REGIONAL GEOLOGY AND TECTONICS

SECOND EDITION

REGIONAL GEOLOGY AND TECTONICS

Volume 2: Phanerozoic Rift Systems and Sedimentary Basins

SECOND EDITION

Edited by

DOMENICO CHIARELLA

Department of Earth Sciences, Royal Holloway University of London, Egham, United Kingdom

NICOLA SCARSELLI

Department of Earth Sciences, Royal Holloway University of London, Egham, United Kingdom

JÜRGEN ADAM

Department of Earth Sciences, Royal Holloway University of London, Egham, United Kingdom

ELSEVIER

Elsevier
Radarweg 29, PO Box 211, 1000 AE Amsterdam, Netherlands
125 London Wall, London EC2Y 5AS, United Kingdom
50 Hampshire Street, 5th Floor, Cambridge, MA 02139, United States

ISBN: 978-0-444-64136-6

For Information on all Elsevier publications
visit our website at https://www.elsevier.com/books-and-journals

Publisher: Candice G. Janco
Acquisitions Editor: Peter Llewellyn
Editorial Project Manager: Ellie Barnett
Production Project Manager: Kumar Anbazhagan
Cover Designer: Mark Rogers

Typeset by MPS Limited, Chennai, India

Contents

List of contributors

Gábor Bada TDE Services, Budapest, Hungary

Dan Bosence Department of Earth Sciences, Royal Holloway University of London, Egham, Surrey, United Kingdom

Edna Rodríguez Calzado Department of Earth and Planetary Sciences, Jackson School of Geosciences, University of Texas at Austin, Austin, TX, United States

Octavian Catuneanu Department of Earth and Atmospheric Sciences, University of Alberta, Edmonton, AB, Canada

Tim Cope Department of Geosciences, DePauw University, Greencastle, IN, United States

Patricia W. Dickerson American Geological Institute and Jackson School of Geosciences, University of Texas at Austin, Austin, TX, United States

Zhiqiang Feng Petroleum Exploration and Production Research Institute, SINOPEC, Beijing, China

Stephan Alan Graham Department of Earth and Planetary Sciences, Stanford University, Stanford, CA, United States

Li-Yuan Hsiao Chevron North America Exploration and Production, Covington LA, United States

Cari Lynne Johnson Department of Geology and Geophysics, University of Utah, Salt Lake City, UT, United States

Samir Khalil Department of Geology, Faculty of Science, Suez Canal University, Ismailia, Egypt

Sean T. Kinney Department of Earth and Planetary Sciences, Rutgers University, Piscataway, NJ, United States; Department of Earth and Environmental Sciences and Lamont-Doherty Earth Observatory of Columbia University, Palisades, NY, United States

Ágnes Király Centre for Earth Evolution and Dynamics, University of Oslo, Norway

Balázs Koroknai Geothermal and Hydrocarbon Exploration, Geomega, Budapest, Hungary

Gábor Kovács Geothermal and Hydrocarbon Exploration, Geomega, Budapest, Hungary

Viktor Lemberkovics Consultant, Budapest, Hungary

Ken McClay Department of Earth Sciences, Royal Holloway University of London, Egham, Surrey, United Kingdom; School of Physics, Chemistry and Earth Sciences, University of Adelaide, Adelaide, South Australia

Paul E. Olsen Department of Earth and Environmental Sciences and Lamont-Doherty Earth Observatory of Columbia University, Palisades, NY, United States

Jonathan D. Paul Department of Earth Sciences, Royal Holloway, University of London, Egham, United Kingdom

Bradley David Ritts Office of Research, University of Idaho, Moscow, ID, United States

Roy W. Schlische Department of Earth and Planetary Sciences, Rutgers University, Piscataway, NJ, United States

Randell Stephenson School of Geosciences, University of Aberdeen, Aberdeen, Scotland

Daniel F. Stockli Department of Earth and Planetary Sciences, Jackson School of Geosciences, University of Texas at Austin, Austin, TX, United States

Sergiy Stovba Institute of Geophysics of National Academy of Sciences of Ukraine, Kyiv, Ukraine

Gábor Tari OMV Energy, Vienna, Austria

Tamás Tóth Geothermal and Hydrocarbon Exploration, Geomega, Budapest, Hungary

Justin Edward Tully Department of Geology and Geophysics, University of Utah, Salt Lake City, UT, United States

Martha Oliver Withjack Department of Earth and Planetary Sciences, Rutgers University, Piscataway, NJ, United States

Foreword

This is the second in a series of books celebrating and documenting basin-scale regional geology. The first volume, published in 2020 and edited by Scarselli et al. (2020) of Royal Holloway, University of London, concentrated on the principles of regional geologic analysis. The same editors have now organized and collated a series of papers focusing on extensional sedimentary basins. The premise behind these excellent chapters is to update and make current geological knowledge that was initially compiled and published through Elsevier by David Roberts and Albert (Bert) Bally in a pioneering series of books in 2012. Many of the authors in the original series have contributed to this volume that is the successor to Volume 1B of Roberts and Bally (2012b).

David and Bert, undoubtedly world leaders in basin analysis, were ambitiously attempting to document the Phanerozoic Geology of the World, and succeeded in publishing three volumes, with the fourth, on compressional basins, unfortunately never completed. David died in 2013, and Bert in 2019. The challenge they set themselves has been taken up by the academic staff at Royal Holloway where David was the visiting professor and an inspirational teacher, and appropriately the editors have dedicated this volume to him. Gabor Tari provided a touching tribute to both Bert and David in the foreword of Volume One, and I would like to take the opportunity here to reflect on my career with BP which was indelibly marked by over 20 years of DGR's wisdom and influence.

David Roberts and I both joined BP Exploration in the same year, 1981, albeit from very different backgrounds. My introduction to BP was at the lower end of the technical spectrum as a green postgraduate geologist fresh from field mapping areas of North Cornwall under contract to the then Institute of Geological Sciences (now BGS). David on the other hand had already established himself as a pioneering marine geoscientist, with an exceptional flair for regional basin understanding and continental margin geology. His early career prior to 1981 had largely been at the Institute of Oceanography (now IOS; Gerdes, 2018), during a period of remarkable growth of scientific understanding as more widespread seismic surveys were being collected by both the exploration industry and academia, complimented by industry and research drilling. His contributions for example in the Initial Reports of DSDP Leg 48 (Bay of Biscay and Rockall) and Leg 81 (Rockall) and related publications include seminal observation–based accounts of passive margin evolution that exponentially increased the understanding of both nonvolcanic (Bay of Biscay) and volcanic (West Rockall) continental margins, and the highly variable nature of the continent-ocean transition. Inherent in the reports are concepts such as differential crustal stretching, low angle detachment faulting, post-rift thermal subsidence, and the influence of structural inheritance on subsequent extensional basin development. His contribution to the understanding of passive margins and their underlying sedimentary basins cannot be underestimated.

David was approached in 1981 by David Jenkins, the then Chief Geologist of BP, who was building a team of highly regarded academic geoscientists to reinvigorate and refresh geoscience in BP Exploration during an era that was becoming dominated by outstanding regional geological description notably by scientists of the Shell Oil Company (e.g., Ziegler, 1982).

David developed and led the Basin Studies Group (BSG) in the Company, and in it BP could support David's thirst for geological insight with its vast library of 2D seismic data and well information from around the globe. One of my earliest recollections is of entering a corridor in Britannic House West to find a 50-ft-long regional seismic section, still smelling of ammonia, laid out along the floor with David crawling over it making noises of appreciation. It was such regional-scale paper data that formed the backbone of David's teaching within the company and subsequently in his academic teaching and coaching career. His mantra that regional geological understanding cannot be replaced by workstation technology, only supplemented by it, is equally relevant today. In BP, exploration training was brought back in-house, most importantly with David's 2-week mandatory residential course called "Seismic Stratigraphy." It was residential due to extended hours and "homework" each day, but it was also a remarkable learning experience moving from reflection seismic interpretation, depositional sequence mapping through chronostratigraphic chart construction to basin analysis, and, ultimately, petroleum play description and estimation. It was competitive with multiple exercises and presentations, but, more importantly, it created a remarkable camaraderie between a generation of explorers that was to last throughout their careers. It was also the entry ticket to David's field-based course in the Western Alps (passive margin extension, through basin inversion, thrust led contraction to ultimately ophiolite and oceanic crust obduction). A truly remarkable geological experience. Among other basin-scale field offerings from David and his team were the rift and passive margin development of the dolomites, the extensional basin structure, facies and development of the Sardinian and Sinai/Gulf of Suez rift basins, and the spectacular Pyrenees and its forelands. These were not "jollies" but serious educational courses with multiple exercises and guest presenters from David's extensive contacts in both academia and the petroleum industry. Thanks to David's influence and teaching, BP from the mid-1980s through to this century developed an exceptional group of regional thinking petroleum exploration scientists. The regional seismic-based approach was published (Hubbard et al.,1985a,b), although at the same time it was developing an even more powerful string to its bow, spear-headed by David's colleague and coauthor, Richard Hubbard. Gross Depositional Environment (GDE) mapping of depositional sequences and their convolution into Common Risk Segment (CRS) maps together with the other elements of the petroleum system to form a Composite CRS (CCRS) map of a play, was a key tool utilized in BP, and built around David's sequence approach. While it is now widespread throughout the industry, for many years BP had a competitive advantage and a common and consistent exploration approach to every basin it chose to explore in, and a methodology to ensure it focused on the lowest risk segments within any particular play.

Meanwhile David still found time to publish extensively from both his academic "backlog" as well as observations from BP's data archive including a particularly valuable contribution documenting extensional basin inversion in and around the British Isles identifying Variscan, Pyrenean and Alpine far field effects (Roberts, 1989). He also founded a new journal dedicated to Marine and Petroleum Geology from the mid-1980s and as editor-in-chief continued to add new geological learnings and insights to his already encyclopedic global knowledge. In the late 1980s, David accepted a position in BP America at a time when exploration was expanding into the deeper waters of the Gulf of Mexico and a whole new geography was beginning to become available to the industry. The science, knowledge and do-ability of deepwater exploration was to rapidly evolve here, and Houston was an ideal placement for him as chief geologist. It was also an opportunity to build a cadre of geoscientists on the other side of the Atlantic through his teaching, particularly after the take-over of Sohio in 1987, and to reinforce the application of the exploration common process throughout the Company. At this time, 3D surveys were becoming standard practice in the Gulf of Mexico, and the learnings from the Gulf were being rapidly exported to new geography on a global scale. David returned to London in this global advisor role which was to be vastly expanded after the merger mania of the late 1990s, and the combined portfolios were providing a wealth of choice. I traveled extensively with David up to his retirement from BP, as he assured a consistency of best practice between companies and site offices in regional studies and opportunity evaluations. At times, I felt I was there just to make sure he got back to the right floor in the right hotel at night as the number of different sites visited in those postmerger years was extraordinary.

David retired from BP in 2003. He was an exceptional scholar, outstanding teacher and to me and many others a remarkable mentor. Although I inherited many of his teaching responsibilities in BP, and later his position, there was no-one who was ever quite able to fill his shoes and inspire a generation of geoscientists in the way that he did.

Ian James Stewart
Integrated Petroleum Exploration Ltd., Woking, United Kingdom

References

Gerdes, K.D., 2018. Professor David Gwyn Roberts – a life in geoscience. In: McClay, K.R., Hammerstein, J.A. (Eds.), Passive Margins: Tectonics, Sedimentation and Magmatism, Volume 476. Geological Society, London, pp. 11–21. , Special Publications, 2020.

Hubbard, R.J., Pape, J., Roberts, D.G., 1985a. Depositional sequence mapping as a technique to establish tectonic and stratigraphic framework and evaluate hydrocarbon potential on a passive continental margin. In: Berg, O.R., Woolverton, D.G. (Eds.), Seismic Stratigraphy II: An Integrated Approach to Hydrocarbon Exploration, 39. AAPG Memoirs, AAPG, Tulsa, USA, pp. 79–91.

Hubbard, R.J., Pape, J., Roberts, D.G., 1985b. Depositional sequence mapping to illustrate the evolution of a passive continental margin. In: Berg, O.R., Woolverton, D.G. (Eds.), Seismic Stratigraphy II: An Integrated Approach to Hydrocarbon Exploration, 39. AAPG Memoirs, pp. 93–115.

Roberts, D.G., 1989. Basin inversion in and around the British Isles. In: Cooper, M.A., Williams, G.D. (Eds.), Inversion Tectonics, 44. Geological Society, London, pp. 131–150. , Special Publications.

Roberts, D.G., Bally, A.W., 2012b. Regional Geology and Tectonics: Principles of Geologic Analysis. Volume 1B.Phanerozoic Rift Systems and Sedimentary Basins. Elsevier, 528pp.

Regional Geology and Tectonics: Principles of Geologic Analysis 2nd Edition. In: Scarselli, N., Adam, J., Chiarella, D. (Eds.), Volume 1: Principles of Geologic Analysis. 2 ed. Elsevier, 894pp.

Ziegler, P.A., 1982. Geological Atlas of Western and Central Europe. Shell Int. Petrol. Mij. B.V. and Elsevier Science Publishers, Amsterdam, 130pp and 40 Plates. ISBN 978-0-444-42084-8.

CHAPTER

1

The Pannonian Basin: regional geology of a back-arc basin

Gábor Tari[1], *Gábor Bada*[2], *Ágnes Király*[3], *Balázs Koroknai*[4], *Gábor Kovács*[4], *Viktor Lemberkovics*[5] *and Tamás Tóth*[4]

[1]OMV Energy, Vienna, Austria [2]TDE Services, Budapest, Hungary [3]Centre for Earth Evolution and Dynamics, University of Oslo, Norway [4]Geothermal and Hydrocarbon Exploration, Geomega, Budapest, Hungary [5]Consultant, Budapest, Hungary

Introduction

This overview paper on the Pannonian Basin aims to highlight the need for a regional geology understanding of large basin complexes, regardless of whether they are rift basins, passive, transform, and active margins, or cratonic basins. While regional geology began through observations in the field (Roberts and Bally, 2012; Scarselli et al., 2020, for a historical review), it has been supplemented during the last few decades by the ever-increasing amount of subsurface data acquired primarily by the hydrocarbon industry and, to a certain extent, the academia. The variety of useful subsurface data includes not only industry-style reflection seismic and borehole data but also crustal/lithospheric-scale deep seismic reflection and refraction profiles, seismic wave tomography, magnetic and gravity data modeling, and magnetotelluric profiling (Stewart, 2020).

The Pannonian Basin, as a classic petroleum province (e.g., Boote et al., 2018), has been explored by the hydrocarbon industry for more than a century (e.g., Tari and Bérczi, 2018). The amount of geological and geophysical data acquired in this large basin complex, spreading across eight countries (listed in a clockwise sense: Austria, Hungary, Slovakia, Ukraine, Romania, Serbia, Croatia, and Slovenia), is staggering and many monographs were published describing its various geographic segments including various aspects of its formation and evolution (e.g., Royden and Horvath, 1988).

This chapter provides a brief description of the Pannonian Basin highlighting the principles of regional geology (Roberts and Bally, 2012; Scarselli et al., 2020) and demonstrating in detail the use of a wide-ranging set of geological and geophysical tools derived from both academic and industry sources.

Besides the demonstration of the "how-to" of regional geological analysis, this paper is intended to provide an analog for back-arc basins worldwide. The Pannonian Basin provides a snapshot in the general evolution of continental extensional basins, which many times leads to the formation of a passive margin and, ultimately, to the opening of an oceanic basin. The Neogene Pannonian Basin complex followed this typical evolution path until about a few million years ago when the process terminated and was replaced by the basin-scale inversion of the entire system. Whereas the ongoing inversion has already a significant footprint on the basin architecture, the original extensional fabric is still well preserved and, again, can be seen as a *locus typicus* for back-arc extensional basins developed on highly extended continental crust (Bally and Snelson, 1980).

Regional Geology and Tectonics: Volume 2
DOI: https://doi.org/10.1016/B978-0-444-64136-6.00005-1

Basin classifications: the Pannonian Basin is the type example of collision-associated back-arc basins on continental crust

The basin classification scheme of Roberts and Bally (2012) lists the Pannonian Basin as the type example of collision-associated back-arc basins situated on continental crust following the earlier designation by Bally and Snelson (1980) to "Pannonian-type" basins.

In a broader sense, back-arc basins like the Pannonian are located within orogenic belts ("episutural" basins of Bally and Snelson, 1980) and associated with continental collisions (i.e., with A-subduction of Bally and Snelson, 1980). These basins are occasionally referred to as "late to postorogenic collapse basins" (Roberts and Bally, 2012). More specifically, a Pannonian-type basin has a continental crust as opposed to, for example, West Mediterranean back-arc basins where extension advanced to the opening of a (proto) oceanic basin (Bally and Snelson, 1980). Other basin classifications, including those of Klemme (1980), Kingston et al. (1983), and John et al. (1984) also account for back-arc basins but did not list the Pannonian Basin specifically as a *locus typicus*.

The Pannonian Basin, located in the Alpine chain of Europe, is superimposed on a collage of previous basin fragments. Almost without exception, all sedimentary basins are composite in nature and origin, reflecting discrete phases of basin evolution, which combine to form the basin. Each of these basin-forming phases can be described in terms of successive tectonostratigraphic megasequences (TSMs) as summarized by Roberts and Bally (2012), following Hubbard et al. (1985).

A TSM is defined as the gross sediment package deposited during a well-defined tectonic phase of basin formation. On its top and base, it is bounded by regional-scale unconformities corresponding to the onset and the end of a basin-forming process (Roberts and Bally, 2012).

For petroleum geologists, preserved sedimentary basins, in general, are not adequately defined by the perceived origin of their inception (Roberts and Bally, 2012). Instead, they are often better defined by their latest phase of a basinal development. In fact, as Roberts and Bally (2012) argued, the most valuable schemes for hydrocarbon exploration are those that classify basins in terms of the latest basin-forming phase that completes the maturation process of source rocks in the basin. This is a practical necessity as hydrocarbon systems are best understood in an evolutionary sense including the latest phase in the evolution of the basin and/or uplift. Thus there is little value in describing a basin only by its early rifting phase TSM if later phases, as defined by superimposed TSMs, really control the complete evolution of hydrocarbon and hydrodynamic systems. Indeed, the latest unconformity associated with basin-scale inversion and uplift of the basin has the most profound effect on the hydrocarbon (re)migration and accumulation in the basin complex.

The aim of this overview paper is to describe the main TSMs of the Neogene to Quaternary Pannonian Basin focusing on the regional geology aspects using decidedly a very broad range of subsurface data examples acquired by different geological and geophysical methods.

The Pannonian Basin in the framework of the Mediterranean basins

During and mainly after Cenozoic continental collisions, regions of extension and subsidence developed within the Alpine-Mediterranean orogenic belt (Fig. 1.1) located behind the contemporaneously active arcs (Horváth and Berckhemer, 1982; Royden, 1993). Mediterranean back-arc basins are characterized by a hot upper mantle, overlain by crust transitional from continental to oceanic, reflecting various stages of basin evolution. In general, back-arc basins form in a subduction setting on the overriding plate behind a volcanic arc due to extensional stresses transferred from the subduction zone (e.g., Király et al., 2016). Both passive (due to stretching) and active asthenospheric upwelling (as part of the subduction-induced flow) lead to high heat flow, weak lithosphere rheology, volcanism, and positive dynamic topography.

The four Mediterranean back-arc basins considered here are, in order of increasing maturity, the Pannonian, Aegean, Alboran-Balearic, and Tyrrhenian Basins. Back-arc extension is not a rigid plate opening often described in models of continental rifting but rather an areal expansion associated with the progressive bending of an adjacent arc formed by an arcuate thrust-fold belt (e.g., Royden and Faccenna, 2018). The most appropriate models suggested for the evolution of Mediterranean back-arc basins imply updoming of the asthenosphere accompanied by lithospheric attenuation, widespread extension observed at different levels of the crust, various modes of stretching, retreating subduction zones, and volcanism. Instead of discussing the very large number of publications devoted to the Mediterranean basins, here we just list some of the key papers dealing specifically with the Pannonian Basin such as those by Royden et al. (1983), Royden and Horvath (1988), Horváth (1993), Tari et al. (1999), Cloetingh et al. (2006), Horváth et al. (2006, 2015, 2018), Balázs et al. (2016, 2018), and Király et al. (2021).

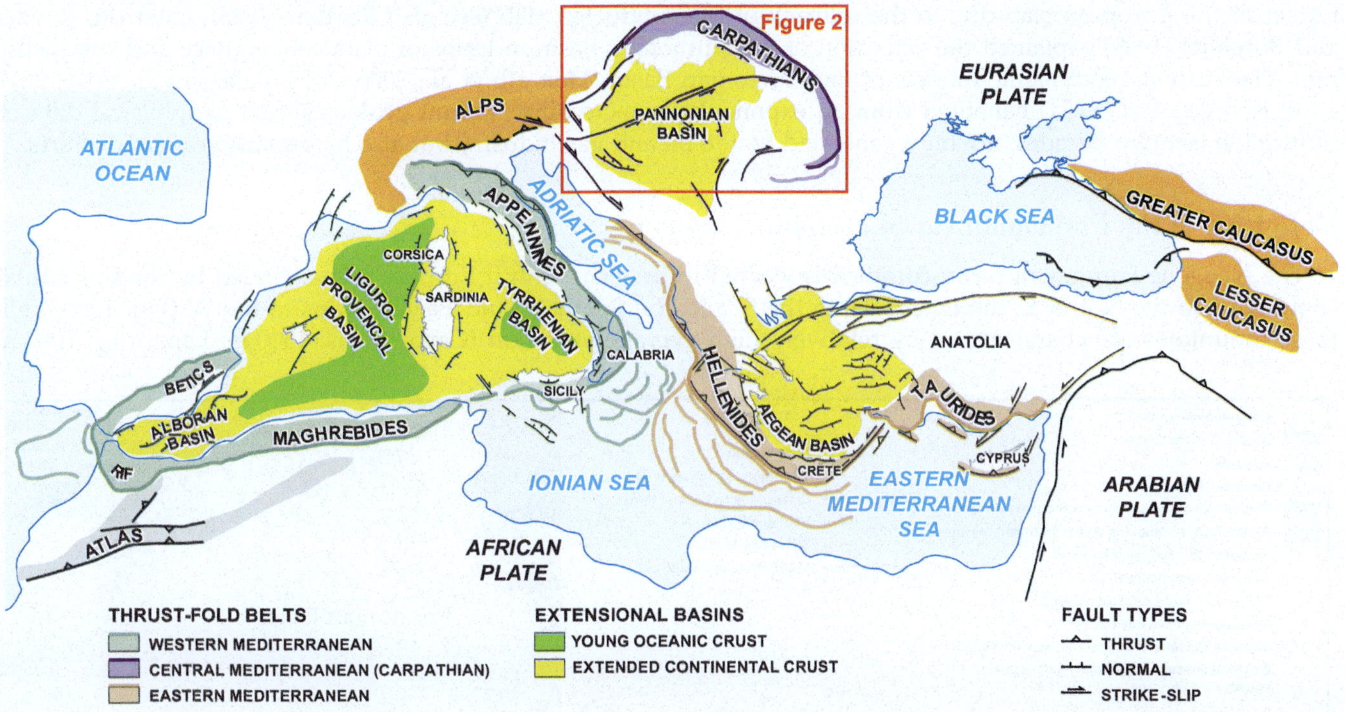

FIGURE 1.1 Simplified overview map of the Mediterranean back-arc basins. Source: *Adapted from (Royden and Faccenna, 2018).*

The Pannonian Basin is a complex of back-arc basin segments superimposed on the Alpine chain filled by Miocene to Quaternary sedimentary and volcanic rocks with a thickness of up to 8000 meters. It is characterized by high heat flow (70–120 mW/m^2, Békési et al., 2018; Lenkey et al., 2021) and attenuated continental crust (20–28 km Moho depth, Tari et al., 1999; Kalmár et al., 2021). The pre-Cenozoic basement of the Pannonian Basin is a collage of two major crustal blocks, the ALCAPA (i.e., ALps-CArpathians-PAnnonian Basin) and Tisza-Dacia terranes, escaped from the Alpine and Dinaric collision zones, respectively (Csontos and Vörös, 2004). Their Early to Middle Miocene escape and extensional collapse were facilitated by the rollback of the oceanic lithosphere in the Carpathian embayment (Royden, 1988) and this Early to Middle Miocene slab retreat led to the formation of the Outer Carpathian accretionary wedge (Horváth et al., 1981). Palinspastic restorations of the preextensional paleogeography in various segments of the Carpathian thrust-fold belt suggested already early on in the 1980s crustal shortening of about 200 km (e.g., Burchfiel, 1980) and more recent works documented even larger numbers around 600 km (e.g., Gagala et al., 2012). Similar magnitude of the crustal extension was inferred in the Pannonian domain either using subsidence modeling (Royden et al., 1983) or interpreting upper-crustal extensional structures (Tari et al., 1992, 1999). Full consumption of the Carpathian slab and "soft" collision of the Pannonian terrane with the East European craton occurred by the late Miocene (Peresson and Decker, 1997). The postrift period of basin evolution has been characterized by rapid subsidence and sedimentation during the Late Miocene and an ongoing period of basin inversion since the start of the Pliocene (Tari, 1994; Horváth, 1995; Horváth and Cloetingh 1996).

The Neogene Pannonian Basin was one of the first areas where McKenzie's (1978) pure shear model of lithospheric extension was tested (Sclater et al., 1980). Analysis of subsidence curves and thermal data suggested that a modified lithospheric extensional model (Royden et al., 1983), whereby large amounts of heat are added to the uppermost mantle during extension, produced good agreement with the observed heat flow, thermal gradients, rates of thermal subsidence, and vitrinite reflectance. In Royden's model, the initial crustal thinning phase occurred during the Middle Miocene (17.5–10.5 Ma), while the subsequent late Miocene to Recent period (10.5–0 Ma) corresponds to the thermal subsidence phase. Crustal extension beneath most parts of the basin is estimated to have been about 50%–120% (Horváth et al., 1988).

Royden et al. (1983) and Royden (1988) made the point that the extension in the Pannonian Basin and coeval compression in the Carpathians were the result of the continued subduction of the European plate, probably driven by negative buoyancy of the subducted slab. In this model, back-arc extension occurs to fill the gap created by the retreat of the European slab due to the rollback of the subducted slab (see also Royden, 1993). Later on, Royden and Burchfiel (1989) explained the extension of the Pannonian Basin in terms of plate convergence and subduction rate. The current geodynamic models of the Pannonian Basin (Horváth et al., 2006, 2015, Balázs et al., 2016, 2017, 2018; Kalmár et al., 2021) benefited from an exponentially increasing amount geological and geophysical datasets during the last two decades, not only generated by the oil and gas industry but also by various academic efforts.

Definition of the Pannonian Basin complex

The Neogene Pannonian Basin complex is located in eastern Central Europe. It is encircled by the Carpathian Mountains to the N and E, the Dinarides to the S, and the Southern and Eastern Alps to the W (Fig. 1.2). While these mountains are characterized by relatively high average elevation (Carpathians: 1500 m; Dinarides: 1000 m;

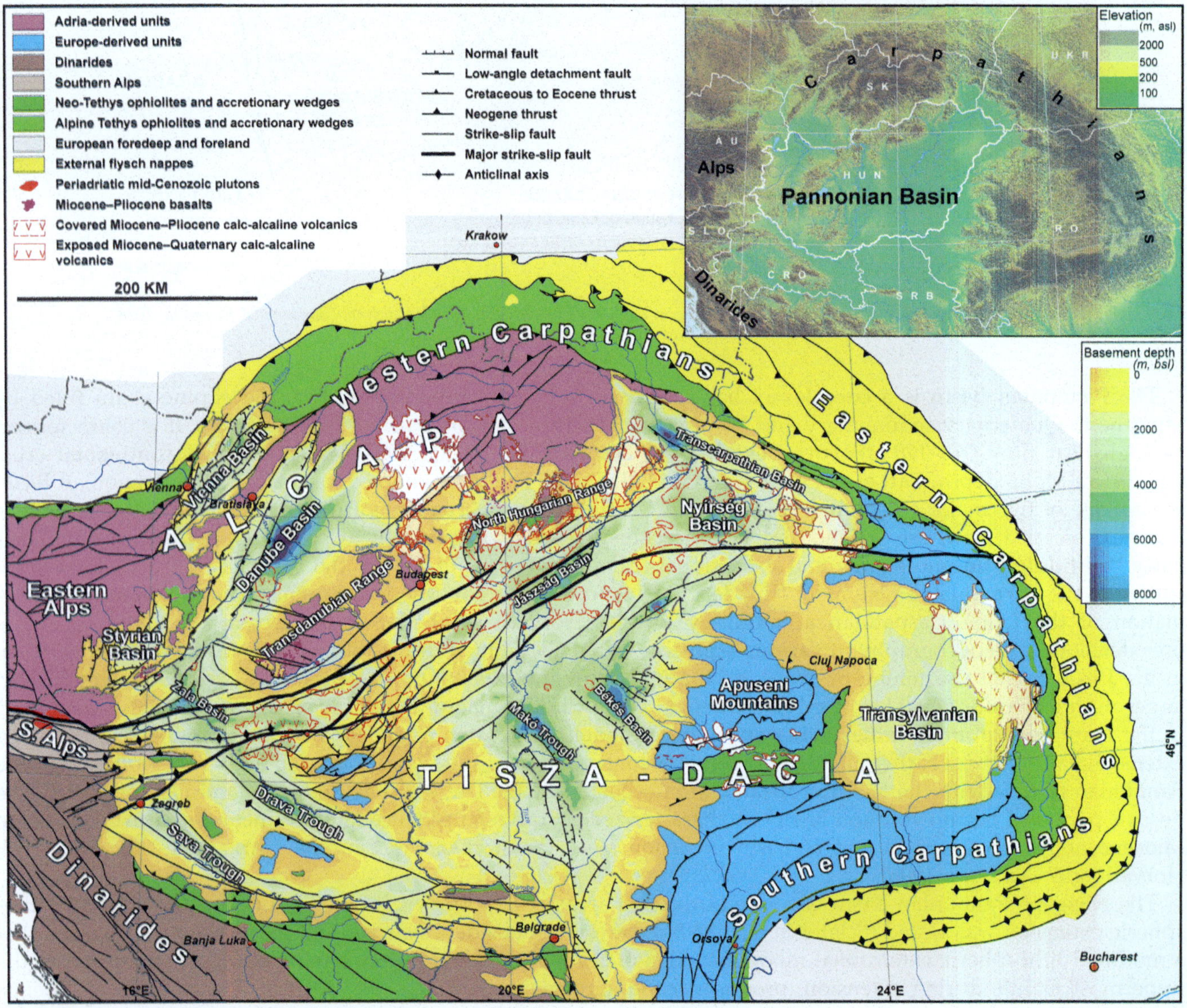

FIGURE 1.2 The Pannonian Basin complex in the context of the Alps, Carpathians, and the Dinarides. The isopach values of the Neogene basin fill outline the most important subbasins. For the complex evolution of the pre-Cenozoic basement of the Pannonian Basin as a collage of two major crustal blocks, the ALCAPA (i.e., ALps-CArpathians-PAnnonian Basin) and Tisza-Dacia terranes, see Vörös and Csontos (2004). Source: *Modified from Horváth, F., Dulić, I., Vranković, A., Koroknai, B., Tóth, T., Wórum, G., et al., 2018. Overview of geologic evolution and hydrocarbon generation of the Pannonian Basin. Interpretation 6, SB111–SB122. https://doi.org/10.1190/int-2017-0100.1; Koroknai, B., Wórum, G., Tóth, T., Koroknai, Z., Fekete-Németh, V. and Kovács, G., 2020. Geological deformations in the Pannonian Basin during the neotectonic phase: New insights from the latest regional mapping in Hungary. Earth-Science Reviews, 211, p.103411.*

Eastern and Southern Alps: 2000 m), the Pannonian Basin s.l. is a lowland with an average elevation of only 100 m above sea level. In this lowland, which is about 400 km wide from N to S and 800 km long from E to W, isolated mountain ranges (or "Inselgebirge" with a German expression) emerge from the plain with elevations up to 1000 m. These surface ranges and many subsurface basin highs subdivide the Pannonian Basin into a number of subbasins in all segments of the basin complex (Fig. 1.3).

The Vienna Basin stretching between Austria and Slovakia is not part of the Pannonian Basin *sensu stricto*. It is located between the eastern termination of the Eastern Alps, the westernmost part of the West Carpathians and the Bohemian Massif (Fig. 1.2). The Danube Basin (or Little Hungarian Plain in strictly geographic terms) of Hungary and Slovakia is bounded by the Eastern Alps and the West Carpathians on the W and N, respectively, while it is bordered by the Transdanubian Range to the S and the E (Fig. 1.2).

Toward the west, the Styrian Basin in Austria is separated by the Zala and Mura basins of Hungary and Slovenia. To the SW the major basins are the Drava and Sava basins located mostly in Croatia.

The Great Hungarian Plain is the largest geographic region of the Pannonian Basin as it occupies the central portion of the intra-Carpathian area stretching between Hungary, Romania, and Serbia. In the subsurface,

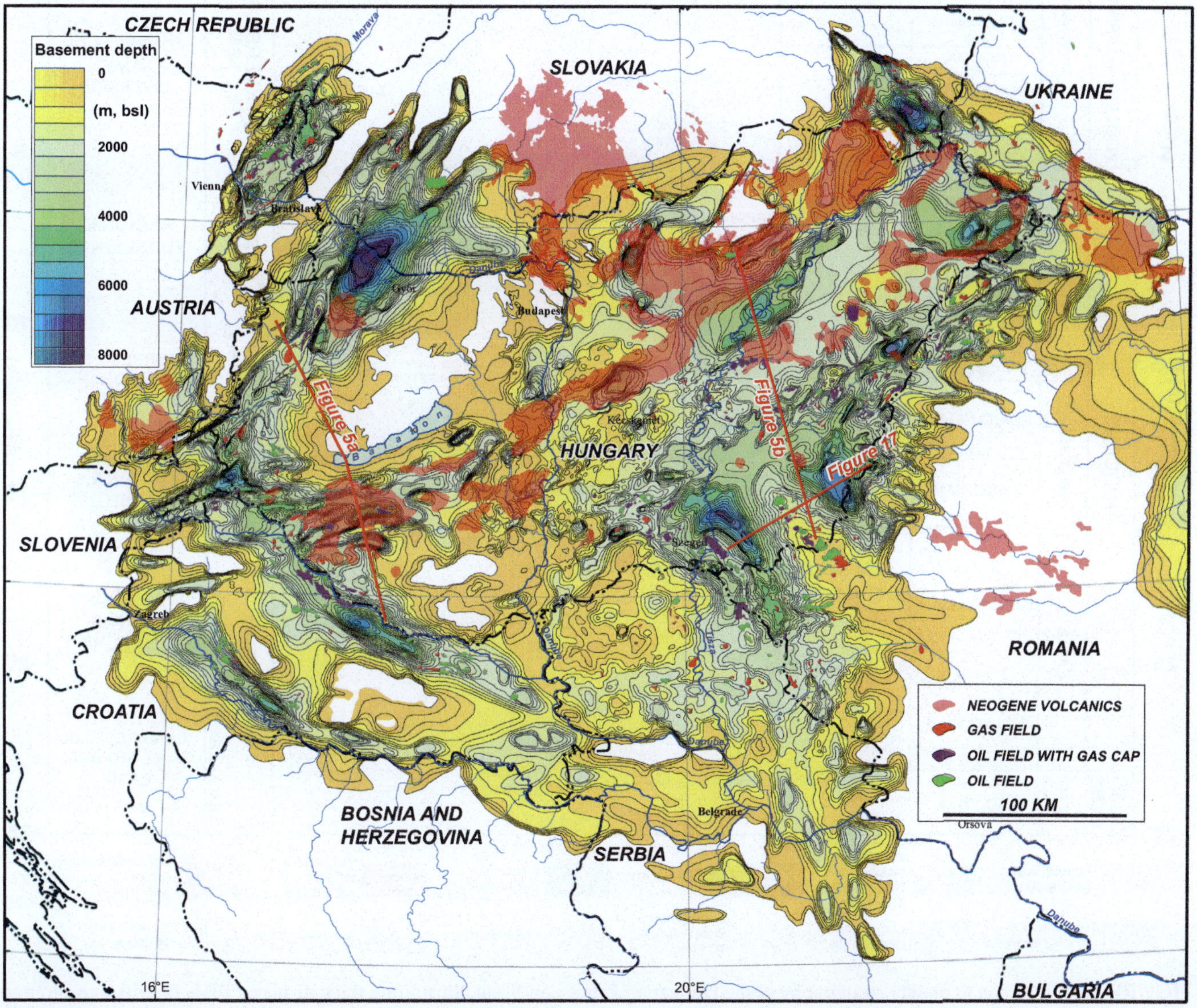

FIGURE 1.3 Depth of the prerift basement in the western and central part of the Pannonian Basin complex. The location of some of the illustrations are indicated in red. Source: *Modified after Horváth, F., Dulić, I., Vranković, A., Koroknai, B., Tóth, T., Wórum, G., et al., 2018. Overview of geologic evolution and hydrocarbon generation of the Pannonian Basin. Interpretation 6, SB111–SB122. https://doi.org/10.1190/int-2017-0100.1.*

however, there are many subbasins in this large region, the most important ones are the Jászság and Nyírség basins on its northern perimeter and the Makó and Békés Troughs in the south. At the NE perimeter of the basin complex, a satellite basin is the Transcarpathian (or East Slovak) basin of Slovakia and the Ukraine. To the East, the Transylvanian Basin in Romania is located between the Apuseni Mts. and the Eastern and Southern Carpathians and it belongs only in a wider sense to the Pannonian Basin.

Largely covered by the basin fill highlighting the need for subsurface datasets, such as seismic and wells (Lemberkovics et al., 2018; Koroknai et al., 2020).

The entire basin system is located in the Central Paratethys region, which has its own chronostratigraphic scheme due to the ever-increasing isolation from the world ocean. For the correlation of the local stages with global ones (Fig. 1.4), the reader is referred to a recent summary by Magyar et al. (2020).

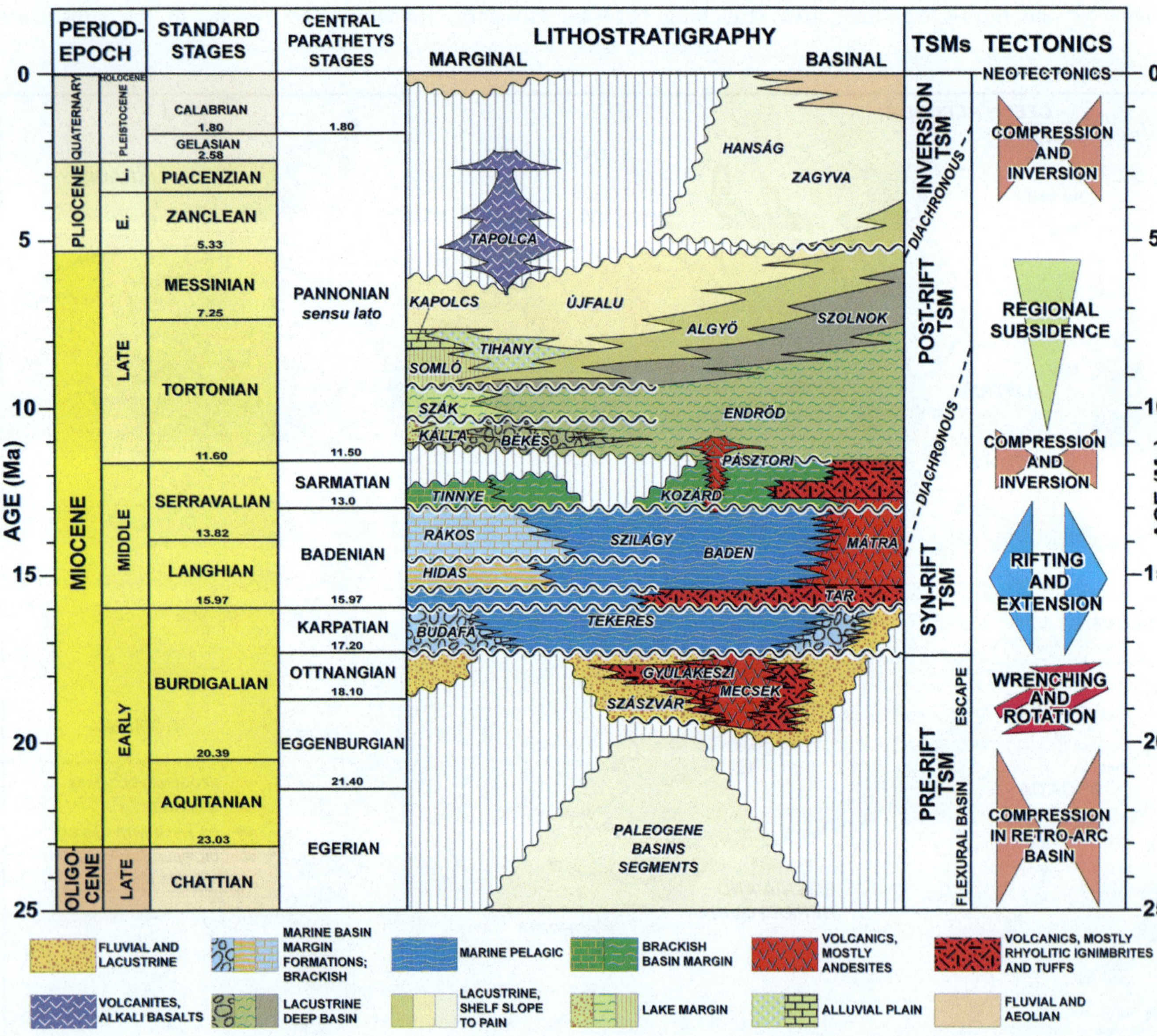

FIGURE 1.4 Miocene to present stratigraphy of the Hungarian Pannonian Basin, illustrating the four major TSMs (prerift, syn-rift, post-rift, and inversion) in the central part of the extensional basin complex manifested by major regional or smaller, local-scale unconformities in the basin fill. *TSM*, Tectonostratigraphic megasequence. Source: *Modified from Horváth, F., Dulić, I., Vranković, A., Koroknai, B., Tóth, T., Wórum, G., et al., 2018. Overview of geologic evolution and hydrocarbon generation of the Pannonian Basin. Interpretation 6, SB111–SB122. https://doi.org/10.1190/int-2017-0100.1.*

Tectonostratigraphic megasequences of the Pannonian Basin based on regional reflection seismic and other subsurface datasets

TSMs are bounded by regional-scale unconformities corresponding to the onset and the end of a basin-forming process with a gross sediment package deposited during the corresponding tectonic phase in the basin evolution (Roberts and Bally, 2012). In their scheme, the "Syn-rift and/or volcanic TSM" is bound at the bottom by the "prerift unconformity," which includes "Igneous and metamorphic basement, including some folded sediments."

In the Pannonian Basin, this unconformity has an early Miocene (intra-Burdigalian) age (Fig. 1.4) separating Lower Miocene to Oligocene sequences below it, which have a different tectonostratigraphic origin such as related to escape tectonics (Royden, 1988) and flexural subsidence in a retroarc basin setting (Tari et al., 1993). In many parts of the basin complex, however, these units are missing and various Mesozoic or Paleozoic units subcrop the prerift unconformity.

The syn-rift and/or volcanic "back-arc TSM" (Roberts and Bally, 2012) spans mostly the Middle Miocene and the top bounding unconformity was variably placed at the middle Badenian (e.g., Tari, 1994), at the top Badenian, or at the top Sarmatian by others (Fig. 1.4). Many of these unconformities are related to significant eustatic sea-level falls within the Paratethys region (Vakarcs et al., 1998) and therefore they are tectonically enhanced by extensional syn-rift faulting (Tari, 1994). More recent work by Balázs et al. (2016) and Horváth et al. (2018), however, emphasized the diachronous nature of the postrift unconformity across the Pannonian Basin complex ranging from the intra-Badenian (c.14 Ma) generally in the west to the early middle Pannonian (c.8 Ma) in the east (Fig. 1.4). We interpret the gradual spatial (west to east) and temporal (middle Miocene to Pliocene) shifting of the top back-arc TSM unconformity as reflecting the gradual shifting of the docking (final thrusting) age of the thrust-fold belt along the Carpathian arc onto the European platform as already observed in his landmark work by Jiřiček (1979).

The syn-rift stratigraphy of the basin from bottom to top is dominated by terrestrial, then shallow marine clastics with occasional neritic limestone formations and eventually deeper marine clastics (Fig. 1.4). Arc volcanism associated with the subduction along the surrounding Carpathians did play a large role in the basin system producing lots of volcanics and volcaniclastics in many subbasins (Harangi and Lenkey, 2007; Seghedi and Downes, 2011).

Many of the unconformities within the syn-rift TSM, subdividing into several subunits (Balázs et al., 2016) are tectonically enhanced in character, that is, the eustatic signal was superposed on the ongoing extensional tectonics (Tari, 1994; Vakarcs et al., 1998).

The postrift (or breakup) unconformity is typically well defined as an erosional unconformity (Fig. 1.4). During the postrift sedimentation period the Pannonian Basin, as part of the Paratethys region, became totally isolated from the world seas and marine conditions were replaced by lacustrine environments in the very large but isolated Pannon Lake (Vakarcs et al., 1994; Magyar et al., 1999, 2013). Lake Pannon was a huge, long-lived, and brackish lake, similar to the modern Caspian Sea.

Compared to other rift basins or passive margins, the relative proportions between the syn-rift versus postrift basin fill are unusual with the dominance of the latter. This peculiar feature of the central Pannonian Basin has been noted by many, including Grow et al. (1994). The reason for the thin syn-rift sequence is the relatively small sediment supply from the Carpathians and Alps during the rifting process. The provenance areas of sediment systems feeding the back-arc region were far during the Middle Miocene, at least a few hundred kilometers away in the surrounding mountain chains.

A prominent intra-Pliocene (Pannonian) southward prograding system across the entire Pannonian Basin from Austria to Serbia corresponds to a general source region to the NW and N (Magyar et al., 2012, 2020 ; Sztanó et al., 2013). It is the various lithofacies units of the postrift prograding Pannonian megasequence (Bérczi, 1988; Juhász, 1994; Radivojević et al., 2022), which provided the most important reservoirs in the Pannonian Basin complex responsible for the largest hydrocarbon reserves (Horváth et al., 2018; Boote et al., 2018; Lemberkovics et al., 2018).

The "Uplift and/or Inversion" TSM of Roberts and Bally (2012) manifests itself as a major erosional unconformity on top of the Pannonian sequence (Fig. 1.4). This unconformity is more pronounced along the basin margins as the regional uplift produced by the basin-scale inversion tended to remove a significant part of the postrift basin fill. Within the basin fill, the unconformity tends to be more subtle. As to the age of this inversion unconformity, different estimations of its maximum age were published since the earliest descriptions in the early 1990s. However, it is generally accepted that inversion started about 4–5 million years ago during the Late Pliocene (e.g., Tari, 1994; Horváth, 1995; Horváth and Cloetingh, 1996; Cloetingh et al., 2006; Uhrin and Sztanó, 2011; Bartha et al., 2018; Ruszkiczay-Rüdiger et al., 2020; Šujan et al., 2021). Since the map view of the inverted structures seems to show propagation from the west to the east across the Pannonian Basin (Tari et al., 2020), the inversion unconformity is a diachronous one, similar to that of the postrift unconformity (Fig. 1.4). Since positive structural inversion is an ongoing neotectonic process (Koroknai et al., 2020), the top surface of the "uplift and/or inversion" TSM coincides with the present-day surface.

Regional geology of the Pannonian Basin based on subsurface datasets

As an illustration of the large-scale basin architecture, regional seismic transects were constructed across the central part of the Pannonian Basin as line drawings (Tari, 1994). This graphical display method is the only practical way to show reflection seismic data on the regional, that is, 100 + km scale, as the reproduction of the actual seismic reflection data is not feasible, not even by vertical exaggeration. For brevity, only two regional transects were reproduced here (Fig. 1.5) spliced together from several 2D seismic profiles and constrained by a large number of exploration wells. These transects were also described by Tari and Horváth (2006) and, therefore, here just the key regional aspects are highlighted with additional subsurface data.

Regional transect A

The ~230-km long transect starts at the NW margin of the Pannonian Basin (Fig. 1.3), at the Austrian-Hungarian border (Fig. 1.5A). This area belongs to the Danube Basin at the flank of the Burgenland basement high separating it from Styrian Basin. Using 2D seismic reflection data, the outcropping greenschist metamorphosed basement rocks in the Penninic Rechnitz Window Group were interpreted by Tari and Bally (1990) as a metamorphic core complex corresponding to extreme upper-crustal extension. The large detachment fault depicted on Transect A could have accommodated horizontal extension on the scale of tens of kilometers (Tari, 1994, 1996). The possible presence of low-angle normal faults in the Pannonian Basin was first mentioned

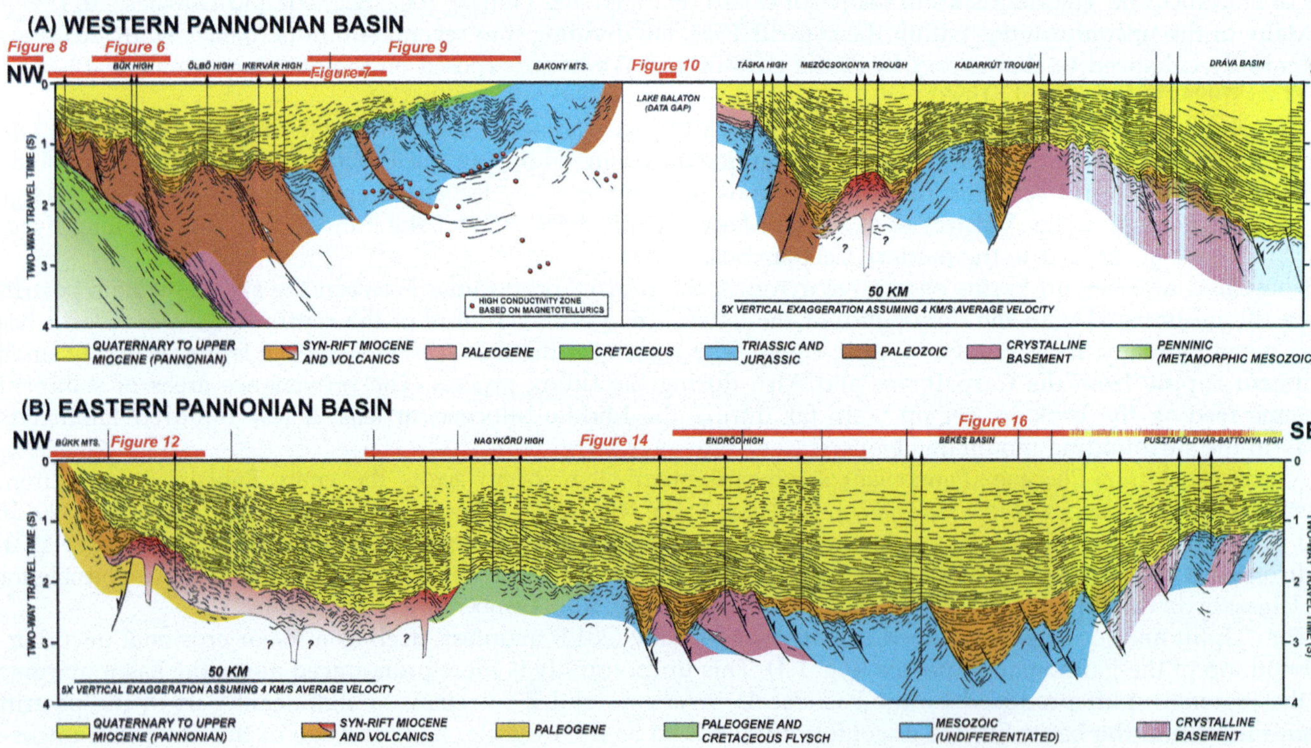

FIGURE 1.5 Line drawing interpretation of regional seismic transects (A) in the western Pannonian Basin and (B) in the eastern Pannonian Basin. For location, see Fig. 1.3. The approximate projections of some of the following figures into these transects are also indicated. Source: *Modified from Tari, G. and Horváth, F., 2006. Alpine evolution and hydrocarbon geology of the Pannonian Basin: an overview. In: J. Golonka and F. Picha (Editors), The Carpathians and their Foreland: Geology and Hydrocarbon Resources. AAPG Memoir, 84, 605–618.*

by Royden (1988) quoting analogs from the Basin and Range Province of the western United States. Indeed, systematic mapping of seismic data did confirm this speculation (Tari et al., 1992; Tari et al., 1999, 2020). A representative 2D seismic example illustrates the low-angle normal fault extensional style, where the Bük basement high forms a hangingwall anticline underlain by a large single detachment fault (Fig. 1.6). The Miocene fault plane itself is interpreted to be inherited from an earlier Cretaceous Alpine overthrust plane, although not in its entire length, similarly to other detachment faults along strike (Tari, 1996). The partial and selective reactivation of preexisting thrust fault planes is basically negative structural inversion as emphasized recently by Tari et al. (2021).

The basin fill in the deepermost part of the Danube Basin, some 100 km along strike to the NE from the basin center depicted in Fig. 1.5A, is up to 8 km thick (Fig. 1.3). The 2D seismic composite section illustrating this part of the basin (Fig. 1.7) is based on legacy 1980s seismic sections (Gjerazi, 2016) from Austria and Hungary.

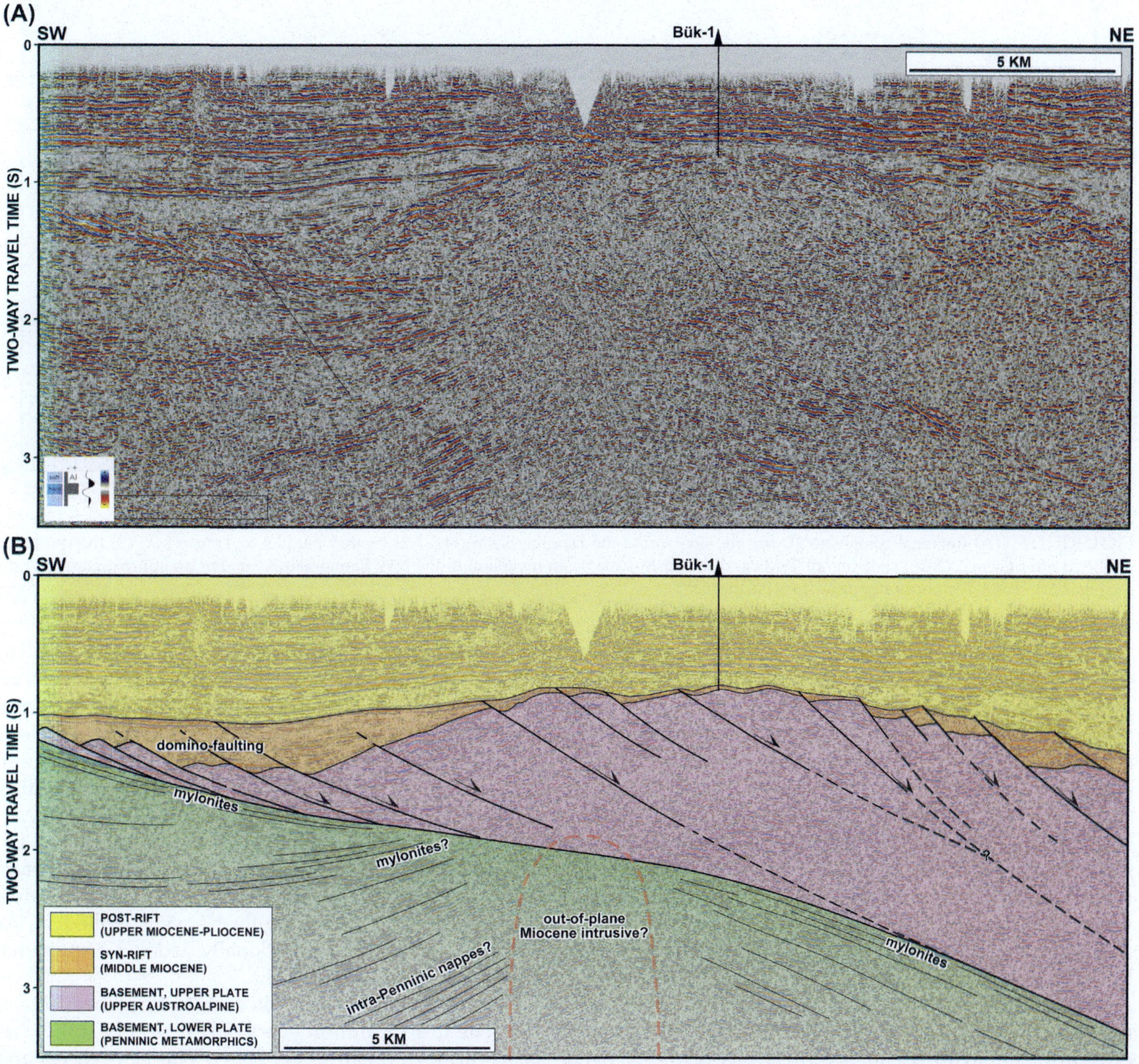

FIGURE 1.6 (A) 2D seismic section across the Bük High in the southern part of the Danube Basin. (B) The interpretation highlights the regional-scale low-angle Rechnitz detachment fault, which is responsible for many tens of kilometers of horizontal extension in this part of the Pannonian Basin. For location, see Fig. 1.5A. Source: *Modified from Tari, G. C., Gjerazi, I., & Grasemann, B., 2020. Interpretation of vintage 2D seismic reflection data along the Austrian-Hungarian border: subsurface expression of the Rechnitz metamorphic core complex. Interpretation, 8(4), SQ73-SQ91.*

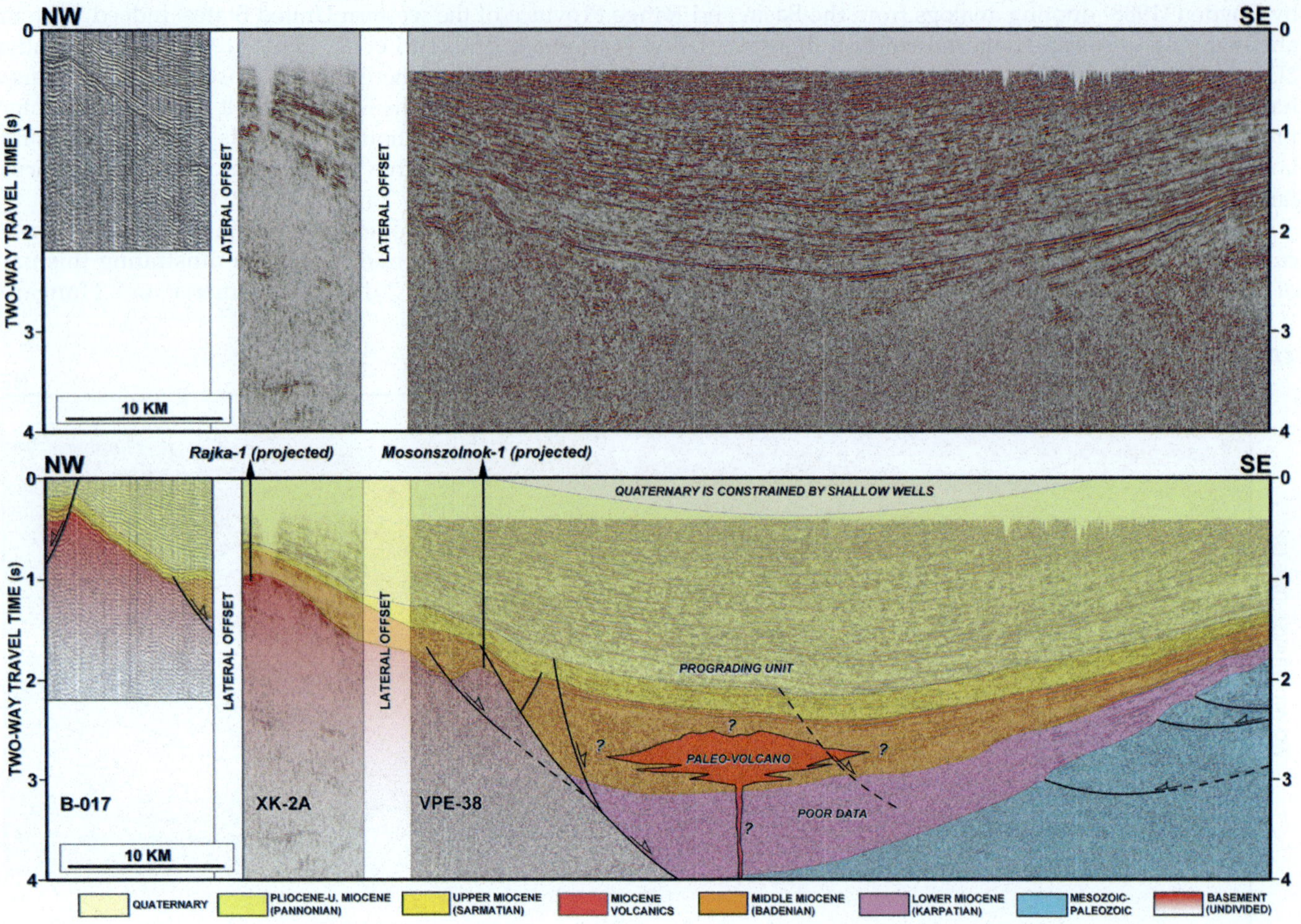

FIGURE 1.7 (A) Regional composite 2D seismic line across the Danube Basin (Fig. 1.2) located parallel to Transect A. (B) Interpretation modified after Gjerazi (2016) showing all TSMs and the basin-scale basin inversion in the NW Pannonian Basin. For an approximate location, see Fig. 1.5A. *TSM*, Tectonostratigraphic megasequence.

The most striking part of the section is the synformal geometry of the Pannonian sequence all the way up to the top of the seismic data truncated at 400 Ms two-way travel time as documented by Tari (1994). Since there is no growth or onlaps associated with differential compaction in this Upper Pannonian (Pliocene) sequence (Wórum et al., 2020), the syncline is the result of large-scale tectonic downwarping of the basin center due to lithospheric buckling during the contractional Pliocene to Recent inversion stage (Tari, 1994; Horváth, 1995; Horváth and Cloetingh, 1996). The adjacent Slovakian part of the Danube Basin shows the same geometry (Sztanó et al., 2016; Šujan et al., 2021).

There is a significant truncation of the Pannonian strata at the NW edge of the Pannonian Basin (Figs. 1.5 and 1.7) seen on many seismic sections in the area. This primarily structural phenomenon, not being fully appreciated before the early 1990s, manifests itself at different scales and in various geological and geophysical datasets. The example we chose to include here is based on shallow (50–200 m deep, on average) "counter-flush" wells drilled in the Austrian side of the area shortly after WWII (Tauber, 1959). The close spacing and exceptionally detailed documentation of these wells allowed a superb lithostratigraphic correlation between them (Fig. 1.8).

With the 20x vertical exaggeration applied, the dramatic truncation of the Upper Miocene to Pliocene strata at the basin margin is obvious and it is attributed to an intra-Pliocene (c.4–5 million years old) period of uplift and erosion of the syn-rift and postrift TSMs of the Neogene basin fill (e.g., Šujan et al., 2021). The isopachous and thin (just a few m thick) Quaternary in this area clearly postdates the regional inversional period. Similar resolution high-frequency seismic data were acquired over the Neusiedler Lake along strike by Loisl et al. (2018) and it displays the same geometry.

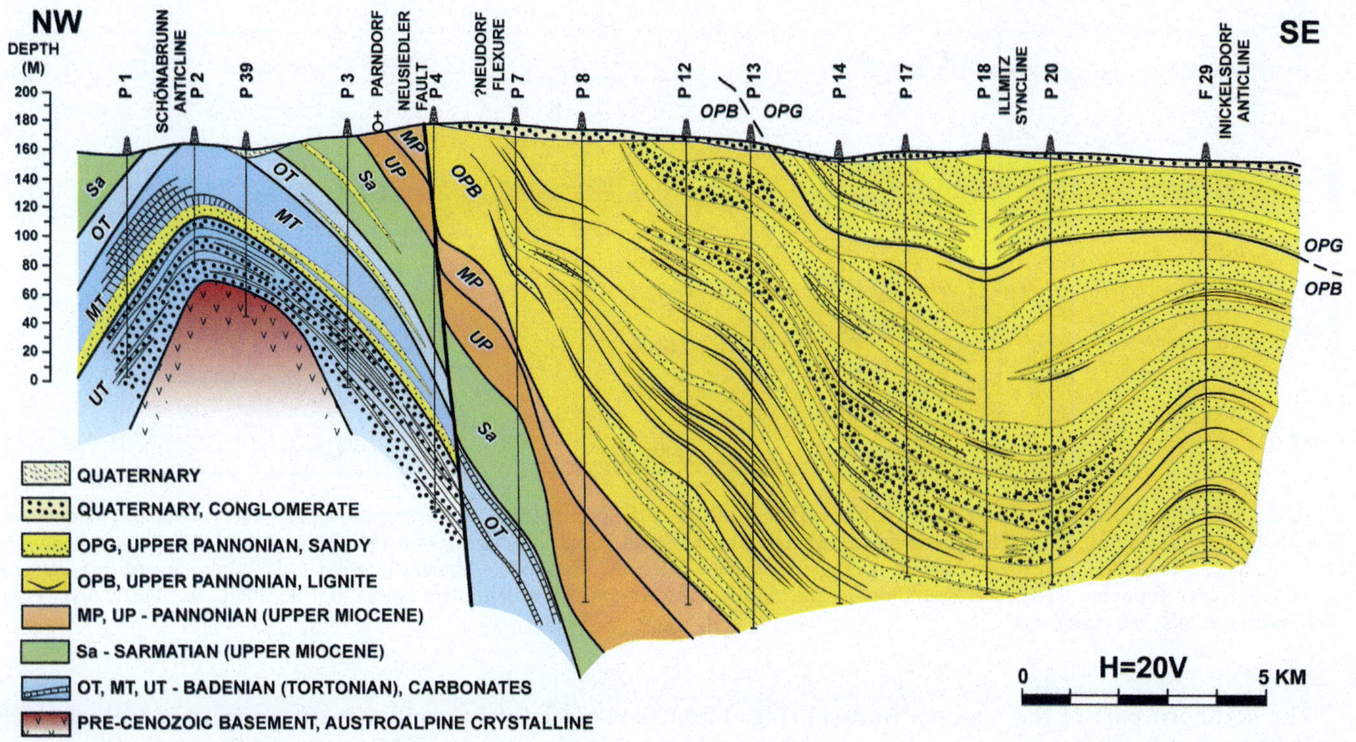

FIGURE 1.8 Detailed geologic cross-section based on shallow drilling to the north of the Neusiedlersee in Austria, redrafted after Tauber (1959). For location, see Fig. 1.5A. Note that the section is 20x overexaggerated. The basement high on the NW end of the section is the NE-oriented downplunge of the basement high exposed in the Leitha Mts. (Fig. 1.2).

Farther along the transect, on the southeastern flank of the Danube Basin and beneath the Transdanubian Range, a low-elevation mountain range, there are very distinct seismic reflector packages in the prerift TSM (Fig. 1.5A). Multiple repetitions of Paleozoic and Mesozoic units were mapped in the subsurface using 2D seismic profiles by Tari (1994), who interpreted them as the result of Alpine (Cretaceous) thrust imbrications. Many of the overthrust planes in between are located at great depths (4–10 km) and, therefore, never penetrated by drilling except in one case (Tari et al., 2019).

An independent geophysical dataset appears to confirm the presence of Alpine overthrust planes in the prerift TSM. Magnetotelluric measurements in the NW Pannonian Basin found an exceptionally high conductivity (c.1–2 Ωm) unit with a large map-view extent (on the order of a few thousand km^2, Ádám et al., 1970). The so-called transdanubian conductivity anomaly (TCA) is an entirely subsurface anomaly occurring at a depth between c.3 and 15 km, with no corresponding outcrops. The pronounced seismic reflector packages attributed to subhorizontally oriented Cretaceous Alpine nappe surfaces and the low resistivity TCA seem to coincide in the subsurface (Fig. 1.9). Therefore, Tari et al. (2019) proposed that the best explanation for the observed extent and geometry of the TCA is the presence of graphite in subhorizontal, tectonically thinned detachment surfaces at the base of the Upper Austroalpine nappe edifice of NW Hungary. In a regional geology sense, the proper analogs for graphite units were found to the west in the Greywacke Zone of the Eastern Alps in Austria, where numerous graphite localities are known historically (Tari et al., 2019).

There is a ~10 km wide data gap in the regional transect (Fig. 1.5A) corresponding to the Lake Balaton, Europe's largest lake (Fig. 1.3). There are no industry-style seismic lines across the lake, but instead, many high-resolution academic datasets were acquired over it during the last two decades (e.g., Sztanó, and Magyar, 2007; Bada et al., 2010; Visnovitz et al., 2015, 2021). An example of these datasets (Fig. 1.10) reveals the faulted and truncated nature of the postrift TSM truncated close to the base of the Holocene mud at the lake bottom. The overall geometry is quite similar to the one depicted in Fig. 1.8 using a very different dataset. Both profiles show the near-surface "uplift and/or inversion" unconformity of Roberts and Bally (2012) on top of the postrift TSM.

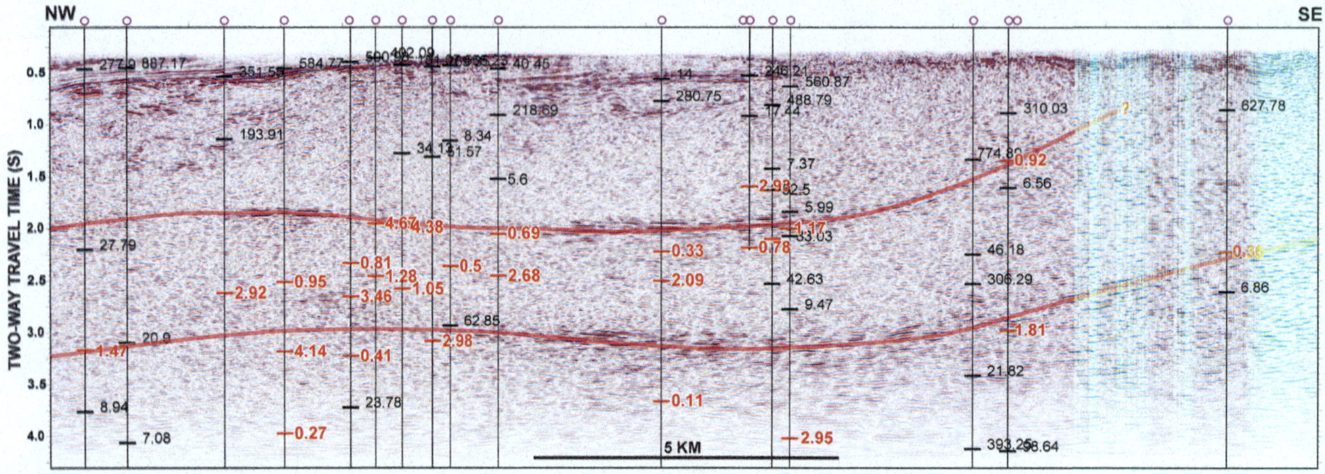

FIGURE 1.9 2D reflection seismic example of the internal structure of the prerift TSM using conventional 2D seismic data and magnetotelluric (MT) data (Tari et al., 2019). High conductivity values (i.e., less than 5 Ωm) are highlighted in red. Note the fairly good coincidence between the MT anomalies and the prominent package of seismic reflectors. These were already inferred to be Alpine overthrust surfaces by Horváth and Rumpler (1984) corresponding to the nappe stack of the Transdanubian Range (for location, see Fig. 1.5A). *TSM*, Tectonostratigraphic megasequence.

The southern part of the regional transect (Fig. 1.5A) shows relatively small syn-rift grabens beneath a thick postrift basin fill. There are several large anticlinal folds within the postrift TSM involving the entire Pannonian sequence. These folds are quite widespread in the SW part of the Pannonian Basin complex as they were mapped by surface measurements as early as the 1930s (Tari and Bérczi, 2018).

Returning to the post-Pontian (Pliocene) neotectonic inversion in the Pannonian Basin complex (Fig. 1.4), it manifests itself on two scales (Tari, 1994; Horváth, 1995; Tari et al., 2020), either producing basin-scale uplifts and erosion on the scale of 50–100 km and/or also structure-scale (5–20 km) inversion features producing lot of late-stage traps in the basin fill (Sacchi et al., 1999; Sacchi and Horváth, 2002; Lemberkovics et al., 2018; Koroknai et al., 2020). While the local-scale inversions are attributed to the reactivation of preexisting extensional normal faults in a compressive stress field, the basin-scale inversions may be due to large-scale folding of the lithosphere reacting to the buildup of the same stress field (Horváth and Cloetingh, 1996; Cloetingh et al., 2006). Since the ultimate cause of the post-Pontian inversion in the Pannonian Basin is the northward movement and counterclockwise rotation of the Adriatic promontory located to the SW of the basin (Vrabec and Fodor, 2006; Bada et al., 2007; Tomljenović et al., 2008), the footprint of the inversion is more pronounced on the western transect (Fig. 1.5A) and more subtle on the eastern one (Fig. 1.5B).

At the SW part of the Pannonian Basin, the Haloze anticline in Slovenia (Fig. 1.11) illustrates the case where the regional-scale basin uplift exposed one of the inverted structures. It is the surface geology here, which constrains the geometry of the structure including the overturned Pliocene sequence along the Ljutomer fault on the northern flank and the large strike-slip zone of the vertical Donat fault on the southern side (Fodor et al., 2002). It is difficult to estimate the amount of missing syn- and postrift TSM strata over this transpressional anticline but their combined thickness could be significantly more than 2000 m. The geologic section constructed by Fodor et al. (2002) has some speculative elements in the subsurface, but it provides an internally consistent picture of a deeply eroded inverted anticline based on the analog structures seen on reflection seismic data farther to the east in the basin itself in Slovenia, Hungary, and Croatia (Dank, 1988; Sacchi et al., 1999; Horváth et al., 2006; Horváth et al., 2018; Tóth and Tari, 2014; Tari et al., 2020).

It is important to emphasize that preexisting syn-rift normal faults tend to be reactivated not only in a positive structural inversion sense (Tari et al., 2020) but also the ongoing neotectonic deformation across the basin complex frequently turns them into strike-slip faults (Koroknai et al., 2020; Wórum et al., 2020). The Donat fault could be seen in this context as well (Fig. 1.11), although the strike-slip deformation on this fault has a prolonged history since the Miocene.

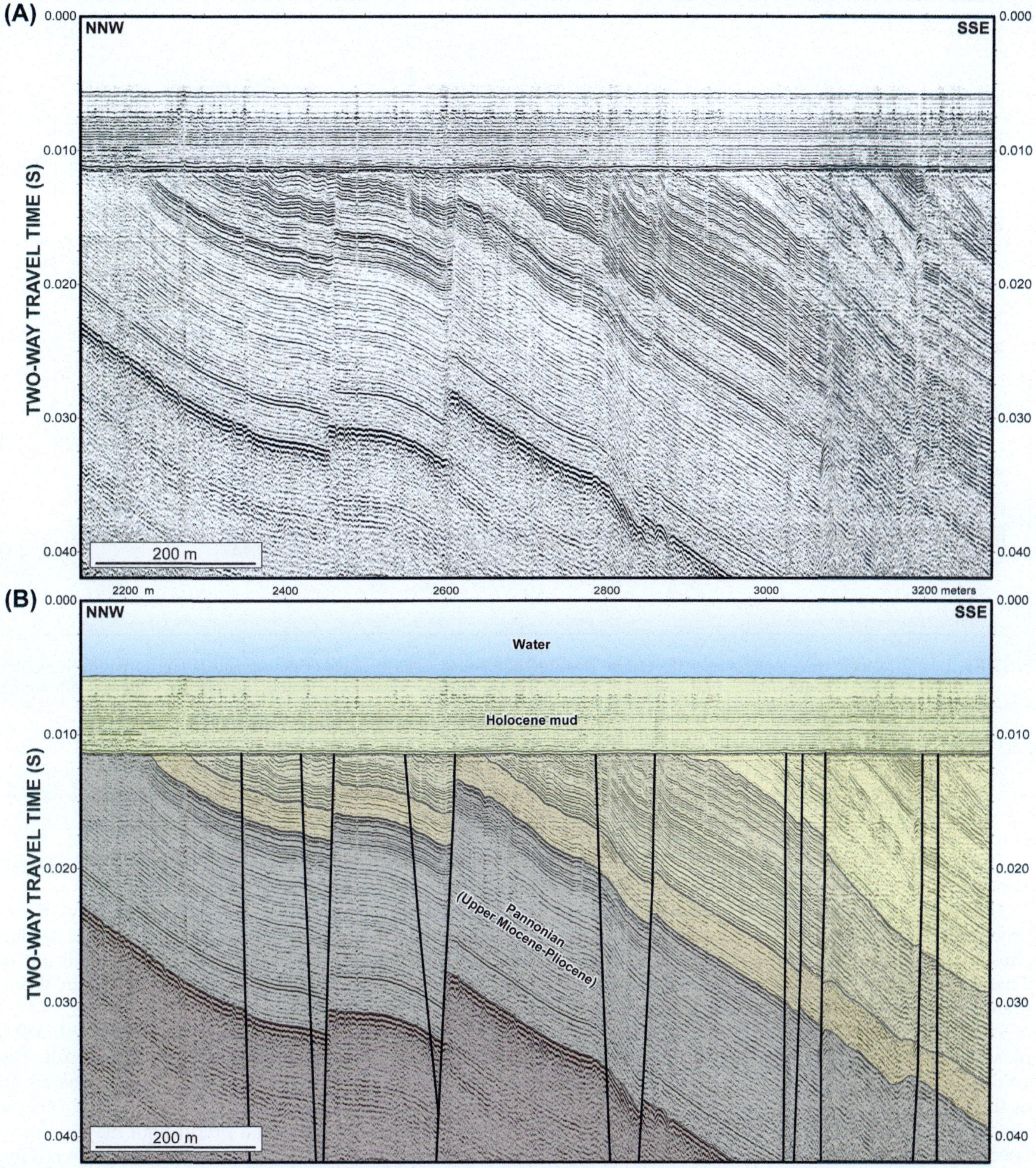

FIGURE 1.10 (A) Ultra-high-resolution seismic example acquired across the Lake Balaton. (B) The interpretation highlights the prominent uplift and inversion unconformity TSM beneath the lake (see Fig. 1.5A for location). The Pannonian strata display an overall steepening toward the SSE (vertical exaggeration is about 20x). *TSM*, Tectonostratigraphic megasequence. Source: *Modified from Visnovitz, F., Horvath, F., Fekete, N., Spiess, V., 2015. Strike-slip tectonics in the Pannonian basin based on seismic surveys at Lake Balaton. International Journal of Earth Sciences (Geol. Rundsch.) 104/8, 2273–2285. https://doi.org/10.1007/s00531-015-1179-x*

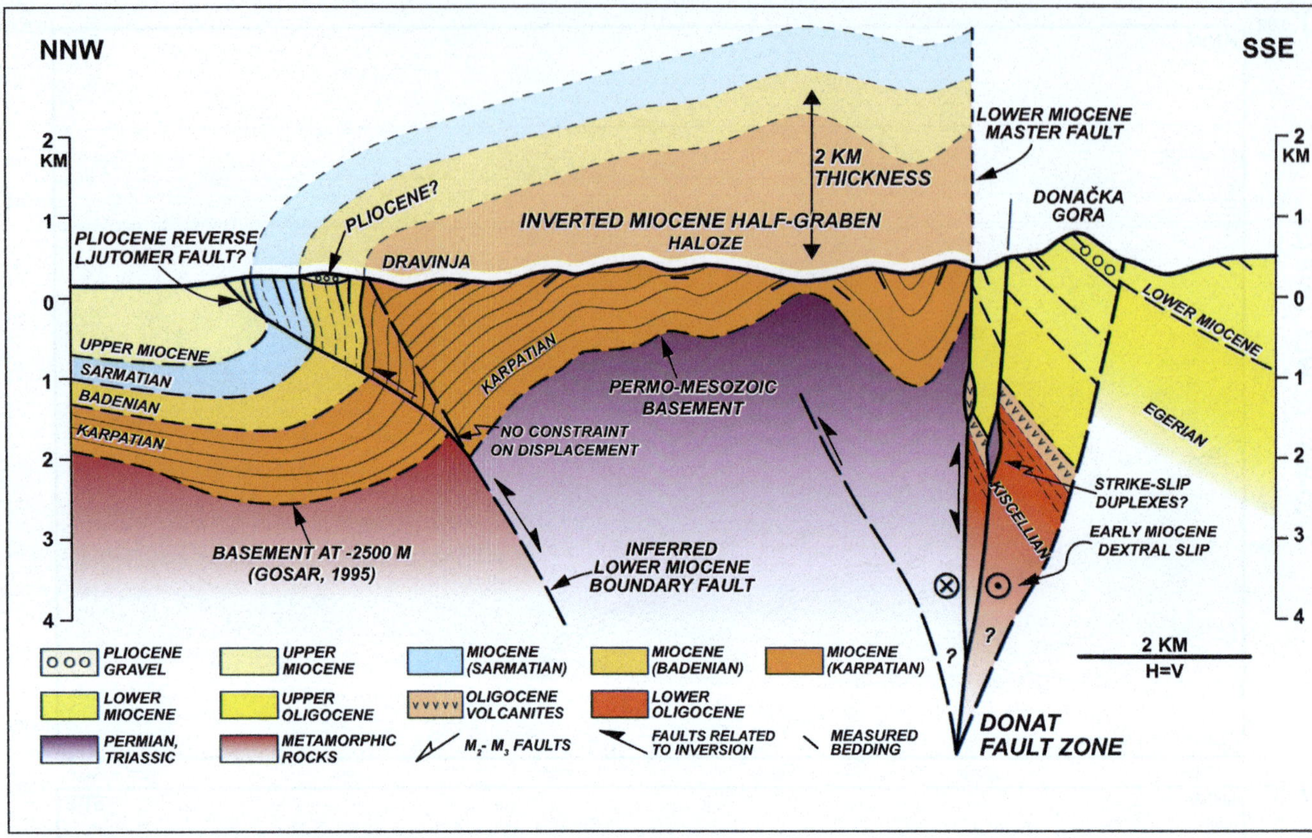

FIGURE 1.11 Geologic cross-section across the Haloze structure in Slovenia as an example of the advanced stage of positive inversion exposing the syn-rift TSM on the surface at the southwestern margin of the Pannonian Basin. For location, see Fig. 1.3. Compare with seismic examples shown in the general area by Sacchi et al. (1999), Fodor et al. (2008), Fodor et al. (2002, 2021), Tóth and Tari (2014), and Tari et al. (2020). *TSM*, Tectonostratigraphic megasequence. Source: *Modified from Fodor, L., Jelen, B., Marton, E., Rifelj, H., Kraljić, M., Kevrić, R., et al., 2002. Miocene to Quaternary deformation, stratigraphy and paleogeography in Northeastern Slovenia and Southwestern Hungary. Geologija, 45, 103–114.*

Regional transect B

The ~220-km-long regional section (Fig. 1.5B) starts in the north from near the outcropping large syn-rift Miocene volcanic centers of the Mátra Mts. and the Triassic prerift carbonates in the Bükk Mts. (Fig. 1.3).

The northern basin margin is tilted and uplifted in this segment of the Pannonian Basin as well (Horváth and Cloetingh, 2006), similar to the ones depicted in Figs. 1.7 and 1.8 (Fig. 1.12). On the particular 2D seismic section chosen to illustrate this late-stage uplift, all TSMs can be observed (Fig. 1.12). The prerift TSM is made up of Triassic carbonates overlain by Paleogene clastics (Petrik, 2016). In the northern part of the Pannonian Basin, there are several Paleogene basin fragments that also belong to the prerift TSM (Tari et al., 1993).

The top of the Paleogene to Lower Miocene strata cannot be differentiated from the base of the syn-rift TSM, mostly because of the presence of Middle Miocene volcanics masking seismically the underlying basin fill. The volcanics can be quite thick, more than 1 km, especially in and around paleo-volcanoes like on the southern end of the section. There is a clastic part of the syn-rift TSM above the Miocene volcanics with well-imaged seismic signature. At the NW end of the seismic profile, the N-dipping normal fault defining a prerift high may be correlated with a regional-scale syn-rift transfer fault described by Tari et al. (1992). The basal part of the overlying postrift TSM has a seismic facies that corresponds to deeper water lacustrine shales deposited in Lake Pannon (Sztanó et al., 2013). This is overlain by a prograding delta sequence with some localized slumping and an incised valley feature at the southern end of the profile (Fig. 1.12). The postrift Pannonian TSM is truncated at the basin margin and a regional unconformity surface, marked by the onlap of several seismic reflectors defines the base of the uplift and/or inversion TSM here. This unconformity is intra-Pannonian in age (Fig. 1.12), possibly post-Pontian (Fig. 1.4), but a more precise age dating is missing in this example.

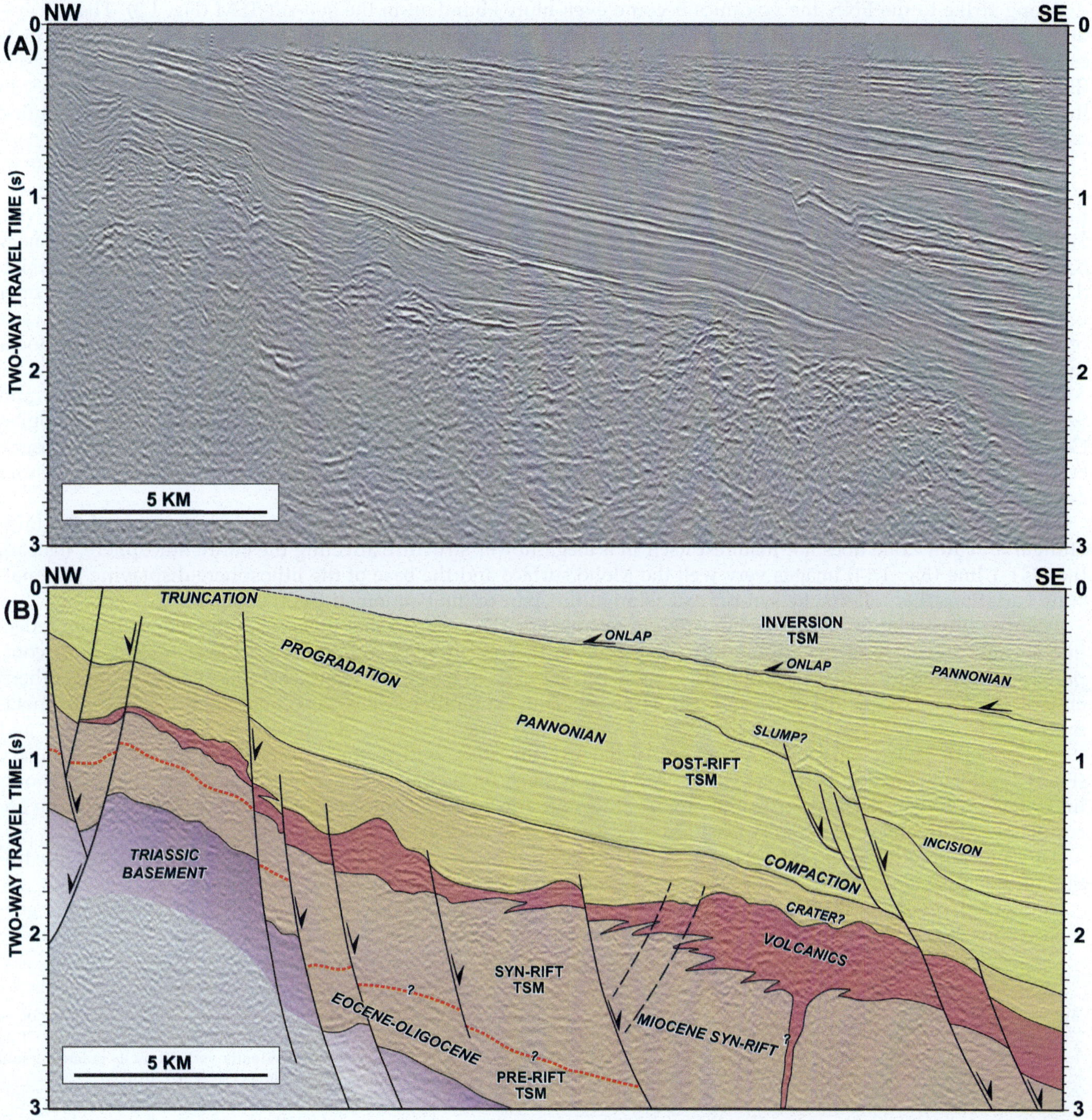

FIGURE 1.12 (A) 2D seismic section along the basin margin. For an approximate location, see Fig. 1.5b. (B) The interpretation illustrates well the regional-scale intra- and post-Pannonian uplift and truncation of the basin fill toward the north. Vertical exaggeration is about five-fold assuming an average seismic velocity of 2000 m/s. Source: *Adapted from Petrik, A. (2016) Cenozoic structural evolution of the southern Bükk foreland. Ph.D. thesis, Eötvös University, Dept. of Applied and Physical Geology, Budapest.*

Farther to the S along the regional transect, the syn-rift Miocene TSM consists of dominantly volcanics, both extrusives and intrusives (Fig. 1.5B). This interpretation is based not only on well-controlled but also on nearby 3D reflection seismic data (Petrik et al., 2018). The deep Jászság subbasin with its volcanics is indicating the zone of the Mid-Hungarian Line, a major Miocene strike-slip zone across the Pannonian Basin (Csontos and Nagymarosy 1998; Fig. 1.1).

Along strike to the ENE, the volcanics become even more abundant in the syn-rift TSM (Fig. 1.3). The integration of well-studied surface and modern analogs into the mapping of these exclusively subsurface features is especially important in areas where only legacy 2D seismic profiles are available. One example of these datasets (Fig. 1.13) illustrates the legacy seismic expression of a Miocene paleo-volcano in the Nyírség Basin (Fig. 1.3) in the back-arc region of the Pannonian Basin, where many of these features within the syn-rift TSM were identified using potential field methods (Kiss et al., 2019).

The Miocene volcanism in the intra-Carpathian region has been studied for more than a century (e.g., Harangi, 2001; Harangi and Lenkey, 2007; Seghedi and Downes, 2011; Harangi and Lukács, 2019 and references therein). There are many volcanic centers and subbasins along the arc and within the back-arc region as well. Since in the Pannonian Basin, most of the Miocene volcanics are buried beneath the Pliocene and Quaternary basin fill (e.g., Zelenka et al., 2004), their mapping requires geophysical interpretation either by seismic reflection data (e.g., Tari, 1994; Petrik et al., 2018) or magnetic field modeling (e.g., Kiss et al., 2019; Kiss, 2021). The proper understanding of igneous and volcaniclastic rocks requires a thorough knowledge of the geology and geophysics of these rocks, especially in the case of subsurface examples and the use of analogs, which are much better documented and understood, can be very useful (e.g., Bischoff et al., 2017).

An exceptionally deep, lithospheric-scale seismic reflection line (Fig. 1.14), published by Posgay and Szentgyörgyi (1991), runs parallel to the Transect B (Fig. 1.5B). The alternative interpretation of the Pannonian Geotraverse (PGT) 1 line is different in many ways from that of Posgay and Szentgyörgyi (1991) and Posgay et al. (1995). These authors identified the Mid-Hungarian Line as a zone of disturbance crosscutting the entire crust and even the mantle.

Our view, however, following that of Tari et al. (1999), is that the wide left-lateral shear zone of the mid-Hungarian Fault Zone does not manifest itself in a mega-flower structure affecting the entire lithosphere. Instead, the PGT-1 line (Fig. 1.14) images very well the Moho surface and the base of the lithosphere displaying regional-scale variations that reflects the syn-rift extension of this part of the Pannonian Basin complex.

Another important aspect of the lithospheric seismic line (Fig. 1.14) is that it captures the regional-scale reflection seismic signature of the Alpine thrust-fold belt in the prerift TSM, although at a lower resolution as compared to the standard industry seismic data, which were used to construct Transect B (Fig. 1.5B). At the southern end of the transect, on the Pusztaföldvár High, several exploration wells encountered medium-grade metamorphics in the pre-Cenozoic basement with stripes of unmetamorphosed Triassic in between (Kovács and Kurucz, 1984). Based on several cases of Alpine overthrusting within the prerift basement of SE Hungary (Fig. 1.15) documented in boreholes (Pap, 1990) and mapped on regional 2D seismic data, (Grow et al., 1994) outlined several N-verging nappes in the area on the southern end of Transect B. Many of these nappe units are correlatable with outcrops in the Apuseni Mts. of Romania, some 200 km to the east (Fig. 1.2).

Tari and Horváth (2006) and Tari et al. (2021) built on these results and here we present yet another illustration of the importance of the Alpine structural heritage in the prerift TSM.

A detailed version of the southern one-third of the regional cartoon transect of Tari et al. (2021) shows the geometric constraints by the borehole penetrations into the basement (Fig. 1.16). The substratum of the Cenozoic

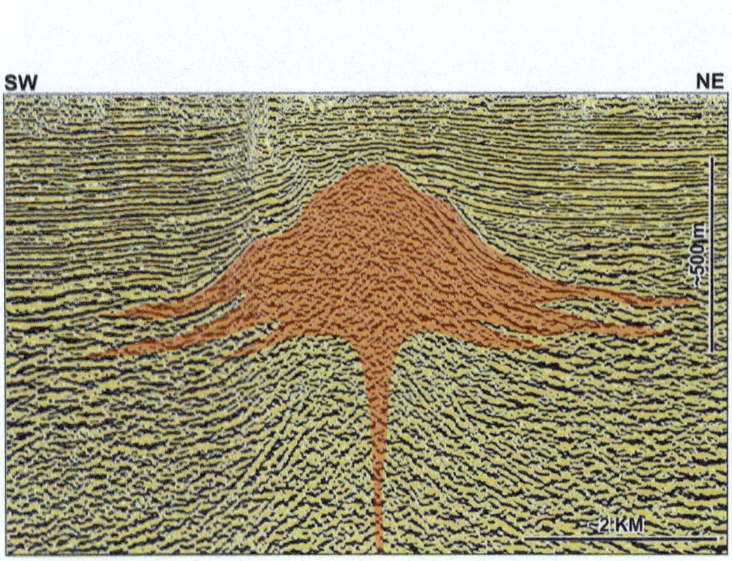

FIGURE 1.13 2D reflection seismic example of a Miocene paleo-volcano in the Nyírség Basin (Fig. 1.2), which has been independently verified by its magnetic field signal. Source: *Adapted from Kiss et al. (2019).*

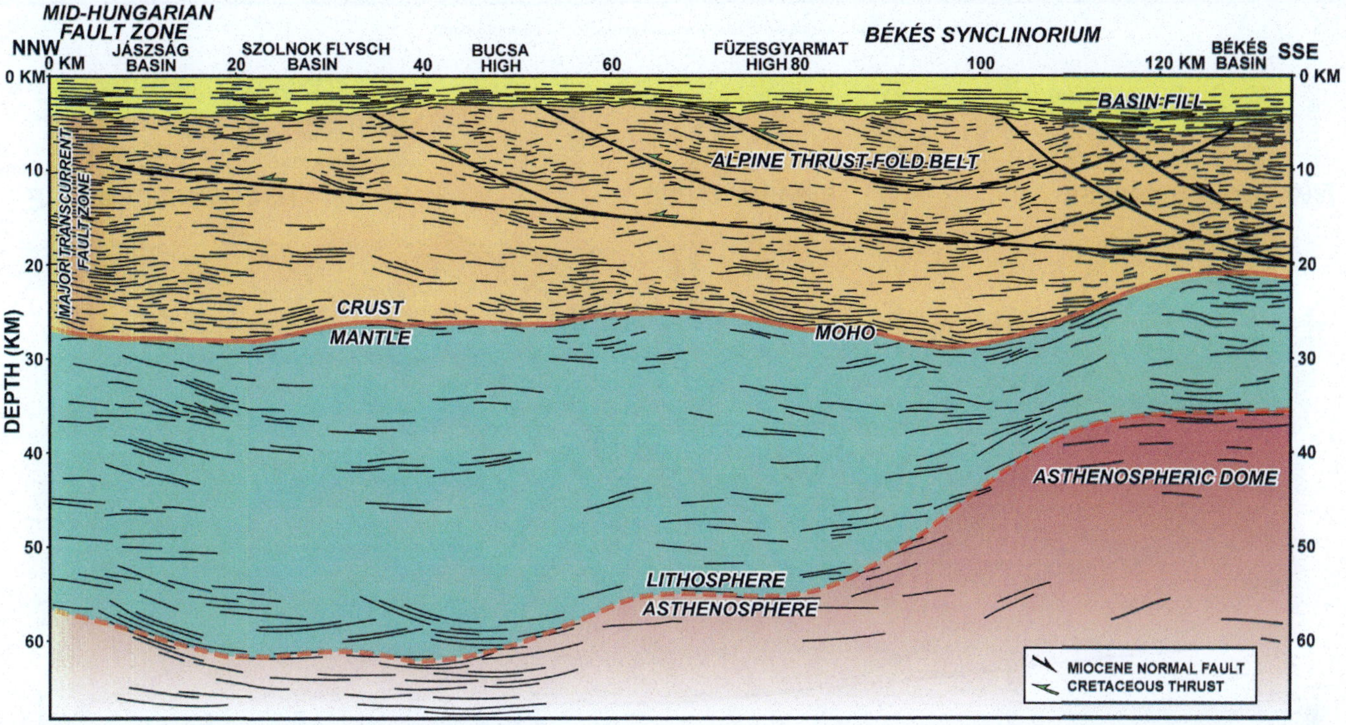

FIGURE 1.14 Example of a lithospheric scale depth-converted seismic section in the Pannonian Basin (PGT-1 of Posgay and Szentgyörgyi, 1991; Posgay et al., 1995), reproduced here as a line drawing reinterpretation. No vertical exaggeration. Source: *Modified after Tari, G., Dövényi, P., Dunkl, I., Horváth, F., Lenkey, L., Stefanescu, M., et al., 1999. Lithosphere structure of the Pannonian basin derived from seismic, gravity and geothermal data. In: Durand, B., Jolivet, L., Horváth, F., Séranne, M. (Eds.), Geological Society London Special Publications, 156, 215–250.*

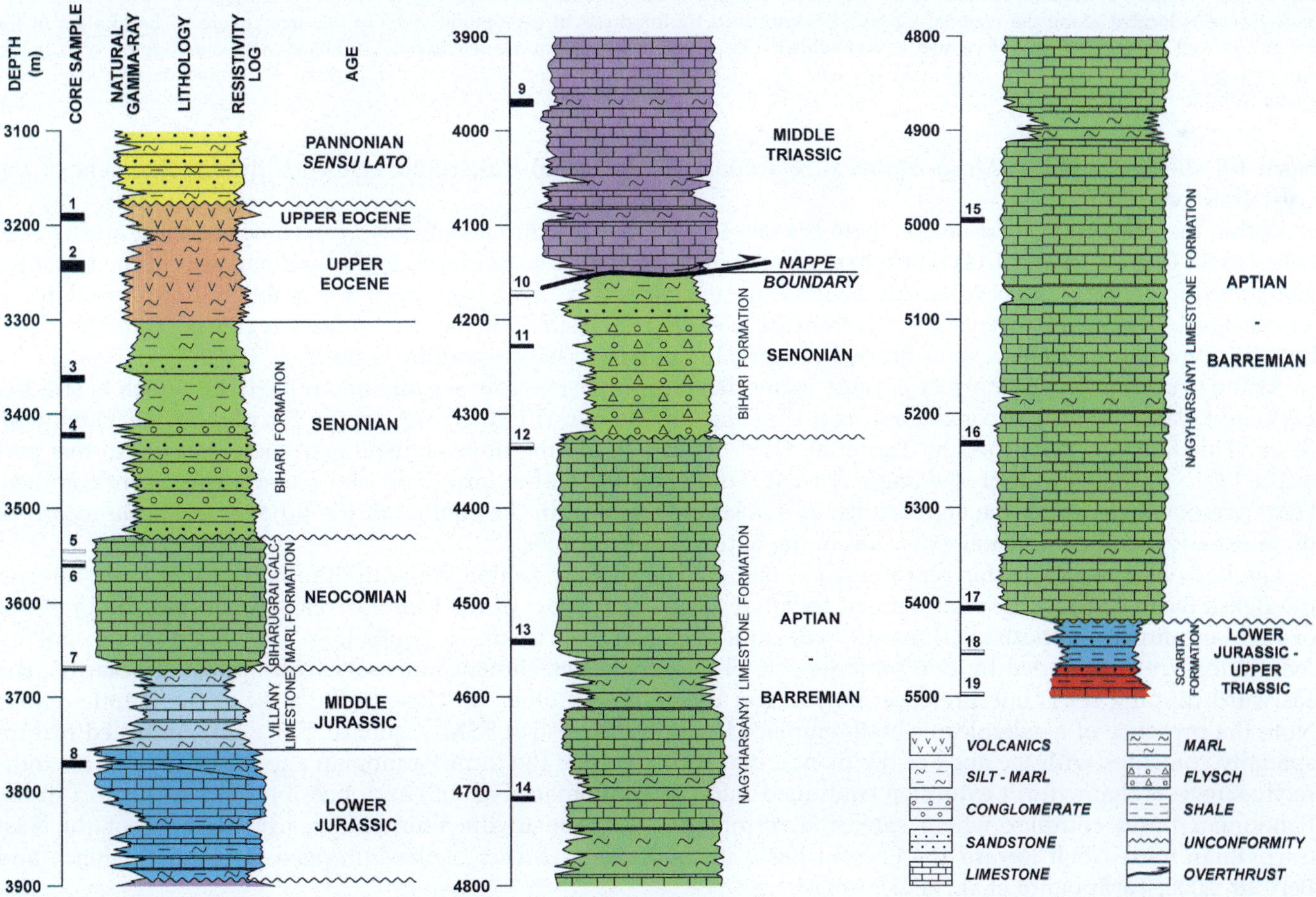

FIGURE 1.15 Exploration well data documenting a large Alpine overthrust plane at 4250 m depth in the Békés-2 well. This exceptional well drilled some 2400 m into the pre-Cenozoic basement beneath the 3180-m-thick basin fill (Pap, 1990). Average well penetrations into the prerift TSM are typically only on the order of 10 s of meters as it is considered the economic basement. *TSM*, Tectonostratigraphic megasequence.

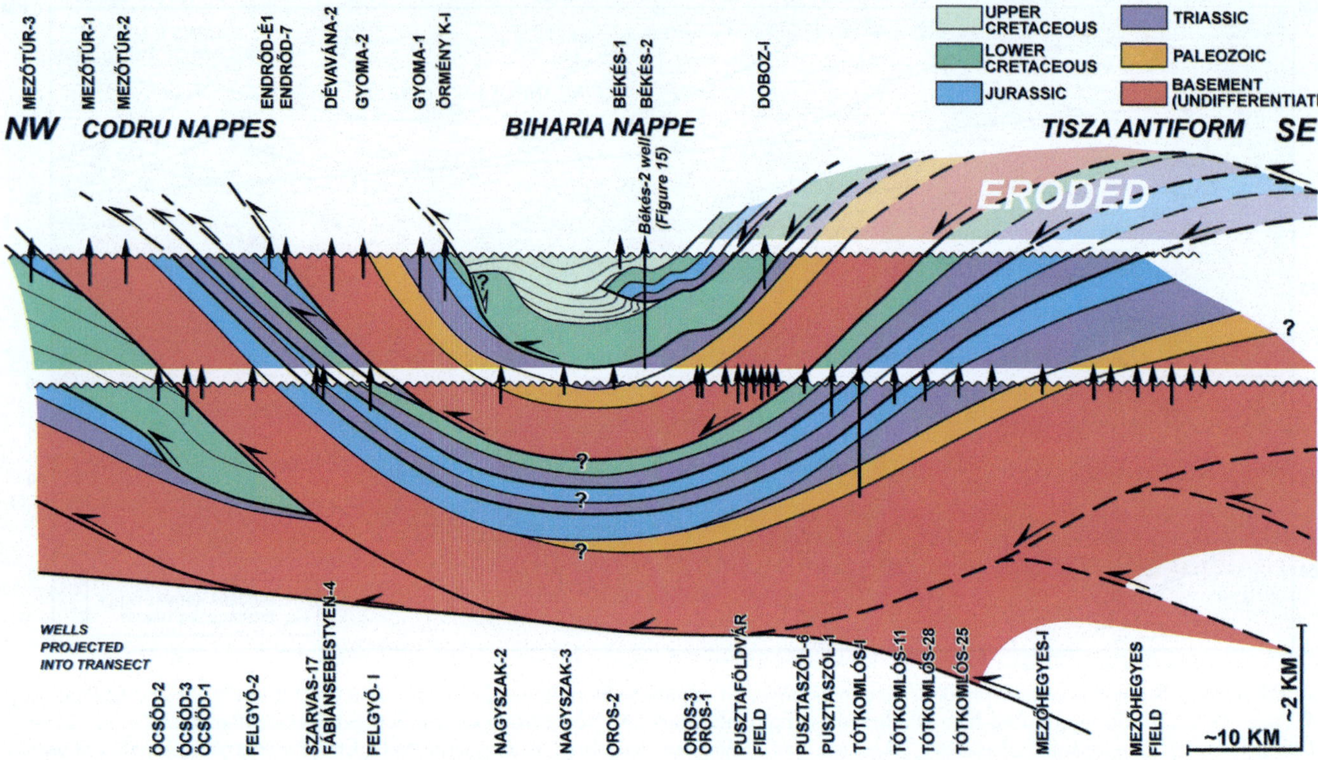

FIGURE 1.16 Simplified and cartoonish summary of the seismic and borehole-based definition of the Alpine folded belt of the prerift TSM beneath the Neogene Pannonian Basin by flattening on the top prerift unconformity as a reference level. This compilation is based on two separate transects located along the regional transect B constrained by hundreds of exploration wells in the area covered. The location of the Békés-2 is well shown in Fig. 1.15, which is highlighted. The nappe units shown are tentatively correlated with outcropping ones in the Apuseni Mts. of Romania, some 200 km to the east (Fig. 1.2), and this correlation is captured here by the name designations. *TSM*, Tectonostratigraphic megasequence.

basin fill shows an Eo- to Meso-Alpine (Cretaceous to Paleogene) thrust-fold belt with the involvement of the crystalline basement.

Within the pre-Cenozoic basement, there is a large synformal structure, the Békés Synform located just north of the Békés Basin (Figs. 1.5B and 1.14). The major antiformal stack, the Tisza Antiform, in the southern part of the section is interpreted to be the result of Meso-Alpine (Paleogene) contraction, involving a deeper, possibly middle crustal décollement level. The development of the antiformal stack is responsible for the pronounced regional northerly dip of both Mesozoic and crystalline units in southeastern Hungary already observed by Grow et al. (1994).

At the southern end of transect B, there is another deep, crustal-scale seismic line (Fig. 1.17), which is striking perpendicular to it and is considered as a dip line for the syn-rift structures in the SE part of the Pannonian Basin. This line was published by Tari et al. (1999) emphasizing the upper crustal extreme extension in this part of the basin in the context of low-angle normal faulting and even the formation of a metamorphic core complex. Here, we develop further the significance of this seismic section in the context of the lithospheric-scale record of the most advanced continental extension in the entire basin complex.

The key observation on this reprocessed version of the seismic section is the updoming of the mantle beneath the Békés Basin outlined by well-defined regional seismic reflectors of the Mohorovičić surface and the presence of high-amplitude subhorizontal mantle reflectors as shallow as 15 km in depth (Fig. 1.17). This diapir-looking mantle feature is bounded by two separate sets of low-angle detachment fault sets with opposing polarity. The eastward dipping set is interpreted to be younger offsetting the other set close to the apex of the mantle diapir. Note the presence of a subvolcanic body emplaced within the postrift TSM sequence. Since this undrilled feature spatially coincides with the apex of the mantle diapir, it suggests the intra-Pannonian age of the latter. This indirectly suggests that syn-rift extension continued into the Pannonian (Fig. 1.5) and, possibly, this is the area in the Pannonian Basin complex where extension terminated last due to the unimpeded advancement of the East Carpathian folded belt toward the Focsani Basin in the Vrancea corner of the European foreland (Matenco and Bertotti, 2000; Tărăpoancă et al., 2003).

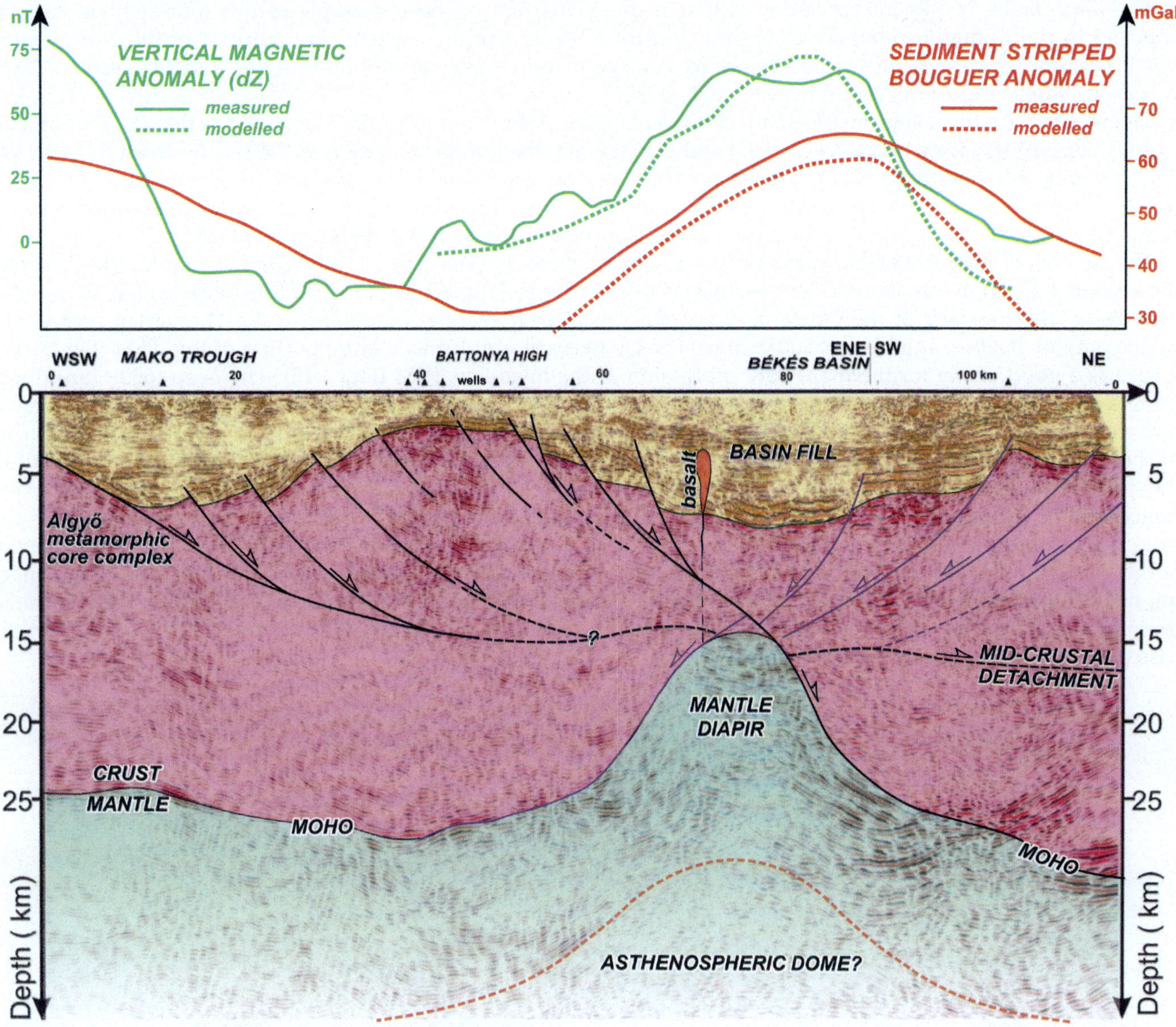

FIGURE 1.17 Crustal-scale depth-converted reflection transect PGT-4 in the SE Pannonian Basin. For approximate location, see Fig. 1.2. Note the orientation that is WSW-ENE, which is considered the dip direction for the Neogene basin structure in the SE Pannonian Basin. No vertical exaggeration. Source: *Adapted from Kiraly, Á., Horvath, F. and Tari, G., 2012. Revisiting the anomalous crustal structure of the Pannonian Basin. Abstract, 29th IAS Meeting of Sedimentology, Schladming, Austria.*

The seismic interpretation of the crustal line has been also confirmed by potential field modeling (Király et al., 2012), assuming serpentinized rocks in the mantle diapir with high magnetic susceptibility. The anomalous nature of both the gravity and magnetic fields beneath the Békés Basin has been already documented by Grow et al. (1996).

The significance of the new interpretation of the crustal-scale line is that shows a distinct deformation mode typical in the evolution of rifted margins as described, among others, by Péron-Pinvidic et al. (2013). The key geometric elements, such as (1) the large-scale detachment faults, (2) the individualization of the H-block (*sensu*, Lavier and Manatschal, 2006; in our case it is the Battonya-Pusztaföldvár High in the middle of the section, Fig. 1.17), (3) the large-scale Moho uplift and the rising mantle, and (4) the coupling of the upper and lower crustal conjugate deformation structures all point to the "thinning and coupling" phase defined in these models. This phase is preceding the next one, which would exhume the mantle at the basin floor by hyperextension. The model briefly outlined here is consistent with the fundamental geodynamic character of "Pannonian-type basins" where back-arc extension did not advance to the opening of an oceanic basin (Bally and Snelson, 1980; Roberts and Bally, 2012).

The importance of this interpretation is that it offers the first onshore example of this deformation stage as opposed to the numerous offshore examples (mostly Atlantic margins) where the footprint of the entire extensional breakup process is preserved in various passive margin segments (Péron-Pinvidic and Manatschal, 2009; Péron-Pinvidic et al., 2013; Péron-Pinvidic et al., 2022).

The reason why the continental breakup process stopped in the Pannonian Basin complex is that the Carpathians "softly" collided (Peresson and Decker, 1997) and docked onto the European margin in the east by about 6–8 ma ago (Fig. 1.2) and, therefore, there was no more space for back-arc extension. It goes beyond the scope of this chapter to speculate how much more horizontal extension would have been required to exhume the mantle and then a proto-oceanic basin in the SE Pannonian Basin, but it could be on the order of "just" another 50–100 km.

The inversion TSM is not indicated on the regional transects (Fig. 1.5), as the topmost few hundred meters thick sequence imaged by industry 2D seismic sections is always muted (e.g., Fig. 1.7). However, there is regional control on the thickness of the Quaternary sequence based on shallow boreholes in the Hungarian part of the basin complex (Gábris and Nádor, 2007). Since the Quaternary unit forms a large portion of this TSM (Fig. 1.4), it serves as a good proxy for the map-view expression of the inversion TSM (Fig. 1.18). The deep (650 + m) depocenters in SE Hungary are the result of relatively fast subsidence. However, these depocenters do not represent a renewed stage of extension in the basin but rather correspond to the crustal-scale buckling across the basin due to the prevailing compressional stress field (Horváth and Cloetingh, 1996; Cloetingh et al., 2006). As a result of this process, existing basins became even deeper (e.g., Fig. 1.7) and the broader structural highs in between experienced uplift (Koroknai et al., 2020; Ruszkiczay-Rüdiger et al., 2020). This neotectonic process also explains the presence of uptilted and eroded basin flanks (Figs. 1.7, 1.8, and 1.12). As Kovács and Telbisz (2013) documented at the western edge of the Pannonian Basin, even the Holocene river terraces show subtle tilting due to the ongoing neotectonic uplift. The interplay between the late-stage subsidence anomalies and the uplifting areas dictated the present-day course of the major rivers across the Pannonian Basin (Gábris and Nádor, 2007), especially in the case of the Tisza River (Fig. 1.18).

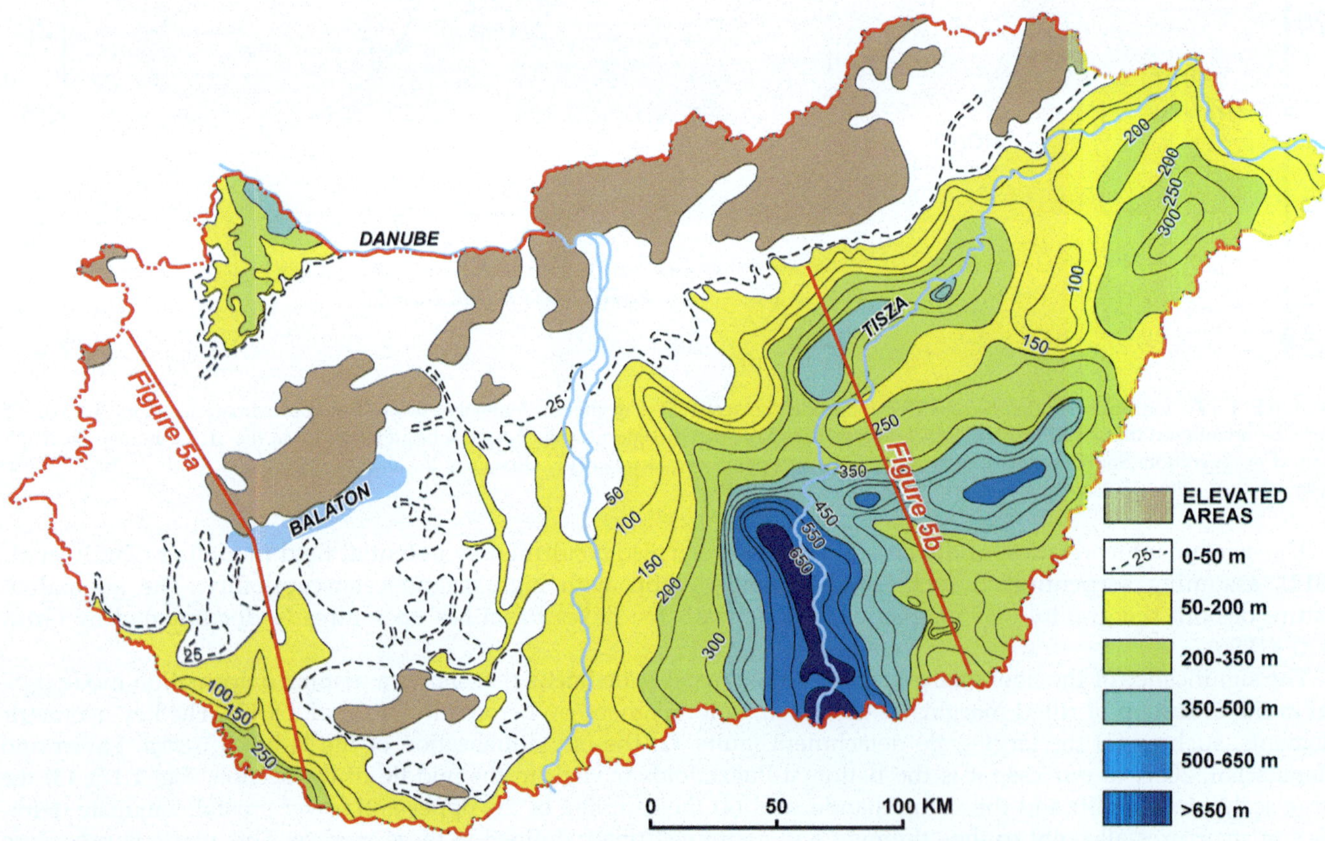

FIGURE 1.18 Isopach map of the Quaternary in the Hungarian part of the Pannonian Basin. This sequence forms a significant portion of the inversion TSM (Fig. 1.4); therefore, it is considered as a reasonable proxy for it. The regional transects shown on Fig. 1.5 are indicated by red lines. *TSM*, Tectonostratigraphic megasequence. Source: *Adapted from Gábris and Nádor (2007).*

Conclusions

The Pannonian Basin system offers exceptional insights to the evolution of back-arc basins due to the very large subsurface datasets, including industry and academic seismic and wells (Lemberkovics et al., 2018; Koroknai et al., 2020). The abundance of these datasets was, and still is, driven by ongoing oil and gas exploration efforts in every segment of the basin complex (Horváth et al., 2018; Boote et al., 2018). Due to the energy transition, many other geo-energy projects are also in progress in this basin complex, evaluating the potential for geothermal energy, carbon dioxide sequestration, hydrogen storage, and natural hydrogen.

The Pannonian Basin, therefore, provides a very useful analog for other basins worldwide, either recent or fossil, where continental extension did not advance to continental breakup and the formation of a passive margin. The Neogene Pannonian Basin complex provides a snapshot in the typical evolution path leading to the formation of an oceanic basin. Whereas the extensive calc-alkaline volcanism during the Early and Middle Miocene is the expression of the strike-slip faulting and retreating subduction zones (Lukács et al., 2018), the Upper Miocene to Pliocene alkaline basalts are related to the asthenospheric updoming beneath the basin complex (Harangi and Lenkey, 2007; Harangi and Lukács, 2019).

Since the back-arc extension terminated only about 6–8 million years ago, the various extensional basin elements are still very well preserved across the basin complex. The superposition of the syn-rift TSM on the prerift TSM was influenced by the preexisting Alpine contractional elements, such as thrust-fold belts of various styles and map view extent. This process occurred frequently, but not exclusively, by negative structural inversion as modern studies showed in the area (Tari et al., 2021). Again, from a regional geology point of view, the Pannonian Basin is a very proper *locus typicus* for back-arc extensional basins developed on highly extended continental crust (Bally and Snelson, 1980).

Another important aspect of this basin is that the extension was replaced by the basin-scale inversion of the entire system about 4–5 million years ago due to the overall convergence between Europe and Africa, which exerted a direct effect by the advancing Adriatic promontory to the SW of the Pannonian Basin. The ongoing inversion as a neotectonic process has already a significant footprint on the basin architecture, including its post-rift TSM, and the frequently observed positive structural inversion tends to utilize the original extensional fabric. The map-view propagation of inversion across the large area provides very useful analogs for the petroleum industry and the inversion TSM had a profound effect on the generation, migration, and selective preservation of hydrocarbons in the basin.

Finally, this chapter serves as an illustration of how the regional geology mindset (Roberts and Bally, 2012) may be used to understand a basin system. Beyond this, it is also important to realize that the geological and geophysical tools available to understand the subsurface is a very broad one. These diverse data input elements should range from raw magnetotelluric- to lithospheric-scale reflection seismic datasets complemented by various interpretation and modeling methods as highlighted in this paper.

Acknowledgments

This paper is dedicated to the memory of Prof. Albert W. Bally (1925–2019) who was the PhD Advisor of the first author at Rice University, Houston, Texas. Many of the ideas highlighted in this paper, such as the basin-scale inversion in the Pannonian region, were already recognized by him in the early 1990s. The authors are pleased to thank Nick Scarselli for inviting this contribution to this book series on regional geology. Discussions about the regional geology of the Pannonian Basin with the late Albert Bally, István Bérczi, David Boote, László Csontos, László Fodor, Zsófia Harald, the late Frank Horváth, Hans-Gert Linzer, Imre Magyar, Árpád Magyari, Béla Márton, Reinhard Sachsenhofer, Árpád Szalay, Orsolya Sztanó, Gábor Vakarcs, and István Vető are also acknowledged. Peter Pernegr kindly drafted some of the figures. Nick Scarselli provided helpful comments on the draft version of this paper.

References

Ádám, A., Horváth, F., Stegena, L., 1970. Investigation of plate tectonics by magnetotelluricanistropy.Annales Universitatis Scientiarum Budapestinensis de Rolando Eötvös Nominatae. sectioGeologica 14, 209–218.

Bada, G., Horváth, F., Dövényi, P., Szafián, P., Windhoffer, G., Cloetingh, S., 2007. Present-day stress field and tectonic inversion in the Pannonian basin. Global and Planetary Change 58, 165–180.

Bada, G., Szafián, P., Vincze, O., Tóth, T., Fodor, L., Spiess, V., Horváth, F., 2010. Theneotectonic habitat of the eastern part of Lake Balaton and its broader environs: inferencesfrom highresolutionseismic profiling. FöldtaniKözlöny. 367-390(in Hungarian) .

Balázs, A., Matenco, L., Magyar, I., Horváth, F., Cloetingh, S.A.P.L., 2016. The link between tectonics and sedimentation in back-arc basins: new genetic constraints from the analysis of the Pannonian Basin. Tectonics 35 (6), 1526–1559.

Balázs, A., Burov, E., Matenco, L., Vogt, K., Francois, T., Cloetingh, S., 2017. Symmetry during the syn-and post-rift evolution of extensional back-arc basins: the role of inherited orogenic structures. Earth and Planetary Science Letters 462, 86–98.

Balázs, A., Matenco, L., Vogt, K., Cloetingh, S., Gerya, T., 2018. Extensional polarity change in continental rifts: inferences from 3-D numerical modeling and observations. Journal of Geophysical Research: Solid Earth 123, 8073–8094.

Bally, A.W., Snelson, S., 1980. Realms of subsidence. In: Miall, A.D. (Ed.), Facts and principles of world petroleum occurrence, 6. Canadian Society of Petroleum Geologists Memoir, pp. 9–94.

Bartha, A., Balázs, A., Szalay, Á., 2018. On the tectono-stratigraphic evolution and hydrocarbon systems of extensional back-arc basins: inferences from 2D basin modelling from the Pannonian basin. Acta Geodaetica et Geophysica 53, 369–394.

Békési, E., Lenkey, L., Limberger, J., Porkoláb, K., Balázs, A., Bonté, D., et al., 2018. Subsurface temperature model of the Hungarian part of the Pannonian Basin. Global and Planetary Change 171, 48–64.

Bérczi, I., 1988. Preliminary sedimentological investigation of a Neogene depression in the Great Hungarian Plain. Am. Assoc. Petrol. Geol. Memoir 45, 107–116.

Bischoff, A.P., Nicol, A., Beggs, M., 2017. Stratigraphy of architectural elements in a buried volcanic system and implications for hydrocarbon exploration. Interpretation 5, SK141–SK159.

Boote, D.R.D., Sachsenhofer, R.F., Tari, G., Arbouille, D., 2018. Petroleum provinces of the Paratethyan region. Journal of Petroleum Geology 41 (3), 247–297.

Burchfiel, B.C., 1980. Eastern European alpine system and the Carpathian orocline as an example of collision tectonics. Tectonophysics 63, 31–61.

Cloetingh, S., Bada, G., Matenco, L., Lankreijer, A., Horváth, F., Dinu, C., 2006. Modes of basin (de)formation, lithospheric strength and vertical motions in the Pannonian-Carpathian system: inferences from thermo-mechanical modelling. Geological Society, London, Memoirs 32 (1), 207–221.

Csontos, L., Nagymarosy, A., 1998. The mid-Hungarian line: a zone of repeated tectonic inversions. Tectonophysics 297 (1–4), 51–71.

Csontos, L., Vörös, A., 2004. Mesozoic plate tectonic reconstruction of the Carpathian region. Palaeogeography, Palaeoclimatology, Palaeoecology 210, 1–56.

Dank, V., 1988. Petroleum geology of the Pannonian basin: an overview. Am. Assoc. Petrol. Geol. Memoir 45, 319–331.

Fodor, L., Jelen, B., Marton, E., Rifelj, H., Kraljić, M., Kevrić, R., et al., 2002. Miocene to Quaternary deformation, stratigraphy and paleogeography in Northeastern Slovenia and Southwestern Hungary. Geologija 45, 103–114.

Fodor, L., Balázs, A., Csillag, G., Dunkl, I., Héja, G., Jelen, B., et al., 2021. Crustal exhumation and depocenter migration from the Alpine orogenic margin towards the Pannonian extensional back-arc basin controlled by inheritance. Global and Planetary Change 201103475.

Fodor, L.I., Gerdes, A., Dunkl, I., Koroknai, B., Pécskay, Z., Trajanova, M., Horváth, P., Vrabec, M., Jelen, B., Balogh, K., Frisch, W., 2008. Miocene emplacement and rapid cooling of the Pohorje pluton at the Alpine-Pannonian-Dinaridic junction, Slovenia. Swiss Journal of Geosciences 101, 255–271.

Gábris, G., Nádor, A., 2007. Long-term fluvial archives in Hungary: response of the Danube and Tisza rivers to tectonic movements and climatic changes during the Quaternary: a review and new synthesis. Quaternary Science Reviews 26, 2758–2782.

Gagala, L., Vergés, J., Saura, E., Malata, T., Ringenbach, J.-C., Werner, P., et al., 2012. Architecture and orogenic evolution of the northeastern Outer Carpathians from cross-section balancing and forward modeling. Tectonophysics 532–535, 223–241.

Gjerazi, I., 2016, Structural re-evaluation of northern Burgenlandand NW Hungary based on subsurface data: M.Sc. thesis, University of Vienna.

Grow, J.A., Mattick, R.E., Bérczi-Makk, A., Péró, Cs, Hajdú, D., Pogácsás, Gy, et al., 1994. Structure of the Békés basin inferred from seismic reflection, well and gravity data. In: Teleki, P.G., Mattick, R.E., Kókai, J. (Eds.), Basin Analysis in Petroleum Exploration. Kluwer Acad. Publ., Dordrecht, pp. 1–38.

Harangi, Sz, 2001. Neogene to Quaternary volcanism of the Carpathian-Pannonian region - a review. Acta Geologica Hungarica 44, 223–258.

Harangi, Sz, Lenkey, L., 2007. Genesis of the Neogene to Quaternary volcanism in the Carpathian-Pannonian region: role of subduction, extension, and mantle plume. In: eds. Beccaluva, L., Bianchini, G., Wilson, M. (Eds.), Cenozoic volcanism in the Mediterranean area, 418. Geological Society of America Special Paper, pp. 67–92.

Harangi, Sz, Lukács, R., 2019. The Neogene to Quaternary volcanism and its geodynamic relations in the Carpathian–Pannonian Region. Földtani Közlöny 149, 197–232.

Horváth, F., 1993. Towards a mechanical model for the formation of the Pannonian basin. Tectonophysics 226, 333–357.

Horváth, F., 1995. Phases of compression during the evolution of the Pannonian Basin and its bearing on hydrocarbon exploration. Marine and Petroleum Geology 12, 837–844. Available from: https://doi.org/10.1016/0264-8172(95)98851-u.

Horváth, F., Berckhemer, H., 1982. Mediterranean backarc basins. Alpine-Mediterranean Geodynamics 7, 141–173.

Horváth, F., Cloetingh, S.A.P.L., 1996. Stress-induced late-stage subsidence anomalies in the Pannonian basin. Tectonophysics 266, 287–300.

Horváth, F., Rumpler, J., 1984. The Pannonian basement: extension and subsidence of an alpine orogene. Acta Geologica Hungarica 27, 229–235.

Horváth, F., Stegena, L., Berckhemer, H., 1981. Models of Mediterranean back-arc basin formation. Philosophical Transactions of the Royal Society of London. Series A, Mathematical and Physical Sciences 300 (1454), 383–402.

Horváth, F., Dövényi, P., Szalay, Á., Royden, L.H., 1988. Subsidence, thermal, and maturation history of the Great Hungarian Plain. Am. Assoc. Petrol. Geol., Memoir 45, 355–372.

Horváth, F., Bada, G., Szafián, P., Tari, G., Ádám, A., Cloetingh, S., 2006. In: Gee, D.G., Stephenson, R.A. (Eds.), Formation and deformation of the Pannonian basin: Constraints from observational data, 32. European Lithosphere Dynamics, Geol. Soc., London, Mem, pp. 191–206.

Horváth, F., Musitz, B., Balázs, A., Végh, A., Uhrin, A., Nádor, A., et al., 2015. Evolution of the Pannonian basin and its geothermal resources. Geothermics 53, 328–352.

Horváth, F., Dulić, I., Vranković, A., Koroknai, B., Tóth, T., Wórum, G., Kovács, G., 2018. Overview of geologic evolution and hydrocarbon generation of the Pannonian Basin. Interpretation 6, SB111–SB122.

Hubbard, R.J., Pape, J., Roberts, D.G., 1985. Depositional sequence mapping as a technique to establish tectonic and stratigraphic framework and evaluate hydrocarbon potential on a passive continental margin. In: Berg, O.R., Wolverton, D.G. (Eds.), Seismic Stratigraphy 11; An Integrated Approach to Hydrocarbon Exploration, vol. 39. American Association of Petroleum Geologists Memoir, Tulsa, pp. 79–92.

Jiřiček, R., 1979. Tectonic development of the Carpathian arc in the Oligocene and Neogene. In: Mahel, M. (Ed.), Tectonic Profiles through the Western Carpathians. Geol. Inst. Dionyz Stur, Bratislava, pp. 205–214.

John St., B., Bally, A.W., Klemme, H.D., 1984. Sedimentary provinces of the world–hydrocarbon productive and nonproductive. American Association of Petroleum Geologists, Tulsa, p. 35, 1 Map.

Juhász, G., 1994. Comparison of the sedimentary sequences in Late Neogenesubbasins in the Pannonian Basin, Hungary. FöldtaniKözlöny 124, 341–365 (in Hungarian).

Kalmár, D., Hetényi, G., Balázs, A., Bondár, I.AlpArray Working Group, 2021. Crustal thinning from Orogen to back-arc basin: the structure of the Pannonian Basin region revealed by P-to-S converted seismic waves. Journal of Geophysical Research: Solid Earth 126 (7) e2020JB021309.

Kingston, D.R., Dishroon, C.P., Williams, P.S., 1983. Global basin classification. American Association of Petroleum Geologists Bulletin 67, 2175–2193.

Király, Á., Horvath, F. and Tari, G., 2012. Revisiting the anomalous crustal structure of the Pannonian Basin. Abstract, 29th IAS Meeting of Sedimentology, Schladming, Austria.

Király, Á., Capitanio, F.A., Funiciello, F., Faccenna, C., 2016. Subduction zone interaction: controls on arcuate belts. Geology 44, 715–718.

Király, Á., Funiciello, F., Capitanio, F.A., Faccenna, C., 2021. Dynamic interactions between subduction zones. Global and Planetary Change 202103501.

Kiss, J., 2021. In search of buried volcanic eruption centres in North-Eastern Hungary. Magyar Geofizika 62, 150–169.

Kiss, J., Vértesy, L., Zilahi-Sebess, L., Takács, E., Gulyás, Á., 2019. Geophysical research of Nyírség area. Magyar Geofizika 60/3, 103–130.

Klemme, H.D., 1980. Petroleum basins–classifications and characteristics. Journal of Petroleum Geology 3, 187–207.

Koroknai, B., Wórum, G., Tóth, T., Koroknai, Z., Fekete-Németh, V., Kovács, G., 2020. Geological deformations in the Pannonian Basin during the neotectonic phase: new insights from the latest regional mapping in Hungary. Earth-Science Reviews 211103411.

Kovács, G., Kurucz, B., 1984. The pre-Mesozoic formations of the southern part of theGreat Hungarian Plain. Hung. Geol. Inst., Budapest, 55 p (in Hungarian).

Kovács, G., Telbisz, T., 2013. The role of tectonic and fluvial forces in the formation of the hilly area between the Kőszeg Mountains and the Rába River. Földtani Közlöny 143, 157–176.

Lavier, L.L., Manatschal, G., 2006. A mechanism to thin the continental lithosphere at magma-poor margins. Nature 440 (7082), 324–328.

Lemberkovics, V., Kissné Pável, E., Badics, B., Lőrincz, K., Rodionov, A., Galimullin, I., 2018. Petroleum system of Miocene troughs of the Pannonian Basin in southern Hungary, based on 3D basin modeling. Interpretation 6 (1), SB37–SB50.

Lenkey, L., Mihályka, J., Paróczi, P., 2021. Review of geothermal conditions of Hungary. FöldtaniKözlöny 151, 65.

Loisl, J., Tari, G., Draganits, E., Zámolyi, A., Gjerazi, I., 2018. High-resolution seismic reflection data acquisition and interpretation, Lake Neusiedl, Austria, northwest Pannonian Basin. Interpretation 6 (1), SB77–SB97.

Lukács, R., Harangi, Sz, Guillong, M., Bachmann, O., Fodor, L., Buret, Y., et al., 2018. Early to Mid-Miocene syn-extensional massive silicic volcanism in the Pannonian Basin (East-Central Europe): eruption chronology, correlation potential and geodynamic implications. Earth-Science Reviews 179, 1–19.

Magyar, I., Geary, D.H., Müller, P., 1999. Paleogeographic evolution of the late Miocene Lake Pannon in Central Europe. Palaeogeography, Palaeoclimatology, Palaeoecology 147, 151–167.

Magyar, I., Krezsek, C., Tari, G., 2020. Clinoforms as paleogeographic tools: Development of the Danube catchment above the deep Paratethyan basins in Central and Southeast Europe. Basin Research 32, 320–331.

Magyar, I., Radivojević, D., Sztanó, O., Synak, R., Ujszászi, K., Pócsik, M., 2013. Progradation of the paleo-Danube shelf margin across the Pannonian Basin during the Late Miocene and Early Pliocene. Global and Planetary Change 103, 168–173.

Matenco, L., Bertotti, G., 2000. Tertiary tectonic evolution of the external East Carpathians (Romania). Tectonophysics 316 (3–4), 255–286.

McKenzie, D., 1978. Some remarks on the development of sedimentary basins. Earth and Planetary science letters 40, 25–32.

Pap, S., 1990. Thrust-faulted sequences in Transtisia. Spec. Publ. Hung. GeolSurv., Budapest,36 p. In Hungarian.

Peresson, H., Decker, K., 1997. Far-field effects of Late Miocene subduction in the Eastern Carpathians: E-W compression and inversion of structures in the Alpine-Carpathian-Pannonian region. Tectonics 16 (1), 38–56.

Péron-Pinvidic, G., Manatschal, G., 2009. The final rifting evolution at deep magma-poor passive margins from Iberia-Newfoundland: a new point of view. International Journal of Earth Sciences 98 (7), 1581–1597.

Péron-Pinvidic, G., Manatschal, G., Osmundsen, P.T., 2013. Structural comparison of archetypal Atlantic rifted margins: a review of observations and concepts. Marine Petroleum Geology 43, 21–47.

Péron-Pinvidic, G., Fourel, L., Buiter, S.J.H., 2022. The influence of orogenic collision inheritance on rifted margin architecture: insights from comparing numerical experiments to the Mid-Norwegian margin. Tectonophysics 828, 229–273.

Petrik, A. (2016) Cenozoic structural evolution of the southern Bükk foreland. Ph.D. thesis, Eötvös University, Dept. of Applied and Physical Geology, Budapest.

Petrik, A., Fodor, L., Bereczki, L., Klembala, Z., Lukács, R., Baranyi, V., et al., 2018. Variation in style of magmatism and emplacement mechanism induced by changes in basin environments and stress fields (Pannonian Basin, Central Europe). Basin Research 31 (2), 380–404.

Posgay, K., Szentgyörgyi, K., 1991. Shear zone penetrating the lithosphere at the eastern part of the Pannonian Basin. Magyar Geofizika 32, 1–15.

Posgay, K., Bodoky, T., Hegedüs, E., Kovácsvölgyi, S., Lenkey, L., Szafián, P., et al., 1995. Asthenospheric structure beneath a Neogene basin in southeast Hungary. Tectonophysics 252, 467–484.

Radivojević, D., Radonjić, M., Katona, T.L., Magyar, I., 2022. Against the tide: southeast to northwest shelf-edge progradation in the southeastern margin of Lake Pannon, Banat (Serbia and Romania). International Journal of Earth Sciences 111 (6), 1–21.

Roberts, D.G., Bally, A.W., 2012. Some remarks on basins and basin classification and tectonostratigraphic megasequences. Regional Geology and Tectonics: Principles of Geologic Analysis. Elsevier, pp. 76–92.

Royden, L. and Horváth, F. (eds), 1988. The Pannonian Basin - a Study in Basin Evolution. Amer. Assoc. Petr. Geol. Mem., 45, p. 394.

Royden, L.H., 1993. The tectonic expression of slab pull at continental convergent boundaries. Tectonics 12, 303–325.
Royden, L.H., Burchfiel, C.B., 1989. Are systematic variations in thrust belt style related to plate boundary processes? (The Western Alps versus the Carpathians). Tectonics 8, 51–61.
Royden, L., Faccenna, C., 2018. Subduction orogeny and the Late Cenozoic evolution of the Mediterranean Arcs. Annual Review of Earth and Planetary Sciences 46, 261–289.
Royden, L.H., Horváth, F., Rumpler, J., 1983. Evolution of the Pannonian basin system: 1. Tectonics. Tectonics 2, 63–90.
Ruszkiczay-Rüdiger, Zs, Balázs, A., Csillag, G., Drijkoningen, G., Fodor, L., 2020. Uplift of the Transdanubian Range, Pannonian Basin: how fast and why? Global and Planetary Change 103263. Available from: https://doi.org/10.1016/j.gloplacha.2020.103263.
Sacchi, M., Horváth, F., 2002. Towards a new time scale for the Upper Miocene continental series of the Pannonian basin (Central Paratethys). EGU Stephan Mueller Special Publication Series 3, 79–94.
Sacchi, M., Horváth, F., Magyari, O., 1999. Role of unconformity-bounded units in the stratigraphy of the continental record: a case study from the Late Miocene of the western Pannonian Basin, Hungary. Geological Society, London, Special Publications 156, 357–390.
Scarselli, N., Adam, J., Chiarella, D., 2020. Regional Geology and Tectonics. Elsevier. Available from: 10.1016/B978-0-444-64134-2.00001-8.
Sclater, J.G., Royden, L.H., Horváth, F., Burchfiel, B.C., Semken, S., Stegena, L., 1980. The formation of the intra-Carpathian basins as determined from subsidence data. Earth Planet. Sci. Lett 51, 139–162.
Seghedi, I., Downes, H., 2011. Geochemistry and tectonic development of Cenozoic magmatism in the Carpathian-Pannonian region. Gondwana Research 20, 655–672.
Stewart, I.J., 2020. Regional geology and tectonics of sedimentary basins. In: Scarselli, N., Adam, J., Chiarella, D., Roberts, D.G., Bally, A.W. (Eds.), Regional Geology and Tectonics, Second Edition Elsevier, pp. 1–6. Available from: 10.1016/B978-0-444-64134-2.00001-8.
Šujan, M., Rybár, S., Kováč, M., Bielik, M., Majcin, D., Minár, J., et al., 2021. The polyphase rifting and inversion of the Danube Basin revised. Global and Planetary Change 196103375.
Sztanó, O., Magyar, I., 2007. Deltaic parasequences on gamma logs, ultra-high resolution seismic images and outcrops of Lake Pannon deposits. Joannea Geologie und Paläontologie 9, 105–108.
Sztanó, O., Szafián, P., Magyar, I., Horányi, A., Bada, G., Hughes, D.W., et al., 2013. Aggradation and progradation controlled clinothems and deep-water sand delivery model in the Neogene Lake Pannon, Makó Trough, Pannonian Basin, SE Hungary. Global and Planetary Change 103, 149–167.
Sztanó, O., Kováč, M., Magyar, I., Šujan, M., Fodor, L., Uhrin, A., et al., 2016. Late Miocene sedimentary record of the Danube/Kisalföld Basin: interregional correlation of depositional systems, stratigraphy and structural evolution. Geologica Carpathica 67, 525–542. Available from: https://doi.org/10.1515/geoca-2016-0033.
Tărăpoancă, M., Bertotti, G., Maţenco, L., Dinu, C., Cloetingh, S.A.P.L., 2003. Architecture of the Focşani depression: a 13 km deep basin in the Carpathians bend zone (Romania). Tectonics 22 (6).
Tari, G., 1994. Alpine tectonics of the Pannonian Basin. PhD thesis, Rice University, Houston, Texas, 501 p.
Tari, G., 1996. Extreme crustal extension in the Rába River extensional corridor (Austria/Hungary). Mitteil. Ges. Geol. Bergbaust. Österreich 41, 1–17.
Tari, G., Bally, A.W., 1990. Metamorphic core complexes at the boundary of the Eastern Alps and the Pannonian basin. Geological Society of America, Abstracts With Programs 22, 97–98.
Tari, G., Bérczi, I., 2018. Simon Papp, a prominent Hungarian petroleum geologist: how to run exploration projects from a prison cell. Geological Society, London, Special Publications 465, SP465-19.
Tari, G., Horváth, F., 2006. Alpine evolution and hydrocarbon geology of the Pannonian Basin: an overview. In: Golonka, J., Picha, F. (Eds.), The Carpathians and Their Foreland: Geology and Hydrocarbon Resources, 84. AAPG Memoir, pp. 605–618.
Tari, G., Horváth, F., Rumpler, J., 1992. Styles of extension in the Pannonian Basin. Tectonophysics 208, 203–219.
Tari, G., Báldi, T., Báldi-Beke, M., 1993. Paleogene retroarc flexural basin beneath the Neogene Pannonian basin: a geodynamic model. Tectonophysics 226, 433–455.
Tari, G., Dövényi, P., Dunkl, I., Horváth, F., Lenkey, L., Stefanescu, M., et al., 1999. Lithospheric structure of the Pannonian basin derived from seismic, gravity and geothermal data. In: Durand, B., Jolivet, L., Horváth, F., Séranne, M. (Eds.), The Mediterranean basins: extension within the Alpine Orogen, 156. Geological Society London Special Publications, pp. 215–250.
Tari, G.C., Németh, V., Horváth, F., Wesztergom, V., 2019. A regional Alpine graphite décollement level beneath the NW Pannonian Basin. Földtani Közlöny 149, 279-279.
Tari, G.C., Gjerazi, I., Grasemann, B., 2020. Interpretation of vintage 2D seismic reflection data along the Austrian-Hungarian border: subsurface expression of the Rechnitz metamorphic core complex. Interpretation 8 (4), SQ73–SQ91.
Tari, G., Bada, G., Beidinger, A., Csizmeg, J., Danišik, M., Gjerazi, I., et al., 2021. The connection between the Alps and the Carpathians beneath the Pannonian Basin: selective reactivation of Alpine nappe contacts during Miocene extension. Global and Planetary Change 197103401.
Tauber, A.F., 1959. Grundzüge der Tektonik des Neusiedlerseegebietes. Wissenschaftliche Arbeiten aus dem Burgenland 23, 26–31.
Tóth, T., Tari, G., 2014. Structural inversions in Western Hungaryand Eastern Slovenia: their impact on hydrocarbon trapping andreservoir quality. AAPG Search and Discovery Article 30387.
Tomljenović, B., Csontos, L., Márton, E., Márton, P., 2008. Tectonic evolution of the northwestern Internal Dinarides as constrained by structures and rotation of Medvednica Mountains, North Croatia, 298. Geological Society, London, pp. 145–167, Special Publications.
Uhrin, A., Sztanó, O., 2011. Water-level changes and their effect on deepwater sand accumulation in a lacustrine system: a case study from the Late Miocene of western Pannonian Basin, Hungary. International Journal of Earth Sciences 101, 1427–1440.
Vakarcs, G., Hardenbol, J., Abreu, V.S., Vail, P.R., Várnai, P., Tari, G., 1998. Oligocene-Middle Miocene depositional sequences of the central Paratethys and their correlation with regional stages. SEPM Special Publication 60, 209–231.
Vakarcs, G., Vail, P.R., Tari, G., Pogacsas, Gy, Mattick, R.E., Szabó, A., 1994. Third-order Middle Miocene-Early Pliocene depositonal sequences in the prograding delta complex of the Pannonian basin. Tectonophysics 240, 81–106.
Visnovitz, F., Horvath, F., Fekete, N., Spiess, V., 2015. Strike-slip tectonics in the Pannonian basin based on seismic surveys at Lake Balaton. International Journal of Earth Sciences (Geol. Rundsch.) 104 (8), 2273–2285. Available from: https://doi.org/10.1007/s00531-015-1179-x.

Visnovitz, F., Jakab, B., Czecze, B., Hámori, Z., Székely, B., Fodor, L., et al., 2021. High resolution architecture of neotectonic fault zones and post-8-Ma deformations in western Hungary: observations and neotectonic characteristics of the fault zone at the Eastern Lake Balaton. Global and Planetary Change 203103540.

Vrabec, M., Fodor, L., 2006. Late Cenozoic tectonics of Slovenia:structural styles at the Northeastern corner of the Adriatic microplate. In: Pinter, N., Grenerczy, G., Weber, J., Medak, D., Stein, S. (Eds.), The Adria microplate: GPS geodesy, tectonics andhazards. Springer, Dordrecht, the Netherlands, pp. 151–168.

Wórum, G., Koroknai, B., Koroknai, Zs, Fekete-Németh, V., Kovács, G., Tóth, T., 2020. Young Geological Deformations in Hungary. 1:500,000 Scale Map. Geomega Ltd, Budapest. Available from: 10.17632/dnjt9cmj87.1.

Zelenka, T., Balázs, E., Balogh, K., Kiss, J., Kozák, M., Nemesi, L., et al., 2004. Buried Neogene volcanic structures in Hungary. Acta Geologica Hungarica 47 (2–3), 177–219.

CHAPTER

2

Structural control on syn-rift sedimentation in rift basins: outcrop analogs for fault architectures and fan-delta geometries, northwestern margin of the Red Sea, Egypt

Samir Khalil[1] *and Ken McClay*[2,3]

[1]Department of Geology, Faculty of Science, Suez Canal University, Ismailia, Egypt [2]Department of Earth Sciences, Royal Holloway University of London, Egham, Surrey, United Kingdom [3]School of Physics, Chemistry and Earth Sciences, University of Adelaide, Adelaide, South Australia

Introduction

Knowledge of the structure and tectonic development of rift basins and the effect of extensional faulting on the depositional patterns of syn-rift sequences has increased immensely during the past few decades. Much of the research studies in extensional settings have evaluated the relative roles of fault evolution and interaction in controlling the deposition, dispersal, and architecture of syn-extensional sediments (Dart et al., 1994; Dawers and Underhill, 2000; Ford et al., 2016; Gawthorpe et al., 1990, 1997, 2017, 2022; Gawthorpe and Hurst, 1993; Gupta et al., 1999; Hardy et al., 1994; Hemelsdael and Ford, 2016; Li et al., 2021; Ma et al., 2021; Moustafa and Khalil, 2017; Purser and Bosence, 1998; Ravnås and Steel, 1998; Scholz et al., 1990; Young et al., 2000). Structural and sedimentological field studies of ancient and active rift basins indicate that the development of normal faults, their physiographic expressions, and variations in fault slip rates are the major factors influencing the spatial distribution and architecture of depositional systems adjacent to the fault zones (cf. Seger and Alexander, 1993; Dart et al., 1994; Gawthorpe and Leeder, 2000; Leeder et al., 2002; Leppard and Gawthorpe, 2006; Ford et al., 2016; Moustafa and Khalil, 2017; Pechlivanidou et al., 2019). In addition, seismic interpretations of rift structures in the East African rift system have shown that the primary control on syn-rift sedimentation is tectonic (Scholz et al., 1990, 1993), and it can be modified by prerift structural fabrics (cf. Versfelt and Rosendahl, 1989; Lezzar et al., 2002) and prerift topography (Crossley, 1984). Similarly, seismic datasets from many basins around the world, for example, central Corinth rift (Hemelsdael and Ford, 2016) and Bohai Bay Basin, NE China (Junqiao et al., 2018; Li et al., 2021 and Ma et al., 2021) have revealed the rule of fault propagation and relay-zone evolution on the rate of sediment supply, sediment delivery and depositional locations of syn-rift sediments.

In the northwestern Red Sea, early syn-rift clastics form coarse-grained fan deltas at key localities (these are recognized as sediment-input sites) along the coastal plain. The location, geometry, and development of these fan-delta systems in relation to the structural architecture and tectonic evolution of the northern Red Sea are still not fully understood. Understanding the close relationship between rift fault architecture and syn-rift sedimentation patterns may help to predict clastic reservoir distribution, which is a key element of many hydrocarbon plays in rift basins. In this chapter, we present three detailed case studies and conceptual models to illustrate the structural control on deposition of the Early Miocene syn-rift coarse clastics and characterize the influence of the

Regional Geology and Tectonics: Volume 2
DOI: https://doi.org/10.1016/B978-0-444-64136-6.00009-9

structures on the localization of sediment-input points in the northwestern margin of the Red Sea. We also present new detailed geological maps and cross-sections for the study area and describe the stratigraphic architecture and sedimentary facies of the Miocene clastic units and categorize the geometrical characteristics of the fan-delta systems. This study is based on a detailed field mapping (scale 1:20,000) and interpretation of Landsat and Spot Images of a 120-km-long segment of the northwestern Red Sea margin, between Safaga and just to the south of Quseir (Fig. 2.1), together with outcrop observations and conventional field mapping methods, including measurements of structural elements, facies description, and section logging.

Stratigraphy and structure of the northwestern Red Sea

The Red Sea is a relatively young and active continental rift system that initiated in the Late Oligocene–Early Miocene (Mckenzie et al., 1970; Le Pichon and Francheteau, 1978; Cochran, 1983; Steckler et al., 1988; Coleman, 1993; Bosworth and Stockli, 2016; Bosworth et al., 2005; 2020). It formed in response to the northeast separation of the Arabian plate from the African plate, and part of the plate movement was taken up by the opening of the Gulf of Suez rift during the Late Oligocene–Early Miocene (Hempton, 1987; Joffe and Garfunkel, 1987; Mckenzie et al., 1970; Girdler and Southren, 1987; Coleman, 1993). Part of the extension was taken by the sinistral strike-slip motion along the Aqaba–Dead Sea transform during the Mid-Miocene–Quaternary times (Freund, 1970; Mckenzie et al., 1970; Quennell, 1984; Morgan, 1990; Abdel Khalek et al., 1993; Bosworth et al., 2005; 2020; Bosworth, 2015; Moustafa and Khalil, 2020; Stockli and Bosworth 2019). In the Safaga–Quseir area, rotated extensional fault blocks expose pre- and syn-rift strata, together with large areas of the Precambrian basement, which form the Red Sea Hills on the rift margins (Fig. 2.1).

Stratigraphy of the northwestern Red Sea margin

Prerift strata

In the northwestern Red Sea, the prerift strata range in age from the Precambrian to the Eocene (Fig. 2.2). The Precambrian basement is heterogeneous in composition and it consists of igneous, metamorphic, and sedimentary rocks deformed during the Pan-African orogeny (Akaad and Noweir, 1980; Stern, 1981, 1994; Kröner; 1984, 1993; Stoeser and Camp, 1985). The basement exhibits a strong fabric, which includes two major shear zones (Hamrawin and El Quwyh shear zones, Fig. 2.1B). These shear zones are oriented WNW, oblique to the NNW rift trend, and were reactivated during the Miocene rifting (Moustafa, 1997; Khalil and McClay 2001; Younes and McClay, 2002). Fracturing is well developed in the basement, and this, together with the lithological heterogeneities, produced easily eroded weak zones in the basement. These acted as source areas for the Early to Mid-Miocene sediments, which were transported to the Red Sea basin through a number of structurally controlled sediment-input sites located along the rift margin.

The Precambrian basement is unconformably overlain by a 500- to 700-m-thick section of Lower Cretaceous–Middle Eocene strata (Fig. 2.2). These sediments occur in a few isolated, fault-bounded subbasins (e.g., Gebel Duwi and Gebel Hamadat; Fig. 2.1), as well as in relatively small, tilted fault blocks along the coastal area (Figs. 2.1 and 2.3). The lower part of this prerift sedimentary section is predominantly siliciclastic, with the lowermost unit consisting of Lower Cretaceous, shallow marine to fluvial sandstones of the Nubia Formation. This is overlain by interbedded shales, sandstones, and limestones with phosphate and oyster thin beds of the Upper Cretaceous to Paleocene Quseir, Duwi, Dakhla, and Esna Formations, respectively (Fig. 2.2) (Youssef, 1957; Abd El-Razik, 1967; Issawi et al., 1969; Said, 1990; Khalil and McClay 2001, 2020). This Upper Cretaceous–Paleocene mixed clastic and carbonate section is overlain by a 120- to 280-m-thick section of chalky and cherty limestones of the Lower to Middle Eocene Thebes Formation (Fig. 2.2).

Syn-rift strata of the northwestern Red Sea margin

The earliest syn-rift sediments of the northwestern Red Sea margin are lacustrine red mudstones, sandstone, breccias, and conglomerates of the Nakheil Formation (Fig. 2.2). The Nakeil Formation is Late Oligocene in age (Akkad and Dardir, 1966; Said, 1990) and is locally preserved in small hanging-wall synclines formed along the Border Fault system (Fig. 2.1). In Wadi El Nakheil (Fig. 2.1), the Nakheil Formation is 60–120 m thick, and it unconformably overlies the prerift Eocene Thebes Formation. At this location, the Nakheil Formation is locally covered by Quaternary alluvium and terraces. The Nakheil breccias and conglomerates consist of reworked chert

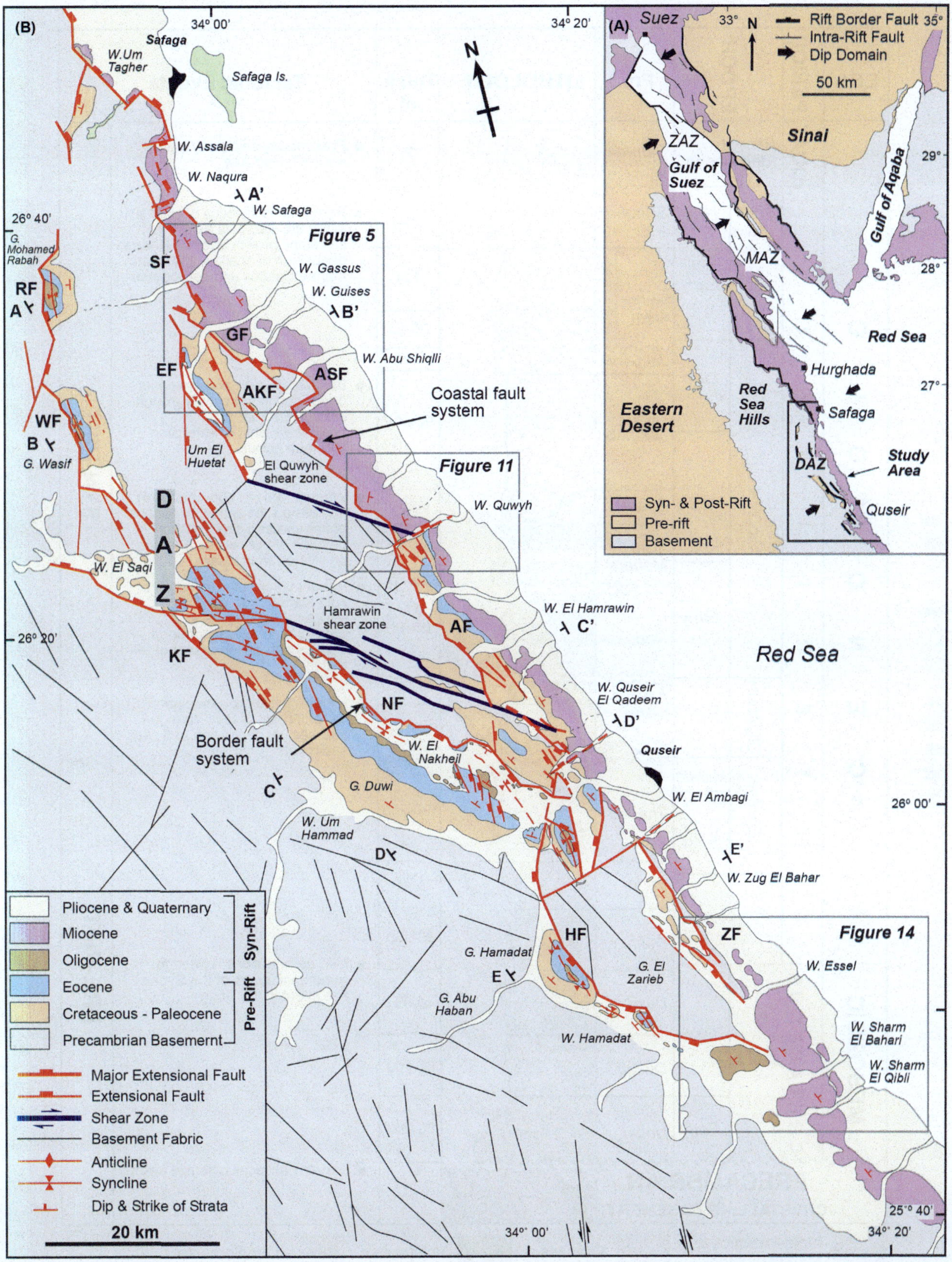

FIGURE 2.1 Surface geologic map of the northwestern Red Sea margin. Locations of cross-sections of Fig. 2.3 are labeled as A–A0 to E–E0. RF, WF, KF, NF, and HF indicate the Rabah, Wasif, Kallahin, Nakheil, and Hamadat segments of the Border Fault system, respectively. SF, EF, GF, AKF, ASF, AF, and ZF indicate the Safaga, Um El Huetat, Gassus, Abu Kherfan, Abu Shiqlli, Anz, and Zug El Bahar segments of the Coastal Fault system, respectively. Inset shows the tectonic setting of the Red Sea and location of the study area. ZAZ, MAZ, and DAZ indicate the Zaafarana, Morgan, and Duwi accommodation zones, respectively.

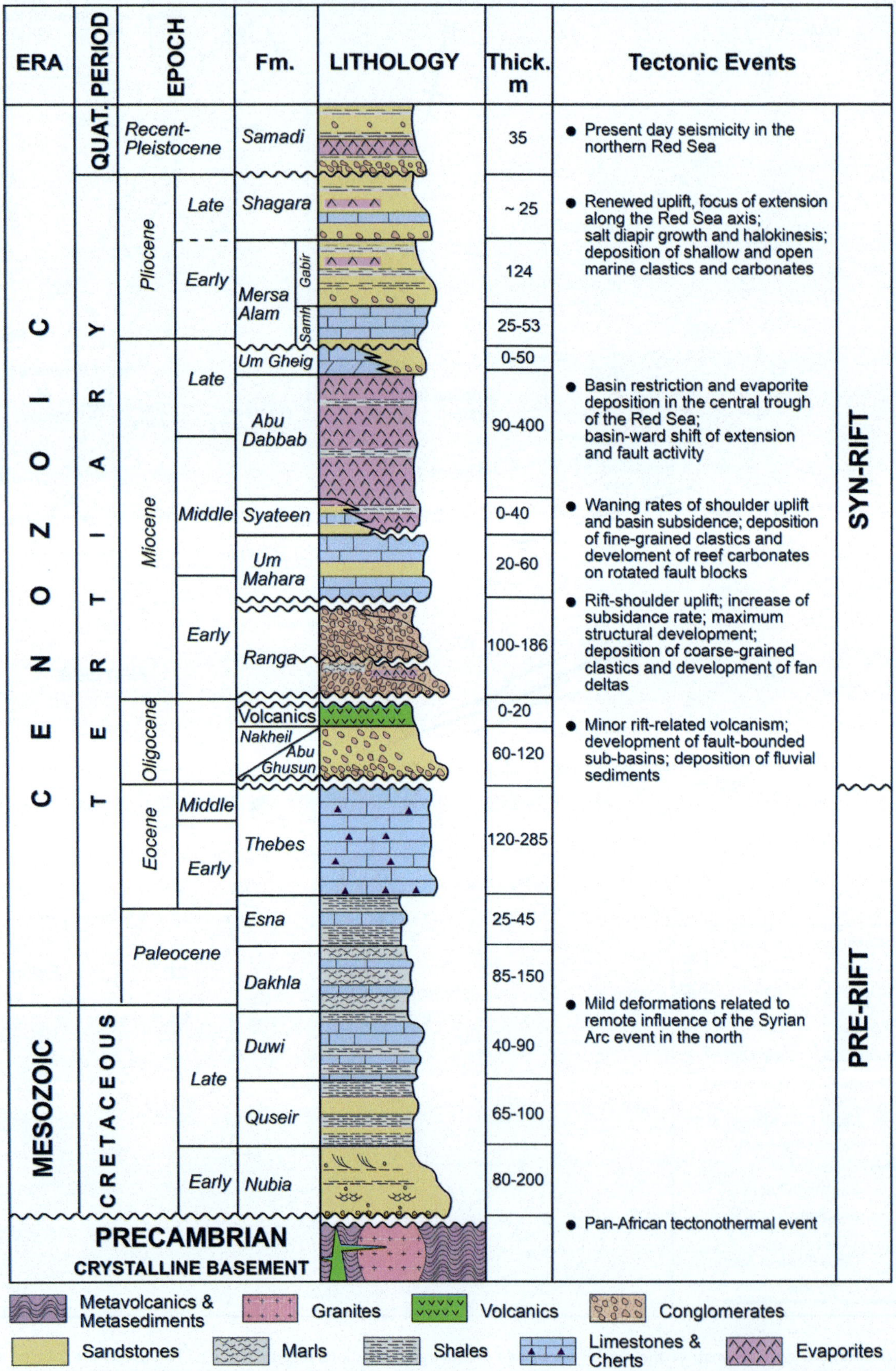

FIGURE 2.2 Summary stratigraphy of the northwestern Red Sea based on the authors' field mapping and data from references cited in the text.

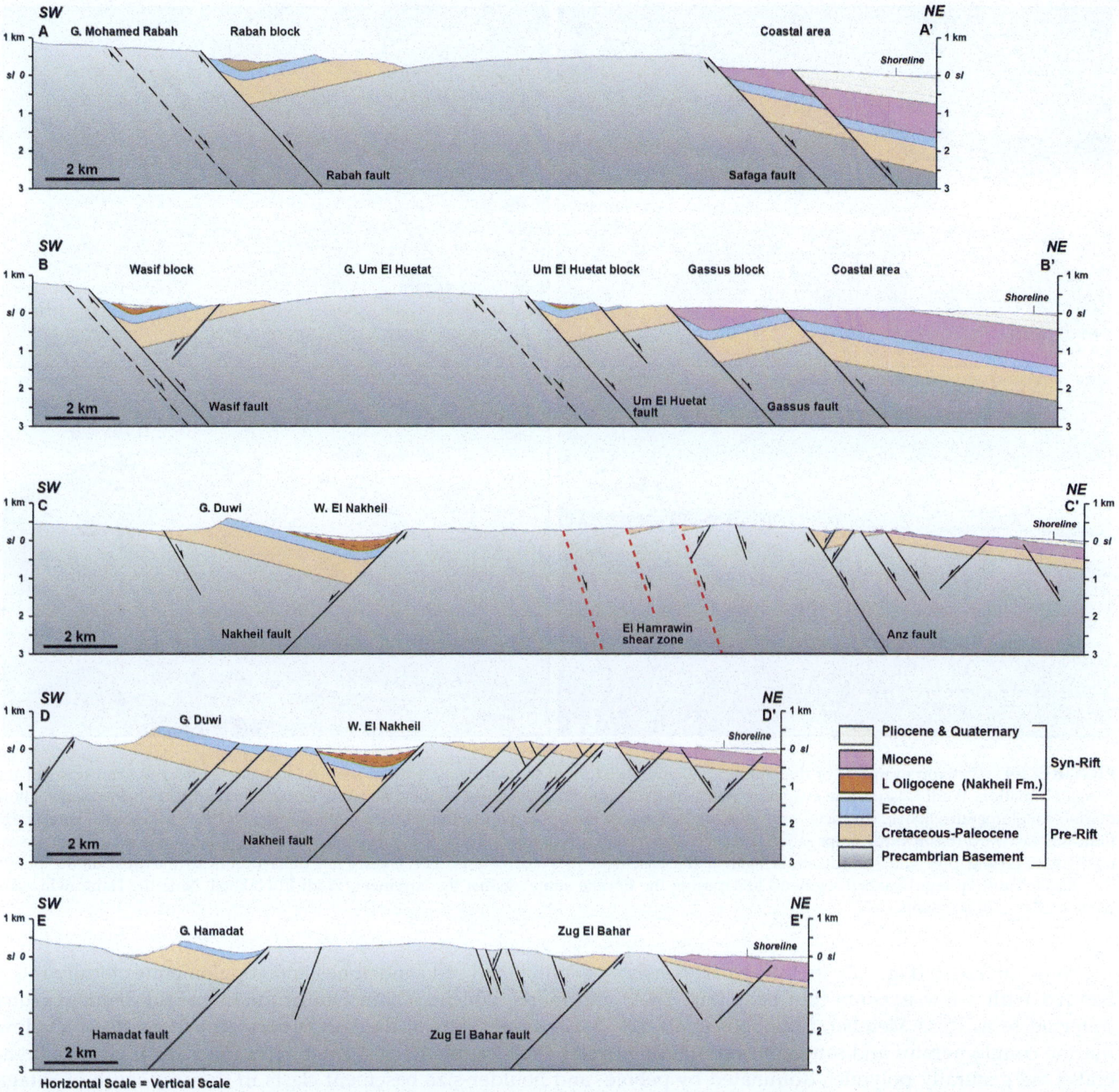

FIGURE 2.3 Cross-sections through the northwestern Red Sea margin, Egypt. The cross-sections illustrate the domino-style geometry of the fault blocks and change of polarity of the major block-bounding faults. Locations of cross-sections are shown in Fig. 2.1.

and limestone clasts derived from the underlying Eocene Thebes Formation. These grade upward into reddish and varicolored sandstones and mudstones interbedded with a few thin laminated carbonate beds. In the southern part of the study area (south Wadi Hamadat and Wadi Sharm El Bahari, Fig. 2.1), these conglomerates change to polymict facies (referred to as Abu Ghusun Formation; El Gezeery and Marzouk, 1974; Phillobbos et al., 1988) containing basement clasts, the proportion of which increases gradually upward, alternating with red sandstones and mudstones locally interbedded with, and overlain by, Upper Oligocene thin basaltic flow (dated at 24.9–26 Ma; El Haddad, 1984; Roussel et al., 1986).

The coarse-grained sandstones and conglomerates of the Aquitanian–Burdigalian age Ranga Formation (El Bassyony, 1982) form a distinct syn-rift unit that unconformably overlies the prerift Precambrian to Eocene units and in places also overlies the Late Oligocene syn-rift red clastics. The Ranga Formation varies from 100 to

FIGURE 2.4 Outcrop examples of the stratigraphy and structure of the northwestern Red Sea. (A) Conglomerates and sandstones of the Ranga Formation, Wadi Gassus. The evaporite interbeds (A, B, and C) of the Rosa Member separate fluvial conglomerates below and shallow marine conglomerates above. (B) Facies and stacking patterns of the upper part of the Ranga conglomerates, North Wadi Gassus. Foreset dip indicates eastward (basinward) propagation direction. Patch reefs are common in the topset beds. (C) Coastal Fault system (Abu Kherfan segment) juxtaposing pre- and syn-rift strata against the basement in Wadi Guesis. The Guesis fan delta occurs to the west of the photo. (D) Border Fault system (Nakheil segment) juxtaposing the Eocene strata against the basement, Wadi El Nakheil. Note the Hamrawin relay between the fault segments.

186 m in thickness (Fig. 2.2). Its lower part consists of continental red sandstones and conglomerates locally interbedded with a few gypsum thin beds (Fig. 2.4A) and marls, which contain Late Aquitanian–Burdigalian fauna (referred to as Rosa Member, Philobbos et al., 1993). The upper part of the Ranga Formation consists of shallow marine conglomerates and sandstones containing oyster shell fragments and patch reefs (Fig. 2.4B). The conglomerates are generally polymict, dominated by pebble- and boulder-size basement clasts in the area south of Safaga (Wadis Gassus, Guesis, and Quwyh, Fig. 2.1) and dominated by chert clasts in the area south of Quseir (Wadi Sharm El Bahari, Fig. 2.1).

Unconformably overlying the Ranga conglomerates are reefal limestones and shallow marine fine-grained clastics of the Late Burdigalian–Langhian Um Mahara Formation (Fig. 2.2) (El Bassyony, 1982; Said, 1990). The thickness of Um Mahara Formation is 20–60 m in the study area, and it increases southwards to about 180 m at Wadi Abu Ghusun (Said, 1990). Along the coastal area, in the region of Safaga, Wadis El Ambagi, and Zug El Bahar, the reefal limestone and sandy limestone of the Um Mahara Formation form local carbonate platforms with associated reef talus. These onlap the eastern slopes of exposed rotated basement blocks, which are topped in places by conglomerates and thin beds of detrital carbonates. The Middle–Late Miocene evaporite sequence of the Abu Dabbab Formation unconformably overlies the older syn-rift and prerift strata (Fig. 2.2). It consists of 90–400 m of massive to poorly bedded gypsum outcropping along the coastal plain, and it is up to 3 km in the offshore area (Tewfik and Ayyad, 1982; Said, 1990).

Overlying the evaporite sequence is a 200- to 300-m-thick succession of Upper Miocene–Quaternary sediments that are exposed along the coastal plain (Figs. 2.1 and 2.2). Offshore, these sediments can be up to 1000 m thick

(Heath et al., 1998; Orszag-Sperber et al., 1998). They include Upper Miocene–Pliocene shallow to open marine sandstones and mudstones of the Mersa Alam Formation, interbedded limestone, coarse sandstones, and conglomerates of the Shagara Formation, as well as Pleistocene conglomerates and coarse sands of the Samadi Formation (Fig. 2.2).

Structure of the northwestern Red Sea margin

The structural architecture of the northwestern margin of the Red Sea consists of two main extensional fault systems, together with a number of NW- and WNW-trending half-graben basins (Figs. 2.1 and 2.3). The western fault system (herein called the Border Fault system) forms the western margin of the rift and juxtaposes the pre-rift Cretaceous–Eocene strata and syn-rift Nakheil Formation against Precambrian basement in the footwall of the extensional faults. The eastern fault system (herein called the Coastal Fault system) borders the Red Sea coastal plain, and it juxtaposes pre- and syn-rift Miocene strata against the Precambrian basement (Figs. 2.1 and 2.3). The Coastal Fault system generally trends NW, subparallel to the shoreline, and delineates the syn-rift exposures along the Red Sea coast (Fig. 2.4C). The Border Fault system is strongly segmented and trends oblique to the rift trend, reflecting the inherited basement fabric (cf. Khalil and McClay, 2002, 2016, 2020). Both of these fault systems consist dominantly of NW, WNW, and NS fault segments that link through breached relay ramps and form a zig-zag fault pattern that bounds rhomboidal fault blocks (Figs. 2.1 and 2.4C and D). The throw of the Border and Coastal fault systems varies from 0.5 to >2 km (Khalil and McClay 2002) (Fig. 2.3).

In the northern part of the study area (between Safaga and the northern part of Gebel Duwi, Fig. 2.1), the Border Fault system dips northeastwards and bounds southwest-dipping half grabens (Mohamed Rabah and Wasif basins), whereas to the south, the Border Faults dip southwestwards and bound northeast-dipping half grabens (Duwi and Hamadat basins) (Figs. 2.1 and 2.3). This change in fault polarity and half-graben-dip direction occurs across a complex accommodation zone in the north of Gebel Duwi—called the Duwi accommodation zone (DAZ) (Moustafa, 1997; Khalil and McClay, 2001; Younes and McClay, 2002). The DAZ accommodates the transfer of displacement between the oppositely dipping fault segments of the Border Fault system and its location is controlled by the reactivation of the WNW trending Precambrian Hamrawin shear zone (Fig. 2.1; Moustafa, 1997; Younes and McClay, 2002). Northeast–southwest cross-sections through the northwestern Red Sea (Fig. 2.3) show that the half grabens are dissected by intra-rift faults into 1- to 3-km wide, domino-style fault blocks. These intra-rift faults are dominantly planar and have throws that range from tens of meters to a few hundred meters (Fig. 2.3).

Stratigraphic architecture and sedimentology of the Miocene syn-rift strata

In the Safaga–Quseir study area (Fig. 2.1), the Lower Miocene syn-rift clastics are localized in a number of structurally controlled sediment-input sites. Of particular importance are four sites that clearly show the interrelationship between the structural architecture and localization and accumulation of syn-rift clastics. From north to south, these are located at Wadi Gassus, Wadi Guesis, Wadi Quwyh, and Wadi Sharm El Bahari (Fig. 2.1). At these localities, the coarse-grained deposits of the Lower Miocene syn-rift Ranga Formation form coarse-grained fan-delta systems, which are dominated by pebble and cobble conglomerates, with minor amounts of pebbly and muddy sandstones. These Lower Miocene clastics unconformably overlie the prerift strata or Late Oligocene early syn-rift strata (Nakhiel and Abu Ghusun Formations) and are overlain by reef carbonates and/or evaporite successions of the Middle to Upper Miocene Um Mahara and Abu Dabbab Formations (Fig. 2.2).

Wadi Gassus–Wadi Guesis area

In the Wadi Gassus and Wadi Guesis areas (Fig. 2.5), Lower Miocene clastics of the Ranga Formation develop kilometer-scale fan-delta systems, formed in the hanging wall of the fault segments of the Coastal Fault system. In these two Wadis, the Ranga Formation consists of conglomerates and sandstones, with moderately to steeply dipping foresets prograding eastward (basinward). The structural setting and geometry of the Gassus and Guesis fan systems are shown in Figs. 2.5–2.8.

The Wadi Gassus fan delta is about 10 km^2 in outcrop area (Figs. 2.5 and 2.6), with a longitudinal length (parallel to the progradation direction) of 2 km and an average strike length of 5 km. The fan delta includes a pre-evaporite clastic section, which is 65–140 m in thickness (Fig. 2.9) and has a thickness/progradation ratio of

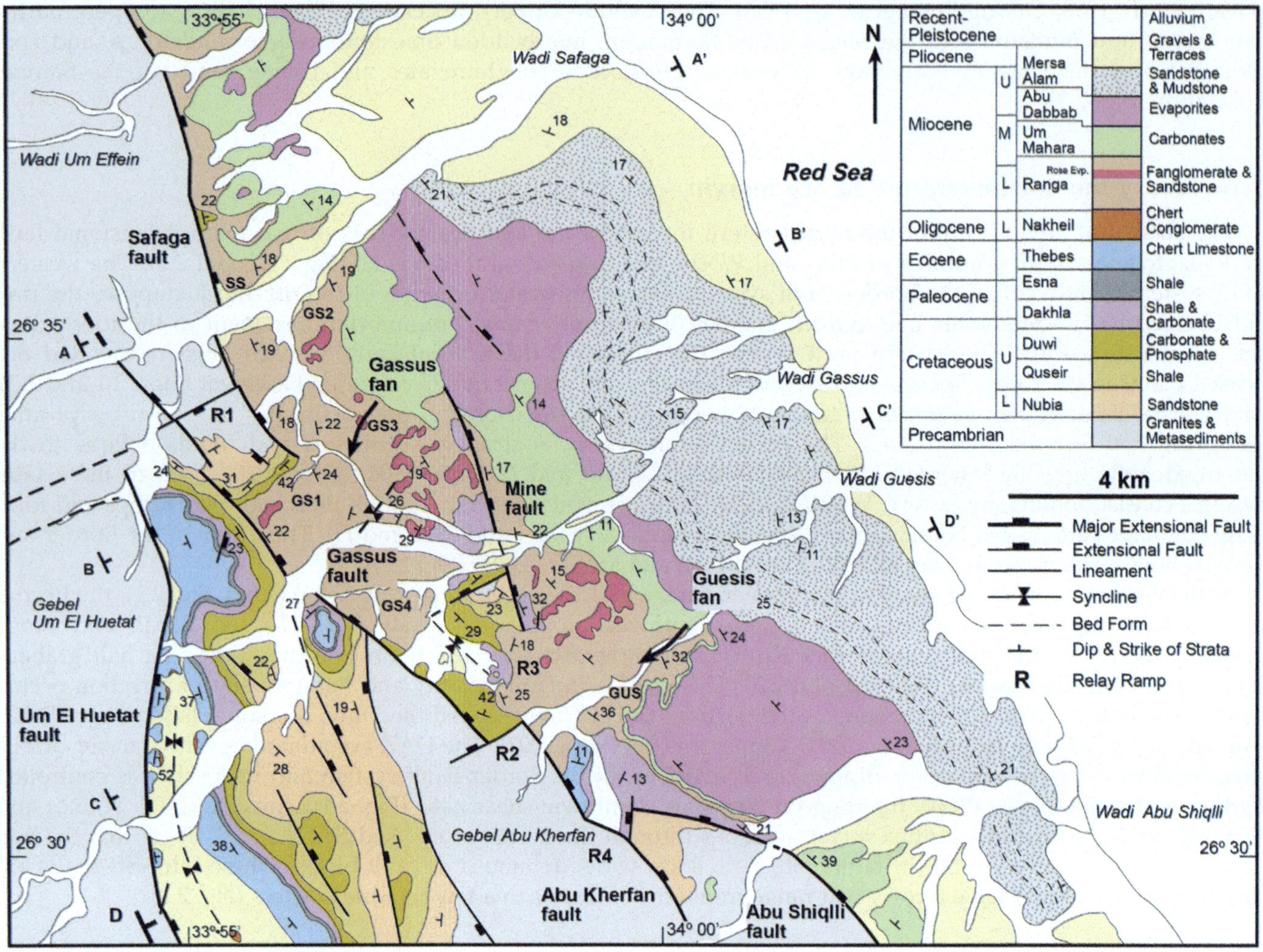

FIGURE 2.5 Detailed geologic map of the Wadi Gassus and Wadi Guesis area. R1, R2, R3, and R4 are relay ramps between the fault segments. GS and GUS indicate the locations of the measured log sections.

0.03–0.07. In Wadi Gassus, the conglomerates of the Ranga Formation are displaced against the Quseir and Duwi Formations, and in places unconformably overlie the Dakhla and Thebes Formations in the hanging-wall of the Gassus and Abu Kherfan fault segment (Fig. 2.4C).

On the northern flank of the Wadi Gassus (Figs. 2.1 and 2.5), the lower 30 m of the Ranga Formation consists of interbedded red siltstones and sandstones, and polymictic conglomerates that include chert and basement clasts (Fig. 2.8B and C). Palaeosols and red fine clastics (Fig. 2.8B) in this part of the section indicate subaerial exposure and deposition in the continental environment. The remainder of the section consists of basement–clast conglomerates and includes sandstones and mudstones in thin beds and lenses. Within the clastic section, there are three to five evaporite beds (Lower Miocene Rosa Member) whose average thickness is 4 m (Fig. 2.9; also Fig. 2.4A). In places, these evaporaite beds overlie thin algal laminates (stromatolites) and shale units which, together with the evaporite beds, represent the restricted transitional facies that mark the Early Miocene marine transgression (Philobbos et al., 1993; Purser and Philobbos, 1993). The basement clasts are dominantly granitic, but they also include metavolcanic and metasedimentary components (Fig. 2.8C). They are subrounded to rounded and vary in size from 3 to 25 cm. Palaeocurrent data (based on conglomerate imbrications, cross-bedding, and the dip directions of foreset beds) indicate NE to ENE palaeoflow directions (Figs. 2.8B and 2.9). The clastic section in Wadi Gassus is unconformably overlain by the Middle Miocene carbonates of the Um Mahara Formation and the Middle–Upper Miocene evaporites of the Abu Dabbab Formation. The thickness of the clastic section decreases from 140 m at the central part of Wadi Gassus to about 100 m in Wadi Safaga in the north and 65 m in the southern flank of Wadi Gassus.

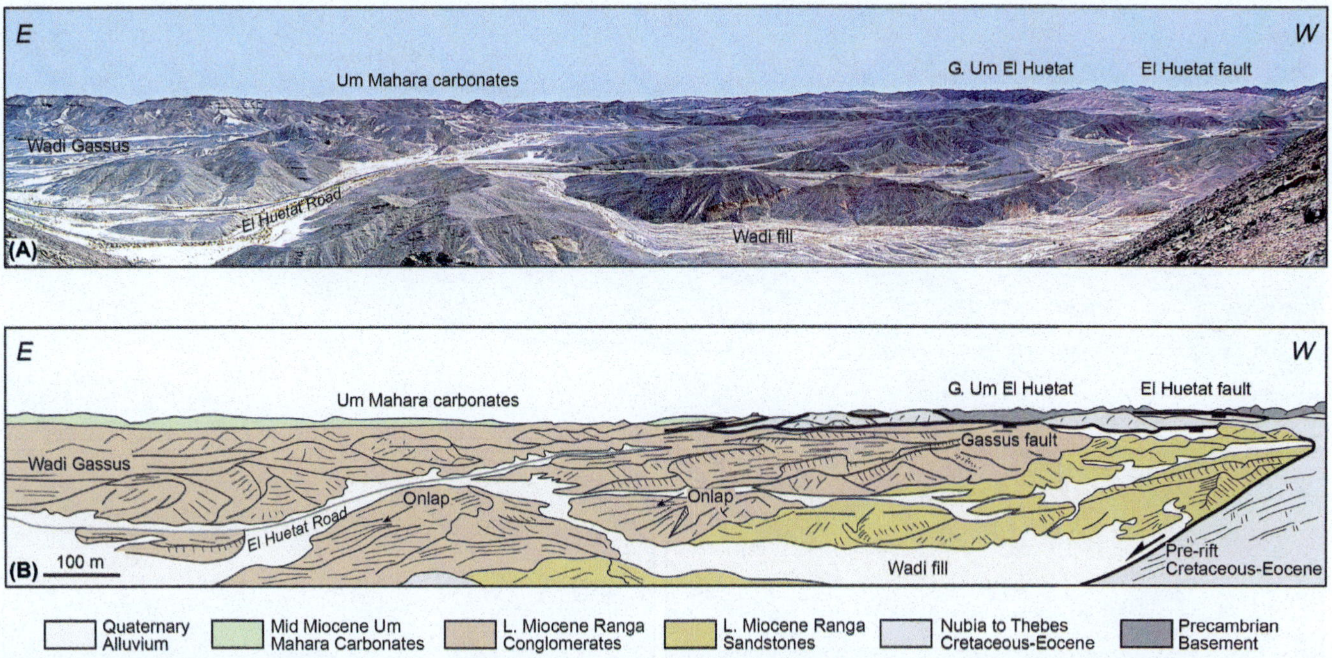

FIGURE 2.6 Field photo (A) and perspective sketch (B) showing the Wadi Gassus fan delta and the Gassus fault segment.

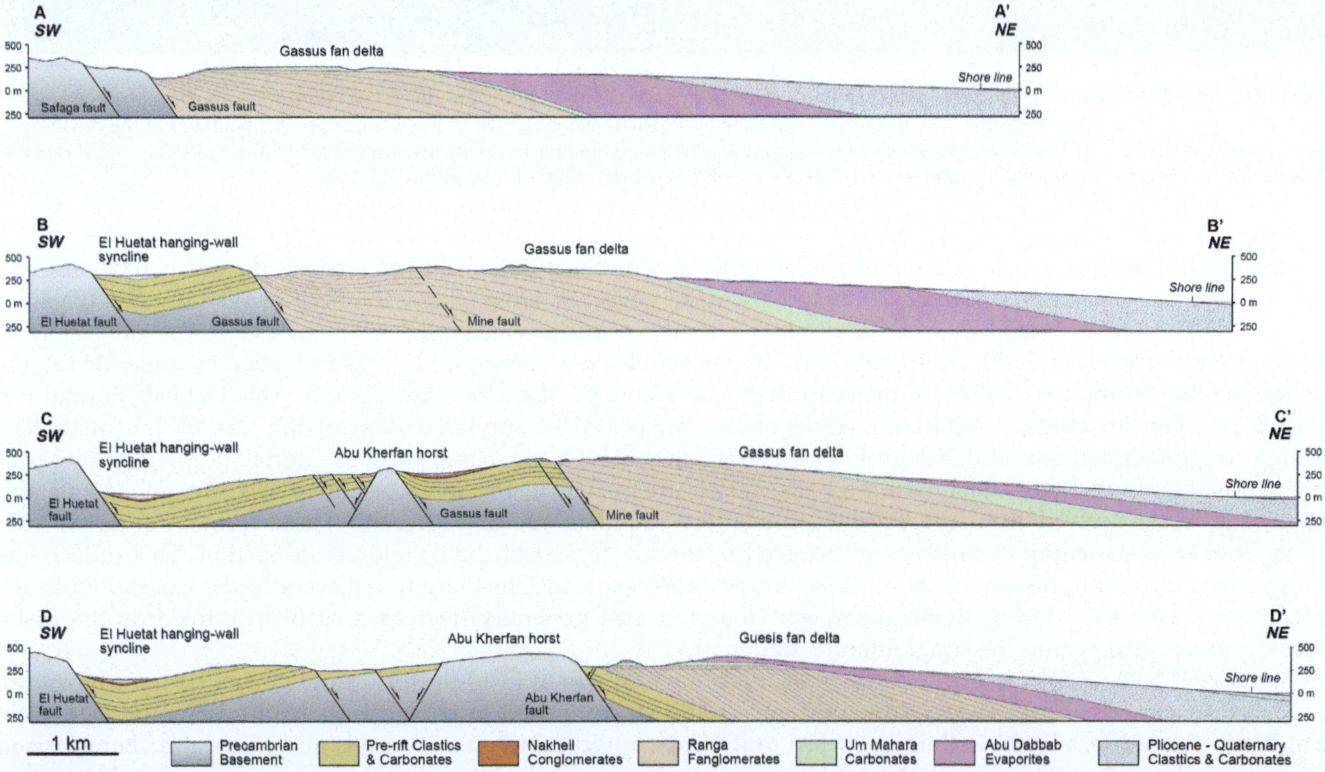

FIGURE 2.7 Cross-sections showing the Wadi Gassus (sections A, B, and C) and Wadi Guesis (Section D) fan delta systems. Note wedge- to sheet-like geometries of the fans and the moderate to steep dip of the foreset beds. Locations of cross-sections are shown in Fig. 2.5.

FIGURE 2.8 Field views showing the geometry and sedimentary facies of the Gassus fan delta. (A) South view showing the geometry of the Gassus fan delta, propagation direction and foreset dip in the immediate hanging wall of the Safaga fault segment and at the downdip of the northern relay R1. (B) Paleosole, red sandstone, and polymictic conglomerate facies in the lower part of the fan delta. (C) Basement-dominated, sand-matrix-supported conglomerate with imbrication fabric indicating NE palaeoflow direction.

The Wadi Guesis fan delta is about 4 km^2 in outcrop area (Fig. 2.5) and it has a sheet-like geometry, with longitudinal and strike lengths of about 2 km and a vertical height of 120 m above the floor of Wadi Guesis (Figs. 2.7 and 2.10). The fan geometry is well exposed in a cliff exposure, which runs near parallel to the progradation direction of the fan (Fig. 2.10). Along this cliff, the fan sequence is composed of a 95-m-thick conglomerate section of the Ranga Formation, which is unconformably overlain by the Um Mahara and Abu Dabbab Formations (Fig. 2.10). The thickness/progradation ratio of the fan is ~0.06, similar to that of the Gassus fan delta. The Ranga conglomerates consist dominantly of granitic and metavolcanic clasts and are interbedded with a few pebbly sandstone, mudstone, and limestone beds, which are 2–3 m in thickness (section GUS, Fig. 2.9). Chert and limestone clasts derived from the prerift sedimentary sequence are also present. These form a minor fraction (<15%) of the clast component and are localized in the lower part of the conglomerate section. This reflects the progressive unroofing history of the uplifted footwall blocks in the rift margin. Adjacent to the Coastal fault, the basement clasts are 0.5–1 m in diameter, and the clast size generally decreases eastwards towards the distal areas, and upward within the conglomerate section.

The topset beds of the fan are subhorizontal, and they form an 8- to 10-m-thick unit that can be traced for hundreds of meters (Fig. 2.10). These beds are thinly stratified, composed of pebble and cobble conglomerates set in sandy matrix and locally show fining-upward units. The contact between the topsets and foresets is characterized by toplap of the foresets against the topset beds, creating typical oblique clinoform geometry similar to that described from other fan-delta systems with comparable extensional setting (cf. Gawthorpe et al., 1990; Dart et al., 1994; Rubi et al., 2018). This geometry may be interpreted to indicate stillstand conditions where erosion occurred above sea level, and the sediments bypassed the break in slope and were deposited basinward on the foreset slope (Gawthorpe et al., 1990).

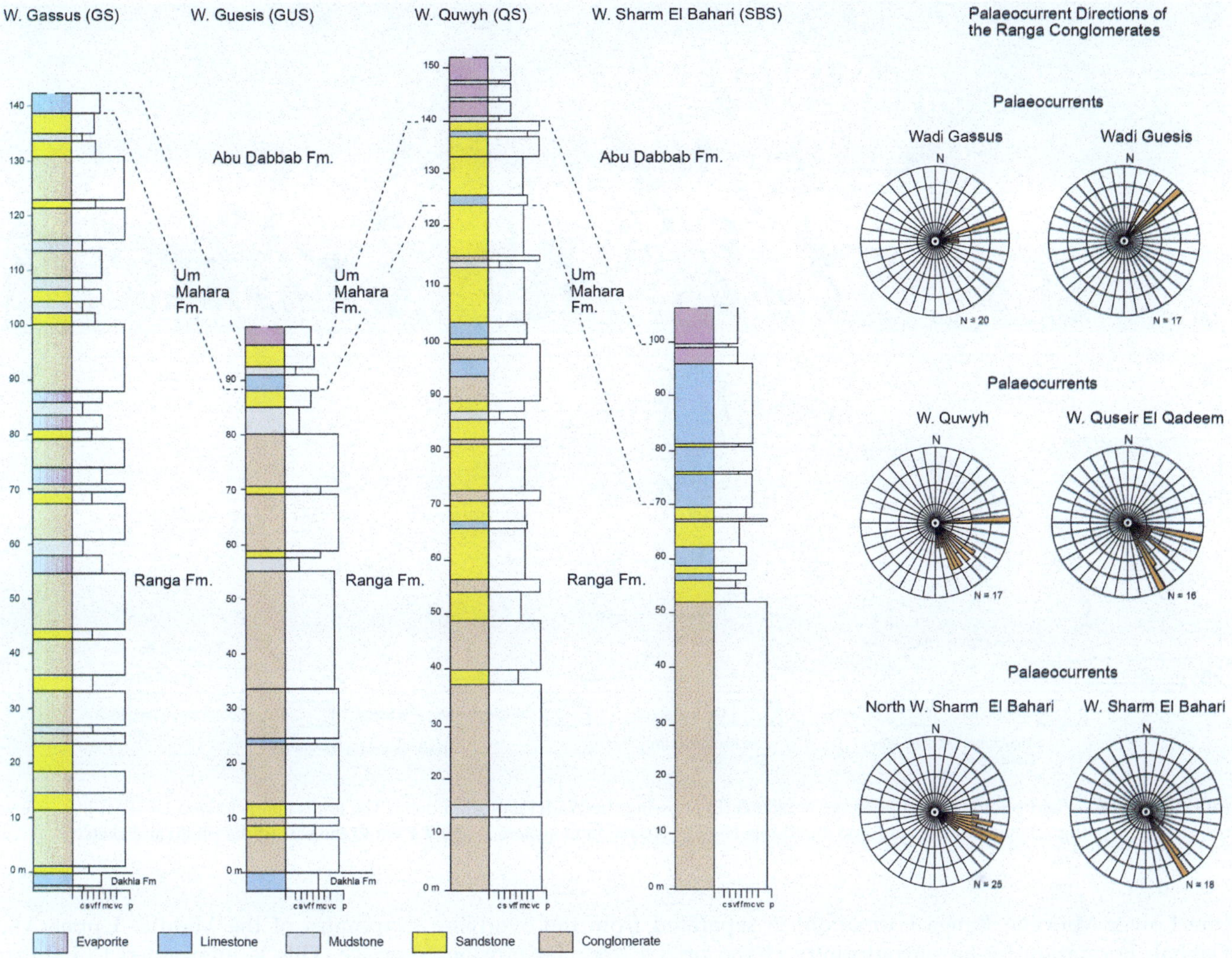

FIGURE 2.9 Lithostratigraphic logs of the Ranga Formation in Wadis Gassus, Guesis, Quwyh, and Sharm El Bahari. Locations of logged sections are shown on the geological maps of Figs. 2.5, 2.11, and 2.14. Rose diagrams show the fanglomerates palaeocurrent directions at the localities cited in the text.

The foreset beds are planar, about 60 m high, and have a depositional dip of 31–38°ENE (Fig. 2.10), decreasing eastwards to ~20° toward the distal part of the fan. The beds are 0.5–1 m in thickness, composed of highly organized, matrix-supported, pebble and cobble conglomerates, which in places show normal grading into medium- to coarse-grained sandstones. The conglomerate clasts are subrounded and show an imbricated fabric which, together with the eastward dip of the foreset beds, indicates a dominant N50°E palaeocurrent direction (Fig. 2.9). Marine oyster fragments and patch reef debris are found to occur in some of the conglomerate and sandstone beds, indicating probable deposition of the conglomerates in a shallow marine environment.

Although the bottomset beds are not well exposed along the cliff exposure of the fan, because of the foreset-bottomset transition zone below the wadi floor level, shallowing of the dip of the foresets in the downdip direction is locally present, probably inferring a gentler dip for the bottomset units.

The overall characteristics of the conglomerate sequence in Wadi Guesis as described above suggest that the depositional system may be interpreted as a "Gilbert-type" fan delta (e.g., Gilbert, 1885; Holmes, 1965; Ethridge and Wescott, 1984; McPherson et al., 1987; Backert et al., 2010; Rubi et al., 2018). This interpretation is similar to that for the syn-rift coarse-grained fan-delta systems described from the eastern margin of the Gulf of Suez (cf. Gawthorpe et al., 1990; McClay et al., 1998; Gupta et al., 1999) and from the Gulf of Corinth (Dart et al., 1994; Ford et al., 2016; Hemelsdael & Ford 2016; Rubi et al., 2018). The fanglomerate sequence in Wadi Guesis

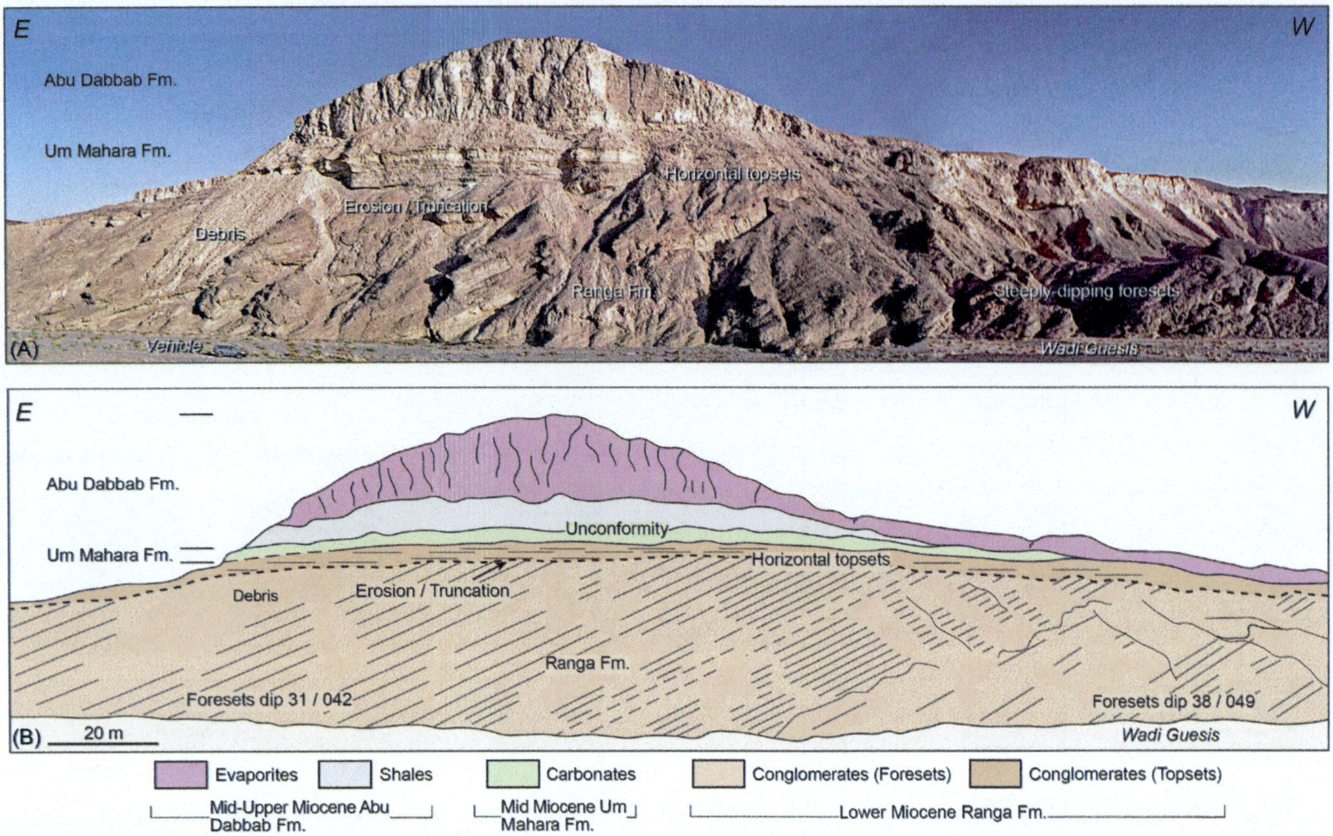

FIGURE 2.10 Field photo (A) and perspective sketch (B) showing the Wadi Guesis fan delta. The fan is characterized by sheet-like geometry, steeply dipping, planar foresets, and horizontal topsets. The coastal fault system is about 1 km to the east of the photo and sketch.

(the Lower Miocene Ranga Formation) is separated from the overlying evaporates of the Middle–Upper Abu Dabbab Formation by an unconformity characterized irregular erosional surface. This is interpreted to indicate the emergence and erosion of much of the fanglomerate topsets before the deposition of the overlying evaporites during the Middle Miocene.

Structural control on Miocene sedimentation in the Gassus and Guesis areas

Detailed structural mapping of the Gassus-Guesis area revealed that the two fan systems are formed in the hanging wall of several linked fault segments (Fig. 2.5). Between Wadi Safaga in the north and Wadi Abu Shiqlli in the south, the Coastal Fault system consists of four main segments from north to south: the Safaga, Gassus, Abu Kherfan, and Abu Shiqlli faults, respectively (Fig. 2.5).

The Safaga fault is oriented N12°W, and it juxtaposes Miocene clastics and carbonates against the Precambrian basement in the north and against Cretaceous–Eocene strata in the south (Figs. 2.5 and 2.8A). The maximum throw of this fault is ~1000 m. The Gassus fault consists of two segments oriented N45°W and has a maximum throw of ~800 m. It juxtaposes Miocene conglomerates down to the east with the Precambrian basement and Cretaceous strata in the footwall (Fig. 2.5). The Abu Kherfan Fault is oriented N35–40°W and has a maximum throw of ~850 m. It throws the Eocene Thebes Formation and Miocene conglomerates down to the east against the basement of Gebel Abu Kherfan (Fig. 2.5). The Abu Shiqlli fault is oriented N50°W and its maximum throw is 870 m. Another fault cuts across the Wadi Gassus (the Mine fault, Fig. 2.5) and is oriented N10°W. The throw of this fault changes along its strike as it juxtaposes Cretaceous–Paleocene strata against Lower Miocene strata. The throw of this fault is a few hundred meters toward the east–northeast.

Calculation of the stratigraphic and topographic offsets shows that the throw of the fault segments decreases along strike as these faults are traced both to the north and to the south. In the map and cross-sections (Figs. 2.5 and 2.7), the change of fault displacement is indicated by the variable juxtaposition of the basement against the prerift and syn-rift sediments along the strike of faults. This along-strike change in fault displacements produced relay ramps that link the fault segments where the fault tips approach and overlap each other (cf. Childs et al., 1995). The deposition of Miocene conglomerates in the Wadi Gassus fan delta was controlled by two relay ramps which served as pathways for clastic discharge and accumulation at the Wadi Gassus sediment-input point. The northern relay ramp (R1) is formed between the Safaga and Gassus fault segments and dips 31–42°SE (Figs. 2.5 and 2.8A), whereas the southern relay ramp (R2) (Fig. 2.5) is formed between Gassus fault and Abu Kherfan fault and dips 36–42°NE (Fig. 2.5). The southern relay (R2) is breached in the footwall (Gebel Abu Kherfan) via a small cross-fault, which strikes approximately parallel to Wadi Guesis (Fig. 2.5). The area between the two relays is a structural low, located opposite to the central part of the Gassus fault and it forms an open syncline that is filled with Miocene clastics (Fig. 2.5). This syncline has a structural relief of ~350 m.

Similar to the Wadi Gassus fan delta, the Wadi Guesis fan delta was controlled by relay structures formed between three fault segments—from north to south, respectively, the Abu Shiqlli, Abu Kherfan, and Mine faults (Fig. 2.5). The throws of these faults decrease as they approach each other as indicated by the juxtaposition of Cretaceous strata against basement, and the Lower Miocene conglomerates against Cretaceous strata at the fault terminations (Fig. 2.5). The northern relay (R3) (Fig. 2.5) occurs between the Abu Kherfan and the Mine faults, and it dips 17–25°ESE; the southern relay (R4) occurs between the Abu Shiqlli and Abu Kherfan faults, and it dips 11–18°NE (Fig. 2.5). The two relay ramps (R3 and R4) served as conduits for the clastic discharge to enter the basin and hence form the Guesis fan-delta system.

Despite the fact that both the Gassus and Guesis fan deltas are formed in the hanging-wall segments of the Coastal Fault system and appear to have been similarly controlled by relay structures, the thickness and outcrop area of the coarse clastics in the proximal part of the Gassus fan system is 10 km^2 as compared with 4 km^2 in the Guesis fan system. This difference in fan size is interpreted to reflect the fact that the Gassus fan was formed almost opposite to the central part of the Gassus fault where this fault had its maximum displacement and, as a consequence, the hanging-wall accommodation space was greater than in the Guesis fan, which was localized close to the tips of the fault segments.

Wadi Quwyh area

In Wadi Quwyh, the Lower Miocene sediments of the Ranga Formation form a fan delta, which is about 4 km^2 in outcrop area (Fig. 2.11). The geological map, perspective sketch, and cross-section (Figs. 2.11–2.13) summarize the structural setting and geometry of this fan delta. The fan delta has a sheet-like geometry, with 2 km of average longitudinal and strike lengths (Figs. 2.11 and 2.12). The fan sequence is 145 m in thickness, which decreases gradually to <50 m in both the longitudinal and strike directions, resulting in a thickness/progradation ratio of 0.25–0.07. It is composed of conglomerates and sandstones interbedded with 3–5 ledge-forming reef limestone units whose average thickness is 2.5 m (Fig. 2.9; log section QS and Fig. 2.13).

Topset beds are generally horizontal to gently dipping, with the dip increasing gradually in the progradation direction to ~8° at the distal areas (Fig. 2.13A). They are crudely stratified and composed of basement-clast-dominated conglomerates and pebbly sandstones (Fig. 2.13B and C). The conglomerate clasts are subrounded and moderately sorted (Fig. 2.13C). The sandstones are medium- to coarse-grained and locally show wave ripple cross-laminations. The limestone interbeds are pink-colored, locally bioturbated, and laminated and contain algal and coral reefs.

Within the fan system, the foreset beds are 20–40 m high and dip 15–20°SE and ESE. They are characterized by oblique to slightly sigmoid clinoforms (Figs. 2.12 and 2.13A). The beds are 0.5–2 m thick and are composed of crudely stratified polymict conglomerates dominated by basement clasts and minor chert and limestone clasts (~90% and 10%, respectively). The clasts are 4–10 cm in diameter, rounded to subrounded with crude imbrications, and in places show normal or inverse grading. The conglomerates in some locations contain patch reefs at higher levels close to the topset-foreset transition.

The conglomerate and sandstone foresets may be interpreted to indicate deposition by fluvial processes, with the coarse sediments transported to the foresets by sheetfloods (Gawthorpe et al., 1990). The dip direction of the foresets and the imbricated clast fabric indicate east and southeast paleocurrent directions (Fig. 2.9). The sandstones within the foresets are interpreted to have been deposited during periods of reduced fluvial input, with the sediment processes mainly dominated by reworking by wave activity (Gawthorpe et al., 1990). The presence of reef limestone and sandstone alternations within the topsets indicates episodic, generally low, fluvial input, allowing coral growth and reef buildup.

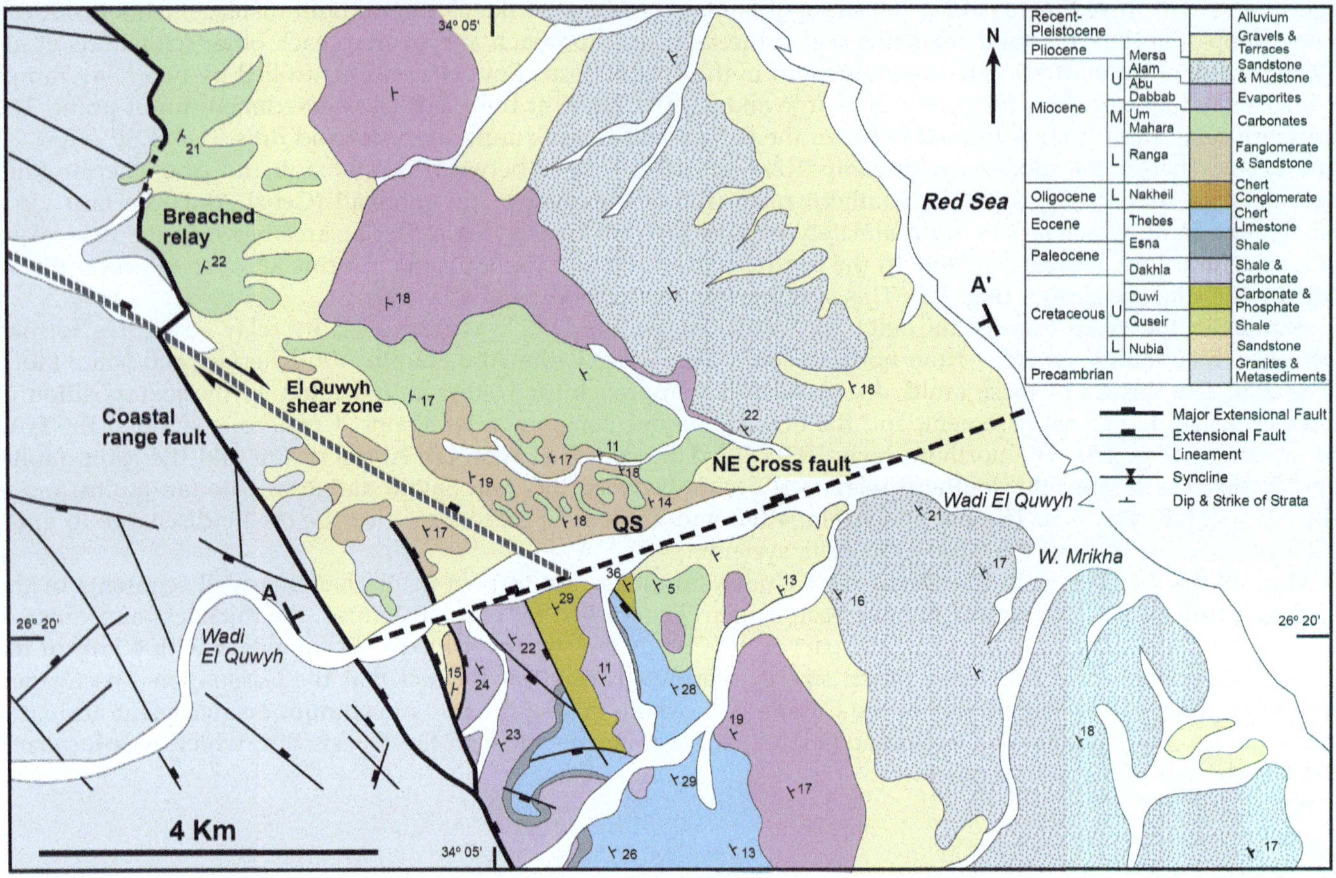

FIGURE 2.11 Detailed geologic map of the Wadi Quwyh area showing the structural setting of the Quwyh fan delta. QS indicates the location of the measured log section.

Structural control on Miocene sedimentation in the Wadi Quwyh area

The map, perspective sketch, and cross-section of the Quwyh fan delta (Figs. 2.11 and 2.12) show that the deposition of the conglomerates was restricted to the northern flank of the Wadi Quwyh rather than along the southern flank. Here, the Coastal Fault system is oriented N25°W and juxtaposes prerift strata (in the southern flank of the wadi) and Miocene strata (in the northern flank of the wadi) against the Precambrian basement in the footwall (Fig. 2.11). Within Wadi Quwyh, a N70°E-oriented cross-fault displaces the Miocene conglomerates down to the north against the Cretaceous strata to the south (Fig. 2.11). The cross-fault interacts with the Quwyh shear zone, which strikes N60°W (Fig. 2.11). This intersection produced a trapdoor structure, which controlled the localization of the conglomerate units (Figs. 2.11 and 2.12). The east and southeastward transportation (as indicated by the palaeocurrents) of the coarse clastics derived from the basement footwall block was controlled by a relay ramp and the Quwyh shear zone. The relay ramp is formed between the coastal fault segments to the north of Wadi Quwyh, and the ramp dips 20–22°ESE (Fig. 2.11). This relay, together with the Quwyh shear zone, which strikes parallel to the downdip direction of the relay, provided a pathway for the coarse clastics to enter the basin (Fig. 2.11).

The influence of cross-faults that controlled the location of Miocene clastics along the rift margin is also seen in Wadi Quseir El Qadeem, about 30 km to the south of Wadi Quwyh (Fig. 2.1). In this Wadi, the Lower Miocene sandstones and conglomerates (90 m thick) develop a local, ENE-trending channel localized by a cross-rift fault that appears to have interacted with the coastal, rift-parallel fault segments. The palaeocurrent data from this channel indicate east and southeast sediment transport directions (Rose diagrams, Fig. 2.9), similar to the palaeocurrent directions of the Wadi Quwyh fan delta.

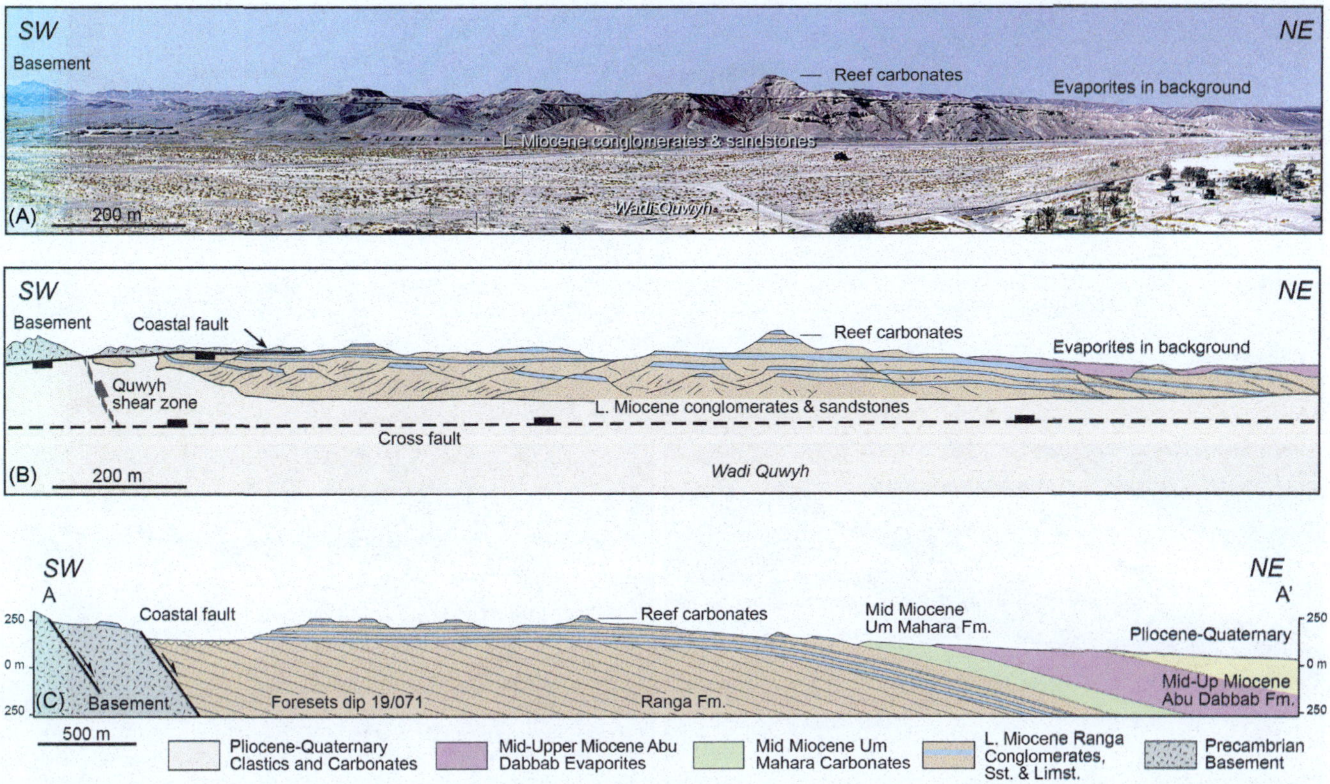

FIGURE 2.12 Field photo (A), perspective sketch (B), and cross-section (C) showing the geometry of the Wadi Quwyh fan delta. See Fig. 2.11 for location of cross-section.

Wadi Sharm El Bahari area

Wadi Sharm El Bahari is located about 30 km south of Quseir (Figs. 2.1 and 2.14). In this Wadi, the conglomerates of the Ranga Formation form a coarse-grained fan delta, which has a wedge-like to a sheet-like geometry and an east to southeastwards progradation direction (Figs. 2.15 and 2.16A). The length of the fan in the progradation direction is 3 km and its average strike length is 2 km. The maximum height of the fan is about 200 m above the wadi floor and it decreases eastwards (at distal areas) to <50 m (Figs. 2.15 and 2.16A). This fan sequence is dominated by lower Miocene coarse clastics of the Ranga Formation, which is 60–100 m in thickness (cf. Fig. 2.9; log section SBS). The average sequence thickness/progradation ratio of the fan is 0.025.

In the northern flank of Wadi Sharm El Bahari, the contact between the Ranga Formation and the older strata is covered by the wadi deposits, whereas in the southern flank, it unconformably overlies the red sandstone and conglomerates of the early syn-rift Oligocene strata, which locally contain basalt flows. The conglomerate sequence is characterized by large-scale foresets, which are 20–40 m high (Figs. 2.15 and 2.16A). The average dip of these foresets is 22°, but this decreases gradually in both updip and downdip directions to <10°, producing sigmoidal to gently oblique clinoforms (Figs. 2.15 and 2.16A). The foreset beds are moderately stratified, with bed thickness ranging from 0.5 to 3 m. These units are composed of chert-dominated conglomerates, which are both clast supported and sand-matrix supported. Clasts are 2–10 cm in diameter, rounded to subrounded, and in places have an imbricated fabric (Fig. 2.16C). When tracing the foreset updip, the beds become sandier, the dipping shallow, and there are shell fragments and patch reefs. These are onlapped in many places by gently dipping to horizontal shallow marine sands and reef (coral and algal) carbonates (Fig. 2.16B) developed on the emergent topography that extends southeastwards to Wadi Sharm El Qibli (Figs. 2.14 and 2.15). The coarse clastic facies of the Wadi Sharm El Bahari fan delta also continue southeastwards to Wadi Sharm El Qibli, where these fanglomerates unite and probably coalesce into a larger fan system.

The conglomerate facies of the fan sequence in Wadi Sharm El Bahari is interpreted to indicate the dominance of fluvial processes, with the coarse sediments transported to the foreset slopes by sheet flows. Foreset bed dips

FIGURE 2.13 Field views showing the bedform geometries and facies characteristics of the Quwyh fan delta. (A) Crudely stratified fanglomerate and sandstone with sheet-like fan geometry characterized by moderately dipping foresets and gently dipping topsets. Note the dominance of carbonate beds in the upper section and at the topset-foreset transition. (B) Medium- to coarse-grained sandstone with tabular cross bedding truncated by paleosoles and polymictic conglomerate. (C) Matrix-supported, basement-clast-dominated conglomerate with rounded to subrounded clasts showing crude imbrications.

and imbricated clast fabrics indicate ESE to SE palaeoflow directions (Fig. 2.9). The finer and sandier facies were deposited when the clastic discharge was low, and the siliciclastic input was reduced enough to permit reef growth and colonization of large areas of the delta topsets.

Structural control on Miocene sedimentation in the Wadi Sharm El Bahari area

The Wadi Sharm El Bahari sediment-input site is located at the southeast corner of the Gebel El Zarieb and Zug El Bahar basement ridges (Fig. 2.14). The localization of the fan delta in Sharm El Bahari appears to have been controlled by the WNW- and NW-trending, WSW-dipping faults (Hamadat and Zug El Bahar faults) that bound the Hamadat and Zug El Bahar half grabens (Figs. 2.1 and 2.14). The basement footwall blocks of these faults (Gebel El Zarieb and Zug El Bahar ridges) are interpreted to have formed barriers against the eastward transportation of the clastics supplied by the basement of the rift shoulder in the west (Figs. 2.1 and 2.14). The Hamadat and Zug El Bahar faults offset and lose displacement as they are traced southeastward (toward the Sharm El Bahari and Sharm El Qibli area). This is indicated by the juxtaposition of the Cretaceous Nubia and Duwi Formations against the basement at the southern termination of the faults (Fig. 2.14) and by the southeastward plunge of the footwall blocks (Zug El Bahar and El Zarieb basement blocks; Philobbos and Purser, 1993; Purser et al., 1987). These basement blocks have their lowermost structural relief at the locality of Sharm El Bahari (Fig. 2.14). Because of these combined structural and topographic elements, the Miocene clastics, instead of being transported eastward toward Zug El Bahar in the coastal area, are considered to have been carried southeastward, toward the Wadi Sharm El Bahari sediment-input point, and farther south to Wadi Sharm El Qibli. In addition, the relay structures that occur between the offset fault segments (e.g., at the southern corner of

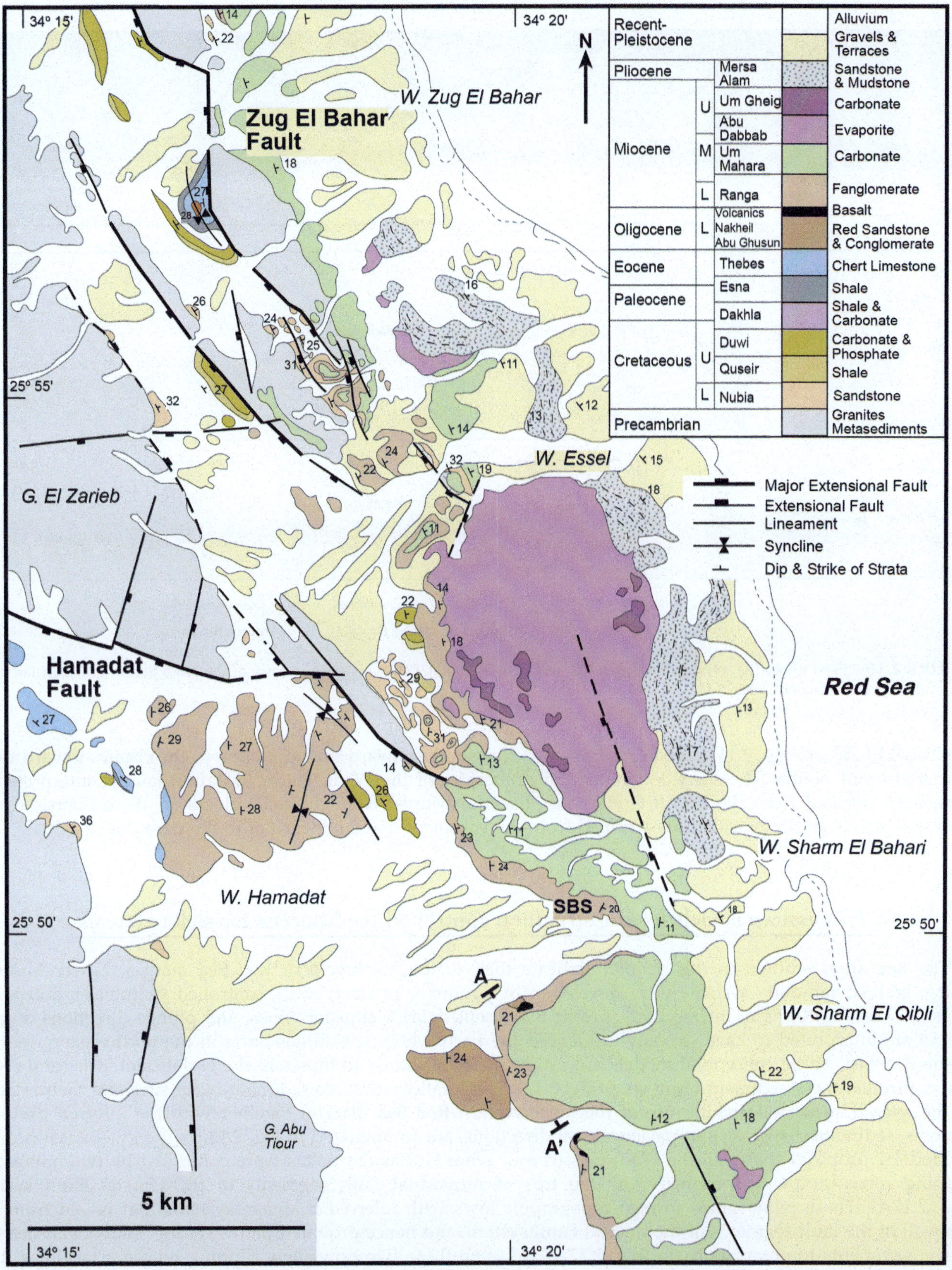

FIGURE 2.14 Detailed geologic map showing the structural setting of the Wadi Sharm El Bahari and Sharm El Qibli area. Note the localization of the Ranga Formation conglomerates at the southern termination of the Hamadat and Zug El Bahar fault segments. At this locality, the basement blocks plunge southeastwards and have low structural relief. SBS indicates the location of the measured section.

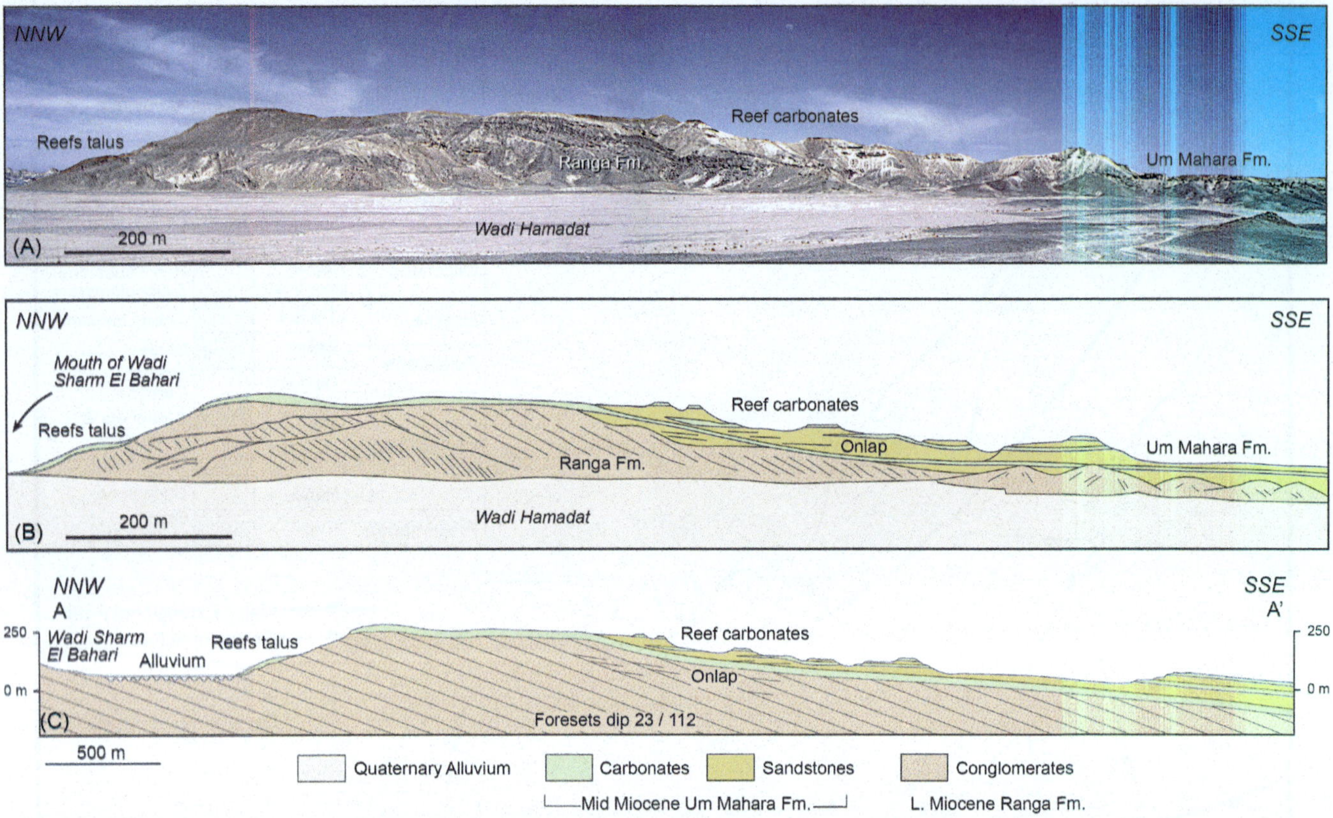

FIGURE 2.15 Field photo (A), perspective sketch (B), and cross-section (C) showing the geometry of the Wadi Sharm El Bahari fan delta. See location of cross-section in Fig. 2.14.

the Gebel El Zarieb, Fig. 2.14) probably augmented this southeastward transportation of the clastics toward these sediment-input points. The chert-dominated conglomerates of the Wadi Sharm El Bahari fan are interpreted to have been derived from the prerift Cretaceous–Eocene stripped from the fault blocks in Wadi Hamadat and probably also from the prerift strata stripped from the basement highlands farther to the west (Fig. 2.14).

Discussion: models for the structural control on the Miocene fan-delta systems

The key areas studied in this chapter indicate that, in the northwestern Red Sea margin, Lower Miocene coarse-grained fan-delta systems were developed in a number of structurally controlled sediment-input points (Figs. 2.17 and 2.18). Fault intersections, prerift (basement) fabrics, transfer zones, and plunge directions of fault blocks are interpreted to have strongly influenced the Early Miocene sedimentation in the northwestern margin of the Red Sea. Three conceptual models are proposed in this study to illustrate the varieties of structural styles that controlled these sediment-input sites (Fig. 2.17). The locations of these sediment-input sites and their relation to the overall structural architecture of the northwestern Red Sea margin, together with the inferred drainage systems, sediment dispersal, and palaeocurrent directions, are summarized in Fig. 2.18.

Model 1 proposes that both the Wadi Gassus and Wadi Guesis fan deltas were controlled by two oppositely dipping relay ramps that formed near the tips of individual fault segments of the Coastal Fault system (Fig. 2.17A). These relay ramps formed topographic lows with inferred drainage systems that issued from the footwall of the fault segments of the Coastal Fault system, and hence provided pathways for the footwall-derived clastic sediments to enter the basin (Fig. 2.17A). Between these two oppositely dipping relays, a local hanging-wall syncline formed adjacent to the central part of the Gassus fault segment, where the fault throw was maximum (Fig. 2.17A), and this provided accommodation space for the accumulation of the conglomerates and development of the fan delta in this locality.

FIGURE 2.16 Field views showing the geometrical characteristics and sedimentary facies of Sharm El Bahari fan delta. (A) Moderately stratified fanglomerate with sigmoidal to gently oblique clinoforms forming wedge-like to sheet-like fan geometry. Note the large-scale, moderately dipping foresets and gently dipping topsets and bottomsets. (B) Closeup showing paleosoles, algal (laminated stromatolites), and coral reef carbonates. (C) Closeup showing clast- and matrix-supported conglomerate with imbricated chert and basement clasts.

In addition to the structural control exerted by the relay system as proposed earlier, the Gassus fan was probably also fed by the long-lived ancestral drainage system of Wadi Gassus (Figs. 2.17A and 2.18), which ran transversely across the Gassus and Um El Huetat fault systems in the west (Fig. 2.18). The large size of the Gassus fan (~ 10 km^2) and the boulder-sized basement clasts probably were derived from the exhumed basement in the footwall of the Gassus and Um El Huetat faults. The role of long-lived antecedent drainage systems in the deposition of coarse-grained sediments is also well known from similar rift basins, for example, Gulf of Corinth rift (Seger and Alexander, 1993; Hemelsdael and Ford, 2016; Ford et al., 2016) and Gulf of Suez rift (Gawthorpe et al., 1990; Khalil, 1998; McClay et al., 1998; Gupta et al., 1999; Leppard and Gawthorpe, 2006). In these examples, drainage incision kept pace with footwall uplift and sediment transport was maintained into the basins in the hanging walls of the extensional faults (Leppard and Gawthorpe, 2006).

Model 2 for the Wadi Quwyh sediment-input point is shown in Fig. 2.17B. This is interpreted to have been controlled by the intersection of the WNW-trending Quwyh shear zone and an ENE-trending cross-fault, together with a relay ramp formed between the NNW-trending segments of the Coastal Fault system (Fig. 2.17B). The shear zone formed a weak basement region that was subjected to severe erosion and, therefore, coarse-grained clastics were shed into the Quwyh sediment-input point. The shear zone, together with the relay ramp between the coastal fault segments, formed a pathway for the sediment influx. In addition, the interaction of the shear zone with the cross-fault formed a "trapdoor" accommodation structure for these footwall-derived clastic sediments (Fig. 2.17B). The cross-fault is also interpreted to have controlled the course of the ancestral Wadi Quwyh, which overprinted the cross-fault and ran perpendicular to the Coastal Fault system (Figs. 2.17A and 2.18). It is interpreted that the ancestral Wadi Quwyh therefore equally sourced the footwall-derived basement conglomerates of the Wadi Quwyh fan-delta system. NE-trending cross-faults are also interpreted to have controlled the

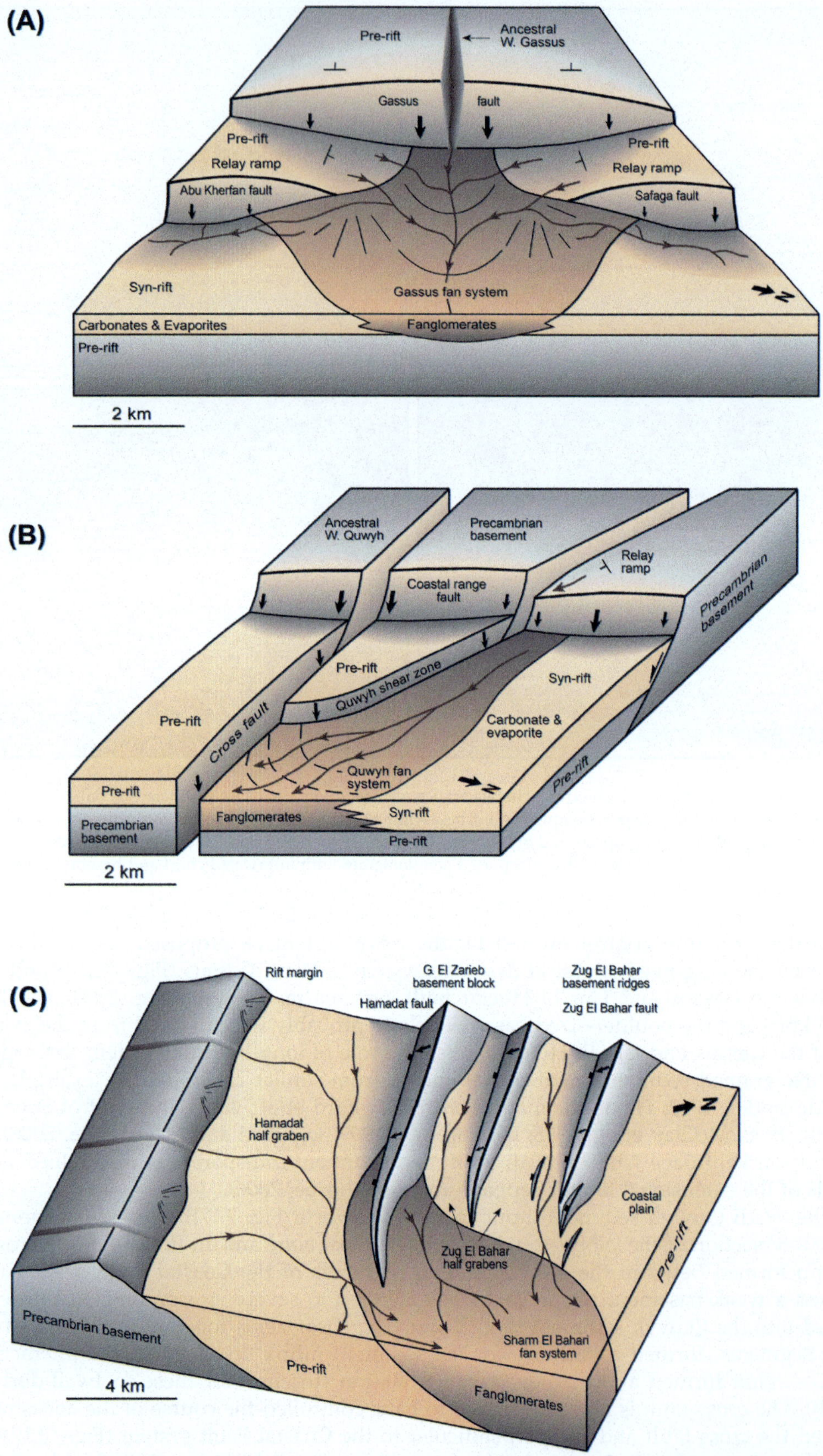

FIGURE 2.17 Summary schematic models illustrating the structural controls on sedimentation of Miocene coarse clastics in the northwestern Red Sea. (A) Wadi Gassus and Wadi Guesis fan deltas, (B) Wadi Quwyh fan delta, and (C) Wadi Sharm El Bahari fan delta.

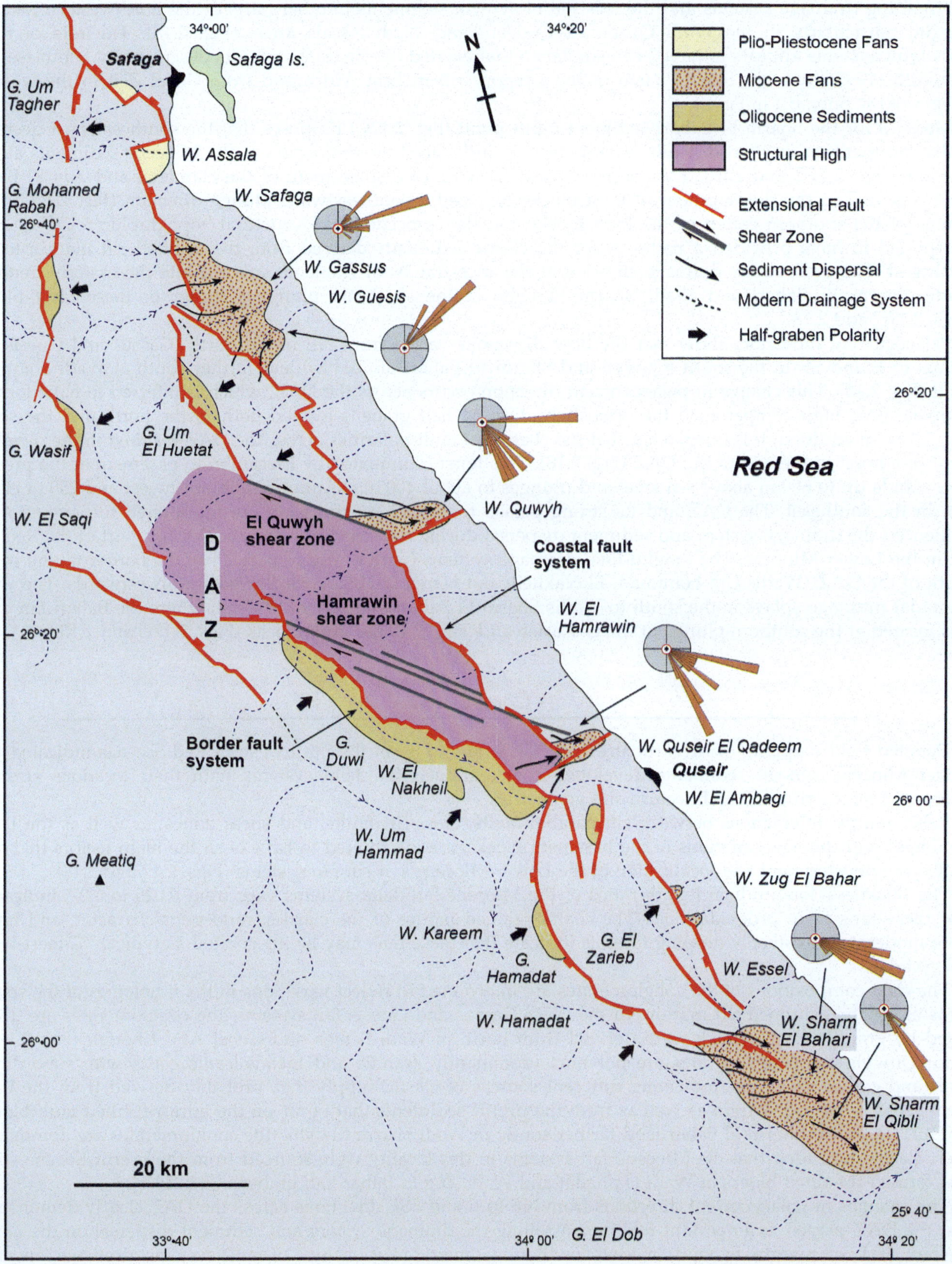

FIGURE 2.18 Summary map showing the major structures, sediment-input sites, drainage systems, and palaeocurrents of the lower Miocene coarse clastics of the Ranga Formation in the northwestern Red Sea.

NE-trending drainage systems and the location of syn-rift sediments in other parts of the northwestern Red Sea margin, particularly in the Wadi Quseir El Qadeem and Wadi Assala areas (Fig. 2.18). The role of these cross-fault systems appears to have been similar to that exerted by the transfer faults described by Lambiase and Bosworth (1995) from the Sugata basin of the Gregory rift in East Africa and from Gebel Zeit in the western margin of the Suez rift in Egypt.

Model 3 for the Sharm El Bahari sediment-input point (Fig. 2.17C) proposes that the southeastward decrease of displacement on the Border Fault system strongly influenced the deposition of the Miocene sediments at this location (Figs. 2.17C and 2.18). The southeastward decrease of displacement of the Hamadat and Zug El Bahar faults resulted in the development of fault blocks that had a southeastward plunge direction (Fig. 2.17C). The plunge of the southern corner of the fault blocks and the associated low structural/topographic relief are thus interpreted to have provided a pathway for the clastic sediments derived from the uplifted rift margin to the northwest. The present-day drainage shows that the ancestral Wadi Hamadat splayed into the easterly trending Wadis Sharm El Bahari and Wadi Sharm El Qibli at the southeast plunging corner of these fault blocks (Figs. 2.17C and 2.18).

Palaeocurrent indicators show that the flow directions varied from NE in the north (Gassus and Guesis fan deltas) to E and SE in the south (Quwyh and Sharm El Bahari fan deltas) and farther south at Wadi Sharm El Qibli (Fig. 2.18). This change in palaeocurrent direction occurs across the DAZ, which is inferred to have formed a topographic high compared to the oppositely dipping half grabens located both to the north and the south (Fig. 2.18). In addition, it is interpreted that the change of fault polarities across the DAZ resulted in the development of a horst to the east of the DAZ (Fig. 2.18). This horst (dominated by Precambrian basement at the present day) stands up to 600 m above sea level and plunges to about 400 m elevation in the northwest and 250 m elevation in the southeast. The DAZ and its strong palaeo topographic expression are interpreted to have strongly influenced the drainage system and sediment dispersal during the structural evolution of this part of the Red Sea rift in the Lower Miocene. The longitudinal drainage systems (with drainage parallel to the border faults) in the south of the DAZ (Wadis Um Hammad, El Nakheil, and Hamadat, Fig. 2.18) flowed southwestwards. The Wadi Hamadat drainage splays in the south to Wadis Sharm El Bahari and Sharm El Qibli, where the Bahari fan delta was located at the southern plunge of the G. Zarieb and Zug El Bahar fault blocks (Figs. 2.17C and 2.18).

Conclusions

Detailed field mapping of rift structures and the stratigraphy in the northwestern Red Sea has indicated that Lower Miocene syn-rift sediments developed coarse-grained fan-delta systems with their locations strongly influenced by the rift fault architecture and geometries.

Relay ramps, interactions between rift-parallel faults, cross-rift faults, and shear zones, as well as the DAZ, combined with plunge variations of the basement blocks, are interpreted to have been the main factors that controlled the development and localization of the Lower Miocene syn-rift fan systems (Figs. 2.17 and 2.18).

The thickness/longitudinal length ratios of the Miocene fan-delta systems vary from 0.025 to 0.07, indicating that these fans were progradational. The coarse-grained nature of the clastics, mud-poor character, and large-scale moderately to steeply dipping foresets indicate that these fans may be interpreted as typical "Gilbert-type" fan systems.

The clast compositions of the conglomerates are interpreted to reflect variations in the lithologies of the source areas along the exhumed rift margin. In the Wadi Gassus and Guesis fan systems, the conglomerates are dominated by granitic clasts, which were derived from north of Wadi Safaga and Gebel Abu Kherfan (Fig. 2.1). In Wadi Quwyh, the conglomerates are polymict (dominantly granitic and metavolcanic clasts with subordinate chert and carbonate clasts), indicating different sources of clastic supply (i.e., probably derived from the basement rocks in the rift margin as well as from the prerift sediments that occur on the intra-rift tilted fault blocks, Fig. 2.1). In Wadis Sharm El Bahari and farther south, in Wadi Sharm El Qibli, the conglomerates are dominantly chert clasts, indicating that the Miocene fan systems in this locality were sourced from the prerift Eocene strata that formed the tilted blocks in Wadi Hamadat and Wadi Zug El Bahar half grabens (Fig. 2.1).

The changes in palaeocurrent directions from NE to E and ESE directions across the DAZ clearly demonstrate that the DAZ played an important role in controlling the drainage system and sediment dispersal on the northwestern Red Sea margin. Synoptic models for these geometric relationships of fault systems, drainage, and fan-delta systems are shown in Figs. 2.17 and 2.18. The models proposed in this chapter for the structural control on Lower Miocene syn-rift fan systems and sediment-input points compare well with the results of similar studies in the Gulf of Suez to the north and in the Corinth rift. These models can be applied to other rift basins where preexisting basement structures have exerted a strong control on rift fault geometries and patterns and affect the inherited paleo-relief and antecedent drainage systems.

Our results provide significant information on the sediment transport routing, depositional pattern, and localization of coarse-grained fan-delta systems and, consequently, benefit future reservoir prediction and hydrocarbon exploration in the Red Sea and in many rift basins throughout the world.

Acknowledgments

The research presented in this paper was supported by the Natural Environment Research Council (NERC) ROPA Grant GR3/R9529 to Ken McClay. Samir Khalil was supported by a research grant from BG International to Ken McClay. Additional support was received from the Fault Dynamics Project, Royal Holloway University of London (sponsored by ARCO British Limited, PETROBRAS U.K. Ltd., BP Exploration, Conoco (U.K.) Ltd., Mobil North Sea Limited, and Sun Oil Britain). The authors also gratefully acknowledge support from ARCO British Limited (now BP Exploration). Bill Bosworth and Marathon Petroleum Egypt are thanked for logistical support and for many fruitful discussions on the Northwestern Red Sea geology. BG Egypt and Maher Ayyad kindly sponsored with additional fieldwork in the northwestern Red Sea.

References

Abd El-Razik, T.M., 1967. Stratigraphy of the sedimentary cover of the Anz-Atshan-south Duwi district. Bulletin of the Faculty of Science. Cairo University. 431, 135–179.

Abdel Khalek, M.L., Abdel Wahed, M., Sehim, A., 1993. Wrenching deformation and tectonic setting of the northwestern part of the Gulf of Aqaba. Geodynamics and sedimentation of the Red Sea-Gulf of Aden Rift system. Geological Survey of Egypt Special Publication. 1, 409–445.

Akaad, M.K., Noweir, A.M., 1980. Geology and lithostratigraphy of the Arabian desert orogenic belt between latitudes 25° 35, and 26°30′. In: Cooray, P.A.T.S. (Ed.), Evolution and Mineralization of the Arabian–Nubian shield. Permagon Press, New York, pp. 127–135.

Akkad, S., Dardir, A.A., 1966. Geology and phosphate deposits of Wasif, Safaga area. Geological Survey of Egypt Paper. 36, 35.

Backert, N., Ford, M., Malartre, F., 2010. Architecture and sedimentology of the Kerinitis Gilbert-type fan delta, Corinth Rift, Greece. Sedimentology, 75, 543–586.

Bosworth, W., 2015. Geological evolution of the Red Sea: historical background, review, and synthesis. In: Rasul, N.M.A., Stewart, I.C.F. (Eds.), The Red Sea, 2015. Springer-Verlag, Berlin Heidelberg, pp. 45–78.

Bosworth, W., Stockli, D.F., 2016. Early magmatism in the greater Red Sea rift: timing and significance, Canadian Journal of Earth Sciences., 53, p. 1158117. Available from: http://doi.org/10.1139/cjes-2016-0019.

Bosworth, W., Huchon, P., McClay, K., 2005. The Red Sea and Gulf of Aden basins. Journal of African Earth Sciences. 43, 334–378.

Bosworth, W., Khalil, S.M., Ligi, M., Stockli, D.F., McClay, K.R., 2020. The northern Red Sea. In: Hamimi, Z., El-Barkooky, A., Martinez- Frías, J., Fritz, H., Abd El-Rahman, Y. (Eds.), The Geology of Egypt. Springer Nature, Switzerland, pp. 343–374.

Childs, C., Watterson, J., Walash, J.J., 1995. Fault overlap zones within developing normal fault systems. Journal of the Geological Society of London 152, 535–549.

Cochran, J.R., 1983. A model for development of the Red Sea. AAPG Bulletin 67, 41–69.

Coleman, R.G., 1993. Geologic evolution of the Red Sea. Oxford Monographs on Geology and Geophysics 24. Oxford University Press, Oxford, p. 186.

Crossley, R., 1984. Controls on sedimentation in Lake Malawi Rift Valley, Central Africa. Sedimentary Geology. 40, 33–40.

Dart, C.J., Collier, R.E.L., Gawthorp, R.L., Keller, J.V., Nichols, G., 1994. Sequence stratigraphy of (?) Pliocene–quaternary syn-rift Gilbert type deltas, northern Peloponnesos, Greece. Marine and Petroleum Geology. 11, 545–560.

Dawers, N.H., Underhill, J.R., 2000. The role of fault interaction and linkage in controlling synrift stratigraphic sequences: Statfjord East area, Northern North Sea. AAPG Bulletin 84, 45–64.

El Bassyony, A.A., 1982. Stratigraphical studies on Miocene and younger exposures between Quseir and Berenice. Red Sea coast. Egypt. Ph.D. Thesis, Ain Shams University, Cairo.

Miocene rock stratigraphy of Egypt. In: El Gezeery, M.N., Marzouk, I.M. (Eds.), Egyptian Journal of Geology., 18. The Stratigraphic Sub-Committee of the National Committee of Geological Sciences of Egypt, pp. 1–59.

El Haddad, A.A., 1984. Sedimentological and geological studies on the Neogene sediments of the Egyptian part of the Red Sea. Ph.D. Thesis, Assiut University.

Ethridge, F.G., Wescott, W.A., 1984. Tectonic setting, recognition and hydrocarbon reservoir potential of fan-delta deposits. In: Koster, E.H., Steel, R.J. (Eds.), Sedimentology of Gravels and Conglomerates, Canadian Society of Petroleum Geologists Memoir, 10. pp. 217–234.

Ford, M., Hemelsdael, R., Mancini, M., Palyvos, N., 2016. Rift migration and lateral propagation: evolution of normal faults and sediment-routing systems of the western Corinth rift (Greece). In: Childs, C., Holdsworth, R.E., Jackson, C.A.-L., Manzocchi, T., Walsh, J.J., Yielding, G. (Eds.), The Geometry and Growth of Normal Faults, 439. Geological Society, London, Special Publications, pp. 131–168. Available from: http://doi.org/10.1144/SP439.15.

Freund, R., 1970. Plate tectonics of the Red Sea and Africa. Nature 228, 453.

Gawthorpe, R.L., Andrews, J.E., Collier, R.E.L., Ford, M., Henstra, G.A., Kranis, H., et al., 2017. Building up or out? Disparate sequence architectures along an active rift margin—Corinth rift, Greece. Geology, Geology 45 (12), 1111–1114.

Gawthorpe, R.L., Fabergas, N., Pechlivanidou, S., Ford, M., Collier, R.E.L., Carter, G.D.O., et al., 2022. Late Quaternary mud-dominated, basin-floor sedimentation of the Gulf of Corinth, Greece: implications for deep-water depositional processes and controls on syn-rift sedimentation. Basin Research. Available from: https://doi.org/10.1111/bre.12671.

Gawthorpe, R.L., Hurst, J.M., 1993. Transfer zones in extensional basins: their structural style and influence on draining development and stratigraphy. Journal of the Geological Society of London 150, 1137–1152.

Gawthorpe, R.L., Hurst, J.M., Sladen, C.P., 1990. Evolution of Miocene footwall-derived coarse grained deltas, Gulf of Suez, Egypt: implications for exploration. AAPG Bulletin 74, 1077–1086.

Gawthorpe, R.L., Leeder, M.R., 2000. Tectono-sedimentary evolution of active extensional basins. Basin Research 12, 195–218.

Gawthorpe, R.L., Sharp, I., Underhill, J.R., Gupta, S., 1997. Linked sequence stratigraphic and structural evolution of propagating normal faults. Geology 25, 795–798.

Gilbert, G.K., 1885. The topographic features of lake shores. USGS Annual Reports 5, 69–123.

Girdler, R.W., Southren, T.C., 1987. Structure and evolution of the northern Red Sea. Nature 330, 716–721.

Gupta, S., Underhill, J.R., Sharp, I.R., Gawthorpe, R.L., 1999. Role of fault interactions in controlling synrift sediment dispersal patterns: Miocene Abu Alaqa Group, Suez Rift, Sinai, Egypt. Basin Research 11, 167–189.

Hardy, S., Dart, C.J., Waltham, D., 1994. Computer modelling of the influence of tectonics on sequence architecture of coarse-grained fan deltas. Marine and Petroleum Geology. 11, 561–574.

Heath, R., Vanstone, S., Swallow, J., Ayyad, M., Amin, M., Huggins, P., et al., 1998. Renewed exploration in the offshore north Red Sea Region – Egypt. Proceedings of the 14th Petroleum Conference, Egyptian General Petroleum Corporation, Cairo, Egypt, pp. 16–34.

Hemelsdael, R., Ford, M., 2016. Relay zone evolution: a history of repeated fault propagation and linkage, central Corinth rift, Greece. Basin Research 28, 34–56. Available from: http://doi.org/10.1111/bre.12101.

Hempton, M., 1987. Constraints on Arabian plate motion and extensional history of the Red Sea. Tectonics 6, 687–705.

Holmes, A., 1965. Principles of Physical Geology, second ed. Thomas Nelson, London, p. 1288.

Issawi, B., Francis, M., El-Hinnawi, M., Mehanna, A., 1969. Contribution to the structure and phosphate deposits of Quseir area. Geological Survey of Egypt 50.

Joffe, S., Garfunkel, Z., 1987. Plate kinematics of the circum Red Sea – a re-evaluation. Tectonophysics 141, 5–22.

Junqiao, L., Haixue, W., Yanfang, L., Tongwen, S., Mengdi, Z., Wei, H.E., et al., 2018. Reservoir controlling differences between consequent faults and antithetic faults in slope area outside of source: A case study of the south-central Wenan slope of Jizhong Depression, Bohai Bay Basin, East China. Petroleum Exploration and Development, 45, 88–98.

Khalil, S., McClay, K., 2001. Tectonic evolution of the northwestern Red Sea-Gulf of Suez rift system. Geological Society, London, Special Publications. 187, 453–473.

Khalil, S.M., 1998. Tectonic Evolution of the Eastern Margin of the Gulf of Suez. PhD thesis, Royal Holloway University of London, Egypt, p. 349.

Khalil, S.M., McClay, K.R., 2002. Extensional fault-related folding, northwestern Red Sea, Egypt. Journal of Structural Geology. 24 (Special Volume), 743–762.

Khalil, S.M., McClay, K.R., 2016. 3D geometry and kinematic evolution of extensional fault-related folds, NW Red Sea, 439. Geological Society, London, Special Publications, Egypt, pp. 109–130. Available from: http://doi.org/10.1144/SP439.11.

Khalil, S.M., McClay, K.R., 2020. Extensional fault-related folding in the northwestern Red Sea Egypt: segmented fault growth, fault linkages, corner folds and basin evolution, 476. Geological Society of London Special Publications, pp. 49–81. Available from: https://doi.org/10.1144/SP476.12.

Kröner, A., 1984. Late Precambrian plate tectonics and orogeny: a need to redefine the term Pan-African. In: Klerkx, J., Michot, J. (Eds.), African Geology. Tervuren, Museé R. l'Afrique Centrale, pp. 23–28.

Kröner, A., 1993. The Pan African belt of northeastern and Eastern Africa, Madagascar, southern India, Sri Lanka and East Antarctica: terrane amalgamation during the formation of the Gondwana supercontinentIn: Thorweihe, U., Schandelmeier, H. (Eds.), Geoscientific Research in Northeast Africa, Balkema, Rotterdam, pp. 3–9.

Lambiase, J.J., Bosworth, W., 1995. Structural controls on sedimentation in continental rifts. In: Lambiase, J.J. (Ed.), Hydrocarbon Habitat in Rift Basins, 80. Geological Society of London, Special Publication, pp. 117–144.

Le Pichon, X., Francheteau, J., 1978. A plate tectonic analysis of the Red Sea-Gulf of Aden area. Tectonophysics 46, 369–406.

Leeder, M.R., Collier, R.E.Ll., Aziz, A.L.H., Trout, M., Ferentinos, G., Papatheodorou, G., et al., 2002. Tectono-sedimentary processes along an active marine/lacustrine halfgraben margin: Alkyonides gulf. E. Gulf of Corinth, Greece. Basin Research 14, 25–41.

Leppard, C.W., Gawthorpe, R.L., 2006. Sedimentology of rift climax deep water systems; lower Rudeis Formation, Hammam Faraun Fault Block, Suez Rift, Egypt. Sedimentary Geology. 191, 67–87.

Lezzar, K.E., Tiercelin, J.J., Le Turdu, C., Cohen, A.S., Reynolds, D.J., Le Gall, B., et al., 2002. Control of normal fault interaction on distribution of minor Neogene sedimentary depocentres, Lake Tanganyika, East Africa. AAPG Bulletin 86, 1027–1059.

Li, J., Liu, Z., Liu, J., Chen, L., Liu, H., Lili, H., et al., 2021. Transformation of sediment delivery and dispersal patterns controlled by relay-ramp evolution along the boundary fault of a lacustrine rift: The Eocene Shahejie formation, Dongying Sag, Bohai Bay Basin, NE China. Marine and Petroleum Geology 128, 1–20. Available from: https://doi.org/10.1016/j.marpetgeo.2021.105044.

Ma, P., Lin, C., Ren, L., Jahren, J., Dong, D., Yu, G., et al., 2021. Linkage and growth of the independent and coherent faults: Insight into the effect of relay ramps on sedimentation patterns in the northern Bonan Sag, Bohai Bay Basin. Marine and Petroleum Geology 127, 104985. Available from: https://doi.org/10.1016/j.marpetgeo.2021.104985.

McClay, K.R., Nicols, G.J., Khalil, S.M., Darwish, M., Bosworth, W., 1998. Extensional tectonics and sedimentation, eastern Gulf of Suez, Egypt. In: Purser, B.H., Bosence, D.W.J. (Eds.), Sedimentation and Tectonics of Rift Basins: Red Sea–Gulf of Aden. Chapman and Hall, London, pp. 223–238.

Mckenzie, D.P., Davies, D., Molnar, P., 1970. Plate tectonics of the Red Sea and east Africa. Nature 226, 243–248.

McPherson, J.G., Shanumugam, G., Moiola, R.J., 1987. Fan-deltas and braided deltas: varieties of coarse-grained deltas. Geological Society of America Bulletin 99, 331–340.

Morgan, P., 1990. Egypt in the tectonic framework of global tectonics. In: Said, R. (Ed.), The Geology of Egypt. Balkema, Rotterdam, pp. 91–111. Chapter 7.

Moustafa, A.R., 1997. Controls on the development and evolution of transfer zones: the influence of basement structure and sedimentary thickness in the Suez rift and Red Sea. Journal of Structural Geology. 19 (6), 755–768.

Moustafa, A.R., Khalil, S.M., 2017. Control of extensional transfer zones on syntectonic and post-tectonic sedimentation: implications for hydrocarbon exploration, Journal of the Geological Society of London, 174. pp. 318–335.

Moustafa, A.R., Khalil, S.M., 2020. Structural setting and tectonic evolution of the Gulf of Suez, NW Red Sea and Gulf of Aqaba rift systems. In: Hamimi, Z., El-Barkooky, A., Martinez- Frías, J., Fritz, H., Abd El-Rahman, Y. (Eds.), The Geology of Egypt. Springer Nature, Switzerland, pp. 295–342.

Orszag-Sperber, F., Purser, B.H., Rioual, M., Plaziat, J.C., 1998. Post Miocene sedimentation and rift dynamics in the southern Gulf of Suez and northern Red Sea. In: Purser, B.H., Bosence, D.W.J. (Eds.), Sedimentation and Tectonics of Rift Basins: Red Sea-Gulf of Aden. Chapman and Hall, London, pp. 427–447.

Pechlivanidou, S., Cowie, P.A., Duclaux, G., Nixon, C.W., Gawthorpe, R.L., Salles, T., 2019. Tipping the balance: shifts in sediment production in an active rift setting. Geology 47 (3), 259–262.

Phillobbos, E.R., El Haddad, A.A., Mahran, T.M., 1988. Comparison between Miocene and Pliocene facies distribution related to syn-rift tectonics along the Egyptian Red Sea coastal area. Egyptian General Petroleum Corporation, 9th Petroleum Exploration and Production Conference. vol. 1, pp. 246–254.

Philobbos, E.R., El Haddad, A.A., Luger, P., Bekir, R., Mahran, T., 1993. Syn-rift Miocen sedimentation arround fault blocks in the Abu Ghusun–Wadi el Gemal area, Red Sea, Egypt. In: Philobbos, E.R., Purser, B.H. (Eds.), Geodynamics and Sedimentation of the Red Sea–Gulf of Aden Rift system, 1. Geological Society of Egypt, Special Publication, pp. 115–142.

Geodynamics and sedimentation of the Red Sea–Gulf of Aden rift system. In: Philobbos, E.R., Purser, B.H. (Eds.), Geological Society of Egypt. Special Publication.

Purser, B.H., Bosence (eds.), D.W.J., 1998. Sedimentation and Tectonics of Rift Basins: Red Sea– Gulf of Aden. Chapman and Hall, London, p.680.

Purser, B.H., Orszag-Sperber, F., Plaziat, J.C., 1987. Sedimentation et rifting: les series neogenes de la marge nord-occidental de la Mer Rouge (Egypte). Notes et Memoires, Compagnie Francaise de Petroles, Paris 21, 111–114.

Purser, B.H., Philobbos, E.R., 1993. The sedimentary expressions of rifting in the NW Red Sea, Egypt. In: Philobbos, E.R., Purser, B.H. (Eds.), Geodynamics and sedimentation of the Red Sea–Gulf of Aden rift system, Geological Society of Egypt, 1. Special Publication, pp. 1–45.

Quennell, A.M., 1984. The western Arabian rift system. In: Dixon, J.E., Robertson, A.H.F. (Eds.), The Geological Evolution of the Eastern Mediterranean. Blackwell Scientific, Oxford, pp. 775–788.

Ravnås, R., Steel, R., 1998. Architecture of marine rift-basin successions. AAPG Bulletin 82, 110–146.

Roussel, N., Purser, B.H., Orszag-Sperber, F., Plaziat, J.C., Soliman, M., El Haddad, E.A., 1986. Geologie de la region de Quseir. Documents et Travaux, 10. Institut Geologique Albert de lapparent, Paris, pp. 129–144.

Rubi, R., Rohais, S., Bourquin, S., Moretti, I., Desaubliaux, G., 2018. Processes and typology in Gilbert-type delta bottomset deposits based on outcrop examples in the Corinth Rift. Marine and Petroleum Geology 92, 193–212.

Said, R. (Ed.), 1990. The Geology of Egypt. Balkema, Rotterdam, p. 734.

Scholz, C.A., Johnson, T.C., McGill, J.W., 1993. Deltaic sedimentation in a rift valley lake: new seismic reflection data from Lake Malawi (Nyasa), East Africa. Geology 21, 395–398.

Scholz, C.A., Rosendahl, B.R., Scott, D.L., 1990. Development of coarse-grained facies in lacustrain rift basins: examples from East Africa. Geology 18, 140–144.

Seger, M.J., Alexander, J., 1993. Distribution of Plio-Pleistocene and Modern coarse-grained deltas south of the Gulf of Corinth, Greece. In: Frostick, L., Steel, R. (Eds.), Tectonic Controls and Signatures in Sedimentary Successions, International Association of Sedimentologists, 20. Special Publication, pp. 37–48.

Steckler, M.S., Berthelot, F., Lyberis, N., Le Pichon, X., 1988. Subsidence in the Gulf of Suez: implications for rifting and plate kinematics. Tectonophysics 153, 249–270.

Stern, R.J., 1981. Petrogenesis and tectonic setting of Late Precambrian ensimatic volcanic rocks, Central Eastern Desert of Egypt. Precambrian Research 16, 195–230.

Stern, R.J., 1994. Arc assembly and continental collision in the Neoproterozoic East African orogen: implications for the consolidation of Gondwanaland. Annual Review of Earth and Planetary Sciences 22, 319–351.

Stockli, D.F., Bosworth, W., 2019. Timing of extensional faulting along the Magma-poor Central and Northern Red Sea rift margin—transition from regional extension to necking along a hyperextended rifted margin. In: Rasul, N.M.A., Stewart, I.C.F. (Eds.), Geological Setting, Palaeoenvironment and Archaeology of the Red Sea. Springer Nature Switzerland, pp. 81–111.

Stoeser, D.B., Camp, V.E., 1985. Pan African microplate accretion of the Arabian shield. Geological Society of America Bulletin 96, 817–826.

Tewfik, N., Ayyad, M., 1982. Petroleum exploration in the Red Sea shelf of Egypt. 6th Exploration Seminar. Egyptian General Petroleum Corporation, Cairo.

Versfelt, J., Rosendahl, B.R., 1989. Relationships between pre-rift structure and rift architecture in Lakes Tanganyika and Malawi, East Africa. Nature 337, 354–357.

Younes, A., McClay, K.R., 2002. Role of basement fabric on rift architecture: Gulf of Suez – Red Sea, Egypt. AAPG Bulletin 86, 1003–1102.

Young, M.J., Gawthorpe, R.L., Sharp, I.R., 2000. Sedimentology and sequence stratigraphy of a transfer zone coarse-grained delta, Miocene Suez Rift, Egypt. Sedimentology 47, 1081–1104.

Youssef, M.I., 1957. Upper Cretaceous rocks in Kosseir area. Bulleten. Del Institutedu Desert d Egypt 7, q67A.

CHAPTER

3

The rifted margin of eastern North America: insights into rifting, igneous activity, and breakup

Martha Oliver Withjack[1], Roy W. Schlische[1], Paul E. Olsen[2] and Sean T. Kinney[1,2]

[1]Department of Earth and Planetary Sciences, Rutgers University, Piscataway, NJ, United States
[2]Department of Earth and Environmental Sciences and Lamont-Doherty Earth Observatory of Columbia University, Palisades, NY, United States

Introduction

Eastern North America (ENA) is a natural laboratory for studying rifting and passive-margin development. It hosts one of the world's largest rift systems, one of the world's oldest intact passive margins, and one of the world's largest igneous provinces (the Central Atlantic Magmatic Province [CAMP]). A growing body of geological and geophysical evidence has shown that the tectonic history of the ENA rift system and margin is complex, differing significantly from the classic rift-to-drift model for passive-margin development. This chapter presents an overview of the key geologic features and history of the ENA rift system and margin, focusing primarily on recent developments and insights about rifting, igneous activity, breakup, and postrift deformation during the rift-drift transition in the southern and central segments of the rift system and margin from the southeastern United States to southeastern Canada (Fig. 3.1).

Influence of prerift tectonics on rifting and breakup

The structure and composition of the ENA lithosphere reflects the long tectonic history that preceded ENA rifting. This includes the assembly, rifting, and breakup of Rodinia; the development of the Laurentian passive margin during the early Paleozoic; and the subsequent Paleozoic orogenic activity that culminated in the assembly of Pangea (e.g., Rast, 1988; Rankin, 1994; Faill, 1997, 1998; Hatcher, 2005; Thomas, 2006). The final orogenic event, the Alleghanian orogeny, was completed by the end of Early Permian time. Paleozoic orogenic activity produced a thickened, amalgamated continental crust with profound preexisting zones of weakness and led to the formation of the Appalachian orogenic belt. Two processes may have thinned the thickened continental crust and reduced the topographic elevation before rift onset: (1) postorogenic gravitational collapse triggered by lithospheric delamination and/or ductile flow (e.g., Nelson, 1992) and (2) a postorogenic cycle of erosion and isostatic rebound/uplift (denudation unloading) initiated by erosion-related transfer of mass away from the mountains (e.g., Jamieson and Beaumont, 1988; Slingerland and Furlong, 1989).

By Late Triassic time, a massive rift zone had developed within the former orogenic belt (e.g., Benson and Doyle, 1988; Manspeizer, 1988; Manspeizer and Cousminer, 1988; Olsen et al., 1989; Schlische, 1993; Withjack

Regional Geology and Tectonics: Volume 2
DOI: https://doi.org/10.1016/B978-0-444-64136-6.00001-4

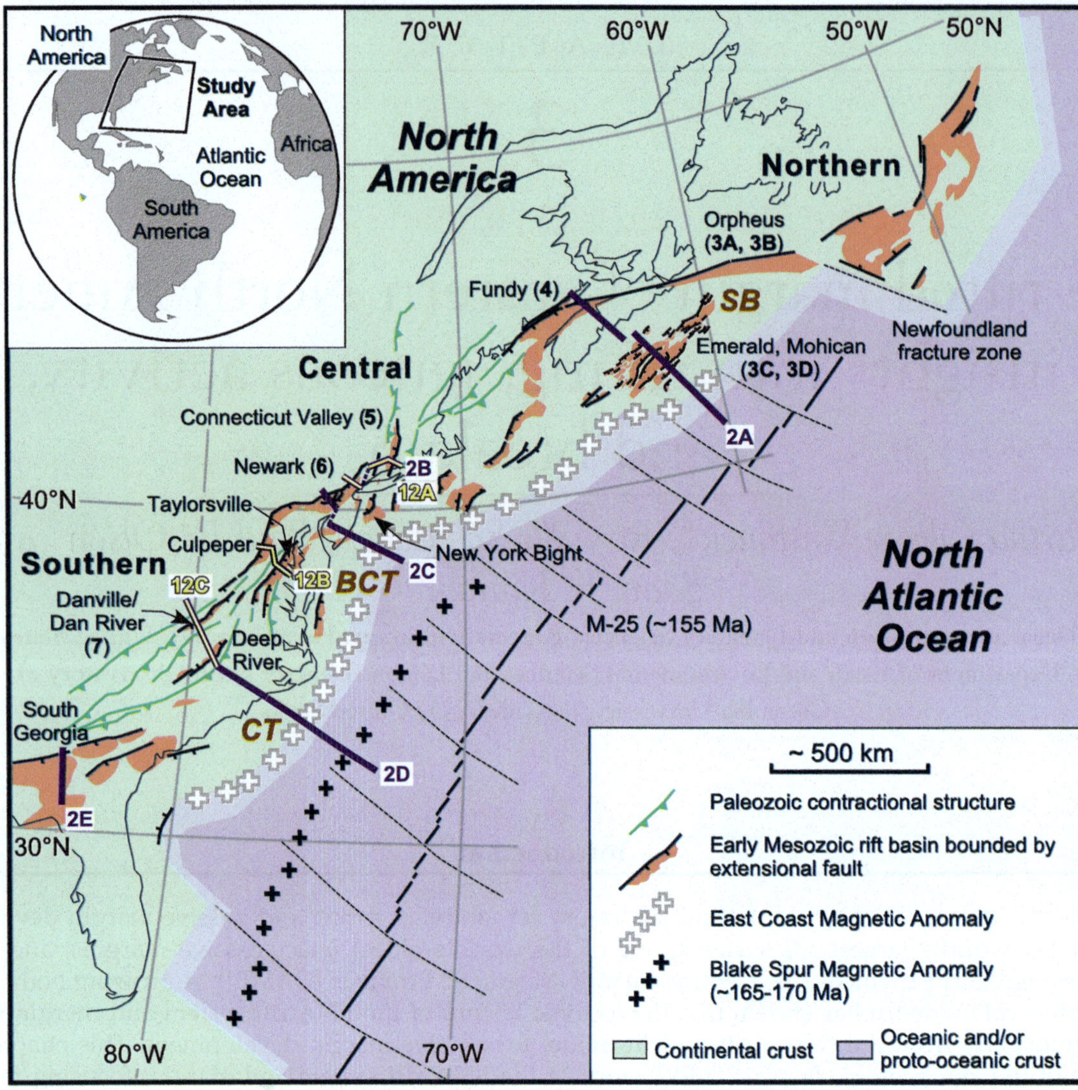

FIGURE 3.1 Map of continental margin of eastern North America (ENA) (see inset for location) showing continental and oceanic crust, major early Mesozoic rift basins, major Paleozoic contractional structures, the East Coast Magnetic Anomaly, the Blake Spur Magnetic Anomaly, and magnetic anomaly M-25 (Klitgord et al., 1988; Olsen et al., 1989; Welsink et al., 1989; Rankin, 1994; Benson, 2003; Wu et al., 2006; Heffner, 2013; Deptuck and Altheim, 2018; Withjack et al., 2020). The exact geometries of the buried rift basins in the southern and central segments are uncertain. SB (Scotian basin), BCT (Baltimore Canyon trough), and CT (Carolina trough) are Mesozoic/Cenozoic postrift basins discussed in the text. Blue and yellow lines show the locations of transects in Figs. 3.2 and 3.12, respectively. Numbers next to rift-basin names indicate figures with detailed map(s) and cross section(s) of rift basin. The ENA rift system consists of deeply eroded rift basins, most offset from the continent-ocean boundary by hundreds of kilometers. It has three segments. The southern segment consists of rift basins in the southeastern United States, the central segment includes rift basins between the southeastern United States and the Newfoundland fracture zone. The northern segment consists of rift basins north of the Newfoundland fracture zone in southeastern Canada. The cessation of rifting in the offset rift basins was diachronous, occurring first in the southern segment (latest Triassic), then in the central segment (Early Jurassic), and finally in the northern segment (Early Cretaceous) (e.g., Withjack et al., 2012).

et al., 1998, 2012). The ENA rift system (i.e., the part of the rift zone preserved on the margin of ENA) is wide, consisting of numerous parallel and en-echelon rift basins (Figs. 3.1 and 3.2). For example, the Danville/Dan River and Deep River rift basins are parallel to each other, whereas the Newark and Connecticut Valley rift basins have an en-echelon arrangement (Fig. 3.1). Most rift basins are deeply eroded (e.g., Withjack et al., 2013, 2020), and most are offset from the continent-ocean boundary by hundreds of kilometers. Examples of offset rift basins include the Fundy and Danville basins, which are located about 500 km (Fig. 3.2A) and 450 km (Fig. 3.2D), respectively, from the continent-ocean boundary. The width of the rift system varies along strike, ranging from about 200 km in the central part of the rift system, where crustal thicknesses (see references in the caption for

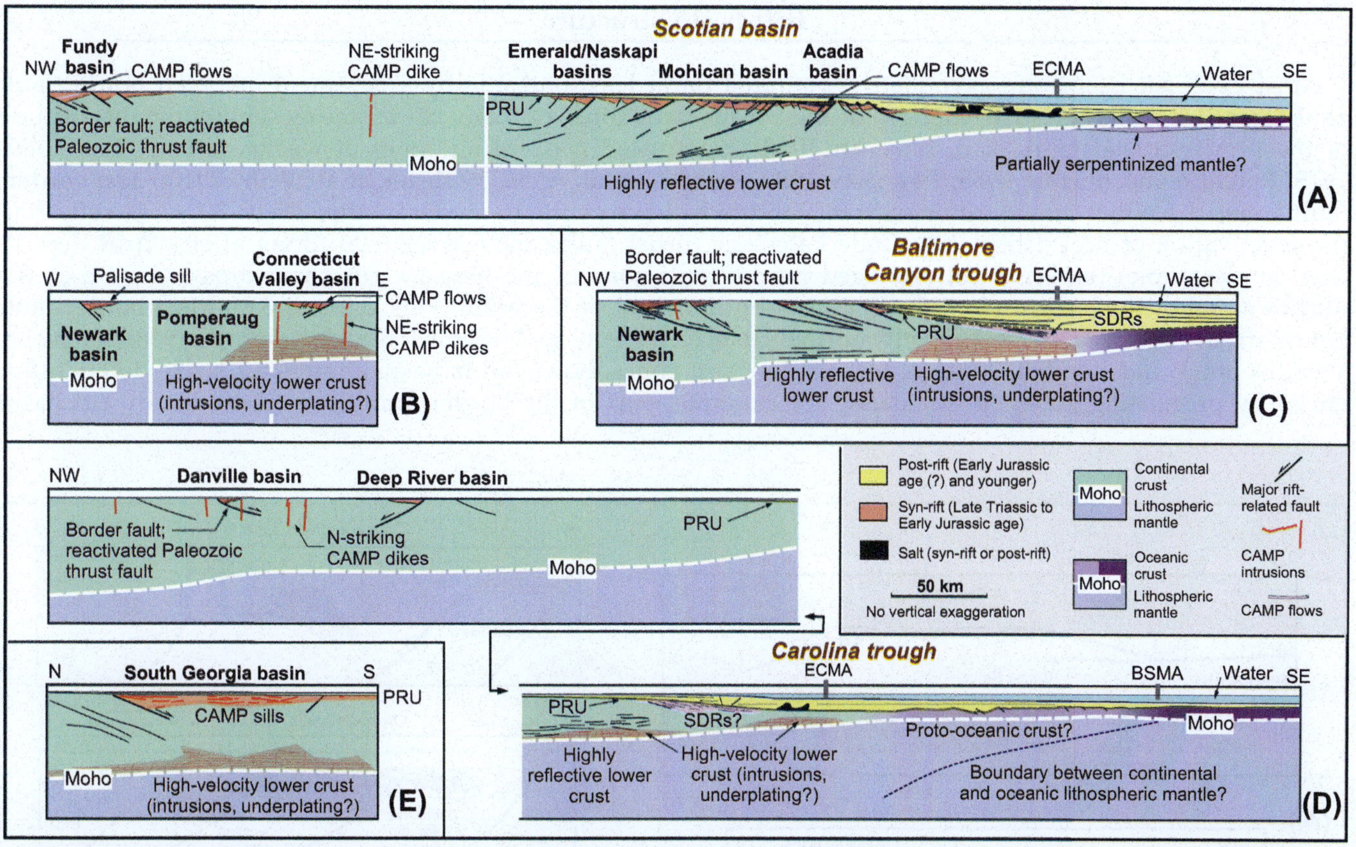

FIGURE 3.2 Transects (no vertical exaggeration) through the ENA rifted margin showing early Mesozoic rift basins, Mesozoic/Cenozoic postrift basins, and major crustal features and architecture. White vertical lines indicate break and/or change in the direction of transect. See Fig. 3.1 for locations. PRU is the postrift unconformity, SDRs are seaward-dipping reflectors, ECMA is the East Coast Magnetic Anomaly, and BSMA is the Blake Spur Magnetic Anomaly. Moho depths are from Hutchinson et al. (1986), Sheridan et al. (1991), Wu et al. (2006), Louden et al. (2013), Labails et al. (2009), Mooney and Kaban (2010), Shen and Ritzwoller (2016), and Li et al. (2018). (A) Transect across southeastern Canada based on data from Keen et al. (1991a, b), Withjack et al. (1995), Wu et al. (2006), Louden et al. (2013), Deptuck and Kendall (2017) and Deptuck and Altheim (2018). (B) Transect through Newark, Pomperaug, and Connecticut Valley rift basins based on data from Wenk (1984), Wise (1992), Resor and eBoer (2005), Burton (2006), Olsen et al. (2016), Withjack et al. (2020), and Gao et al. (2020). Note the high-velocity lower crust beneath the Connecticut Valley rift basin. (C) Transect crossing southern Newark rift basin and northern Baltimore Canyon trough based on seismic line NB-1 (Withjack et al., 2013), interpretation of USGS seismic line 25 (Grow et al., 1988; Labails et al., 2009), and data from LASE (Large Aperture Seismic Experiment) (LASE Study Group, 1986). Note the high-velocity lower crust interpreted as intruded/underplated continental crust and/or initial oceanic crust (e.g., Talwani et al., 1995). Prerift fabric and lower crustal laminations between seismic lines NB-1 and USGS 25 are based on seismic-reflection data from Sheridan et al. (1991). (D) Transect through Danville and Deep River rift basins and Carolina trough based on geological data (Olsen et al., 1989) and geophysical data (Shuck et al., 2019; Bécel et al., 2020). Note possible proto-oceanic crust and continental lithospheric mantle between the ECMA and BSMA. (E) Transect through the southeastern USA based on geophysical data from McBride (1991) and Marzen et al. (2019, 2020). Note the abundant sills within the South Georgia rift basin and the high-velocity lower crust beneath the basin.

Fig. 3.2) sharply decrease toward the continent-ocean boundary (Fig. 3.2C) to more than 500 km in the north (Fig. 3.2A) and 400 km in the south (Fig. 3.2D), where crustal thicknesses more gradually decrease toward the continent-ocean boundary.

Many of the characteristics of the ENA rift system and margin are likely related to the orogenic activity that preceded rifting. (1) Lang et al. (2020) propose that the juxtaposition of crustal blocks with differing dimensions and rheologies during the assembly of Pangaea contributed to the observed along-strike variations in crustal stretching and thinning on the ENA margin. (2) The preexisting zones of weakness created during the orogenic activity strongly influenced the location and distribution of the ENA rift basins. For example, numerical modeling suggests that wide rift systems with offset rift basins, like the ENA rift system, are more likely to develop, where weak crustal layers decouple laterally offset, preexisting zones of weakness within the upper crust from those within the upper mantle (e.g., Harry and Sawyer, 1992; Chenin and Beaumont, 2013). (3) Withjack et al. (2020) propose that denudation unloading related to relic, prerift orogenic crustal thickening and elevated topography caused considerable postrift uplift and erosion on the ENA rifted margin.

Rift-basin structure

Most ENA rift basins are asymmetric, bounded on at least one side by a basement-involved border-fault zone (Figs. 3.1–3.7). The strike and dip of the border-fault zones of many rift basins closely mimic the attitude of the prerift crustal fabric and, thus, are likely reactivated preexisting zones of weakness (e.g., Lindholm, 1978; Ratcliffe and Burton, 1985; Swanson, 1986; Ratcliffe et al., 1986; Withjack et al., 2009, 2010). The border-fault zones dip either landward or seaward, and most have gentle to moderate dips (Figs. 3.1–3.7), reflecting the reactivation of preexisting, low-angle Paleozoic thrust faults and/or regional tilting after synrift deposition. For example, the seismically imaged border-fault zone of the Newark rift basin dips ~45° SE at the northern end of the basin and ~25° SE at the southern end of the basin (Figs. 3.6 and 3.8). Most border-fault zones strike NE-SW and have mostly normal displacement (e.g., Hutchinson and Klitgord, 1988; Schlische, 1993; Deptuck and Kendell, 2017). A few border-fault zones, however, have an anomalous strike, likely influenced by preexisting zones of weakness. For example, part of the border-fault zone of the Fundy rift basin

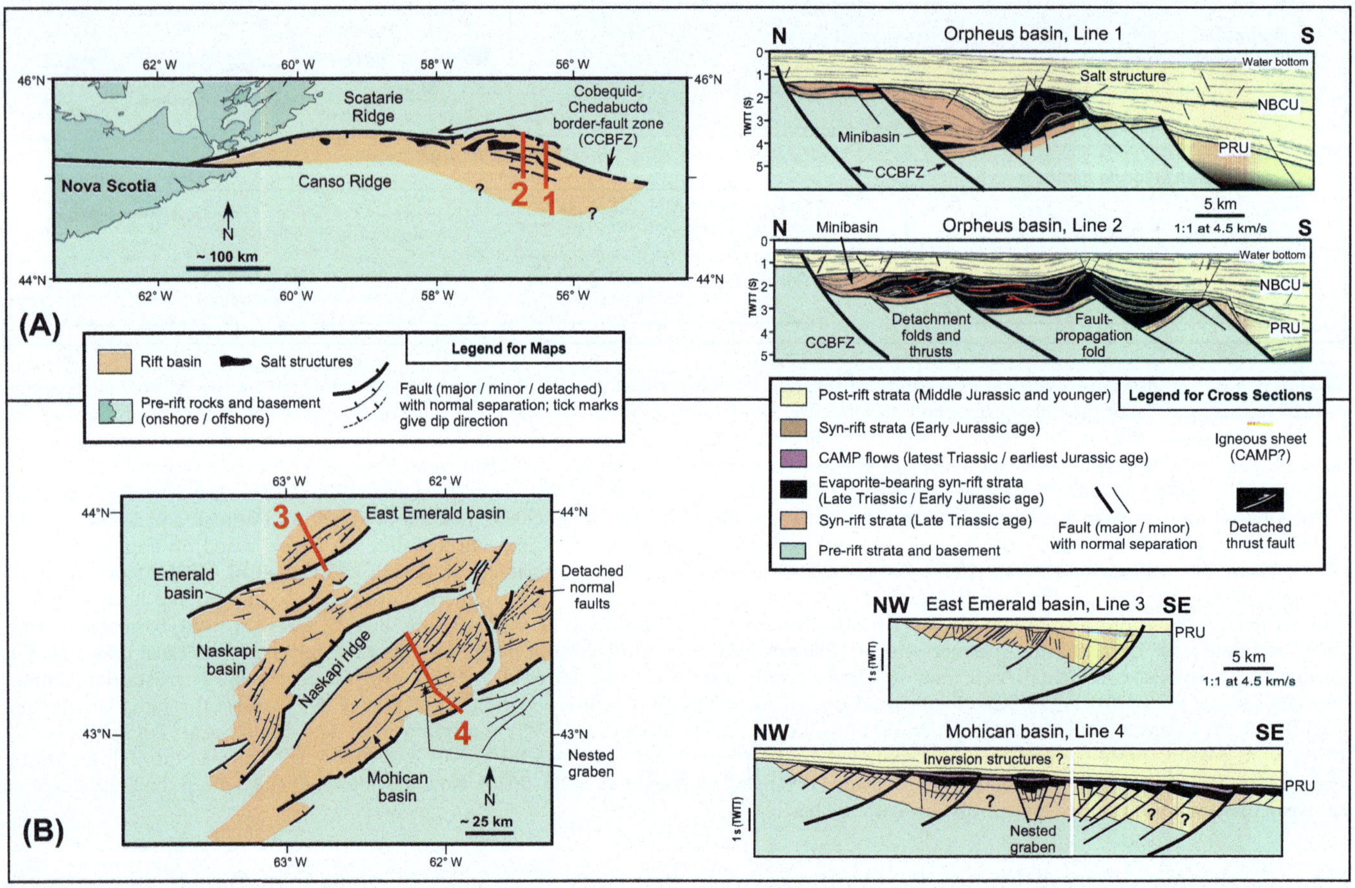

FIGURE 3.3 (A) Map view (left) and cross-sectional views (right) of Orpheus rift basin (based on Syamsir et al., 2010; Hanafi, 2013; Deptuck and Altheim, 2018; Hanafi et al., 2022). PRU is postrift unconformity; NBCU is near base Cretaceous Unconformity. The Cobequid-Chedabucto border-fault zone (CCBFZ) bounds the Orpheus rift basin on the north and continues westward, bounding part of the Fundy rift basin (see Fig. 3.4). Lines 1 and 2 cross a relay zone between two, right-stepping, overlapping border-fault segments. On line 1, the southern segment has more displacement than the northern segment. On line 2, the northern segment has more displacement than the southern segment. Loading associated with minibasin formation has forced evaporites to flow outward, producing adjacent shortening-related structures (Hanafi et al., 2022). On line 1, the evaporite-bearing unit contains few interbedded shales, and salt stocks and walls are common. On line 2, the evaporite-bearing unit contains more interbedded shales, and detachment folds and thrusts are common (Hanafi et al., 2022). Note the presence of likely CAMP igneous sheets in the basin. (B) Top basement structure map (left) and cross-sectional views (right) of East Emerald and Mohican rift basins on the LaHave platform, offshore Nova Scotia (after Deptuck and Altheim, 2018). Most faults dip toward the northwest (i.e., landward). Evaporite-bearing rocks are far more common toward the southeast (i.e., seaward) (line 4 vs line 3). Note the presence of CAMP flows within the evaporite-bearing synrift strata. Possible inversion structures postdate the PRU.

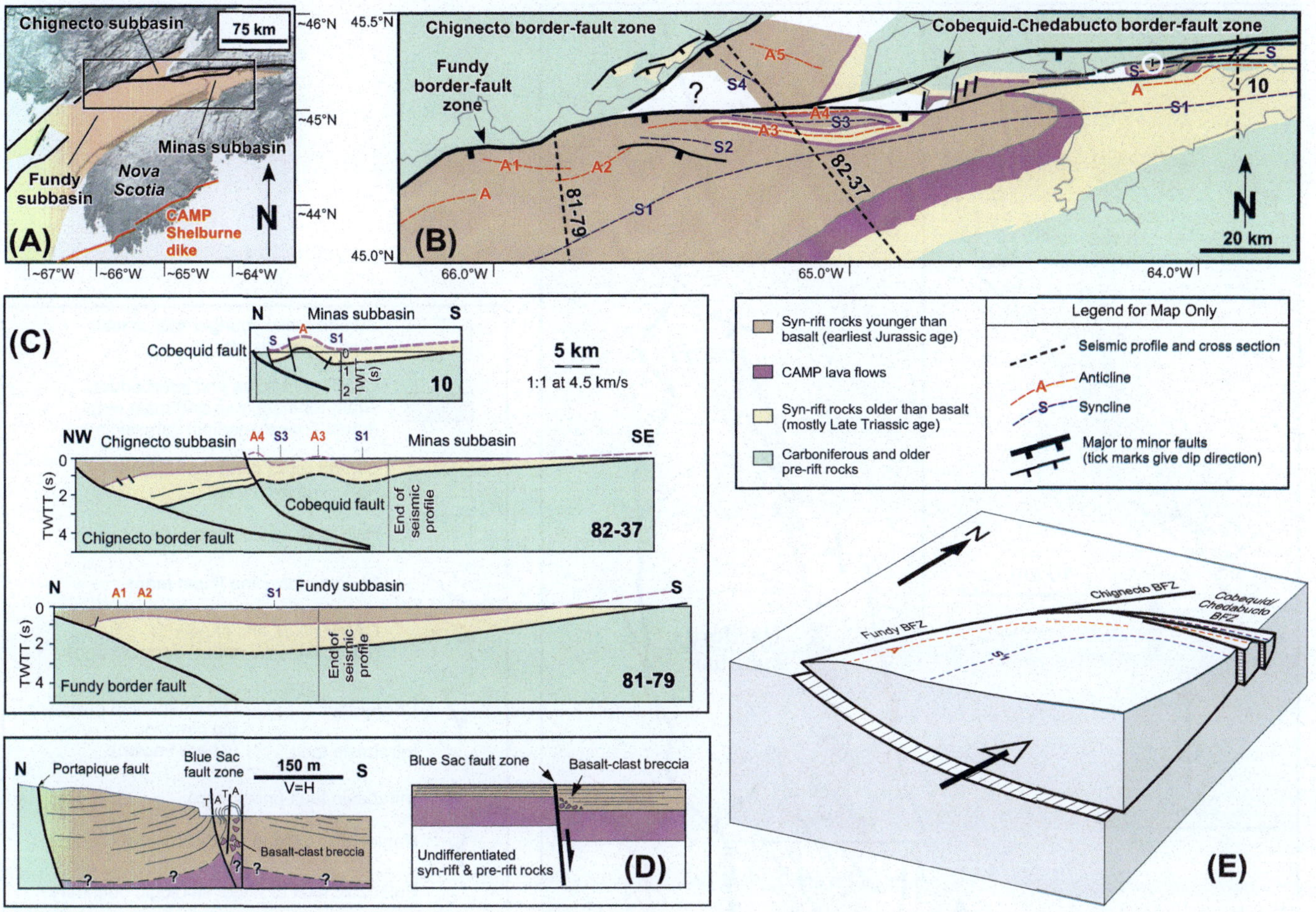

FIGURE 3.4 Geology of the Fundy rift basin . See Fig. 3.1 for location. (A) Shaded-relief map of the Bay of Fundy region, showing location of the Fundy rift basin (orange) consisting of Fundy, Chignecto, and Minas subbasins. (B) Regional geologic map of the area outlined by rectangle in (A) based on field and seismic-reflection data. The Fundy border-fault zone (FBFZ) and Chignecto border-fault zone (CBFZ) trend NE-SW. The Cobequid-Chedabucto border-fault zone (CCBFZ) trends E-W to ENE-WSW. Note that the synrift strata within the basin are folded near the border-fault zones. (C) Line drawings of interpreted seismic profiles with projections from onshore geology; see (B) for line locations. The profiles provide evidence of two phases of deformation. Lines 81–79 and 82–37 show that the prebasalt synrift units thicken toward the border faults (i.e., the faults were active during deposition). Subsequently, the synrift strata were folded. Line 10 shows a prominent anticline flanked by synclines in the hanging wall of the CCBFZ. Line 82–37 shows a train of folds in the hanging wall of the CCBFZ. Line 81–79 shows a broad, gentle anticline and syncline adjacent to the FBFZ. (D) Geologic cross section (left) and simplified restoration to late rifting geometry (right) of the Blue Sac area (see white circle in B). The Blue Sac fault zone is steeply dipping and exhibits left-lateral strike-slip slickenlines. T and A indicate toward and away. Tight folds in and adjacent to the fault zone are locally overturned and involve basalt-clast breccias, which likely accumulated adjacent to fault scarps during late rifting. Thus, this cross section shows evidence of growth faulting during late rifting followed by postrift deformation. (E) Block diagram showing kinematics of postrift deformation. The hanging-wall block of the composite border-fault system moved to the NNE, producing broad buttress folds adjacent to the FBFZ and tight folds and partitioned strike-slip faults along the CCBFZ. The CCBFZ had both normal dip-slip and left-lateral strike-slip components during rifting, and reverse dip-slip and left-lateral strike-slip after rifting. The FBFZ had mostly normal dip-slip during rifting and left-lateral strike-slip and reverse dip-slip components after rifting. Source: *Modified from Withjack, M.O., Schlische, R.W. and Baum, M.S., 2009. Extensional development of the Fundy rift basin, southeastern Canada. Geological Journal, 44(6), pp. 631–651; Withjack, M.O., Baum, M.S. and Schlische, R.W., 2010. Influence of preexisting fault fabric on inversion-related deformation: a case study of the inverted Fundy rift basin, southeastern Canada. Tectonics, 29(6).*

strikes ENE-WSW (Fig. 3.4) and had combined normal dip-slip and left-lateral strike-slip during rifting (Olsen and Schlische, 1990; Withjack et al., 1995, 2009). Most border-fault zones are composed of individual fault segments that are either soft-linked by relay ramps (i.e., the right-stepping border faults of the Orpheus rift basin, Fig. 3.3A) or hard-linked by breached relay ramps (e.g., the right-stepping border faults of the Newark rift basin, Fig. 3.6). Generally, the synrift strata within the ENA rift basins dip gently to moderately

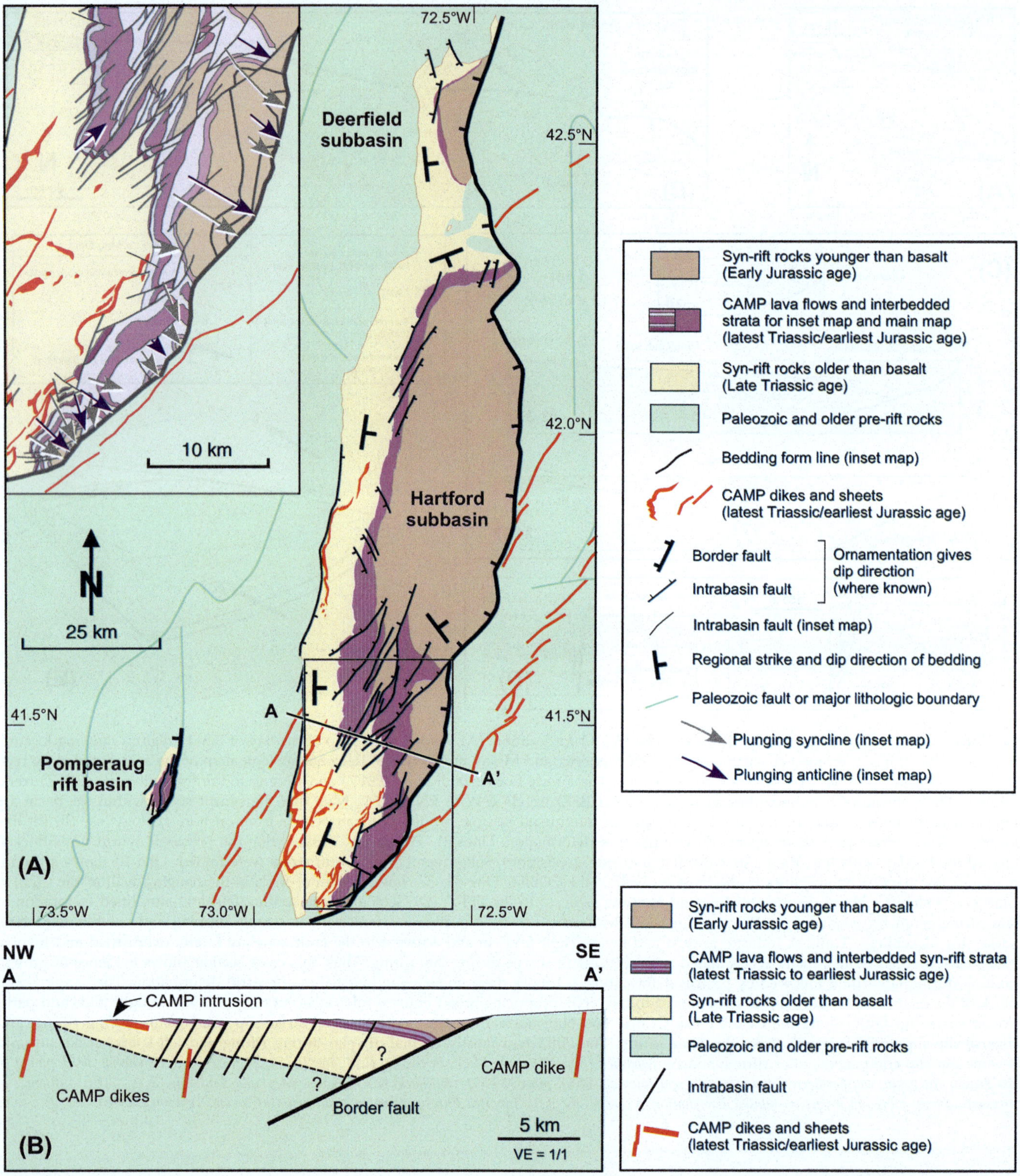

FIGURE 3.5 (A) Geologic map of the Connecticut Valley rift basin complex (consisting of Hartford and Deerfield subbasins) and Pomperaug rift basin . See Fig. 3.1 for location. In both rift basins, the border-fault zone is on the eastern side, and strata generally dip toward the east. The strike of most intrabasin faults in the Connecticut Valley rift basin is oblique to the basin trend but similar to that of CAMP diabase dikes in and near the basin. The Hartford and Deerfield subbasins are broad synclinal structures separated by a narrow intervening anticlinal structure. Inset shows southeastern Hartford subbasin with multiple transverse folds. (B) Geologic cross section of the southern Hartford basin, showing that most intrabasin faults and the border fault dip toward the west. Based on Withjack et al. (2020), which honors

(*Continued*)

toward the border-fault zones (Figs. 3.2–3.7). As discussed below, some of this dip developed during synrift deposition but much of it developed after synrift deposition.

A great variety of smaller-scale extensional structures developed within the ENA rift basins during rifting. Intrabasin faults are common in most rift basins. In the Newark rift basin, the intrabasin faults have variable strikes, ranging from highly oblique to subparallel to the border-fault zone (Fig. 3.6). These faults are basement-involved and developed primarily during the later stages of rifting (Withjack et al., 2013, 2020). In the Connecticut Valley rift basin, most intrabasin faults are oblique to the border-fault zone and developed primarily after synrift deposition (e.g., Wise, 1992) (Fig. 3.5). In the Mohican rift basin, nested graben structures, whose bounding faults strike subparallel to the border-fault zone, developed during the late stages of rifting (Deptuck and Kendell, 2017) (Fig. 3.3D). A series of anticlines and synclines, commonly called transverse folds, also deform the synrift strata in the hanging walls of the border-fault zones and intrabasin faults of many ENA rift basins (e.g., Wheeler, 1939; Sanders, 1963; Schlische, 1992, 1995; Wise, 1992) (Figs. 3.5 and 3.6). The preferential intrusion of diabase along the axial traces of some folds (e.g., the Jacksonwald syncline; Fig. 3.6C) and the subtle thinning and thickening of the synrift strata within the anticlines and synclines, respectively, suggest that some of this folding occurred during synrift deposition, possibly as a result of along-strike variations in fault displacement (e.g., Wheeler, 1939; Lucas et al., 1988; Manspeizer, 1988; Wenk, 1990; Schlische, 1993, 1995; Withjack et al., 2002). As discussed below, however, much of this folding occurred after synrift deposition.

The presence of synrift salt profoundly affected the structural development of many ENA rift basins in southeastern Canada (Fig. 3.3). Tectonic activity, regional tilting, and/or differential sediment loading triggered salt flow, producing a variety of salt structures during and after rifting. For example, salt diapirs, detached normal faults, detached thrust faults, detachment folds, and minibasins formed within the Orpheus rift basin (MacLean and Wade, 1992; Syamsir et al., 2010; Zulfitriadi, 2011; Hanafi, 2013; Hanafi et al., 2022) (Fig. 3.3A) and the eastern rift basins on the LaHave platform of offshore Nova Scotia (e.g., Deptuck and Kendell, 2017) (Fig. 3.2A and B) during and after rifting. Hanafi et al. (2022) showed that the style of detached deformation in the salt-rich Orpheus rift basin depends on the relative abundance of interbedded shales within the evaporites (Fig. 3.3B). If evaporite-bearing units have few interbedded shales, then salt stocks and walls are common. Alternatively, if evaporite-bearing units have numerous interbedded shales, then detachment folds and thrusts are common. Forced folds developed above many deep-seated intrabasin faults in the Orpheus rift basin (Hanafi et al., 2022) (Fig. 3.3A). Scaled experimental models and geologic examples from several salt-rich basins (e.g., Vendeville et al., 1995; Withjack and Callaway, 2000) show that deep-seated normal faults cannot propagate upward through thick salt. Instead, forced folds form in the sedimentary cover above the subsalt faults.

Timing of rifting in the southern and central segments

Strata deposited during extensional deformation (i.e., extensional growth strata) exhibit thickness changes, fanning geometries (i.e., increasing bedding dips with depth), and facies changes near the faults and folds that influenced their deposition (e.g., Ravnas and Steel, 1998; Gawthorpe and Leeder, 2000; Withjack et al., 2002; Withjack and Schlische, 2006; Serck and Braathen, 2019). As described below, growth strata are common throughout the ENA rift system and provide critical information about the timing of rift-related deformation.

Alluvial-fan conglomeritic facies in strata of Late Triassic age are present near the border-fault zones of the ENA rift basins, indicating a local source of relief adjacent to the border-fault zones during the Late Triassic (e.g., LeTourneau, 1985; Arguden and Rodolfo, 1986; Olsen and Schlische, 1990; Smoot, 1991) (e.g., Fig. 3.6A). Geological and geophysical data show that strata of Late Triassic age thicken and fan toward the border-fault zones in many ENA rift basins, including the East Emerald and Mohican rift basins (e.g., Deptuck and Kendell, 2017) (Fig. 3.3B), the Fundy rift basin (e.g., Withjack et al., 1995, 2009, 2010) (Fig. 3.4C), and the Newark rift basin

◄ surface geology from Resor and deBoer (2005) and depth to basement from Wenk (1984). Source: *(A) Adapted from Schlische, R. W., 1993, Anatomy and evolution of the Triassic-Jurassic continental rift system, eastern North America: Tectonics, 12, p. 1026–1042; Schlische, R. W., Withjack, M. O., and Olsen, P. E., 2003, Relative timing of CAMP, rifting, continental breakup, and basin inversion: tectonic significance, in Hanes, W. E., McHone, J. G., Renne, P. R., and Ruppel, C., eds., The Central Atlantic Magmatic Province, Insights from Fragments of Pangea: Washington, D. C., American Geophysical Union, Geophysical Monograph 136, p. 33–60; and Burton, W., 2006, Bedrock map of the early Mesozoic Pomperaug basin and surrounding basement rocks, Litchfield and New Haven counties, CT: U.S. Geological Survey Open-File Report 2006-1011, https://doi.org/10.3133/ofr20061011; (B) Modified from Withjack, M., Malinconico, M., and Durcanin, M., 2020. The "passive" margin of eastern North America: rifting and the influence of prerift orogenic activity on postrift development. Lithosphere, 2020(1).*

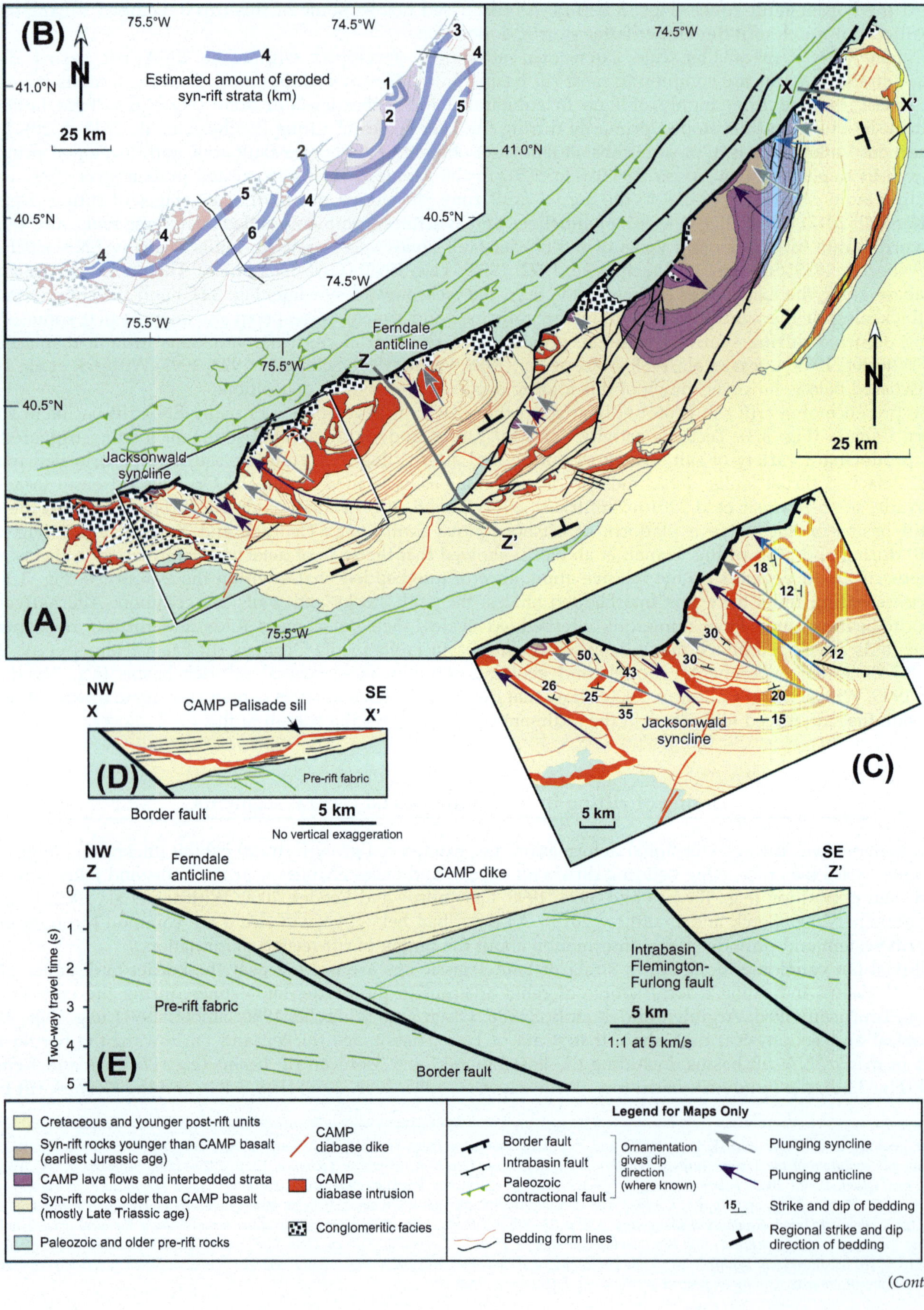
(A)
(B)
(C)
(D)
(E)
Estimated amount of eroded syn-rift strata (km)
25 km
Ferndale anticline
Jacksonwald syncline
NW X
SE X'
CAMP Palisade sill
Pre-rift fabric
Border fault
5 km
No vertical exaggeration
NW Z
SE Z'
CAMP dike
Two-way travel time (s)
Pre-rift fabric
Intrabasin Flemington-Furlong fault
1:1 at 5 km/s
Border fault
Cretaceous and younger post-rift units
Syn-rift rocks younger than CAMP basalt (earliest Jurassic age)
CAMP lava flows and interbedded strata
Syn-rift rocks older than CAMP basalt (mostly Late Triassic age)
Paleozoic and older pre-rift rocks
CAMP diabase dike
CAMP diabase intrusion
Conglomeritic facies
Legend for Maps Only
Border fault
Intrabasin fault
Paleozoic contractional fault
Ornamentation gives dip direction (where known)
Bedding form lines
Plunging syncline
Plunging anticline
Strike and dip of bedding
Regional strike and dip direction of bedding

(Continued)

(e.g., Schlische, 1992; Olsen et al., 1996a; Schlische and Withjack, 2005; Withjack et al., 2013, 2020) (Fig. 3.6D and E). In the Newark rift basin, growth packages near the base of the synrift section are narrow (<10 km wide) and exhibit marked thickening toward the border-fault zone. Higher in the synrift section, growth packages are wide (50–100 km), thicken gradually toward the border-fault zone, and have great lateral continuity (Fig. 3.6D and E). Without ample core and outcrop data, it would be easy to mistake these latter growth packages—with their great width, subtle thickness variations, and lateral continuity—for prerift or postrift strata (e.g., Faill, 1973, 1988, 2003; Schlische and Withjack, 2005). Thus, the presence of growth strata indicates that rifting was underway throughout the ENA rift system by Late Triassic time. Rifting may have begun earlier in some ENA rift basins. For example, Permian and Middle Triassic strata crop out in the Fundy rift basin (Olsen, 1997; Sues and Olsen, 2015; Olsen and McDonald, 2022). Coeval strata on the conjugate margin of Morocco display growth geometries consistent with rifting, suggesting that these older Fundy outcrops also may be synrift strata (Olsen et al., 2000; Kent and Clemmensen, 2021).

The age of the youngest, unambiguously dated synrift strata varies from south to north along the ENA rift system (Figs. 3.1 and 3.9). In the southern segment, the youngest dated synrift strata are in the Taylorsville rift basin (Figs. 3.1 and 3.9) and are Late Triassic in age, ~10 Myr older than the Triassic-Jurassic boundary. Modeling studies based on thermal maturation indices (e.g., Malinconico, 2003, 2015) indicate that, although more strata accumulated in the Taylorsville rift basin than are currently preserved, synrift subsidence most likely ceased by the earliest Jurassic. If these results are applicable to other southern-segment basins, then synrift subsidence ceased in the south by the earliest Jurassic.

In the central segment, the youngest dated synrift strata include basalt flows associated with the CAMP (201.5 Ma) and interbedded and/or overlying sedimentary rocks of the latest Triassic/earliest Jurassic to Early Jurassic age (e.g., Olsen et al., 1989; Olsen, 1997) (Fig. 3.9). In the Newark rift basin, alluvial-fan conglomeritic facies are present in strata interbedded with and succeeding CAMP lava flows, and some of these units contain clasts of CAMP basalt (e.g., Manspeizer, 1980; Olsen, 1980) (Fig. 3.6A). In the Fundy rift basin, outcrop, drill-hole, and seismic data indicate that synrift units of the Early Jurassic age generally thicken toward the border-fault zone (e.g., Olsen et al., 1989; Withjack et al., 1995, 2009, 2010). In addition, the sedimentary unit above CAMP lava flows in the Fundy rift basin contains paleo-talus slope deposits composed predominantly of basalt clasts, indicating active fault-controlled subsidence along the border-fault zone during Early Jurassic time (Olsen and Schlische, 1990; Tanner et al., 1991; Withjack et al., 2009, 2010) (Fig. 3.4D). Thus, the structural and stratigraphic information indicate that, unlike the rift basins in the southern segment, the rift basins in the central segment of the ENA rift system remained active into Early Jurassic time. Because of considerable postrift erosion, it is unclear how long deposition continued in the central-segment rift basins. In the Newark rift basin, deposition likely ended in Early Jurassic time (~198 Ma), based on published sedimentation rates and chronostratigraphy of the youngest preserved synrift Jurassic formations (Olsen et al., 1996b; Kent and Olsen, 2008), isotopic ages of CAMP igneous rocks (Blackburn et al., 2013), and estimates of missing synrift Jurassic overburden (Malinconico, 2002, 2010).

FIGURE 3.6 (A) Geologic map of Newark rift basin showing the location of cross sections in (D) and (E) See Fig. 3.1 for location. The border-fault zone on the northwest side of the basin is subparallel to Paleozoic contractional structures. The strike of most large intrabasin faults differs from that of the border-fault zone. Synrift strata generally dip to the northwest (toward the border-fault zone). (B) Contours showing the estimated amount of eroded synrift strata based on analyses of vitrinite-reflectance and sonic transit-time data. Erosion locally exceeds 6 km. (C) Enlarged view of southwestern part of Newark rift basin showing segmented, right-stepping border-fault zone with breached relay-ramps and transverse folds. (D) Line drawing of depth-migrated version of Sandia seismic line through northern end of Newark rift basin (after Olsen et al., 2016; Withjack et al., 2020). The CAMP Palisade sill mostly cuts across bedding. (E) Line drawing of time-migrated seismic line NB-1 through southern Newark rift basin (after Withjack et al., 2013). The border-fault zone dips about 25° and is subparallel to Paleozoic contractional structures. Only the oldest synrift strata show obvious growth toward the border-fault zone. The large intrabasin fault (i.e., the Flemington-Furlong fault) is mostly postdepositional. The Ferndale anticline, associated with a restraining offset in the border-fault system during postdepositional left-lateral strike-slip, affects the synrift strata near the border-fault zone. Source: *(A) Modified from Schlische, R. W., 1992, Structural and stratigraphic development of the Newark extensional basin, eastern North America: implications for the growth of the basin and its bounding structures: Geological Society of America Bulletin, 104, p. 1246–1263; Withjack, M.O., Schlische, R.W., Malinconico, M.L. and Olsen, P.E., 2013. Rift-basin development: lessons from the Triassic–Jurassic Newark Basin of eastern North America. Geological Society, London, Special Publications, 369(1), pp.301–321; (B) Modified from Withjack, M., Malinconico, M., and Durcanin, M., 2020. The "passive" margin of eastern North America: rifting and the influence of prerift orogenic activity on postrift development. Lithosphere, 2020(1); (C) Modified from Schlische, R. W., 1992, Structural and stratigraphic development of the Newark extensional basin, eastern North America: implications for the growth of the basin and its bounding structures: Geological Society of America Bulletin, 104, p. 1246–1263; Schlische, R. W., 1995, Geometry and origin of fault-related folds in extensional settings, American Association of Petroleum Geologists Bulletin, 79, p. 1661–1678.*

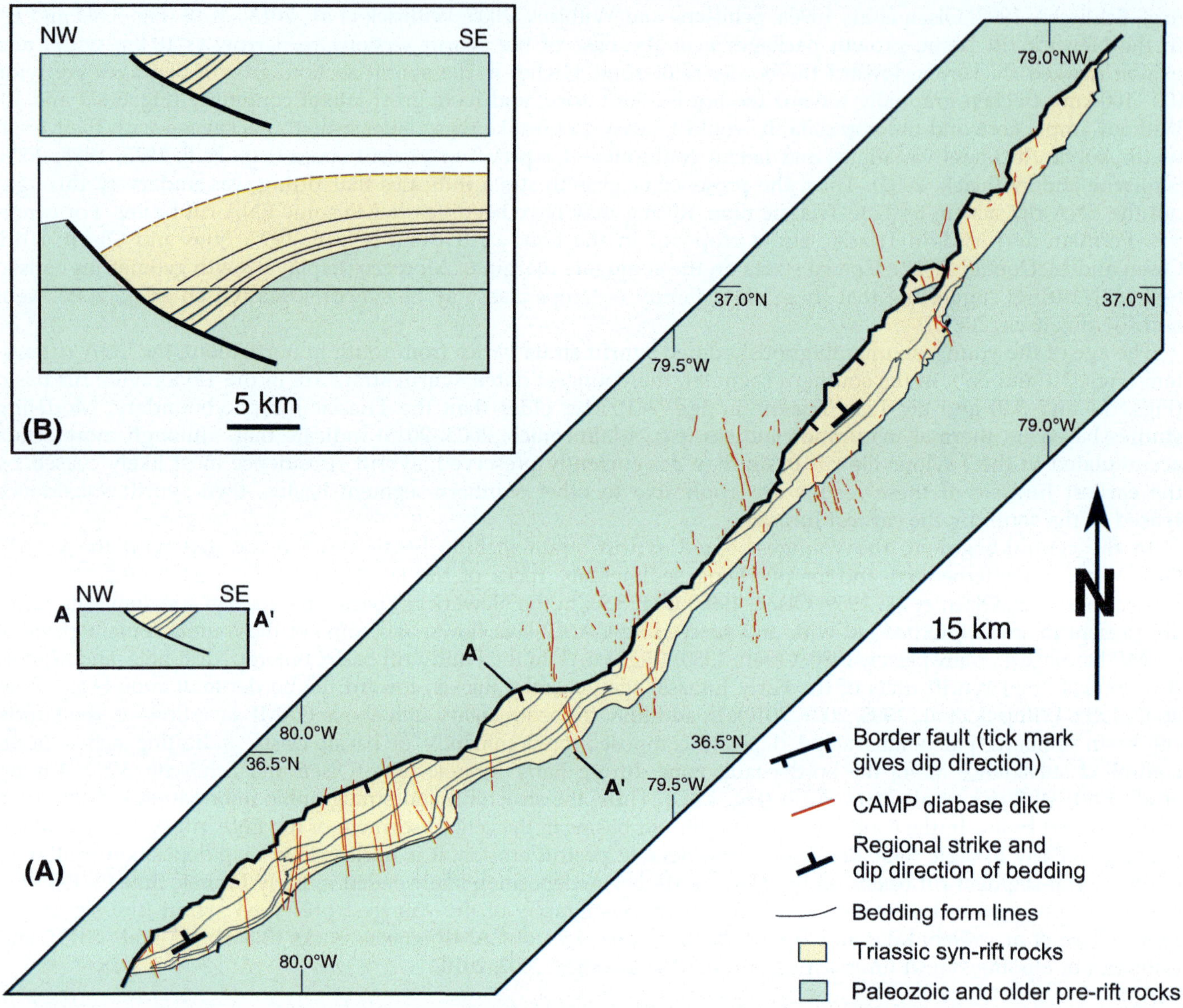

FIGURE 3.7 Geology of the Danville/Dan River rift basin. See Fig. 3.1 for location. (A) Simplified geologic map and cross section, showing that the border-fault zone dips to the southeast; synrift strata generally dip to the northwest; the basin is exceptionally narrow for its length; and N- to NW-trending CAMP diabase dikes cut the synrift strata and cross the border-fault zone without any apparent offset. (B) Enlarged version of cross section (top) and schematic restoration, showing that the basin has undergone considerable postrift erosion and was originally much wider. Mineralogical evidence and vitrinite-reflectance analyses indicate up to 8 km of erosion after synrift deposition (Warrell et al., 2018). Source: *(A) Modified from Olsen, P.E., Reid, J.C., Taylor, K.B., Whiteside, J.H. and Kent, D.V., 2015. Revised stratigraphy of Late Triassic age strata of the Dan River Basin (Virginia and North Carolina, USA) based on drill core and outcrop data. Southeastern Geology, 51(1), pp.1–31.*

Rift-basin evolution: evidence from the Newark Rift Basin

Recent studies of the Newark rift basin using field, seismic-reflection, core, borehole, vitrinite-reflectance, and sonic transit-time data provide a more detailed portrait of rift-basin evolution in the central segment of the ENA rift system (Withjack et al., 2013, 2020). These studies show that rifting had two distinct phases. During the main rifting phase (~230 to 201.5 Ma), the Newark rift basin widened through time. By the end of this phase, the basin was very wide (locally >100 km) with little, if any, intrabasin faulting (Fig. 3.8A and B). Its northern end plausibly connected with the Connecticut Valley and New York Bight rift basins to the northeast and east, respectively (Fig. 3.8B). The basin was asymmetric, bounded on the northwest by a major border-fault zone (Figs. 3.6A and 3.8A).

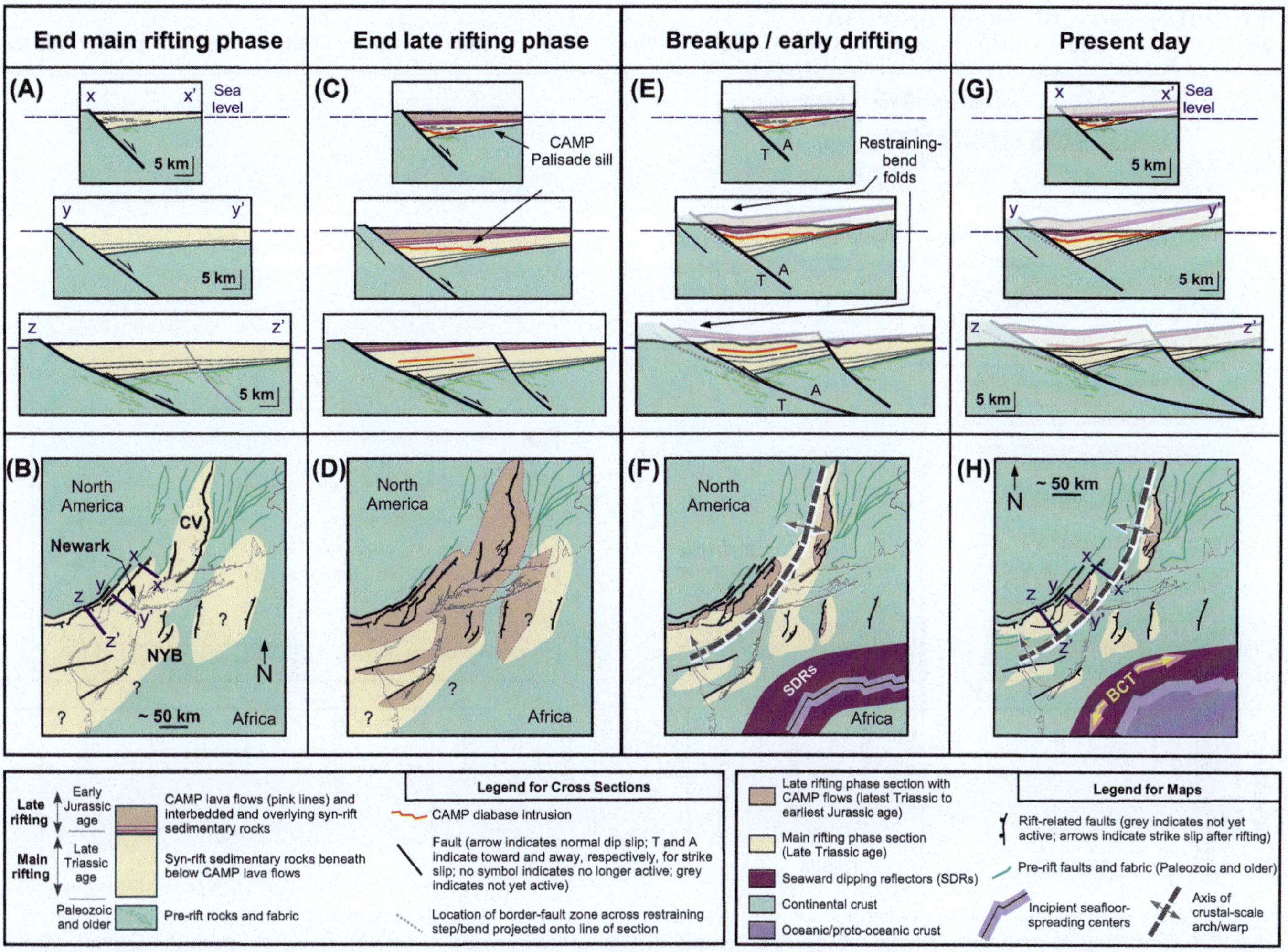

FIGURE 3.8 Evolution of Newark rift basin and surrounding region (after Withjack et al., 2020) in cross-sectional view (A, C, E, and G) and map view (B, D, F, and H). Cross-section locations shown in B and H. Cross sections X-X′ and Z-Z′ are same as those in Fig. 3.6. (A) and (B) End of main rifting phase. The Newark rift basin was asymmetric with a border-fault zone on its northwest side and had become wide and deep, especially in the south. The basin likely connected with the Connecticut Valley rift basin (CV) in the northeast and the New York Bight rift basin (NYB) in the east. (C) and (D) End of late rifting phase. During this phase, the border-fault zone remained active and intrabasin faults formed, linking with the border-fault zone and dissecting the rift basin. CAMP lava flows filled the rift basin, and igneous sheets intruded the basin fill. (E) and (F) Breakup and early drifting. Seaward-dipping reflectors (SDRs) were emplaced near the site of continental breakup and incipient seafloor-spreading centers developed forming oceanic or proto-oceanic crust. The postrift unconformity developed adjacent to the site of breakup. Far from the site of breakup, postrift deformation included the reactivation of the border-fault zone of the Newark rift basin with left-lateral strike-slip. In addition, the Newark rift basin underwent NW tilting, uplift, and erosion, whereas the Connecticut Valley and New York Bight rift basins underwent E to SE tilting, uplift and erosion, leading to the formation of a broad, faulted crustal arch. (G) and (H) Present day. Only erosional remnants of the ENA rift basins remain on the margin. The offshore postrift Baltimore Canyon trough (BCT) filled with sediments shed from the adjacent onshore region.

Cumulative heave on the border-fault zone was greatest in the southern Newark rift basin, exceeding 19 km with corresponding extension rates averaging about 0.7 km/Myr (Withjack et al., 2020). The thickness of the basin fill in the southern Newark rift basin was greater than 9 km near the border-fault zone with compacted accumulation rates averaging about 0.3 km/Myr (Withjack et al., 2020), similar to values reported by Olsen and Kent (1996). The magnitude of the dip of the strata deposited during the main rifting phase gradually increases downward, reflecting tilting (up to 5° NW) toward the border-fault zone during the main rifting phase.

Tectonic activity changed profoundly during the late rifting phase (201.5 to ~198 Ma) (Fig. 3.8C and D). CAMP basalt flows episodically filled the basin, massive igneous sheets (e.g., the Palisade sill) intruded the basin fill, and mostly NE-striking dikes intruded the basin fill and the surrounding basement rocks. The border-fault

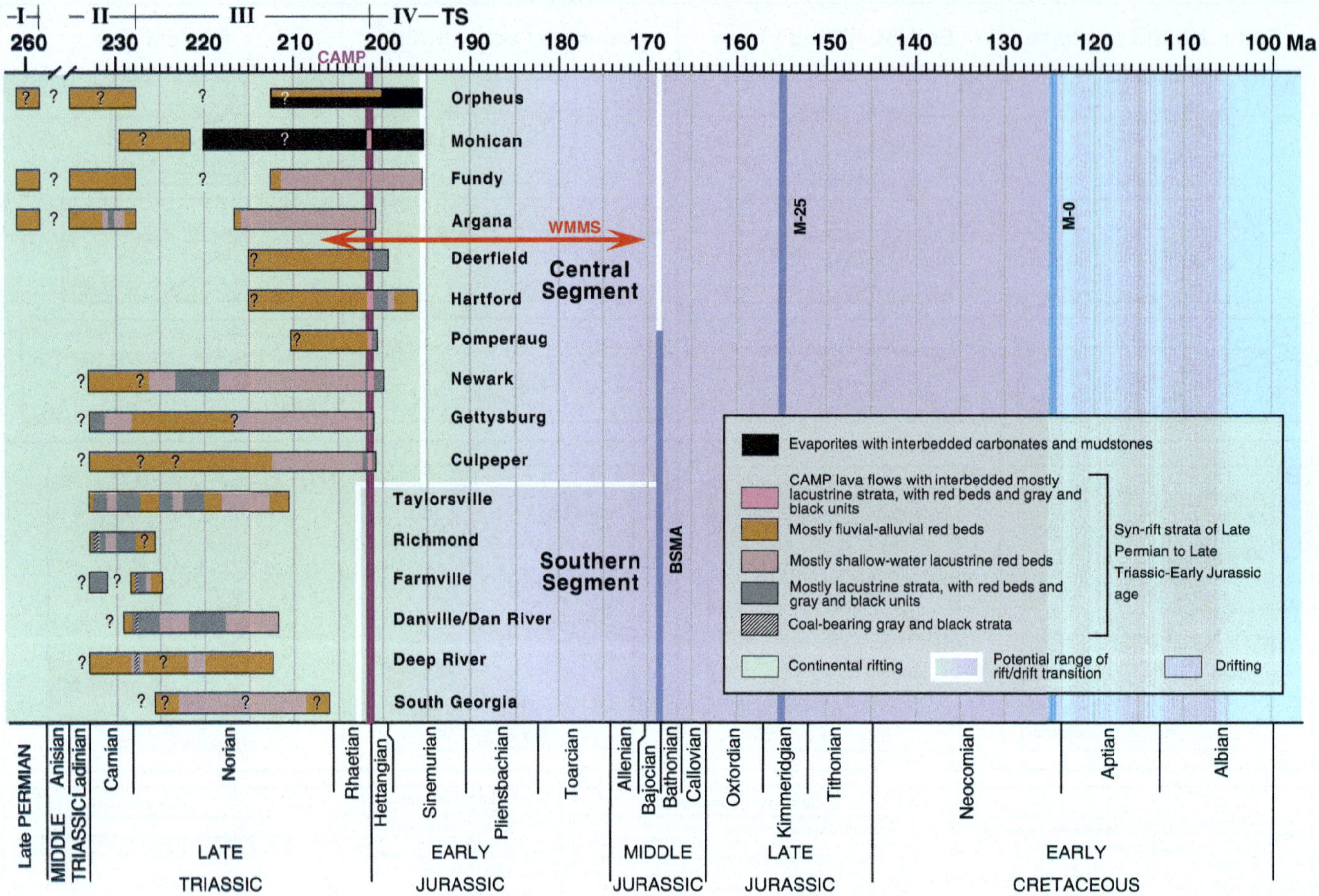

FIGURE 3.9 Ages and basic facies of synrift strata within select rift basins of the ENA rift system and the Argana rift basin of Morocco relative to the timing of continental rifting and drifting, the rift/drift transition, magmatic activity associated with the Central Atlantic Magmatic Province (CAMP) and the White Mountain Magma Series (WMMS), the formation of the Blake Spur Magnetic Anomaly (BSMA), and the formation of magnetic anomalies, M25 and M0. TS are tectonostratigraphic sequences I through IV described in the text. Time scale from Cohen et al. (2013) and Gradstein et al. (2020). Ages and facies based on Olsen et al. (1989), Olsen (1997), Weems and Olsen (1997), Tanner and Brown (2003), Withjack et al. (2012), Olsen et al. (2015), Sues and Olsen (2015), Deptuck and Altheim (2018), MacRae and Rankin (2018), O'Connor et al. (2018), Sues et al. (2021), Hanafi et al. (2022), and Kinney et al. (2022). The age of the youngest, unambiguously dated synrift strata varies along the ENA rift system. In the southern segment, only strata of Late Triassic age are present. In the central segment, synrift strata of Late Triassic and Early Jurassic age are present, including CAMP basalt flows and interbedded and/or overlying sedimentary rocks of latest Triassic to Early Jurassic age. Note the long potential range of the rift/drift transition based on Shuck et al. (2019) and Bécel et al. (2020).

zone remained active and, for the first time, major intrabasin faults developed, linking with the border-fault zone and dissecting the basin. Extension rates increased, reaching values of 1 km/Myr or more (Withjack et al., 2020). Compacted accumulation rates also increased significantly to about 1 km/Myr in the central and northern parts of the basin (Withjack et al., 2020). These compacted accumulation rates are similar to those determined by Olsen et al. (1996b). Schlische et al. (2003) described a similar increase in accumulation rates during the earliest Jurassic for other ENA rift basins in the central segment. By the end of the late rifting phase, the linked border and intrabasin faults remained active, but little sediment accumulated within their hanging walls, signaling an end to widespread synrift deposition.

Depositional patterns

Continental conditions prevailed throughout ENA during the Late Triassic (Carnian) (e.g., Olsen, 1997) (Fig. 3.9). The western rift basins in the southern and central segments remained continental throughout their entire preserved depositional history. By latest Triassic (Norian) time, however, encroachment of the Tethys Sea from the north created evaporitic marine conditions in the northeastern rift basins of the central segment (e.g., the Orpheus and Mohican rift basins, Figs. 3.3 and 3.9). Evaporites (mostly halite) filled these rift basins during Late Triassic and Early Jurassic times (Figs. 3.3 and 3.9). Sulfur isotopic evidence from the oldest evaporites suggests only a moderate marine contribution of highly evolved brines (Holser et al., 1988). Coeval basins on the conjugate Moroccan margin, however, have progressively more marine influence toward the east (Et-Touhami, 2000; Olsen et al., 2003), and in eastern Morocco, the sedimentary units interbedded with CAMP basalt flows are carbonate sedimentary rocks with abundant marine mollusks and echinoids (Olsen et al., 2003; Whiteside et al., 2007).

The evolving geometry of the rift basins also influenced the depositional patterns in the ENA rift system. Tectonostratigraphic sequences bounded by unconformities in the southern and central segments of the ENA rift system and the conjugate Moroccan margin (Olsen, 1997) share a similar stratigraphic succession known as a tripartite stratigraphy (e.g., Schlische and Olsen, 1990). Generally, near the basin depocenter, each tectonostratigraphic sequence (TS I to TS IV, Fig. 3.9) consists of a basal fluvial unit followed upward by lacustrine strata that rapidly reach maximum paleodepth and then slowly shallow upward, sometimes returning to fluvial deposits. The lacustrine strata tend to be strongly cyclical (Olsen et al., 1989; Olsen and Kent, 2000; Olsen et al., 2019; Olsen and McDonald, 2022), reflecting Milankovitch climate-cycle control of lake depth (Olsen, 1986; Olsen and Kent, 1996). Exceptions include TS III in some rift basins, such as the Connecticut Valley rift basin, that consist only of fluvial strata, as well as the basal part of TS IV that can lack a basal fluvial unit in areas with a correlative conformity with TS III (Olsen, 1997). Facies patterns tend to be centripetal with fine-grained lacustrine strata at the basin depocenter. Asymmetrically distributed fluvial and marginal lacustrine strata are present along the basin margins; these tend to be extensively developed along hinge margins and at the lateral ends of the basins and are restricted to much narrower, but commonly very coarse-grained bands, on the border-fault margins (Fig. 3.10) (Olsen, 1997). Both the vertical sequence of facies and their asymmetric distribution suggest that each tectonostratigraphic sequence originated with a major extensional pulse that produced an asymmetrically subsiding basin that widened through time (Schlische and Olsen, 1990; Schlische, 1991; Contreras et al., 1997; Olsen, 1997).

Climate profoundly influenced the depositional patterns in the ENA rift basins. The clearest climatic effect is the arrangement of facies relative to the paleoequator during Carnian time. Coals and deep-water lacustrine deposits accumulated near the paleoequator (the southern basins in Figs. 3.9 and 3.10C), while cyclical perennial lacustrine and playa deposits and bioturbated red beds accumulated at 10° to the north and south of the paleoequator (the Newark basin in Figs. 3.9 and 3.10B) (Kent and Olsen, 1997; Olsen, 1997; Olsen and Kent, 2000). Broadly contemporaneous deposits at 30° N paleolatitude in Greenland (Clemmensen, 1980, Clemmensen et al., 1998; Mau et al., 2022) are composed of red mudstones, eolian sand dunes, and evaporite beds, while farther north on the conjugate margin in the Barents Sea, deltaic coals and black mudstone again dominate (Van Veen et al., 1992). This conforms to a simple zonal pattern, with a narrow equatorial humid zone and an arid belt mostly south of 30°, passing northward into hothouse humid temperate climates. Although nonzonal elements such as orography and an enhanced monsoon may have been important elements of the Earth's climate system during Carnian time, such are not required by the observations (Kent and Tauxe, 2005).

As Pangea drifted northward, the vertical sequence of climate-sensitive facies within the individual basins changed as the basins passed from one climate zone to another. Thus, in the rift basins of ENA (and Morocco), the transition from Carnian- through Norian-age strata is characterized by apparent drying with shallow-water cyclical lacustrine strata predominating in the southern and central basins (Olsen, 1997; Olsen and Kent, 2000) (Figs. 3.9 and 3.10A). Conversely, in Greenland and offshore Norway, the vertical Late Triassic sequence within individual basins is from red beds and evaporites upward into black lacustrine, paludal, and paralic shales (Clemmensen, 1980; Jacobsen and van Veen, 1984). This vertical pattern within all of these basins fits the same basic geographic pattern consistent with a northward drift of central Pangea.

Igneous activity during rifting

Igneous activity was limited within the ENA rift system before CAMP magmatism. Potential pre-CAMP rift-related igneous activity includes magmatism associated with the Coastal New England (CNE) province and the White Mountain Magma Series (WMMS) (e.g., McHone and Butler, 1984), both located within the northern part of the central segment of the ENA rift system. The CNE province consists predominantly of basaltic dikes and

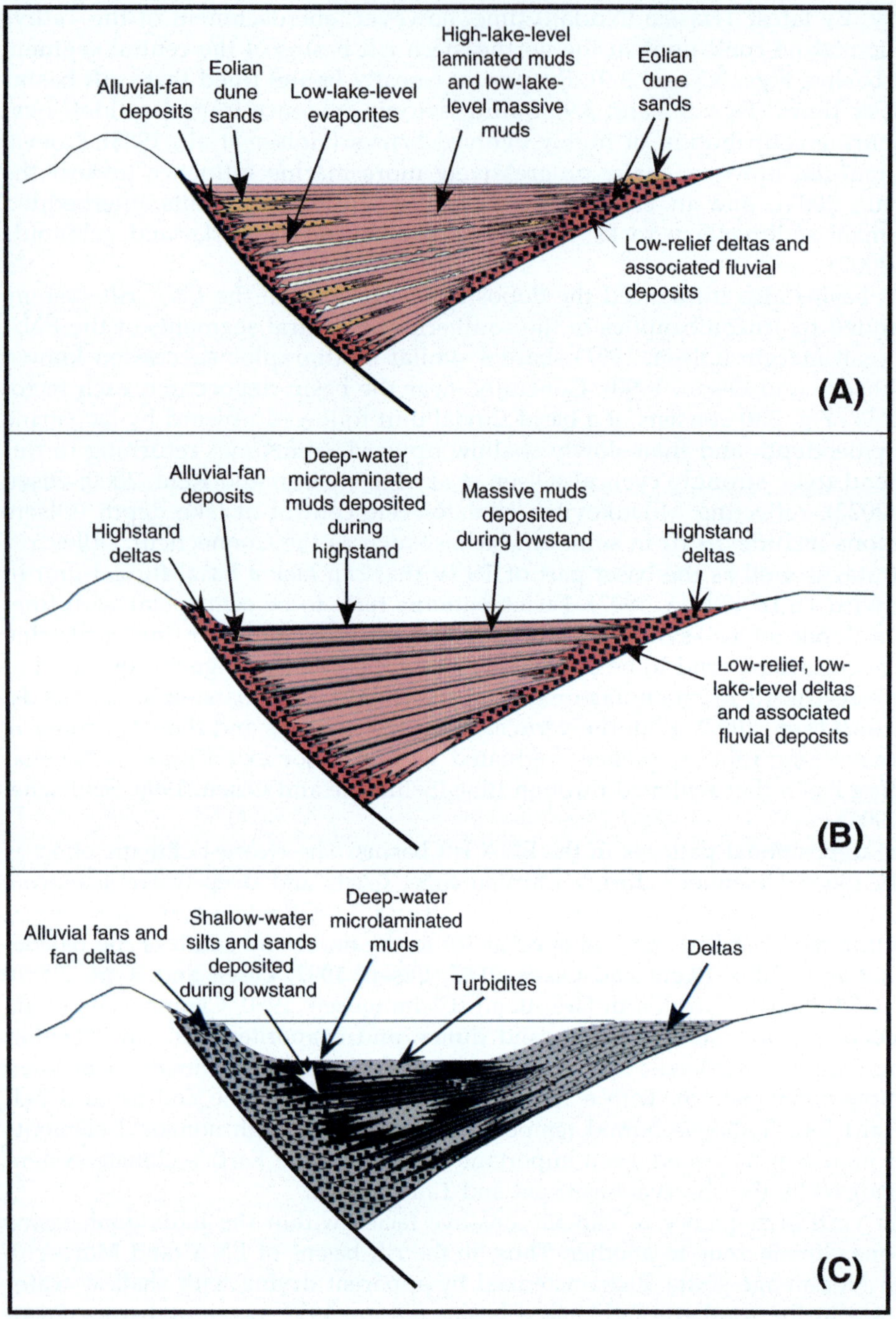

FIGURE 3.10 Cross sections showing rift-basin lacustrine architecture (after Olsen, 1990). Colors in the figure reflect the general color of rocks. (A) Arid type (e.g., Fundy basin during Norian time). (B) Intermediate type (e.g., Newark basin during Carnian time). (C) Humid type (e.g., Richmond basin during Carnian time).

minor silicic plutons, including the intrusive Agamenticus complex dated at ~239 Ma with zircon U-Pb geochronology (Hussey et al., 2016). Although most WMMS activity occurred after CAMP magmatism (see discussion below), some activity predated CAMP magmatism by more than 5 million years (Kinney et al., 2022) (Fig. 3.9).

CAMP magmatism affected a broad region (>10 million km^2), encompassing the entire ENA rift system as well as parts of northern South America, northwestern Africa, and southwestern Europe (Fig. 3.11, inset). CAMP igneous rocks include flood basalts and intrusive dikes and sheets within the upper crust (e.g., Weigand and Ragland, 1970; Sutter, 1988; Dunning and Hodych, 1990; Ragland et al., 1992; Olsen, 1999; Hames et al., 2000, 2003; Schlische et al., 2003; Olsen et al., 2003; Beutel et al., 2005; Schoene et al., 2010; Blackburn et al., 2013; Davies et al., 2017; Marzoli et al., 2018) as well as potential intrusions within the lower crust (Marzen et al., 2019, 2020; Gao et al., 2020). CAMP magmatism was short-lived, lasting for no more than a few million years (Marzoli et al., 1999;

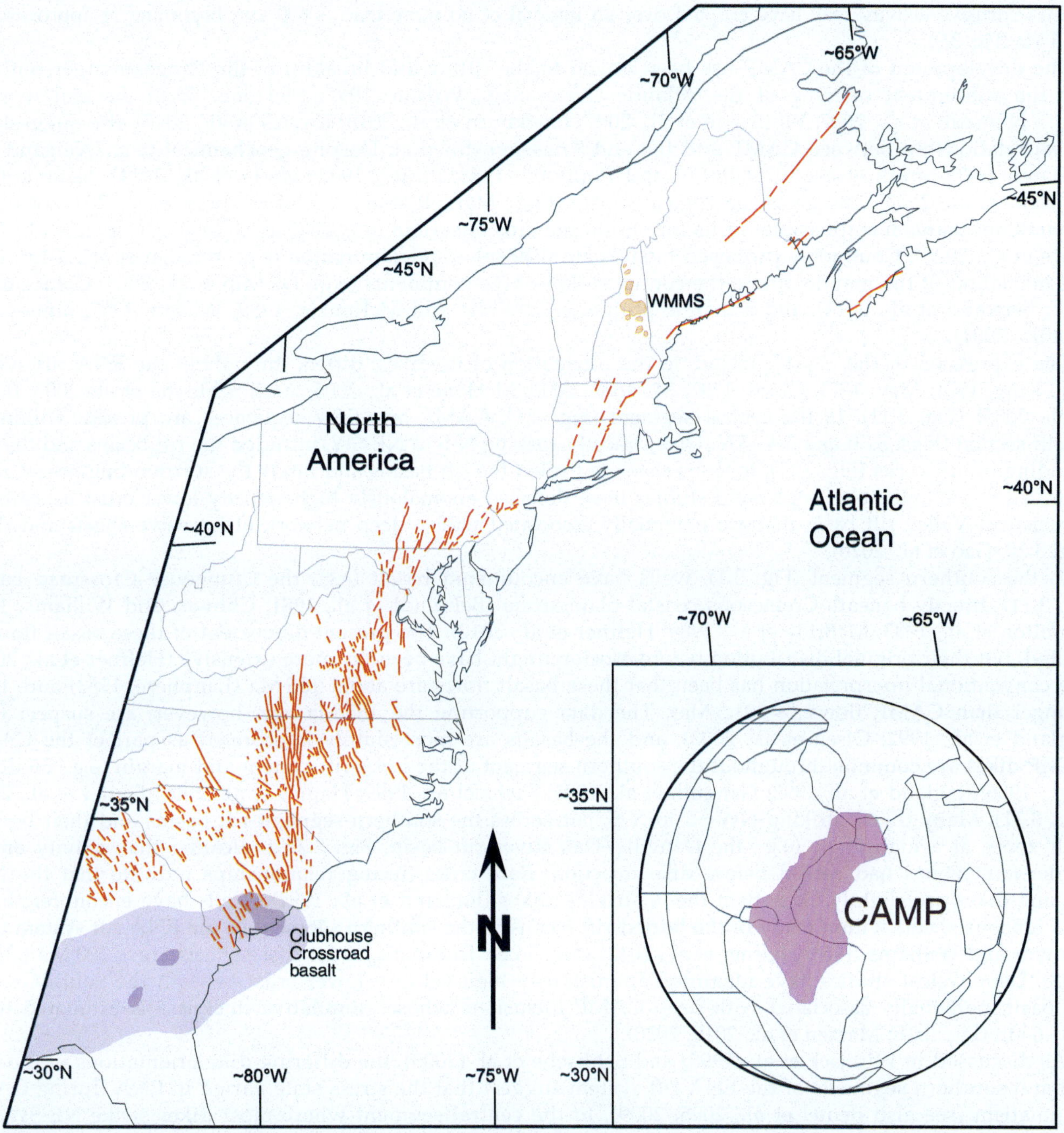

FIGURE 3.11 Map showing the distribution of CAMP dikes (thin red lines) in eastern North America, plutons of the Triassic-Jurassic White Mountain Magma Series, WMMS (light orange) (modified from McHone, 2000; McHone et al., 2005, 2022), well penetrations of basalt (dark purple) including the Clubhouse Crossroad basalt, and well penetrations of diabase (light purple) in the southeastern United States (modified from Chowns and Williams, 1983; Heffner et al., 2012). Inset shows the approximate distribution of CAMP (magenta), based largely on the distribution of lava flows, dikes, and igneous sheets, on the reconstruction of Pangaea at ~200 Ma. Source: *Modified from Olsen, P. E., 1997, Stratigraphic record of the early Mesozoic breakup of Pangea in the Laurasia-Gondwana rift system: Annual Reviews of Earth and Planetary Science, 25, p. 337–401.*

Hames et al., 2000; McHone, 2003; Marzoli et al., 2018), likely over a virtually geologically instantaneous time interval (Olsen et al., 2003; Schoene et al., 2010; Marzoli et al., 2011; Blackburn et al., 2013). For the ENA margin and the conjugate Moroccan margin, zircon U-Pb ages (e.g., Dunning and Hodych, 1990; Schoene et al., 2010; Blackburn et al., 2013) coupled with cyclostratigraphy (Olsen et al., 1996b; Olsen et al., 2003; Blackburn et al., 2013) indicate that the entirety of

CAMP extrusive activity may have erupted over an interval of no more than ~600 kyr, beginning at approximately 201.5 Ma (Fig. 3.9).

The development of the CAMP has been linked to the rifting and breakup of the Pangean supercontinent and the subsequent opening of the Atlantic Ocean (e.g., Wilson, 1997; McHone, 2000). Several workers (e.g., ; Nomade et al., 2007; Whiteside et al., 2007; Blackburn et al., 2013; Davies et al., 2017) also suggested a causal relationship between CAMP and the end Triassic extinction. Despite geochemical (e.g., Weigand and Ragland, 1970; Smith et al., 1975; Puffer and Philpotts, 1988; Puffer, 1992; Merle et al., 2014), tectonic (e.g., Schlische et al., 2003), and geochronological studies (e.g., Marzoli et al., 1999, Blackburn et al., 2013), the geodynamic mechanism responsible for its formation remains uncertain (e.g., McHone, 2000; McHone et al., 2005; Peace et al., 2020). Commonly proposed models are lithospheric delamination (e.g., Whalen et al., 2015), thermal insulation of the mantle due to the configuration of the continents (e.g., De Min et al., 2003; Coltice et al., 2007; Bertrand et al., 2014), and a mantle plume (e.g., White and McKenzie, 1989; Wilson, 1997; Janney and Castillo, 2001).

The expression of the CAMP, including the orientation of its dikes, differs throughout the ENA rift system (e.g., King, 1971; May, 1971; Olsen, 1997; McHone, 2000; McHone et al., 2005, 2022; Schlische et al., 2003; Beutel et al., 2005) (Fig. 3.11). In the central segment (Fig. 3.1), CAMP lava-flow sequences are present within the synrift section (Figs. 3.3B and 3.4–3.6). Dikes, mostly striking N-S to NE-SW, intruded the rift basins and the surrounding prerift rocks (Fig. 3.11). Igneous sheets intruded the rift basins and rarely the surrounding prerift rocks (Figs. 3.3A, 3.5, and 3.6). Geophysical studies have detected anomalously high-velocity lower crust beneath the Connecticut Valley rift basin that are potentially associated with a deep network of intrusive sheets and dikes (Fig. 3.2B, Gao et al., 2020).

In the southern segment (Fig. 3.1), wells have encountered basalt (e.g., the Clubhouse Crossroad basalt, Fig. 3.11) directly beneath Cretaceous coastal-plain strata (Behrendt et al., 1981; Chowns and Williams, 1983; Hamilton et al., 1983; McBride et al., 1989; Heffner et al., 2012). The present-day extent of these basalt flows is limited, but their original distribution before erosion might have been far more extensive (Heffner et al., 2012). The conventional interpretation has been that these basalt flows are about 185 Ma (Lanphere, 1983) and, thus, younger than CAMP flows (~201.5 Ma). The data supporting the 185 Ma age, however, are suspect (e.g., Ragland et al., 1992; Olsen et al., 2003), and the basalts are most simply interpreted as part of the CAMP. CAMP dikes are common throughout the southern segment of the ENA rift system and margin (e.g., Stoddard et al., 1986; Ragland et al., 1992; Ganguli, et al., 1995; Sachs et al., 1999; Hames et al., 2000; Beutel et al., 2005) (Fig. 3.11). Many dikes, striking NW-SE to N-S, cut across the southern-segment rift basins and their border-fault zones at a high angle (e.g., the Danville/Dan River rift basin, Fig. 3.7), indicating that activity on the border-fault zones had ceased before dike injection. Rare cross-cutting relationships indicate that the NW-striking dikes are slightly older than the N-striking dikes (Ragland et al., 1992). Wells have encountered diabase igneous sheets within many of the buried rift basins in the southern segment of the ENA rift system (e.g., Chowns and Williams, 1983; Heffner et al., 2012) (Fig. 3.2E). Isotopic ages for these sheets are ~200 Ma (Olsen, 1998). Geophysical studies have identified anomalously high-velocity lower crust beneath the South Georgia rift basin potentially associated with deep CAMP intrusions whose cumulative thickness is estimated to be ~3–6 km (Fig. 3.2E; Marzen et al., 2019, 2020).

As discussed in Withjack et al. (1998) and Schlische et al. (2003), the differing dike orientations in the central and southern segments of the ENA rift system suggest that the stress state varied in ENA during CAMP magmatism (see also Beutel et al., 2005, 2009). In the central segment where most dikes strike NE-SW, the maximum horizontal stress (S_{Hmax}) was oriented NE-SW, and the minimum horizontal stress (S_{Hmin}) was oriented NW-SE. In the southern segment where most dikes strike NW-SE, S_{Hmax} was NW-SE, and S_{Hmin} was NE-SW. The younger N-striking dike set indicates that S_{Hmax} was N-S and S_{Hmin} was E-W. Thus, the stress state differed for the central and southern segments of the ENA rift system during CAMP magmatism, and it changed between the emplacement of the NW-striking set and the younger N-striking set in the southern segment. Because the boundary between the central and southern segments is diffuse (Fig. 3.1), the stress state should be transitional between NE-SW-oriented S_{Hmin} in the south and NW-SE-oriented S_{Hmin} in the north. In fact, dikes in the transitional region strike ~ N-S, suggesting that S_{Hmin} was oriented ~ E-W. The stress state indicated by the CAMP dikes in the central segment is consistent with continued NW-SE extension and activity on the NE-striking border-fault zones during CAMP magmatism. However, the stress state indicated by the dikes in the southern segment (S_{Hmax} oriented NW-SE, S_{Hmin} oriented NE-SW) is incompatible with continued NW-SE extension and activity on the NE-striking border-fault zones during CAMP magmatism.

Timing of the transition from rifting to drifting

The transition from rifting to drifting is the time interval encompassing the cessation of continental rifting, the breakup of the continental lithosphere, and the onset of seafloor spreading and the formation of oceanic crust. Based on the age of the youngest synrift strata in the offset ENA rift basins (Fig. 3.9) and the stress states indicated by the CAMP dikes (Fig. 3.11), Withjack et al. (1998, 2012) and Schlische et al. (2003) proposed that the cessation of rifting was diachronous, occurring first in the southern segment (latest Triassic, before CAMP magmatism 201.5 Ma) and then in the central segment (Early Jurassic, ~195 Ma). It is unclear, however, whether the cessation of rifting for the offset rift basins represents the cessation of rifting for the entire ENA rift system. For example, it is possible that the locus of extension shifted eastward away from the offset rift basins toward the eventual site of continental breakup during the final stages of rifting.

The timing of the onset of seafloor spreading also remains unclear. (1) The ages of the oldest magnetic anomalies on the ENA margin, that is, those between the East Coast Magnetic Anomaly (ECMA) and the Blake Spur Magnetic Anomaly (BSMA, ~170 Ma) (Fig. 3.1), are poorly constrained (e.g., Greene et al., 2017). (2) The oldest, dated postrift strata on the ENA margin are Middle Jurassic in age, and no wells have penetrated the older, seismically imaged, postrift strata near the continent-ocean boundary, generally interpreted to be Early Jurassic in age (e.g., Klitgord et al., 1988; Poag and Valentine, 1988; Benson, 2003; Weston et al., 2012). (3) The timing of the breakup of the continental crust might differ from that of the underlying lithospheric mantle. Shuck et al. (2019) and Bécel et al. (2020) proposed that, for the ENA margin, the breakup of the continental crust occurred well before the complete rupture of the continental lithosphere, leading to the formation of proto-oceanic crust between the ECMA and the BSMA (Fig. 3.2D). If their model is correct, then the transition from rifting to drifting was prolonged, lasting 20–30 Myr (Fig. 3.9). Thus, the exact timing and duration of breakup for the ENA margin and the early seafloor-spreading history of the Atlantic Ocean remain unclear, leading to a variety of plate reconstructions (e.g., compare those of Bird et al., 2007; Schettino and Turco, 2009; Labails et al., 2010; Kneller et al., 2012; Biari et al., 2017; Greene et al., 2017).

Magmatic and tectonic activity during transition from rifting to drifting

Several magmatic and tectonic events characterize the transition from rifting to drifting on the offshore and onshore parts of the ENA margin.

Predominantly offshore

East Coast Margin Igneous Province

The East Coast Margin Igneous Province (ECMIP, Keleman and Holbrook, 1995) developed during breakup and the earliest stages of seafloor spreading near the boundary between the continental crust and the oceanic and/or proto-oceanic crust (Fig. 3.2C and D). It includes seaward-dipping reflectors (SDRs), presumably of volcanic or volcaniclastic origin, and is associated with the ECMA (e.g., Hinz, 1981; Klitgord et al., 1988; Austin et al., 1990; Holbrook and Keleman, 1993; Sheridan et al., 1993; Keleman and Holbrook, 1995; Benson, 2003) (Figs. 3.1 and 3.2C and D). The age of the SDRs is poorly constrained because no rocks associated with the SDRs have been directly sampled and, as previously discussed, because of uncertainties in the age of the oldest postrift sediments above the SDRs and the age of the oldest oceanic crust. Proposed ages for the SDRs range from c.170 to 200 Ma (e.g., Oh et al., 1995; Benson, 2003; Schlische et al., 2003; Labails et al., 2010; Davis et al., 2018; Bécel et al., 2020). Anomalously high crustal velocities and reflectivity beneath the SDRs (Fig. 3.2C and D) support the presence of lower crustal intrusions and/or magmatic underplating related to the ECMIP (e.g., LASE Study Group, 1986; Keleman and Holbrook, 1995; Shuck et al., 2019; Bécel et al., 2020). The ENA rifted margin, from the southeastern United States to southeastern Canada, with ECMIP extrusions and intrusions is considered a magma-rich margin (Fig. 3.2C and D), whereas the remainder of the margin northward into the Grand Banks region lacks SDRs and is, thus, considered magma-poor (Fig. 3.2A) (Keen and Potter, 1995; Wu et al., 2006; Louden et al., 2013). Recent work by Jian et al. (2021), however, indicates that magmatic activity accompanied breakup even on parts of the magma-poor margin.

Postrift unconformity and postrift basins

A series of broad basins (e.g., the Scotian basin, Fig. 3.2A; the Baltimore Canyon trough, Fig. 3.2C; and the Carolina trough, Fig. 3.2D), associated with tectonic subsidence and sediment loading, formed after continental rifting along the ENA margin. These postrift basins developed mostly seaward of the basement hinge zone where basement depths abruptly increase and crustal thicknesses decrease.

A prominent regional unconformity, termed the postrift unconformity (PRU), separates the eroded prerift and synrift rocks from the overlying strata in the postrift basins (e.g., Grow et al., 1988; Klitgord et al., 1988; Deptuck and Kendell, 2017) (Fig. 3.2A, C, and D, 3.3A and B). The PRU has several distinctive characteristics. (1) It covers a broad region (~ 100 km wide) that stretches from the continent-ocean boundary landward across the basement hinge zone. (2) It is a time-transgressive surface. Near the continent-ocean boundary, uplift and erosion ceased and postrift subsidence began soon after continental breakup. Near the basement hinge zone well landward of the continent-ocean boundary, erosion persisted for tens of millions of years after continental breakup. (3) The development of the PRU was followed by substantial subsidence, locally exceeding 10 km near the continent-ocean boundary, leading to the formation of the postrift basins.

Researchers have proposed a variety of mechanisms to explain the style of uplift and erosion associated with the development of a breakup/postrift unconformity like the PRU. Several mechanisms would produce a more transient phase of uplift and erosion near the continent-ocean boundary followed by significant subsidence. These mechanisms include depth-dependent stretching with excess thinning of the lithospheric mantle near the site of breakup (e.g., Royden and Keen, 1980) and delamination and/or dripping of the lower lithospheric mantle near the site of breakup (Esedo et al., 2012). Other mechanisms would produce uplift and a more sustained phase of erosion flanking the continent-ocean boundary, including deep lithospheric necking during rifting and breakup (e.g., Braun and Beaumont, 1989), flexural uplift caused by mechanical unloading during rifting and breakup (Weissel and Karner, 1989), and peripheral bulging that accompanied offshore postrift subsidence and deposition (Poag and Sevon, 1989). Several of the above mechanisms likely contributed to the development of the PRU on the ENA margin.

Predominantly onshore

White Mountain Magma Series

The WMMS, situated north of the Connecticut Valley rift basin, west of the Fundy rift basin, and within the region of known CAMP dikes (Fig. 3.11), consists of a variety of volcanic and plutonic complexes, predominately ferroan (A1-type) granites (Eby, 1992; Frost and Frost, 2011) and minor silica-undersaturated units (e.g., Creasy, 1989; Henderson et al., 1989). The largest igneous suite in the WMMS is a composite series of intrusions and volcanic rocks forming the White Mountain Batholith. Early geochronology from primarily whole rock and mineral K-Ar and Ar-Ar measurements (Eby et al., 1992; Foland and Faul, 1977; Foland et al., 1971) indicated that its formation occurred over the entire c.50-Myr duration (Eby et al., 1992; Foland and Faul, 1977; Foland et al., 1971) of Jurassic magmatism in the WMMS.

Using a combination of both CA-ID-TIMS and LA-ICP-MS zircon U-Pb geochronologic techniques, Kinney (2021) refined the geochronology and constrained the timing and duration of WMMS magmatism and its relationship with CAMP magmatism, rifting, and breakup (Fig. 3.9). The new dates show that small-scale WMMS magmatism began as early as ~207 Ma (i.e., predating CAMP activity by about 5 Myr) with major discrete pulses of igneous activity occurring at about 195 Ma (i.e., shortly after CAMP activity) and 180 Ma, continuing sporadically until the emplacement of the youngest dated unit at ~170 Ma (Kinney, 2021). For the White Mountain Batholith itself, the new dates show that the oldest units were emplaced no later than ~198.5 Ma, and the youngest units were emplaced at ~182 Ma, more than 25 Myr older than previously published K-Ar and Ar-Ar ages would suggest (Kinney et al., 2022). Thus, the new dates indicate that WMMS magmatism began during rifting; occurred before, during, and after CAMP magmatic activity; and continued into the Early and Middle Jurassic after synrift deposition had ceased in the nearby Connecticut Valley and Fundy rift basins (Fig. 3.9).

Uplift and erosion landward of the basement hinge zone

Most ENA rift basins, well landward of the basement hinge zone, underwent significant tilting, uplift, and erosion after synrift deposition (e.g., Wheeler, 1937; Sanders, 1963; Faill, 1973; Pratt et al., 1988; Wise, 1992; Roden-Tice and Wintsch, 2002; Malinconico, 2003, 2010, 2015; Srogi et al., 2017; Withjack et al., 2013, 2020). For example, Wise (1992) identified about 15° of postdepositional eastward tilting in the Connecticut Valley rift basin, which led to significant erosion along the western side of the basin (Fig. 3.5A). Srogi et al. (2017) recognized about 20° of northwest tilting after CAMP igneous activity in the southernmost Newark rift basin, which led to about 8 km of erosion on the eastern side of the basin. Withjack et al. (2013, 2020) reported northwest tilting in the Newark rift basin that caused differential erosion ranging from 1 km in the northwest to 6 km in the southeast (Figs. 3.6B and 3.8E and G). Vitrinite-reflectance analyses by Malinconico (2003, 2015) showed that the Taylorsville rift basin underwent about 5° of southeast tilting and several kilometers of erosion after synrift deposition. Mineralogical evidence and vitrinite-reflectance analyses indicate that the Danville/Dan River rift basin underwent up to 8 km of exhumation after rifting (Warrell et al., 2018) (Fig. 3.7B).

For many of the western offset rift basins (e.g., the Fundy, Newark, Culpeper, and Danville/Dan River rift basins), the postdepositional tilting was toward the west and northwest. For many of the eastern offset rift basins (e.g., the Connecticut Valley, the New York Bight, Taylorsville, and Deep River rift basins), the postdepositional tilting was toward the east or southeast. Together, the opposing NW and SE tilting defined broad, crustal-scale arches and warps (Fig. 3.12). For example, a faulted arch, about 150 km wide, developed between the northern Newark rift basin and the Connecticut Valley rift basin (Wheeler, 1937; Sanders, 1963; Withjack et al., 2020), and broad asymmetric arches developed between the Culpeper and Taylorsville rift basins and the Danville/Dan

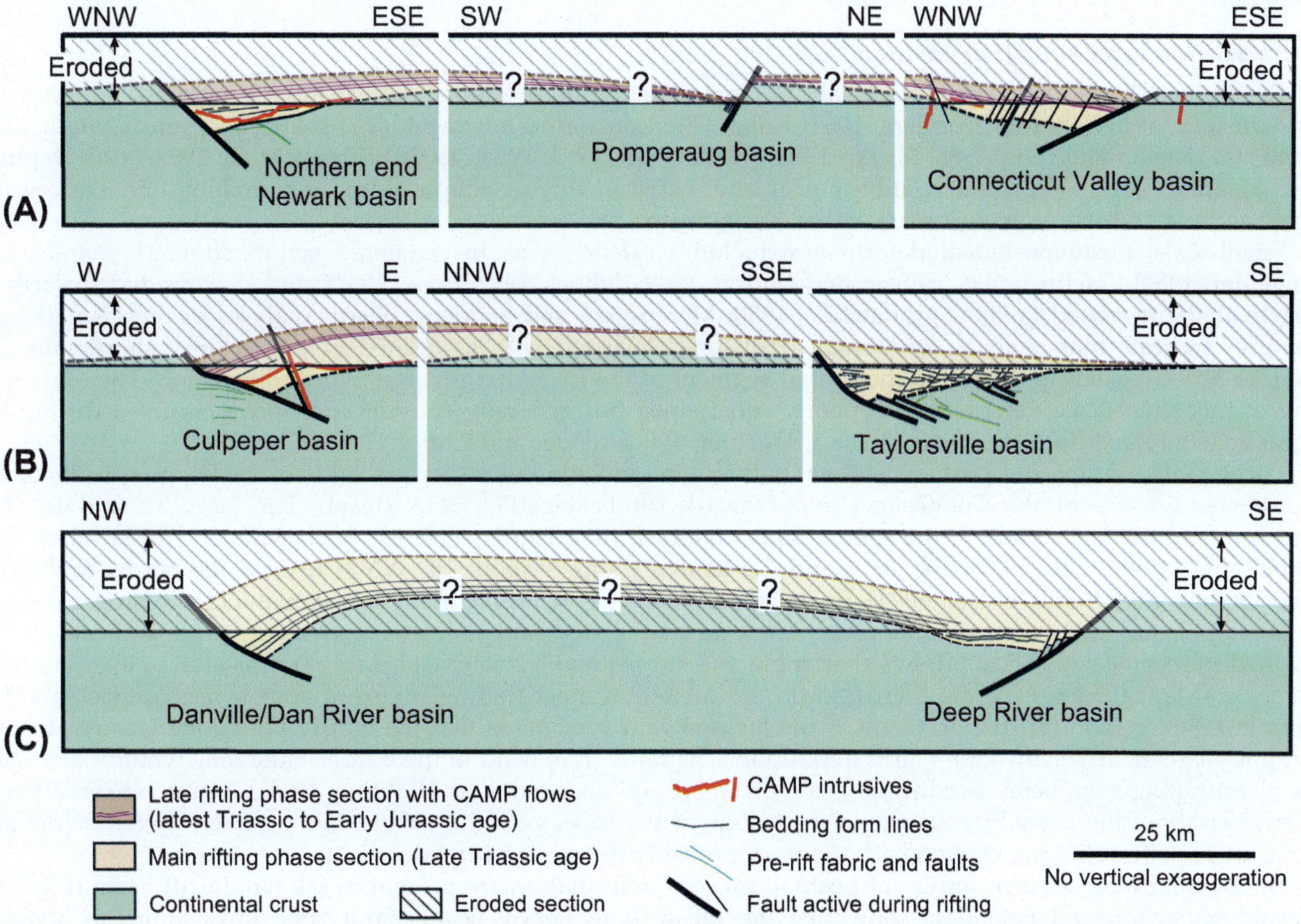

FIGURE 3.12 Regional cross sections across ENA margin with restored eroded material. See Fig. 3.1 for locations. (A) Section across Newark, Pomperaug, and Connecticut Valley rift basins based on data from Wenk (1984), Wise (1992), Resor and deBoer (2005), Burton (2006), Olsen et al. (2016), and Withjack et al. (2020). Postdepositional tilting toward the northwest in the Newark rift basin and toward the east in the Pomperaug and Connecticut Valley rift basins define a broad, deeply eroded, crustal arch. Faulting during arch formation led to the preservation of the Pomperaug rift basin. (B) Section across Culpeper and Taylorsville rift basins based on data from LeTourneau (2003), Malinconico (2003, 2015), and Lyttle et al. (2017) and reprocessed seismic-reflection lines (supervised by John Costain and Cahit Coruh) published in Ryan et al. (2007). Postdepositional tilting toward the west in the Culpeper rift basin and toward the east in the Taylorsville rift basin define a broad, deeply eroded, crustal arch. (C) Section across the Danville/Dan River and Deep River rift basins based on Schlische (1993) and Olsen et al. (2015) and erosion estimates from mineralogical evidence and vitrinite-reflection analyses (Warrell et al., 2018). Post depositional tilting toward the west in the Danville/Dan River rift basin and toward the east in the Deep River basin define a broad, deeply eroded, crustal arch.

River and Deep River rift basins. Because of this significant postdepositional tilting, uplift, and erosion, most of the offset rift basins on the ENA margin today are erosional remnants of once much larger rift basins (e.g., Fig. 3.8B vs H).

As described by Withjack et al. (2020), the style of uplift and erosion landward of the basement hinge zone has characteristics that are distinctly different from those associated with the offshore PRU. Uplift and erosion are associated with crustal-scale arching/warping that affected the offset ENA rift basins and the surrounding rocks landward of the basement hinge zone. The axis of the crustal-scale arching/warping is subparallel to the prerift and synrift crustal fabric, not the continent-ocean boundary. For example, the axis of the crustal-scale arch between the Newark rift basin on the west and the Connecticut Valley and New York Bight rift basins on the east veers sharply northward, following the prerift/synrift fabric but not the continent-ocean boundary (Fig. 3.8F and H). Uplift and erosion locally exceeded 6 km and were long-lived; they likely began during the waning phase of rifting and continued well into the drifting stage (i.e., Late Jurassic to earliest Early Cretaceous). Postrift subsidence was minimal. Based on these characteristics, Withjack et al. (2020) proposed that denudation unloading associated with relic, Paleozoic, orogenic crustal thickening and elevated topography contributed to the postrift tilting, uplift, and erosion. In addition, synrift processes (e.g., crustal thinning, lower crustal intrusion, and/or magmatic underplating) would have influenced the magnitude and distribution of the postrift denudation unloading.

Deformation after synrift deposition

Researchers have long recognized small- to large-scale, postdepositional deformation throughout the ENA rift system (e.g., Wheeler, 1937; Sanders, 1963; Faill, 1973; Lomando and Engelder, 1984; deBoer and Clifton, 1988; Venkatakrishnan and Lutz, 1988; Manspeizer, 1988; deBoer, 1992; Wise, 1992; Withjack et al., 1995, 2010; Deptuck and Kendell, 2017). The exact relationship, if any, between this postdepositional deformation and the crustal-scale arching and warping, described earlier, remains unclear.

Small-scale, postdepositional deformation includes calcite twins in cemented syn-ift strata (Lomando and Engelder, 1984), fractures (e.g., deBoer and Clifton, 1988; deBoer, 1992; Wise, 1992), and slickensided surfaces of reactivated rift-related faults (e.g., Ratcliffe, 1980; Merguerian and Sanders, 1994). Analyses of these small-scale features indicate that regional, horizontal, left-lateral shearing (with associated N-S to NE-SW shortening and W-E to NW-SE extension) affected the central segment of the ENA margin after synrift deposition. The exact timing of this small-scale deformation is poorly constrained but generally considered to have occurred during the Jurassic (e.g., Ratcliffe, 1980; Lomando and Engelder, 1984; deBoer and Clifton, 1988).

Large-scale examples of postdepositional deformation include the transverse folds in the hanging walls of the border-fault zones of the Connecticut and Newark rift basins (Fig. 3.5A (inset), Fig. 3.6A, C, and E). The Jacksonwald transverse fold in the southernmost Newark rift basin is an 8- to 10-km wide and 20-km long syncline (e.g., Lucas et al., 1988) (Fig. 3.6A and C). Its axis plunges about 15° WNW toward the border-fault zone and its limb dips locally exceed 50° (Lucas et al., 1988) (Fig. 3.6C). Like many transverse folds, it is near a major right-step/bend in the border-fault zone. Analyses of the fold and its associated deformation indicate that it formed in response to N-S to NE-SW shortening and subsequent left-lateral shearing (Lucas et al., 1988). The relatively constant thickness of the folded strata indicates that most folding occurred after synrift deposition. The simplest explanation for the formation of the Jacksonwald syncline is that the border-fault zone was reactivated with left-lateral strike-slip after synrift deposition. The right-step/bend in the border-fault zone would have acted as a restraining step/bend, producing local N-S to NE-SW shortening and folding. Based on the preservation of synrift strata of the Early Jurassic age in the trough of the Jacksonwald syncline, folding likely began before significant postrift uplift and erosion had occurred, probably during the Jurassic.

Spectacular large-scale examples of postdepositional deformation are present in the Fundy rift basin (Fig. 3.4). Seismic-reflection and field data show that the NE-striking Fundy border-fault zone links with the ENE- to E-striking Cobequid-Chedabucto border-fault zone, forming a composite border-fault system that bounds the basin on the northwest and north, respectively (Withjack et al., 1995, 2010; Baum et al., 2008) (Fig. 3.4A and B). Broad, NE-striking anticlines developed after synrift deposition in the hanging wall of the NE-striking northwestern border-fault zone (Fig. 3.4B). In contrast, tight ENE- to E-trending synclines and anticlines and E-striking, left-lateral strike-slip faults formed after synrift deposition (Fig. 3.4D) within the ENE-to-E-striking northern border-fault zone (Fig. 3.4B and C). Kinematic analyses for the composite border-fault system indicate that NE-directed displacement of the hanging wall relative to the footwall is necessary to produce the observed deformational patterns on the individual border-fault zones (Fig. 3.4E). With this displacement direction, the NE-striking border-fault zone would have been reactivated with a minor reverse dip-slip and major left-lateral strike-slip, producing the broad buttress folding in its hanging wall. The E-striking border-fault zone would have been reactivated with reverse dip-slip and left-lateral strike-slip components, explaining the tight buttress folding and the left-lateral strike-slip faulting (Withjack et al., 2010).

The postdepositional deformation in the central segment of the ENA rift system,is consistent with the proposal by deBoer and Clifton (1988) that a "shifting" phase occurred on the ENA margin, likely during Jurassic time. During this phase, easternmost North America shifted toward the NE relative to the North American craton. In response, the ENA margin underwent regional, horizontal, left-lateral shearing, resulting in penetrative deformation within broad zones and the reactivation of many rift-related faults. Reactivated, NE-striking structures had mostly left-lateral strike-slip, and reactivated, E-striking structures had both left-lateral strike-slip and reverse dip-slip components. The reactivation of these rift-related faults resulted in the formation of restraining-bend folds and buttress folds (Figs. 3.4–3.6). Reactivated, N-striking structures had mostly normal dip-slip, promoting rift-basin preservation (e.g., the Pomperaug rift basin, Figs. 3.5A and 3.12A). The cause of this enigmatic "shifting" phase remains unclear but may involve strain partitioning (i.e., coeval margin-normal extension near the site of continental breakup and margin-parallel shearing landward of the breakup site) associated with oblique divergence during the transition from rifting to drifting.

Evolution of Eastern North America

Paleozoic orogenic activity (Fig. 3.13A)

Orogenic activity associated with subduction, accretion, and collision occurred throughout ENA during much of Paleozoic time. The orogenic activity produced a thickened, amalgamated continental crust with profound preexisting zones of weakness and led to the formation of the Appalachian orogenic belt. The final collisional event welded the North American and African continents, creating the Pangaean supercontinent. Gravitational collapse and a postorogenic cycle of erosion and isostatic rebound/uplift (denudation unloading) reduced the thickness of the continental crust.

Late Triassic rifting (Fig. 3.13B)

In response to likely NW-SE regional extension, a wide zone of rifting developed within the former orogenic belt by the Late Triassic as Pangaea began to split apart. The preexisting zones of weakness strongly influenced the distribution and location of the ENA rift basins with many developing far from the eventual site of continental breakup (i.e., offset rift basins). Many of the border-fault zones of the offset rift basins are Paleozoic structures reactivated with normal dip-slip or oblique-slip during rifting. Initially, the rift basins were narrow. As rifting progressed, the rift basins widened and deepened considerably and became increasingly interconnected. Fluvial and lacustrine deposits filled most rift basins, but marine encroachment from the north led to the deposition of evaporitic sequences (mostly halite) in the northeastern basins of the central segment of the ENA rift system. WMMS magmatism began in the northern part of the central segment.

Latest Triassic/earliest Jurassic (Fig. 3.13C)

In the central segment of the ENA rift system, CAMP magmatism resulted in the emplacement of mostly N- to NE-trending diabase dikes and intrusive sheets and the eruption of basalt flows that filled the rift basins. Dike trends suggest that extension was NW-SE during CAMP magmatism. Rifting continued in the central segment of the ENA rift system during and for several million years after CAMP magmatism. The rift basins became wider, deeper, and more interconnected as rifting progressed. Continental (fluvial and lacustrine) deposits filled most rift basins, and deposition of evaporitic sequences (mostly halite) continued in the northeastern basins. Subsidence and extension rates increased significantly, and major intrabasin faults formed in several rift basins. WMMS magmatism continued in the northern part of the central segment with a major pulse occurring soon after CAMP magmatism.

In the southern segment of the ENA rift system, the rift basins became inactive before CAMP magmatism. N- to NW-trending dikes cut across the rift basins, indicating that NW-SE extension had ceased before their emplacement. Lava flows (related to CAMP or younger igneous activity) may have covered the region. Extension became focused near the eventual site of continental breakup with the possible formation of SDRs and incipient

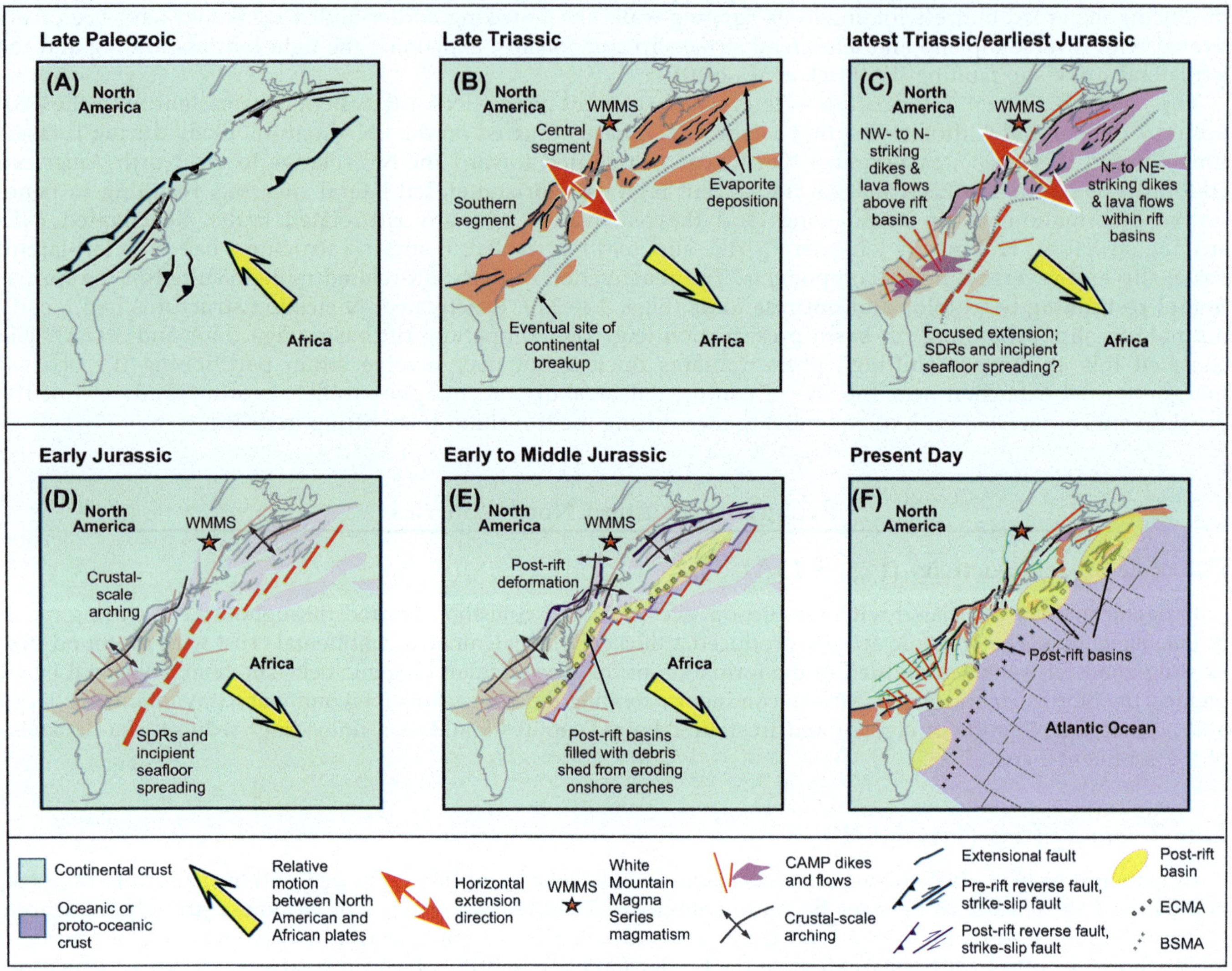

FIGURE 3.13 Schematic evolution for southern and central segments of the ENA rift system and margin. Inactive features are partially transparent, except in F. (A) Late Paleozoic. Orogenic activity created profound zones of weakness within the continental crust. Postorogenic collapse and denudation unloading thinned the thickened crust. (B) Main rifting phase (Late Triassic). In response to the NW-SE extension, a wide rift zone formed within the orogenic belt. Preexisting zones of weakness strongly influenced the distribution and location of the rift basins with most rift basins offset from the eventual site of continental breakup. WMMS magmatic activity began in the northwestern part of the central segment. (C) CAMP magmatic activity (latest Triassic/earliest Jurassic). In the central segment, CAMP magmatism included the intrusion of N- to NE-striking dikes and the emplacement of lava flows and intrusive sheets within the rift basins. The rift basins remained active, and WMMS magmatism continued. In the southern segment, NW- and N-striking dikes cut across the rift basins at a high angle, indicating that the basins had become inactive. Extension became focused near the site of continental breakup with the possible formation of SDRs and incipient seafloor-spreading centers. (D) Beginning of transition from rifting to drifting (Early Jurassic). Deposition ceased in all rift basins, and extension became focused near the site of continental breakup with the formation of SDRs and incipient seafloor-spreading centers. The development of the PRU began near the site of continental breakup. Crustal-scale arching began onshore, and WMMS magmatism continued. (E) Continued transition from rifting to drifting (Early to Middle Jurassic). Formation of oceanic or proto-oceanic crust and postrift basins had begun. The BSMA formed about 170 Ma, reflecting a ridge jump or complete lithospheric rupture. Onshore, crustal-scale arching caused massive erosion with debris shed into the postrift basins. Postdepositional deformation led to the reactivation of many rift-related faults with left-lateral strike-slip and to the formation and/or amplification of transverse and buttress folds. WMMS magmatism continued. (F) Present day. Erosion continued onshore through the Jurassic and into the Early Cretaceous, leaving behind erosional remnants of once massive rift basins. Eroded debris filled the deepening offshore postrift basins.

seafloor-spreading centers associated with the formation of oceanic or proto-oceanic crust. If the model of Shucks et al. (2019) and Bécel et al. (2020) is correct, then the continental crust may have begun to rupture in the southern segment of the ENA rift system, but the underlying lithospheric mantle remained intact.

Early Jurassic (Fig. 3.13D)

Deposition ceased in all offset rift basins of the southern and central segments of the ENA rift system during Early Jurassic time. Instead, the extension became focused near the eventual site of a continental breakup with the formation of SDRs (and the future ECMA) and incipient seafloor-spreading centers associated with oceanic or proto-oceanic crust development. The PRU developed in a wide region surrounding the site of continental breakup. Crustal-scale arching began onshore, and WMMS magmatism continued with a major pulse during late Early Jurassic time. If the model of Shuck et al. (2019) and Bécel et al. (2020) is correct, then the lithospheric mantle had not yet ruptured, and the North American and African plates remained connected at depth.

Early to Middle Jurassic (Fig. 3.13E)

Oceanic or proto-oceanic crust formed along the entire ENA margin. Erosion associated with the PRU persisted near the basement hinge zone. Near the breakup site, however, subsidence and deposition lead to the formation of postrift basins, likely beginning during Early Jurassic time. Simultaneously, landward of the basement hinge zone, continued crustal-scale arching caused massive erosion with much of the debris shed into the growing offshore postrift basins. In addition, deformation associated with the NE shifting of easternmost North America relative to the North American craton, resulted in the reactivation of many rift-related faults in the central segment of the ENA rift system. Some NE-striking segments of border-fault zones were reactivated with mostly left-lateral strike-slip; some E-striking segments were reactivated with reverse dip-slip and left-lateral strike-slip; and some N-striking segments were reactivated with normal dip-slip. WWMS magmatism continued until about 170 Ma.

At about 170 Ma, the BSMA formed in response to a ridge jump to the east (Vogt, 1973), a significant change in the rate and direction of seafloor spreading (Labails et al., 2010), or to the final rupture of the lithospheric mantle and the formation of true oceanic crust (Shucks et al., 2019; Bécel et al., 2020).

Present day (Fig. 3.13F)

The onshore rift basins continued to erode and narrow through the Jurassic and into the Early Cretaceous, leaving behind erosional remnants of the once massive rift basins. The eroded debris filled the deep and wide offshore postrift basins.

Summary

Prerift Paleozoic orogenic activity created profound preexisting zones of weakness within the ENA lithosphere. These zones of weakness strongly influenced the location and distribution of the ENA rift basins and led to the formation of a wide rift zone composed of numerous parallel and en-echelon rift basins, many of which are offset from the continent-ocean boundary by hundreds of kilometers. Rift-basin development began by the Late Triassic throughout the ENA rift system but ceased diachronously, occurring first in the southern segment before the latest Triassic and then in the central segment during the Early Jurassic. The resultant rift basins were deep (locally >5 km), wide (locally >100 km), and interconnected. Fluvial and lacustrine sediments filled most rift basins, although evaporites (mostly halite) filled the basins in the northeastern part of the central segment.

The timing and duration of the ENA rift-drift transition are poorly constrained. The transition occurred after the offset rift basins became inactive (i.e., after latest Triassic in the southern segment and during Early Jurassic in the central segment). If the model proposed by Shuck et al. (2019) and Bécel et al. (2020) is correct, then the transition was a prolonged process with the continental breakup and the formation of proto-oceanic crust followed by complete lithospheric rupture and the formation of true oceanic crust during the formation of the BSMA (~170 Ma).

Several major tectonic events affected the ENA margin during the rift-drift transition and early stages of seafloor spreading. Uplift and erosion led to the development of the PRU mostly seaward of the basement hinge zone. Crustal-scale arching/warping occurred mostly landward of the basement hinge zone, causing significant erosion of the ENA rift basins. The eroded debris filled the postrift basins developing offshore. The ENA margin also underwent regional, horizontal, left-lateral shearing, resulting in penetrative deformation within broad zones and the reactivation of many rift-related faults and the formation of large-scale restraining-bend folds and buttress folds.

Three major igneous events affected the ENA margin during rifting and the rift-drift transition. (1) CAMP magmatism was widespread and short-lived, occurring throughout ENA in the latest Triassic/earliest Jurassic (~201.5 Ma). It includes flood basalts, dikes, and sheets within the shallow crust and potential intrusions within the lower crust. (2) WMMS magmatism was local and long-lived, occurring in the northwestern part of the central segment of the ENA margin. It began during rifting; occurred before, during, and after CAMP magmatism; and continued into the Early and Middle Jurassic after synrift deposition had ceased in the surrounding ENA rift basins. (3) The ECMIP developed during breakup and the earliest stages of seafloor spreading near the continent-ocean boundary. It includes SDRs and lower crustal intrusions and/or magmatic underplating. The age of ECMIP is unknown.

Acknowledgments

The authors gratefully acknowledge the encouragement of Albert Bally and Dave Roberts during the initial project, almost 20 years ago. The authors also thank their colleagues Gail Ashley, Mark Baum, Michael Durcanin, Jennifer Elder Brady, Etikha, Michael Flite, Bari Hanafi, Greg Herman, Claude Herzberg, Dennis Kent, MaryAnn Malinconico, Don Mondeverde, Chris Potter, Cesar Sequeira, Robert Sheridan, Ian Sinclair, LeeAnn Srogi, Beatriz Elena Serano Suarez, Zulfitriadi Syamsir, Jill VanTongeren, and Kathleen Warrell for their thought-provoking comments, insights, and discussions.

References

Arguden, A.T., Rodolfo, K.S., 1986. Sedimentary facies and tectonic implications of lower Mesozoic alluvial-fan conglomerates of the Newark basin, northeastern United States. Sedimentary Geology 51 (1–2), 97–118.

Austin, J.A., et al., 1990. Crustal structure of the Southeast Georgia embayment-Carolina trough: preliminary results of a composite seismic image of a continental suture (?) and a volcanic passive margin. Geology 18, 1023–1027.

Baum, M.S., Withjack, M.O., Schlische, R.W., 2008. The ins and outs of buttress folds: examples from the Fundy rift basin, Nova Scotia and New Brunswick, Canada. Central Atlantic Conjugate Margins, Program and Extended Abstracts, 53–61.

Bécel, A., Davis, J.K., Shuck, B.D., Van Avendonk, H.J., Gibson, J.C., 2020. Evidence for a prolonged continental breakup resulting from slow extension rates at the eastern North American volcanic rifted margin. Journal of Geophysical Research: Solid Earth 125 (9)e2020JB020093.

Behrendt, J.C., Hamilton, R.M., Ackermann, H.D., Henry, V.J., 1981. Cenozoic faulting in the vicinity of the Charleston, South Carolina, 1886 earthquake. Geology 9, 117–122.

Benson, R.N., 2003. Age estimates of the seaward-dipping volcanic wedge, earliest oceanic crust, and earliest drift-stage sediments along the North Atlantic continental margin. In: Hanes, W.E., McHone, J.G., Renne, P.R., Ruppel, C. (Eds.), The Central Atlantic Magmatic Province, Insights from Fragments of Pangea, 136. American Geophysical Union, Geophysical Monograph, Washington, D. C., pp. 61–75.

Benson, R.H., Doyle, R.G., 1988. Early Mesozoic rift basins and the development of the United States middle Atlantic continental margin. In: Manspeizer, W. (Ed.), Triassic-Jurassic Rifting, Continental Breakup and the Origin of the Atlantic Ocean Passive Margins, Part A. Elsevier, New York, pp. 99–127.

Bertrand, H., Fornari, M., Marzoli, A., García-Duarte, R., Sempere, T., 2014. The Central Atlantic magmatic province extends into Bolivia. Lithos 188, 33–43.

Beutel, E.K., 2009. Magmatic rifting of Pangaea linked to onset of South American plate motion. Tectonophysics 468 (1–4), 149–157.

Beutel, E.K., Nomade, S., Fronabarger, A.K., Renne, P.R., 2005. Pangea's complex breakup: a new rapidly changing stress field model. Earth and Planetary Science Letters 236 (1–2), 471–485.

Biari, Y., Klingelhoefer, F., Sahabi, M., Funck, T., Benabdellouahed, M., Schnabel, M., et al., 2017. Opening of the central Atlantic Ocean: implications for geometric rifting and asymmetric seafloor spreading and continental breakup. Tectonics 36, 1229–1150.

Bird, D.E., Hall, S.A., Burke, K., Sawyer, D.S., 2007. Early central Atlantic Ocean seafloor spreading history. Geosphere 3, 282–298.

Blackburn, T.J., Olsen, P.E., Bowring, S.A., McLean, N.M., Kent, D.V., Puffer, J., et al., 2013. Zircon U-Pb geochronology links the end-Triassic extinction with the Central Atlantic Magmatic Province. Science 340 (6135), 941–945.

Braun, J., Beaumont, C., 1989. A physical explanation of the relation between flank uplifts and the breakup unconformity at rifted continental margins. Geology 17, 760–764.

Burton, W., 2006. Bedrock map of the early Mesozoic Pomperaug basin and surrounding basement rocks, Litchfield and New Haven counties, CT: U.S. Geological Survey Open-File Report 2006-1011.

Chenin, P., Beaumont, C., 2013. Influence of offset weak zones on the development of rift basins: activation and abandonment during continental extension and breakup. Journal of Geophysical Research: Solid Earth 118, 1698–1720.

Chowns, T., and Williams, C., 1983, Pre-Cretaceous rocks beneath the Georgia coast plain – regional implications. In: Gohn, G.S. (Ed.), Studies related to the Charleston, South Carolina earthquake of 1886 – Tectonics and seismicity. U.S. Geological Survey Professional Paper 1313, pp. L1-L42.

Clemmensen, L.B., 1980. Triassic rift sedimentation and palaeogeography of central East Greenland. Bulletin Grønlands Geologiske Undersøgelsek 136, 1–72.

Clemmensen, L.B., Kent, D.V., Jenkins, F.A., 1998. A Late Triassic lake system in East Greenland; facies, depositional cycles and palaeoclimate. Palaeogeography, Palaeoclimatology, Palaeoecology 140, 135–159.

Cohen, K.M., Finney, S.C., Gibbard, P.L., Fan, J.-X., 2013. (updated February, 2022), The ICS International Chronostratigraphic Chart. Episodes 36, 199–204.
Coltice, N., Phillips, B.R., Bertrand, H., Ricard, Y., Rey, P., 2007. Global warming of the mantle at the origin of flood basalts over supercontinents. Geology 35 (5), 391–394.
Contreras, J., Scholz, C.H., King, G.C.P., 1997. A general model of rift basin evolution: constraints of first order stratigraphic observations. Journal of Geophysical Research 102, 7673–7690.
Creasy, J.W., 1989. Geology and geochemistry of the Rattlesnake Mountain igneous complex, Raymond and Casco, Maine. Studies in Maine Geology, Maine Geological Survey Publication 66, 63–78.
Davies, J.H.F.L., Marzoli, A., Bertrand, H., Youbi, N., Ernesto, M., Schaltegger, U., 2017. End-Triassic mass extinction started by intrusive CAMP activity. Nature Communications 8 (1), 1–8.
Davis, J.K., Bécel, A., Buck, W.R., 2018. Estimating emplacement rates for seaward-dipping reflectors associated with the U.S. East Coast Magnetic Anomaly. Geophysical Journal International 215 (3), 1594–1603.
De Min, Marzoli, A., Bellieni, G., Renne, P.R., Ernesto, M., et al., 2003. The Central Atlantic Magmatic Province (CAMP) in Brazil: petrology, geohemistry, 40Ar/39Ar ages, paleomagnetism and geodynamic implications. In: Hames, W., McHone, G., Renne, P.R., Ruppel, C. (Eds.), The Central Atlantic Magmatic Province: Insights from Fragments of Pangea, 136. American Geophysical Union, Geophysical Monograph Series, pp. 91–128.
deBoer, J.Z., 1992. Stress configurations during and following emplacement of ENA basalts in the norlthern Appalachians. In: Puffer, J.H., Ragland, P.C. (Eds.), Eastern North American Mesozoic Magmatism, 268. Geological Society of America Special Paper, pp. 361–378.
deBoer, J.Z., Clifton, A.E., 1988. Mesozoic tectogenesis: development and deformation of 'Newark' rift zones in the Appalachians (with special emphasis on the Hartford basin, Connecticut). In: Manspeizer, W. (Ed.), Triassic-Jurassic Rifting, Continental Breakup and the Origin of the Atlantic Ocean Passive Margins, Part A. Elsevier, New York, pp. 275–306. [Clifton was incorrectly published as Clifford.].
Deptuck, M.E., Kendell, K.L., 2017. A review of Mesozoic-Cenozoic salt tectonics along the Scotian margin, eastern Canada. In: Soto, J.I., Flinch, J.F., Tari, G. (Eds.), Permo-Triassic Salt Provinces of Europe, North Africa and the Atlantic Margins. Elsevier, pp. 287–312.
Deptuck, M.E., Altheim, B., 2018. Rift basins of the central LaHave Platform, offshore Nova Scotia. CNSOPB Geoscience Open File Report 2018-001MF, 54p.
Dunning, G.R., Hodych, J.P., 1990. U-Pb zircon and baddeleyite age for the Palisade and Gettysburg sills of northeast United States: implications for the age of the Triassic-Jurassic boundary. Geology 18, 795–798.
Eby, G.N., 1992. Chemical subdivision of the A-type granitoids: petrogenetic and tectonic implications. Geology 20 (7), 641–644.
Eby, G.N., Krueger, H.W., Creasy, J.W., 1992. Geology, geochronology, and geochemistry of the White Mountain batholith, New Hampshire. Special Papers, Geological Society of America 379–397.
Esedo, R., van Wijk, J., Coblentz, D., Meyer, R., 2012. Uplift prior to continental breakup: Indication for removal of mantle lithosphere? Geosphere 8, 1078–1085.
Et-Touhami, M., 2000. Lithostratigraphy and depositional environments of lower Mesozoic evaporites and associated red beds, Khemisset basin, northwestern Morocco. In: Bachmann, G., Lerche, I. (Eds.), Epicontinental Triassic, Volume 2, VIII. Zentralblatt fur Geologie und Palaontologie, pp. 1217–1241.
Faill, R.T., 1973. Tectonic development of the Triassic Newark-Gettysburg basin in Pennsylvania. Geological Society of America Bulletin 84, 725–740.
Faill, R.T., 1988. Mesozoic tectonics of the Newark basin, as viewed from the Delaware River. In: Husch, J.M., Hozik, M.J. (Eds.), Geology of the central Newark basin, field guide and proceedings: Fifth Meeting of the Geological Association of New Jersey. Rider College, Lawrenceville, pp. 19–41.
Faill, R.T., 1997. A geologic history of the north-central Appalachians; part 1, orogenesis from the Mesoproterozoic through the Taconic orogeny. American Journal of Science 297 (6), 551–619.
Faill, R.T., 1998. A geologic history of the north-central Appalachians; part 3, the Alleghany orogeny. American Journal of Science 298 (2), 131–179.
Faill, R.T., 2003. The early Mesozoic Birdsboro central Atlantic margin basin in the Mid-Atlantic region, eastern United States. Geological Society of America Bulletin 115, 406–421.
Foland, K.A., Faul, H., 1977. Ages of the White Mountain intrusives; New Hampshire, Vermont, and Maine, USA. American Journal of Science 277 (7), 888–904.
Foland, K.A., Quinn, A.W., Giletti, B.J., 1971. K-Ar and Rb-Sr Jurassic and Cretaceous ages for intrusives of the White Mountain magma series, northern New England. American Journal of Science 270 (5), 321–330.
Frost, C.D., Frost, B.R., 2011. On ferroan (A-type) granitoids: their compositional variability and modes of origin. Journal of Petrology 52 (1), 39–53.
Ganguli, P.M., Kunk, M.J., Wintsch, R.P., Dorais, M.J., Sacks, P.E., 1995. High precision sanidine 40Ar/39Ar results from Mesozoic rhyolite dikes near Lake Gaston, North Carolina and Virginia. Geological Society of America, Abstract with Programs, 27, p. 45.
Gao, H., Yang, X., Long, M.D., Aragon, J.C., 2020. Seismic evidence for crustal modification beneath the Hartford rift basin in the northeastern United States. Geophysical Research Letters 47 (17)e2020GL089316.
Gawthorpe, R.L., Leeder, M.R., 2000. Tectono-sedimentary evolution of active extensional basins. Basin Research 12, 195–218.
Gradstein, F.M., Ogg, J.G., Schmitz, M.D., Ogg, G.M. (Eds.), 2020. Geologic time scale 2020. Elsevier.
Greene, J.A., Tominaga, M., Miller, N.C., Hutchinson, D.R., Karl, M.R., 2017. Refining the formation and early evolution of the Eastern North American Margin: new insights from multiscale magnetic anomaly analyses. Journal of Geophysical Research: Solid Earth 122 (11), 8724–8748.
Grow, J.A., Klitgord, K.D., Schlee, J.S., 1988. Structure and evolution of the Baltimore Canyon Trough. In: Sheridan, R.E., Grow, J.A. (Eds.), The Atlantic Continental Margin, I-2. Boulder, Colorado, Geological Society of America, Geology of North America, U.S., pp. 269–290.
Hames, W.E., Renne, P.R., Ruppel, C., 2000. New evidence for geologically instantaneous emplacement of earliest Jurassic Central Atlantic magmatic province basalts on the North American margin. Geology 28, 859–862.
Hames, W., McHone, J.G., Renne, P., Ruppel, C. (Eds.), 2003. The Central Atlantic Magmatic Province, Insights from Fragments of Pangea. In: Geophysical Monograph, 136. American Geophysical Union, Washington, D. C., p. 267.

Hamilton, R.M., Behrendt, J.C., Ackermann, H.D., 1983. Land multichannel seismic-reflection evidence for tectonic features near Charleston, South Carolina, in Gohn, G.S., (Ed.), Studies Related to the Charleston, South Carolina, Earthquake of 1886—Tectonics and Seismicity: U.S. Geological Survey Professional Paper 1313, p. I1-I18.

Hanafi, B.R., 2013. The influence of basin architecture and synrift salt on structural evolution during and after rifting: A case study of the Orpheus Rift Basin, offshore Nova Scotia and Newfoundland, M.S. thesis. Rutgers The State University of New Jersey, New Brunswick, New Jersey.

Hanafi, B.R., Withjack, M.O., Durcanin, M.A., Schlische, R.W., 2022. The development of the eastern Orpheus rift basin, offshore eastern Canada: a case study of the interplay between rift-related faulting and salt deposition and flow. Marine and Petroleum Geology 105629.

Harry, D.L., Sawyer, D.S., 1992. A dynamic model of extension in the Baltimore Canyon trough region. Tectonics 11, 420–436.

Hatcher Jr., R.D., 2005. Southern and central Appalachians. In: Selley, R.C., Cocks, L.R.M., Plimer, I.R. (Eds.), Encyclopedia of Geology: Amsterdam. Elsevier, pp. 72–81.

Heffner, D.M., 2013. *Tectonics of the South Georgia rift* (Doctoral dissertation, University of South Carolina).

Heffner, D.M., Knapp, J.H., Akintunde, O.M., Knapp, C.C., 2012. Preserved extent of Jurassic flood basalt in the South Georgia Rift: a new interpretation of the J horizon. Geology 40 (2), 167–170.

Henderson, C.M.B., Pendlebury, K., Foland, K.A., 1989. Mineralogy and petrology of the Red Hill alkaline igneous complex, New Hampshire, USA. Journal of Petrology 30 (3), 627–666.

Hinz, K., 1981. A hypothesis on terrestrial catastrophes; wedges of very thick, oceanward-dipping layers beneath passive continental margins. Geologisches Jahrbuch E22, 3–38.

Holbrook, W.S., Keleman, P.B., 1993. Large igneous province on the US Atlantic margin and implications for magmatism during breakup. Nature 364, 433–436.

Holser, W.T., Clement, G.P., Jansa, L.F., Wade, J.A., 1988. Evaporite deposits of the North Atlantic Rift. In: Manspeizer, W. (Ed.), Triassic-Jurassic Rifting: Continental Breakup and the Origin of the Atlantic Ocean and Passive Margins, Part A, Developments in Geotectonics 22. Elsevier, Amsterdam, pp. 525–556.

Hussey, A., Bothner, W., Thompson, P., 2016. Bedrock geology of the Kittery 1: 100 000 quadrangle, southwestern Maine and southeastern New Hampshire. Maine Geological Survey Bulletin 45, 99p.

Hutchinson, D.R., Klitgord, K.D., 1988. Evolution of rift basins on the continental margin off southern New England. In: Manspeizer, W. (Ed.), Triassic-Jurassic Rifting, Continental Breakup and the Origin of the Atlantic Ocean Passive Margins, Part A. Elsevier, New York, pp. 81–98.

Hutchinson, D.R., Klitgord, K.D., Detrick, R.S., 1986. Rift basins of the Long Island platform. Geological Society of America Bulletin 97 (6), 688–702.

Jacobsen, V.W., van Veen, P., 1984. The Triassic offshore Norway north of 62N. In: Spencer, A.M. (Ed.), Petroleum Geology of the North European Margin. Norwegian Petroleum Society, London, Graham & Trotman, pp. 317–327.

Jamieson, R.A., Beaumont, C., 1988. Orogeny and metamorphism: a model for deformation and pressure-temperature-time paths with applications to the Central and Southern Appalachians. Tectonics 7, 417–445.

Janney, P.E., Castillo, P.R., 2001. Geochemistry of the oldest Atlantic oceanic crust suggests mantle plume involvement in the early history of the central Atlantic Ocean. Earth and Planetary Science Letters 192 (3), 291–302.

Jian, H., Nedimović, M.R., Canales, J.P., Lau, K.H., 2021. New insights into the rift to drift transition across the northeastern Nova Scotian margin from wide-angle seismic waveform inversion and reflection imaging. Journal of Geophysical Research: Solid Earth 126 (12)e2021JB022201.

Keen, C.E., Potter, D.P., 1995. The transition from a volcanic to a nonvolcanic rifted margin off eastern Canada. Tectonics 14, 359–371.

Keen, C.E., Kay, W.A., Keppie, J.D., Marillier, F., Pe-Piper, G., Waldron, J.W.F., 1991a. Deep seismic reflection data from the Bay of Fundy and Gulf of Maine: Tectonic implications for the northern Appalachians. Canadian Journal of Earth Sciences 28, 1096–1111.

Keen, C.E., Kay, W.A., MacLean, B.C., 1991b. A deep seismic reflection profile across the Nova Scotia continental margin, offshore eastern Canada. Canadian Journal of Earth Sciences 28, 1112–1120.

Keleman, P.B., Holbrook, W.S., 1995. Origin of thick, high-velocity igneous crust along the U.S. east coast margin. Journal of Geophysical Research 100, 10077–10094.

Kent, D.V., Olsen, P.E., 1997. Magnetostratigraphy and paleopoles from the Late Triassic Dan River-Danville basin: interbasin correlation of continental sediments and a test of the tectonic coherence of Newark rift basins in eastern North America. Geological Society of America Bulletin 109 (3), 366–377.

Kent, D.V., Tauxe, L., 2005. Corrected Late Triassic latitudes for continents adjacent to the North Atlantic. Science 307 (5707), 240–244.

Kent, D.V., Olsen, P.E., 2008. Early Jurassic magnetostratigraphy and paleolatitudes from the Hartford continental rift basin (eastern North America): testing for polarity bias and abrupt polar wander in association with the central Atlantic magmatic province. Journal of Geophysical Research: Solid Earth 113 (B6).

Kent, D.V., Clemmensen, L.B., 2021. Northward dispersal of dinosaurs from Gondwana to Greenland at the mid-Norian (215–212 Ma, Late Triassic) dip in atmospheric pCO2. Proceedings of the National Academy of Sciences 118 (8).

King, P.B., 1971. Systematic pattern of Triassic dikes in the Appalachian region—second report. U.S. Geological Survey Professional Paper 750-D, pp. D84-D88.

Kinney, S.T., 2021, Re-evaluating the timescale of rift and post-rift magmatism on the Eastern North American Margin via zircon U-Pb geochronology, Ph.D. dissertation, Columbia University, New York, New York.

Kinney, S.T., MacLennan, S.A., Szymanowski, D., Keller, C.B., VanTongeren, J.A., Setera, J.B., et al., 2022. Onset of long-lived silicic and alkaline magmatism in eastern North America preceded Central Atlantic Magmatic Province emplacement. Geology 50 (11), 1301–1305.

Klitgord, K.D., Hutchinson, D.R., Schouten, H., 1988. U.S. Atlantic continental margin; structural and tectonic framework. In: Sheridan, R.E., Grow, J.A. (Eds.), The Geology of North America, v. I-2, The Atlantic Continental Margin. Geological Society of America, U.S., pp. 19–56.

Kneller, E.A., Johnson, C.A., Karner, G.D., Einhorn, J., Queffelec, T.A., 2012. Inverse methods for modeling non-rigid plate kinematics: application to mesozoic plate reconstructions of the Central Atlantic. Computers & Geosciences 49, 217–230.

Labails, C., Olivet, J.-L.The Dakhla study group, 2009. Crustal structure of the SW Moroccan margin from wide-angle and reflection seismic data (the Dakhla experiment). Part B – The tectonic heritage. Tectonophysics 468, 83–97.

Labails, C., Olivet, J.-L., Aslanian, D., Roest, W.R., 2010. An alternative early opening scenario for the Central Atlantic Ocean. Earth and Planetary Science Letters 297, 355–368.

Lanphere, M.A., 1983. $^{40}Ar/^{39}Ar$ ages of basalt from Clubhouse Crossroads test hole #2, near Charleston, South Carolina. In: Gohn, G.S., (Ed.), Studies Related to the Charleston, South Carolina, Earthquake of 1886—Tectonics and Seismicity. U. S. Geological Survey Professional Paper 1313, pp. B1-B8.

Lang, G., ten Brink, U.S., Hutchinson, D.R., Mountain, G.S., Schattner, U., 2020. The role of premagmatic rifting in shaping a volcanic continental margin: An example from the Eastern North American Margin. Journal of Geophysical Research: Solid Earth 125 (11), p.e2020JB019576.

LASE Study Group, 1986. Deep structure of the US East Coast passive margin from large aperture seismic experiments (LASE). Marine and Petroleum Geology 3, 234–242.

LeTourneau, P.M., 1985. Alluvial fan development in the Lower Jurassic Portland Formation, central Connecticut: implications for tectonics and climate. In: Proceedings of the Second U.S. Geological Survey Workshop on the Early Mesozoic Basins of the Eastern United States, 946. U.S. Geological Survey Circular, pp. 17–26.

LeTourneau, P., 2003. Tectonic and climatic controls on the stratigraphic architecture of the Late Triassic Taylorsville basin, Virginia and Maryland. In: LeTourneau, P.M., Olsen, P.E. (Eds.), The Great Rift Valleys of Pangea in Eastern North America, Volume 2, Sedimentology, Stratigraphy, and Paleontology. Columbia University Press, New York, pp. 12–58.

Li, C., Gao, H., Williams, M.L., Levin, V., 2018. Crustal thickness variation in the northern Appalachian Mountains: implications for the geometry of 3-D tectonic boundaries within the crust. Geophysical Research Letters 45, 6061–6070.

Lindholm, R.C., 1978. Triassic-Jurassic faulting in eastern North America—a model based on pre-Triassic structures. Geology 6, 365–368.

Lomando, A.J., Engelder, T., 1984. Strain indicated by calcite twinning: implications for deformation of the early Mesozoic northern Newark basin, New York. Northeastern Geology 6, 192–195.

Louden, K., Wu, Y., Tari, G., 2013. Systematic variations in basement morphology and rifting geometry along the Nova Scotia and Morocco conjugate margins. Geological Society, London, Special Publications 369 (1), 267–287.

Lucas, M., Hull, J., Manspeizer, W., 1988. A foreland-type fold and related structures in the Newark rift basin. In: Manspeizer, W. (Ed.), Triassic-Jurassic Rifting, Continental Breakup and the Origin of the Atlantic Ocean Passive Margins, Part A. Elsevier, New York, pp. 307–332.

Lyttle, P.T., Aleinikoff, J.N., Burton, W.C., Crider, E.A., Jr., Drake, A.A., Jr., Froelich, A.J., et al., 2017. Geologic map of the Washington West 30′ × 60′ quadrangle, Maryland, Virginia, and Washington D.C. U.S. Geological Survey Open-File Report 2017–1142, 1 sheet, scale 1:100,000.

MacLean, B.C., Wade, J.A., 1992. Petroleum geology of the continental margin south of the islands of St. Pierre and Miquelon, offshore eastern Canada. Bulletin of Canadian Petroleum Geology 40, 222–253.

MacRae, R. A., Rankin, S. K., 2018. A palynological Triassic–Jurassic boundary section in the salt-bearing Argo F-38 well, Orpheus Graben, offshore Nova Scotia, Canada. Paper presented at the Central Atlantic Conjugate Margins Conference.

Malinconico, M.L., 2002. Lacustrine organic sedimentation, organic metamorphism, and thermal history of selected early Mesozoic Newark Supergroup basins, Eastern U.S.A., Ph. D. dissertation. Columbia University, New York, New York, 419 p.

Malinconico, M.L., 2003. Paleo-maximum thermal structure of the Triassic Taylorsville (Virginia) basin: evidence for border fault convection and implications for duration of syn-rift sedimentation and long-term elevated heat flow. In: LeTourneau, P.M., Olsen, P.E. (Eds.), The Great Rift Valleys of Pangea in Eastern North America, Volume 1, Tectonics, Structure, and Volcanism. Columbia University Press, New York, pp. 80–103.

Malinconico, M.L., 2010. Synrift to early postrift basin-scale groundwater history of the Newark basin based on surface and borehole vitrinite reflectance data. In: Herman, G.C., Serfes, M.E. (Eds.), Contributions to the Geology and Hydrogeology of the Newark Basin, 77. New Jersey Geological Survey Bulletin, pp. C1–C38.

Malinconico, M.L., 2015. Triassic Taylorsville Basin, Virginia, USA: comparative thermal history and organic facies within the Early Mesozoic eastern North American lacustrine rift basin system. In: Post, P.J., Coleman Jr., J.L., Rosen, N.C., Brown, D.E., Roberts-Ashby, T., Kahn, P., Rowan, M. (Eds.), Proceedings, Petroleum Systems in "Rift" Basins, 34th Annual Foundation GCSSEPM Bob F. Perkins Research Conference. Houston, Texas, U.S.A, Gulf Coast Section SEPM, Houston, pp. 147–173.

Manspeizer, W., 1980. Rift tectonics inferred from volcanic and clastic structures. In: Manspeizer, W. (Ed.), Field studies of New Jersey geology and guide to field trips. New York State Geological Association, 52nd Annual Meeting, Rutgers University, Newark, New Jersey, pp. 314–350.

Manspeizer, W., 1988. Triassic–Jurassic rifting and opening of the Atlantic: an overview. In: Manspeizer, W. (Ed.), Triassic-Jurassic Rifting, Continental Breakup and the Origin of the Atlantic Ocean Passive Margins, Part A. Elsevier, New York, pp. 41–79.

Manspeizer, W., Cousminer, H.L., 1988. Late Triassic-Early Jurassic synrift basins of the U.S. Atlantic margin. In: Sheridan, R.E., Grow, J.A. (Eds.), The Geology of North America, v. I-2, The Atlantic Continental Margin. Geological Society of America, U.S., pp. 197–216.

Marzen, R.E., Shillington, D.J., Lizarralde, D., Harder, S.H., 2019. Constraints on Appalachian orogenesis and continental rifting in the southeastern United States from wide-angle seismic data. Journal of Geophysical Research: Solid Earth 124 (7), 6625–6652.

Marzen, R.E., Shillington, D.J., Lizarralde, D., Knapp, J.H., Heffner, D.M., Davis, J.K., et al., 2020. Limited and localized magmatism in the Central Atlantic Magmatic Province. Nature Communications 11 (1), 1–8.

Marzoli, A., Renne, P.R., Piccirillo, E.M., Ernesto, M., Bellieni, G., deMin, A., 1999. Extensive 200-million-year-old continental flood basalts of the Central Atlantic Magmatic Province. Science 284, 616–618.

Marzoli, A., Jourdan, F., Puffer, J.H., Cuppone, T., Tanner, L.H., Weems, R.E., et al., 2011. Timing and duration of the Central Atlantic magmatic province in the Newark and Culpeper basins, eastern USA. Lithos 122, 175–188.

Marzoli, A., Callegaro, S., Dal Corso, J., Davies, J.H., Chiaradia, M., Youbi, N., et al., 2018. The Central Atlantic magmatic province (CAMP): a review. The Late Triassic World 91–125.

Mau, M., Bjerrum, C.J., Clemmensen, L.B., 2022. Late Triassic paleowinds from lacustrine wave ripple marks in the Fleming Fjord Group, central East Greenland. Palaeogeography, Palaeoclimatology, Palaeoecology 586110776.

May, P.R., 1971. Pattern of Triassic-Jurassic diabase dikes around the North Atlantic in the context of the predrift configuration of the continents. Geological Society of America Bulletin 82, 1285–1292.

McBride, J.H., 1991. Constraints on the structure and tectonic development of the early Mesozoic South Georgia rift, southeastern United States; seismic reflection data processing and interpretation. Tectonics 10 (5), 1065–1083.
McBride, J.H., Nelson, K.D., Brown, L.D., 1989. Evidence and implications of an extensive early Mesozoic rift basin and basalt/diabase sequence beneath the southeast Coastal Plain. Geological Society of America Bulletin 101, 512–520.
McHone, J.G., 2000. Non-plume magmatism and rifting during the opening of the central Atlantic Ocean. Tectonophysics 316, 287–296.
McHone, J.G., 2003. Volatile emissions from Central Atlantic Magmatic Province basalts: mass assumptions and environmental consequences. In: The Central Atlantic Magmatic Province: Insights from Fragments of Pangea, Hames, W., McHone, J.G, Renna, P.R., Ruppel, C. (Eds.), Geophysical monograph, 136. American Geophysical Union, pp. 241–254.
McHone, J.G., Butler, J.R., 1984. Mesozoic igneous provinces of New England and the opening of the North Atlantic Ocean. Geological Society of America Bulletin 95 (7), 757–765.
McHone, J.G., Anderson, D.L., Beutel, E.K., Fialko, Y.A, 2005. Giant dikes, rifts, flood basalts, and plate tectonics: A contention of mantle models. In: Foulger, G.R., Natland, J.H., Presnall, D.C., Anderson, D.L. (Eds.), Plates, Plumes, and Paradigms. Geological Society of America Special Papers, 388, New York, pp. 401–420.
McHone, J.G., Barr, S.M., Jourdan, F., 2022. Petrology and age of the Lepreau River Dyke, southern New Brunswick, Canada: source of the end-Triassic Fundy Group basalts. Canadian Journal of Earth Sciences 59 (1), 12–28.
Merguerian, C., Sanders, J.E., 1994. Post-Newark folds and faults: implications for the geologic history of the Newark basin. In: Hanson, G.N. (Ed.), Geology of Long Island and Metropolitan New York. State University of New York at Stony Brook, New York, pp. 1–165.
Merle, R., Marzoli, A., Reisberg, L., Bertrand, H., Nemchin, A., Chiaradia, M., et al., 2014. Sr, Nd, Pb and Os isotope systematics of CAMP tholeiites from Eastern North America (ENA): evidence of a subduction-enriched mantle source. Journal of Petrology 55 (1), 133–180.
Mooney, W.D., Kaban, M.K., 2010. The North American upper mantle: density, composition, and evolution. Journal of Geophysical Research 115, B12424.
Nelson, K.D., 1992. Are crustal thickness variations in old mountain belts like the Appalachians a consequence of lithospheric detachment?:. Geology 20, 498–502.
Nomade, S., Knight, K.B., Beutel, E., Renne, P.R., Verati, C., Féraud, G., et al., 2007. Chronology of the Central Atlantic Magmatic Province: implications for the Central Atlantic rifting processes and the Triassic–Jurassic biotic crisis. Palaeogeography, Palaeoclimatology, Palaeoecology 244 (1–4), 326–344.
O'Connor, D., Silva, R.L., Wach, G., 2018. Earliest sedimentological evidence for marine ingressions in the eastern North American rift system. Central Atlantic Margin. Lithosphere 10 (6), 783–791.
Oh, J., Austin, J.A., Phillips, J.D., Coffin, M.F., Stoffa, P.L., 1995. Seward-dipping reflectors offshore the southeastern United States: Seismic evidence for extensive volcanism accompanying sequential formation of the Carolina trough and Blake Plateau basin. Geology 23, 9–12.
Olsen, P.E., 1980. The lastest Triassic and Early Jurassic formations of the Newark basin (eastern North America, Newark Supergroup): stratigraphy, structure, and correlation. New Jersey Academy of Science Bulletin 25, 25–51.
Olsen, P.E., 1986. A 40-million-year lake record of early Mesozoic climatic forcing. Science 234, 842–848.
Olsen, P.E., 1990. Tectonic, climatic, and biotic modulation of lacustrine ecosystems: examples from the Newark Supergroup of eastern North America. In: Katz, B. (Ed.), Lacustrine Basin Exploration: Case Studies and Modern Analogs, American Association Petroleum Geologists Memoir 50, pp. 209–224.
Olsen, P.E., 1997. Stratigraphic record of the early Mesozoic breakup of Pangea in the Laurasia-Gondwana rift system. Annual Reviews of Earth and Planetary Science 25, 337–401.
Olsen, P.E., 1998, Geological interpretation of the Rattlesnake Ridge Project - McNair et al. # 1 well. Project report to Rattlesnake Ridge Joint Venture 1986, +21p.
Olsen, P.E., 1999. Giant lava flows, mass extinctions, and mantle plumes. Science (New York, N.Y.) 284, 604–605.
Olsen, P.E., McDonald, N.G., 2022. The stratigraphy and paleontology of the Newark Supergroup in its global context. In: Farlow, J.O. (Ed.), Dinosaur State Park. Indiana University Press, in revision.
Olsen, P.E., Schlische, R.W., 1990. Transtensional arm of the early Mesozoic Fundy rift basin: penecontemporaneous faulting and sedimentation. Geology 18, 695–698.
Olsen, P.E., Kent, D.V., 1996. Milankovitch climate forcing in the tropics of Pangea during the Late Triassic:. Palaeogeography, Palaeoclimatology, and Palaeoecology 122, 1–26.
Olsen, P.E., Kent, D.V., 2000. High resolution early Mesozoic Pangean climatic transect in lacustrine environments. In: Bachmann, G., Lerche, I. (Eds.), Epicontinental Triassic, 3. Zentralblatt fur Geologie und Palaontologie, pp. 1475–1496. , VIII.
Olsen, P. E., Schlische, R. W., and Gore, P. J. W., (Eds.), 1989, Tectonic, depositional, and paleoecological history of early Mesozoic rift basins, eastern North America: International Geological Congress Field Trip T351, American Geophysical Union, Washington, D. C., 174 p.
Olsen, P.E., Kent, D.V., Cornet, B., Witte, W.K., Schlische, R.W., 1996a. High-resolution stratigraphy of the Newark rift basin (early Mesozoic, eastern North America). Geological Society of America Bulletin 108, 40–77.
Olsen, P.E., Schlische, R.W., and Fedosh, M.S., 1996b, 580 kyr duration of the Early Jurassic flood basalt event in eastern North America estimated using Milankovitch cyclostratigraphy. In: Morales, M. (Ed.), The Continental Jurassic. Museum of Northern Arizona 60, p. 11–22.
Olsen, P.E., Kent, D.V., Fowell, S.J., Schlische, R.W., Withjack, M.O., LeTourneau, P.M., 2000. Implications of a comparison of the stratigraphy and depositional environments of the Argana (Morocco) and Fundy (Nova Scotia, Canada) Permian-Jurassic basins. In: Oujidi, M., Et-Touhami, M. (Eds.), Le Permien et le Trias du Maroc. Actes de la Premièr Réunion su Groupe Marocain du Permien et du Trias, Oujda, Hilal Impression, pp. 165–183.
Olsen, P.E., Kent, D.V., Et-Touhami, M., Puffer, J., 2003. Cyclo-, magneto-, and bio-stratigraphic constraints on the duration of the CAMP event and its relationship to the Triassic-Jurassic boundary. In: Hanes, W.E., McHone, J.G., Renne, P.R., Ruppel, C. (Eds.), The Central Atlantic Magmatic Province, Insights from Fragments of Pangea, 136. American Geophysical Union, Geophysical Monograph, Washington, D. C., pp. 7–32.
Olsen, P.E., Reid, J.C., Taylor, K.B., Whiteside, J.H., Kent, D.V., 2015. Revised stratigraphy of Late Triassic age strata of the Dan River Basin (Virginia and North Carolina, USA) based on drill core and outcrop data. Southeastern Geology 51 (1), 1–31.

Olsen, P.E., Kinney, S.T., Zakharova, N.V., Schlische, R.W., Withjack, M.O., Kent, D.V., et al., 2016. New insights on rift basin development and the geological carbon cycle, mass extinction, and carbon sequestration from outcrops and new core, drill holes, and seismic lines from the northern Newark Basin (New York and New Jersey). In 88th Annual New York State Geological Field Conference Guidebook, pp. 190–274.

Olsen, P.E., Laskar, J., Kent, D.V., Kinney, S.T., Reynolds, D.J., Sha, J., et al., 2019. Mapping solar system chaos with the Geological Orrery. Proceedings of the National Academy of Sciences 116 (22), 10664–10673.

Peace, A.L., Phethean, J.J., Franke, D., Foulger, G.R., Schiffer, C., Welford, J.K., et al., 2020. A review of Pangaea dispersal and Large Igneous Provinces—in search of a causative mechanism. Earth-Science Reviews 206102902.

Poag, C.W., Valentine, P.C., 1988. Mesozoic and Cenozoic stratigraphy of the United States Atlantic continental shelf and slope. In: Sheridan, R.E., Grow, J.A. (Eds.), The Geology of North America, I-2. Geological Society of America, The Atlantic Continental Margin, U.S., pp. 67–85.

Poag, C.W., Sevon, W.D., 1989. A record of Appalachian denudation in postrift Mesozoic and Cenozoic sedimentary deposits of the U.S. Middle Atlantic continental margin. Geomorphology 2 (1–3), 119–157.

Pratt, L.M., Shaw, C.A., Burruss, R.C., 1988. Thermal histories of the Hartford and Newark Basins inferred from maturation indices of organic matter. In: Froelich, A.J., Robinson, G.R. (Eds.), Geological Survey Bulletin, 1776. Studies of the Early Mesozoic Basins of the Eastern United States, United States, pp. 58–62.

Puffer, J.H., 1992. Eastern North American flood basalts in the context of the incipient breakup of Pangaea. In: Puffer, J.H. Ragland, P.C. (Eds.), Eastern North American Magmatism, Geological Society of America Special Paper 268, pp. 95–119.

Puffer, J.H., Philpotts, A.R., 1988. Eastern North American quartz tholeiites: geochemistry and petrology. Manspeizer, W. (Ed.),Triassic-Jurassic Rifting, Continental Breakup and the Origin of the Atlantic Ocean Passive Margins, Part AElsevier, pp. 579–605.

Ragland, P.C., Cummins, L.E., Arthur, J.D., 1992. In: Puffer, J.H., Ragland, P.C. (Eds.), Eastern North American Mesozoic Magmatism, Geological Society of America Special Paper 268, pp. 301–331.

Rankin, D.W., 1994. Continental margin of the eastern United States: past and present. In: Speed, R.C. (Ed.), Phanerozoic Evolution of North American Continent-Ocean Transitions. Geological Society of America, pp. 129–218.

Rast, N., 1988. Variscan-Alleghanian orogen. In: Manspeizer, W. (Ed.), Triassic-Jurassic Rifting, Continental Breakup and the Origin of the Atlantic Ocean Passive Margins, Part A. Elsevier, New York, pp. 1–27.

Ratcliffe, N.M., 1980. Brittle faults (Ramapo fault) and phyllonitic ductile shear zones in the basement rocks of the Ramapo seismic zone (NY and NJ), and their relationship to current seismicity. In: W. Manspeizer (Ed.), Field studies of New Jersey geology and guide to field trips: New York State Geological Association, 52nd Annual Meeting, Rutgers University, Newark, New Jersey, pp. 278–311.

Ratcliffe, N.M., Burton, W.C., 1985. Fault reactivation models for origin of the Newark basin and studies related to eastern U.S. seismicity, 946. Geological Survey Circular, States: U.S, pp. 36–45.

Ratcliffe, N.M., Burton, W.C., D'Angelo, R.M., Costain, J.K., 1986. Low-angle extensional faulting, reactivated mylonites, and seismic reflection geometry of the Newark Basin margin in eastern Pennsylvania. Geology 14, 766–770.

Ravnas, R., Steel, R.J., 1998. Architecture of marine rift-basin successions. American Association of Petroleum Geologists Bulletin 83, 110–146.

Resor, P.G., deBoer, J.Z., 2005. Hartford Basin Cross Sections – Southington to Portland, Connecticut. McHone, N.W., Peterson, M.J. (Eds.), Guidebook For Fieldtrips in Connecticut 8, 177–189.

Roden-Tice, M.K., Wintsch, R.P., 2002. Early Cretaceous normal faulting in southern New England: evidence from apatite and zircon fission-track ages. Journal of Geology 110, 159–178.

Royden, L., Keen, C.E., 1980. Rifting process and thermal evolution of the continental margin of Eastern Canada determined from subsidence curves. Earth and Planetary Science Letters 51, 343–361.

Ryan, M.P., Pierce, H.A., Johnson, C.D., Sutphin, D.M., Daniels, D.L., Smoot, J.P., et al., 2007. Reconnaissance borehole geophysical, geological, and hydrological data from the proposed hydrodynamic compartments of the Culpeper basin in Loudoun, Prince William, Culpeper, Orange, and Fairfax Counties, Virginia [Version 1.0], U. S. Geological Survey Open-File Report 2006-12-3, 43 p.

Sachs, P.E., Stoddard, E.F., Berquist, R., Newton, C., 1999. A field guide to the geology of the fall zone region, North Carolina and Virginia state line: road log for CGS field trip. Geology of the Fall Zone Region along the North Carolina. Guidebook for the 1999 Meeting of the Carolina Geological Society, Emporia, Virginia, Virginia State Line.

Sanders, J.E., 1963. Late Triassic tectonic history of northeastern United States. American Journal Science 261, 501–524.

Schettino, A., Turco, E., 2009. Tectonic history of the western Tethys since the Late Triassic. Geological Society of America Bulletin 123, 89–105.

Schlische, R.W., 1991. Half-graben basin filling models: new constraints on continental extensional basin evolution. Basin Research 3, 123–141.

Schlische, R.W., 1992. Structural and stratigraphic development of the Newark extensional basin, eastern North America: implications for the growth of the basin and its bounding structures. Geological Society of America Bulletin 104, 1246–1263.

Schlische, R.W., 1993. Anatomy and evolution of the Triassic-Jurassic continental rift system, eastern North America. Tectonics 12, 1026–1042.

Schlische, R.W., 1995. Geometry and origin of fault-related folds in extensional settings. AAPG Bulletin 79, 661–1678.

Schlische, R.W., Olsen, P.E., 1990. Quantitative filling model for continental extensional basins with applications to early Mesozoic rifts of eastern North America. Journal of Geology 98, 135–155.

Schlische, R.W., Withjack, M.O., 2005. The early Mesozoic Birdsboro central Atlantic margin basin in the mid-Atlantic region, eastern United States: discussion. Geological Society of America Bulletin 117 (5–6), 823–828.

Schlische, R.W., Withjack, M.O., Olsen, P.E., 2003. Relative timing of CAMP, rifting, continental breakup, and basin inversion: tectonic significance. In: Hanes, W.E., McHone, J.G., Renne, P.R., Ruppel, C. (Eds.), The Central Atlantic Magmatic Province, Insights from Fragments of Pangea, 136. American Geophysical Union, Geophysical Monograph, Washington, D. C., pp. 33–60.

Schoene, B., Guex, J., Bartolini, A., Schaltegger, U., Blackburn, T.J., 2010. Correlating the end-Triassic mass extinction and flood basalt volcanism at the 100 ka level. Geology 38 (5), 387–390.

Serck, C.S., Braathen, A., 2019. Extensional fault and fold growth: impact on accommodation evolution and sedimentary infill. Basin Research 31, 967–990.

Shen, W., Ritzwoller, M.H., 2016. Crustal and uppermost mantle structure beneath the United States. Journal of Geophysical Research: Solid Earth 121, 4306–4342.

Sheridan, R.E., Olsson, R.K., Miller, J.J., 1991. Seismic reflection and gravity study of proposed Taconic suture under the New Jersey coastal plain: implications for continental growth. Geological Society of America Bulletin 103 (3), 402–414.

Sheridan, R.E., Musser, D.L., Glover III, L., Talwani, M., Ewing, J.I., Holbrook, W.S., et al., 1993. Deep seismic reflection data of EDGE U.S. Atlantic continent-margin experiment: implications for Appalachian sutures and Mesozoic rifting and magmatic underplating. Geology 21, 563–567.

Shuck, B.D., Van Avendonk, H.J., Bécel, A., 2019. The role of mantle melts in the transition from rifting to seafloor spreading offshore eastern North America. Earth and Planetary Science Letters 525115756.

Slingerland, R., Furlong, K.P., 1989. Geodynamic and geomorphic evolution of the Permo-Triassic Appalachian Mountains. Geomorphology 2, 23–37.

Smith, R.C., Rose, A.W., Lanning, R.M., 1975. Geology and geochemistry of Triassic diabase in Pennsylvania. Geological Society of America Bulletin 86 (7), 943–955.

Smoot, J.P., 1991. Sedimentary facies and depositional environments of early Mesozoic Newark Supergroup basins, eastern North America. Palaeogeography, Palaeoclimatology, Palaeoecology 84 (1–4), 369–423.

Srogi, L., Martinson, P.J., Willis, K.V., Lutz, T., Kulp, R., Pollock, M., 2017. The 201-Ma magma plumbing system in the western Newark Basin: How and (maybe) why it's different from the east. In: Bosbyshel., H. (Ed.), The Piedmont: Old Rocks, New Understanding. Geological Association of New, Jersey, pp. 30–48.

Stoddard, E.F., Delorey, C.M., McDaniel, R.D., Dooley, R.E., Ressetar, R., Fullagar, P.D., 1986. A new suite of post-orogenic dikes in the eastern North Carolina Piedmont: part I. Occurrence, petrography, paleomagnetics, and Rb/Sr geochronology. Southeastern Geology 27, 1–12.

Sues, H.D., Olsen, P., 2015. Stratigraphic and temporal context and faunal diversity of Permian-Jurassic continental tetrapod assemblages from the Fundy rift basin, eastern Canada. Atlantic Geology 51, 139–205.

Sues, H.-D., Olsen, P., Fedak, T., Schoch, R., 2021. Diverse assemblage of Middle Triassic continental tetrapods from the Newark Supergroup of Nova Scotia (Canada). Journal of Vertebrate Paleontology 41 (4), e2023168.

Sutter, J.F., 1988. Innovative approaches to the dating of igneous events in the early Mesozoic basins of the eastern United States. In: Froelich, A.J., Robinson Jr., G.R. (Eds.), Studies of the Early Mesozoic Basins of the Eastern United States, 1776. U.S. Geological Survey Bulletin, pp. 194–199.

Swanson, M.T., 1986. Preexisting fault control for Mesozoic basin formation in eastern North America. Geology 14, 419–422.

Syamsir, Z., Withjack, M.O., Durcanin, M.A., Schlische, R.W. Monteverde, D.H., 2010. The Mesozoic Orpheus rift basin, offshore Nova Scotia and Newfoundland, Canada: Synrift and early postrift evolution of a well-imaged North Atlantic rift basin. In: II Central and North Atlantic Conjugate Margins Conference, Lisbon, Portugal, Extended Abstracts, 29, pp. 279–283.

Talwani, M., Ewing, J., Sheridan, R.E., Holbrook, W.S., Glover, L., 1995. The EDGE experiment and the U. S. East Coast Magnetic Anomaly. In: Banda, E., Torne, M., Talwani, M. (Eds.), Rifted Ocean-Continent Boundaries, 463. Springer, NATO ASI Series, Dordrecht, The Netherlands, pp. 155–181.

Tanner, L.H., Hubert, J.F., 1991. Basalt breccias and conglomerates in the Lower Jurassic McCoy Brook Formation, Fundy Basin, Nova Scotia; differentiation of talus and debris-flow deposits. Journal of Sedimentary Research 61 (1), 15–27.

Tanner, L.H., Brown, D.E., 2003. Tectonostratigraphy of the Orpheus graben, Scotian basin, offshore eastern Canada, and its relationship to the Fundy rift basin. In: LeTourneau, P.M., Olsen, P.E. (Eds.), The Great Rift Valleys of Pangea in Eastern North America, Volume 2, Sedimentology, Stratigraphy, and Paleontology. University Press, New York, Columbia, pp. 59–68.

Thomas, W.A., 2006. Tectonic inheritance at a continental margin. GSA Today 16, 4–11.

Van Veen, P.M., Skjold, L.J., Kristensen, S.E., Rasmussen, A., Gjelberg, J., Stølan, T., 1992. In: Vorren, T.O., Bergsager, Ø.A., Dahl-Stamenesm Holter, E., Johansen, B., Lie, E., Lund, T.B. (Eds.), Arctic Geology and Petroleum Potential, 2. NPF Special Publication, pp. 515–538.

Vendeville, B.C., Ge, H., Jackson, M.P.A., 1995. Scale models of salt tectonics during basement-involved extension. Petroleum Geoscience 1 (2), 179–183.

Venkatakrishnan, R., Lutz, R., 1988. A kinematic model for the evolution of the Richmond basin. In: Manspeizer, W. (Ed.), Triassic-Jurassic Rifting, Continental Breakup and the Origin of the Atlantic Ocean Passive Margins, Part A. Elsevier, New York, pp. 445–462.

Vogt, P.R., 1973. Early events in the opening of the North Atlantic. In: Tarling, D.H., Runcorn, S.K. (Eds.), Implications of Continental Drift to the Earth Sciences. Academic Press, New York, pp. 693–712.

Warrell, K.F., Potter, C.J., Malinconico, M.A., 2018. Insights into the geometry and evolution of the Chatham fault and Dan River basin near Chatham. Virginia, Geological Society of America, Abstracts with Programs 50 (3).

Weems, R.E., Olsen, P.E., 1997. Synthesis and revision of groups within the Newark Supergroup, eastern North America. Geological Society of America Bulletin 109 (2), 195–209.

Weigand, P.W., Ragland, P.C., 1970. Geochemistry of Mesozoic dolerite dikes from eastern North America. Contributions to Mineralogy and Petrology 29 (3), 195–214.

Weissel, J.K., Karner, G.K., 1989. Flexural uplift of rift flanks due to mechanical unloading of the lithosphere during extension. Journal of Geophysical Research 94, 13,919–13,950.

Welsink, H.J., Dwyer, J.D., Knight, R.J., 1989. Tectono-stratigraphy of the passive margin off Nova Scotia. In: Tankard, A.J., Balkwill, H.R. (Eds.), Extensional Tectonics and Stratigraphy of the North Atlantic Margins, AAPG Memoir 46, pp. 215–231.

Wenk, W.J., 1984. Seismic refraction model of depth of basement in the Hartford rift basin, Connecticut and Massachusetts. Northeastern Geology 6, 196–202.

Wenk, W.J., 1990. Syndepositional timing of transverse warps in Mesozoic rocks of the Hartford rift basin, southern New England. Northeastern Geology 12, 132–137.

Weston, J.F., MacRae, R.A., Ascoli, P., Cooper, M.K.E., Fensome, R.A., Shaw, D., et al., 2012. A revised biostratigraphic and well-log sequence-stratigraphic framework for the Scotian Margin, offshore eastern Canada. Canadian Journal of Earth Sciences 49 (12), 1417–1462.

Whalen, L., Gazel, E., Vidito, C., Puffer, J., Bizimis, M., Henika, W., et al., 2015. Supercontinental inheritance and its influence on supercontinental breakup: the Central Atlantic Magmatic Province and the breakup of Pangea. Geochemistry, Geophysics, Geosystems 16 (10), 3532–3554.

Wheeler, G., 1937. The west wall of the New England Triassic lowland. Connecticut Geological and Natural History Survey Bulletin 58, 73.
Wheeler, G., 1939. Triassic fault-line deflections and associated warping. Journal of Geology 47, 337–370.
White, R., McKenzie, D., 1989. Magmatism at rift zones: the generation of volcanic continental margins and flood basalts. Journal of Geophysical Research: Solid Earth 94 (B6), 7685–7729.
Whiteside, J.H., Olsen, P.E., Kent, D.V., Fowell, S.J., Et-Touhami, M., 2007. Synchrony between the Central Atlantic magmatic province and the Triassic–Jurassic mass-extinction event? Palaeogeography, Palaeoclimatology, Palaeoecology 244 (1–4), 345–367.
Wilson, M., 1997. Thermal evolution of the Central Atlantic passive margins: continental break-up above a Mesozoic super-plume. Journal of the Geological Society 154 (3), 491–495.
Wise, D.U., 1992. Dip domain method applied to the Mesozoic Connecticut Valley rift basins. Tectonics 11, 1357–1368.
Withjack, M.O., Callaway, S., 2000. Active normal faulting beneath a salt layer: an experimental study of deformation patterns in the cover sequence. AAPG Bulletin 84, 627–651.
Withjack, M.O., Schlische, R.W., 2006. Geometric and experimental models of extensional fault-bend folds. In: Buiter, S.J.H., Schreurs, G. (Eds.), Analogue and Numerical Modelling of Crustal-Scale Processes, Geological Society, London, Special Publications 253, pp. 285–305.
Withjack, M.O., Olsen, P.E., Schlische, R.W., 1995. Tectonic evolution of the Fundy rift basin, Canada: evidence of extension and shortening during passive margin development. Tectonics 14, 390–405.
Withjack, M.O., Schlische, R.W., Olsen, P.E., 1998. Diachronous rifting, drifting, and inversion on the passive margin of central eastern North America: an analog for other passive margins. AAPG Bulletin 82, 817–835.
Withjack, M.O., Schlische, R.W., Olsen, P.E., 2002. In: Renaut, R., Ashley, G.M. (Eds.), Continental Rift Basin Sedimentology, SEPM Special Publication 73, pp. 57–81.
Withjack, M.O., Schlische, R.W., Baum, M.S., 2009. Extensional development of the Fundy rift basin, southeastern Canada. Geological Journal 44 (6), 631–651.
Withjack, M.O., Baum, M.S., Schlische, R.W., 2010. Influence of preexisting fault fabric on inversion-related deformation: a case study of the inverted Fundy rift basin, southeastern Canada. Tectonics 29 (6).
Withjack, M.O., Schlische, R.W., Olsen, P.E., 2012. Development of the passive margin of eastern North America: Mesozoic rifting, igneous activity, and breakup. In: Bally, A.W., Roberts, D.G. (Eds.), Principles of Phanerozoic Regional Geology, Volume 1. Elsevier, Amsterdam, pp. 301–335.
Withjack, M.O., Schlische, R.W., Malinconico, M.L., Olsen, P.E., 2013. Rift-basin development: lessons from the Triassic–Jurassic Newark Basin of eastern North America. Geological Society, London, Special Publications 369 (1), 301–321.
Withjack, M., Malinconico, M., Durcanin, M., 2020. The "passive" margin of eastern North America: rifting and the influence of prerift orogenic activity on postrift development. Lithosphere 2020 (1).
Wu, Y., Louden, K.E., Funck, T., Jackson, H.R., Dehler, S.A., 2006. Crustal structure of the central Nova Scotia margin off Eastern Canada. Geophysical Journal International 166 (2), 878–906.
Zulfitriadi, Z., 2011. The Mesozoic Orpheus rift basin, offshore Nova Scotia and Newfoundland, Canada: synrift and early postrift evolution of a well-imaged North Atlantic rift basin (M.S. thesis), Rutgers the State University of New Jersey, New Brunswick, New Jersey.

CHAPTER

4

Neoproterozoic-Early Paleozoic rifting in central southern Laurentia—Zircon U-Pb geochronological evidence from the Devils River Uplift, West Texas, United States

Patricia W. Dickerson[1], *Daniel F. Stockli*[2] *and Edna Rodríguez Calzado*[2]

[1]American Geological Institute and Jackson School of Geosciences, University of Texas at Austin, Austin, TX, United States [2]Department of Earth and Planetary Sciences, Jackson School of Geosciences, University of Texas at Austin, Austin, TX, United States

Introduction

Supercontinent interactions, from Grenvillian orogenesis and assembly of Rodinia to its fragmentation by rifting to the amalgamation of Pangaea, have shaped the central southern Laurentian margin. Surface structures and strata record the Grenvillian and Appalachian-Ouachita episodes, but between the southern Oklahoma and New Mexico aulacogens, the Neoproterozoic to early Cambrian rift complex lies buried beneath kilometers of younger rock (Fig. 4.1). However, a coherent and extensive metaigneous and metasedimentary sequence is preserved in the Devils River Uplift (DRU) in the subsurface of West Texas (Val Verde, Kinney, and Edwards counties; Fig. 4.2). That section was cored in a petroleum exploration borehole (Shell #1 Stewart), and whole-rock Rb/Sr and K/Ar ages were determined (R. E. Denison *in* Nicholas and Rozendal, 1975; Nicholas and Waddell, 1989). Among those geochronological data were Neoproterozoic to early Cambrian ages relevant to Rodinia rifting. Zircons in that range are abundant in metaigneous and metasedimentary strata of the DRU; new U/Pb analyses (LA-ICP-Ms) expand on the earlier work and refine the timing of the opening of Iapetus.

Regional tectonic setting

Supplementing outcrop observations, regional spatial relations among Mesoproterozoic-early Paleozoic elements have been interpreted from aeromagnetic, gravity, seismic, and borehole data (King and Flawn, 1953; Muehlberger et al., 1967; Denison et al., 1971; Keller and Peeples, 1985; Mosher, 1998; Nicholas and Rozendal, 1975; Reese et al., 2000). Increasingly available geochronological analyses help to define the temporal relations among those older provinces.

In aeromagnetic data for the study area, the subsurface DRU structure is coextensive with a northwesterly linear positive anomaly of similar intensity to that of the Llano Uplift of Central Texas. In Bouguer gravity data, the DRU-positive anomaly resembles that of the southern periphery of the Llano Uplift but differs from the Llano central core (Figs. 4.3, 4.4). Metaigneous and metasedimentary rocks of the Grenville

Regional Geology and Tectonics: Volume 2
DOI: https://doi.org/10.1016/B978-0-444-64136-6.00010-5

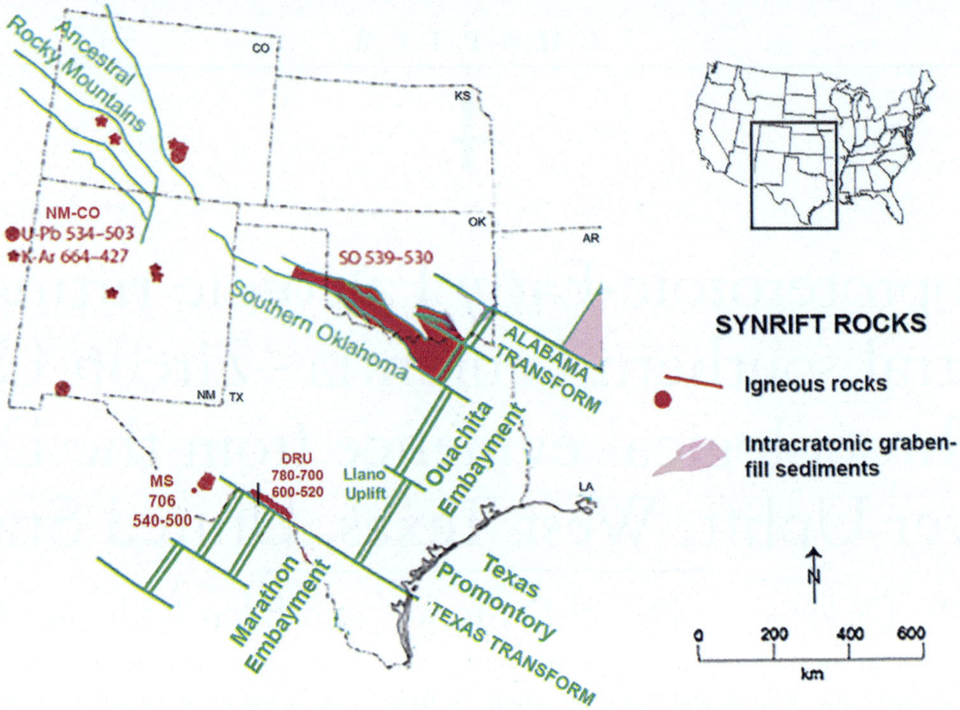

FIGURE 4.1 Region of investigation, conterminous United States, central southern Laurentia. *DRU*, Devils River Uplift; *Ms*, Marathon/Solitario basin. Cryogenian to Early Cambrian synrift igneous rocks (red): Outcrops—Southern Oklahoma and New Mexico aulacogens (U/Pb & K/Ar ages); NM-CO—New Mexico and Colorado data. Subsurface—DRU (U/Pb ages). Detritus in Marathon/Solitario basin—Ms (U/Pb ages). Source: *Modified from Thomas, W. A., 2014, A mechanism for tectonic inheritance at transform faults of the Iapetan margin of Laurentia: Geological Association of Canada, Geoscience Canada, Harold Williams Series, 41, p. 321–344. http://doi.org/10.12789/geocanj.2014.41.048.*

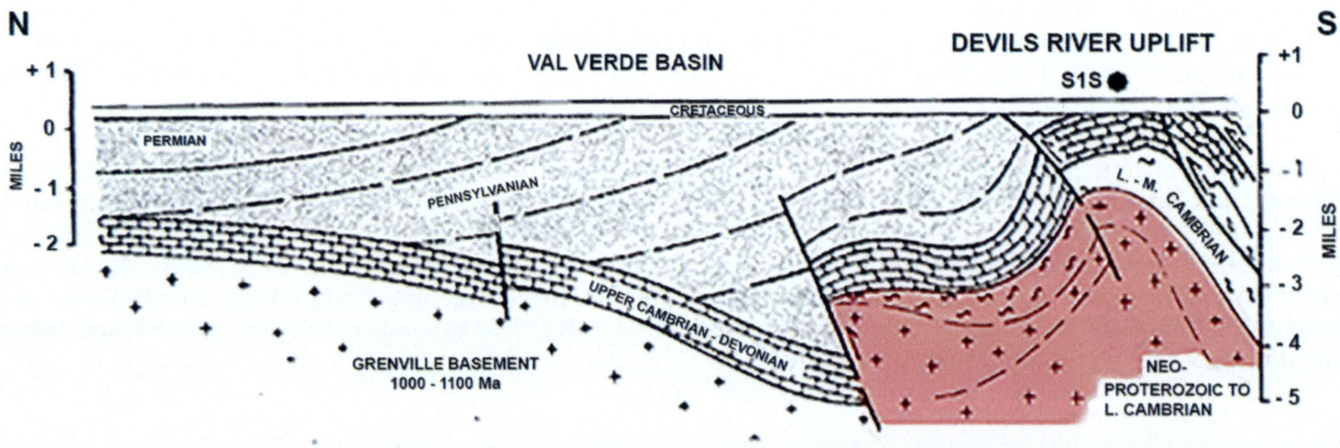

FIGURE 4.2 Devils River Uplift cross-section, location on Fig. 4.1. S1S —Shell #1 Stewart well. Whole-rock Rb/Sr ages indicative of Rodinia rifting were determined for metarhyolites (711 ± 23 Ma) and for other metavolcanics within metasediments (524 ± 32 and 529 ± 31 Ma) in that borehole. See text for zircon U/Pb data (LA-ICP-Ms, this study) for those intervals. Source: *Modified from Nicholas, R. L., and Rozendal, R. A., 1975, Subsurface positive elements within Ouachita foldbelt in Texas and their relation to Paleozoic cratonic margin: American Association of Petroleum Geologists Bulletin, 59, p. 193–216.*

orogenic belt are exposed in the Llano Uplift; deformed between 1200 and 1100 Ma, the complex was then intruded by late orogenic to postorogenic granite plutons at c.1100 Ma. Outcrop, petrographic, geochemical, and zircon U/Pb geochronological data are discussed in detail in Mosher (1998), Reese et al. (2000), and references therein. Analogous Mesoproterozoic underpinnings of the Llano and Devils River basement uplifts are geochronologically corroborated; the 1230-Ma orthogneisses are notably similar. The positive

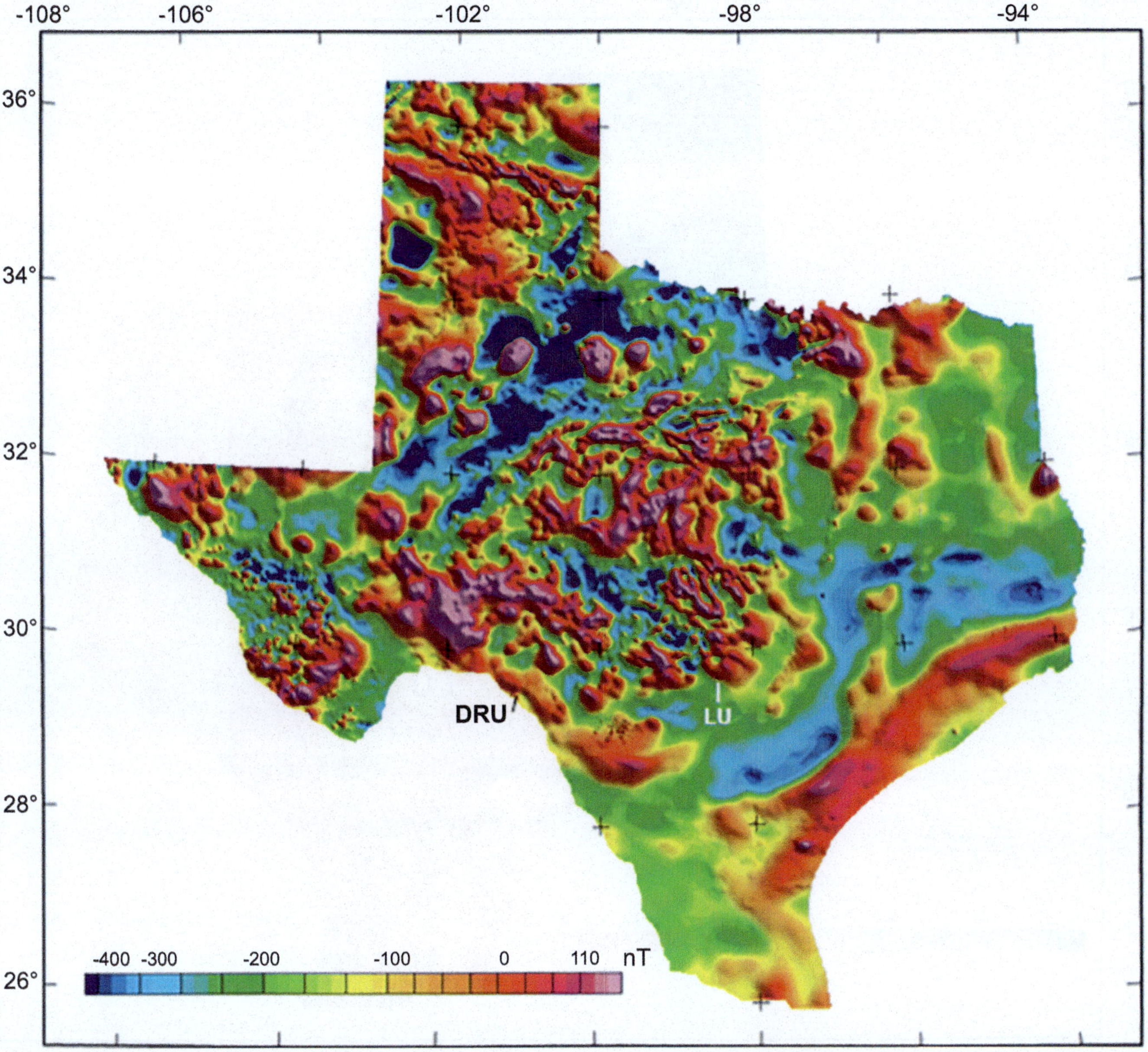

FIGURE 4.3 Aeromagnetic map, Texas. DRU—the northwesterly linear positive anomaly is coextensive with Devils River Uplift. See text for comparison of magnetic expressions of DRU and Llano Uplift, LU. *DRU*, Devils River Uplift. Source: *Modified from Bankey, V., 2016, Texas magnetic and gravity maps and data: a website for distribution of data: U.S. Geological Survey, Data Series 232, 2 maps. [Page last modified: Monday, November 28, 2016] https://pubs.usgs.gov/ds/2006/232/.*

magnetic and gravity anomalies for DRU are both likely influenced by massive metabasalt at depth (not cored in the Shell #1 Stewart well but penetrated in a nearby borehole; seismic interpretation in Nicholas and Rozendal, 1975, Fig. 4.8).

During the Neoproterozoic to early Cambrian breakup of Rodinia, the Grenvillian foundation underwent rifting (geochronological constraints in this paper). The western Iapetus seaway opened, and a vast Cambrian-Ordovician carbonate platform developed along the southern Laurentian margin. Gondwana encroached and Iapetus was closed in Late Carboniferous to Early Permian time. Ouachita-Marathon collisional deformation drove the Laurentian DRU back onto the old rifted margin. Based on lithologic data and on tectonostratigraphic setting, the ~125-km-long subsurface structure is interpreted as an external massif of the Ouachita orogen, analogous to the Blue Ridge and Green Mountains basement uplifts of the

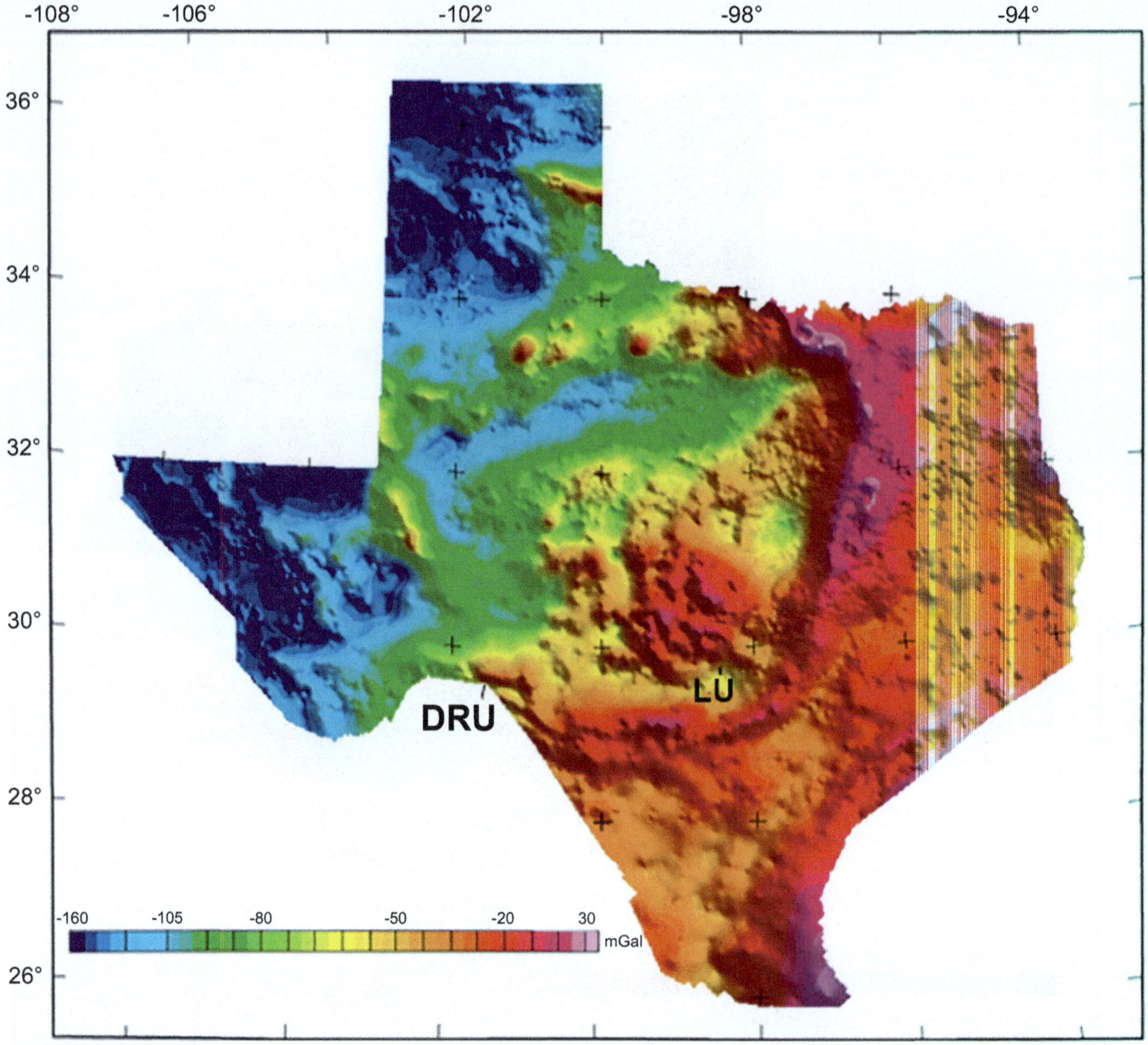

FIGURE 4.4 Bouguer gravity anomalies, Texas. The west-trending DRU-positive anomaly is similar to that of the southern periphery of the Llano Uplift, LU, but not to the central core. Metabasalt at depth and the thicker crust, owing to late Pz overthrusting, contribute to the DRU gravity high. *DRU*, Devils River Uplift. Source: *Modified from Bankey, V., 2016, Texas magnetic and gravity maps and data: a website for distribution of data: U.S. Geological Survey, Data Series 232, 2 maps. [Page last modified: Monday, November 28, 2016] https://pubs.usgs.gov/ds/2006/232/.*

Appalachians (Flawn et al., 1961; Nicholas and Waddell, 1989). Tectonic thickening of the crust in that zone contributes to the positive gravity response.

Zircon U/Pb geochronology

Before the breakup of Rodinia, abundant detritus in prerift metasediments of the DRU was derived from the Yavapai-Mazatzal (1800–1600 Ma), Granite-Rhyolite (1500–1350 Ma), and Grenville (older—1230 Ma; younger—1070 Ma) provinces (Fig. 4.5; Karlstrom et al., 2004). In addition to those populations, the syn-rift strata also

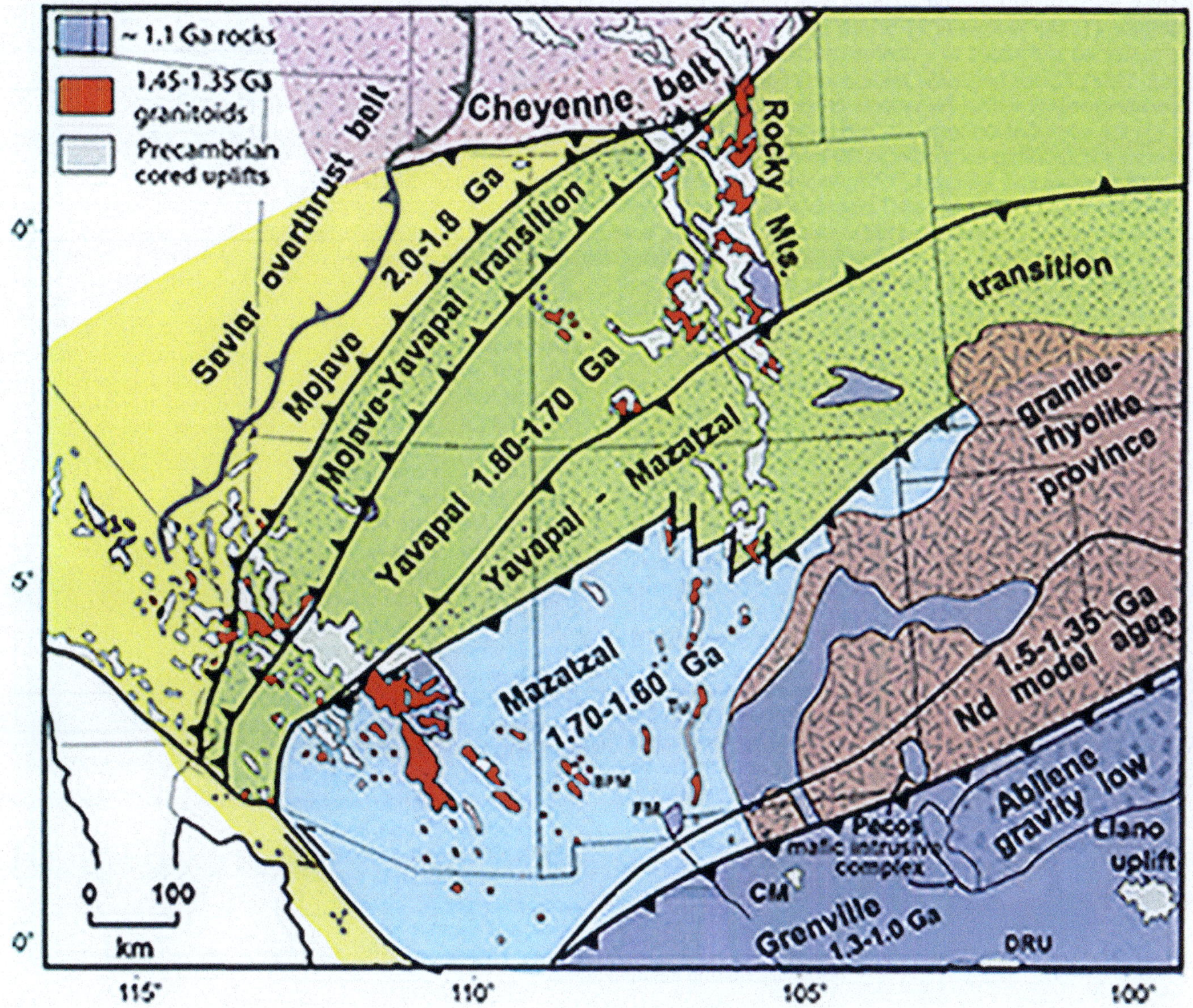

FIGURE 4.5 Proterozoic tectonostratigraphic terranes of southwestern and central southern Laurentia. Grenvillian of Devils River (DRU) and Llano (LU) uplifts. Marathon Basin and the Solitario, on Grenvillian basement, are just south of the map boundary. *CM*, Carrizo Mountains, exposures of Granite-Rhyolite Province strata and Grenvillian thrust faults. *FM*, Franklin Mountains, Grenvillian. Pecos mafic intrusive complex = 1090–1070 Ma (Kargi and Barnes, 1995). BFM, Burro-Florida Mountains and Tu, Tularosa area—localities for early Cambrian synrift igneous rocks of the New Mexico aulacogen. Source: *Modified from Karlstrom, K. E., J. M. Amato, M. L. Williams, M. Heizler, C. A. Shaw, A. S. Read, and et al., 2004, Proterozoic tectonic evolution of the New Mexico region: a synthesis, in G. H. Mack and K. A. Giles, eds., The geology of New Mexico: A geologic history: Socorro, New Mexico, New Mexico Bureau of Geology, p. 1–34.*

contain magmatic zircons that record Cryogenian and early Cambrian rift pulses. These zircon U/Pb determinations aid in refining ~1 Ga of the pre-Pangean history of the DRU and of the region.

Analytical methods

Zircon U/Pb geochronological analyses (LA-ICP-Ms) of Mesoproterozoic and Cambrian units in the DRU were performed by the UTChron Laboratory, University of Texas-Austin. Details of sample ages/types, procedures, reference materials, instruments, and data processing/display are provided in Table 4.1.

TABLE 4.1 Analytical information—Zircon U/Pb Data—LA-ICP-Ms. UTChron Laboratory, Jackson School of Geosciences, University of Texas at Austin.

Laboratory and Sample Preparation	
Laboratory name	UTChron, Department of Geological Sciences, University of Texas at Austin
Sample type/mineral	Detrital zircons
Sample preparation	Conventional mineral separation, grain mount on double-sided tape
Laser ablation system	
Make, model, and type	PhotonMachines Analyte.G2 ATLex 300si ArF Excimer laser
Ablation cell and volume	Helex 9 sample cell, volume c.30 cm^3
Laser wavelength	193 nm
Pulse width	≤ 4 ns
Energy	4 mJ, laser
Fluence	1.5–1.6 $J.cm^{-2}$
Repetition rate	10 Hz
Spot size	25–30 μm
Sampling mode	4 preablation shots, 30 sec of baseline data collection, 300 ablation shots, 30 sec washout
Sample cell washout time (s)	<0.5 sec
Carrier gas	He: 1.0 L/min (ultrapure)
Ablation duration/depth	30 secs at ~0.5 μm/sec for depth of 15–17 μm
ICP-Ms Instrument	
Make, model, and type	ThermoFisher Element2 double-focusing magnetic sector ICP-MS
Sample introduction	Ablation aerosol (dry plasma)
RF power	1175–1240 W
Cooling gas	Ar: 16 L/min
Auxiliary gas	Ar: 0.9 L/min
Sample gas	Ar: 0.85–0.9 L/min
Detection system	Secondary electron multiplier (SEM)
Masses measured	202, 204, 206, 207,208, 232, 235, 238
Dwell time	5 Ms (202, 204, 206, 238); 10 Ms (207, 235), 2 Ms (208, 232)
Total integration time	50 Ms per pass; 410 runs with 3 passes per run
Data processing	
Gas blank	25 sec on-peak zero subtracted
Calibration strategy	GJ1 used as primary reference material, run every 5 unknowns, 91500 and Plesovice used as secondary reference material
Reference material info	GJ1 $^{206}Pb/^{238}U$ age 601.7 ± 1.3 Ma (Jackson et al., 2004); Plesovice $^{206}Pb/^{238}U$ age 337.13 ± 0.37 Ma (Slama et al., 2008); 91500 $^{207}Pb/^{206}Pb$ age 1065.4 ± 0.6 Ma (Wiedenbeck et al., 1995)
Data processing package used	Iolite3 (Paton et al., 2011) with VisualAge (Petrus and Kamber, 2012)
Mass discrimination	$^{206}Pb/^{238}U$, $^{207}Pb/^{235}U$ and $^{208}Pb/^{232}U$ normalized to reference material
Common-Pb correction	None applied
Uncertainty level and propagation	Ages are quoted at 2 sigma absolute error, propagation is by quadratic addition. Reproducibility and age uncertainty of reference material are propagated.

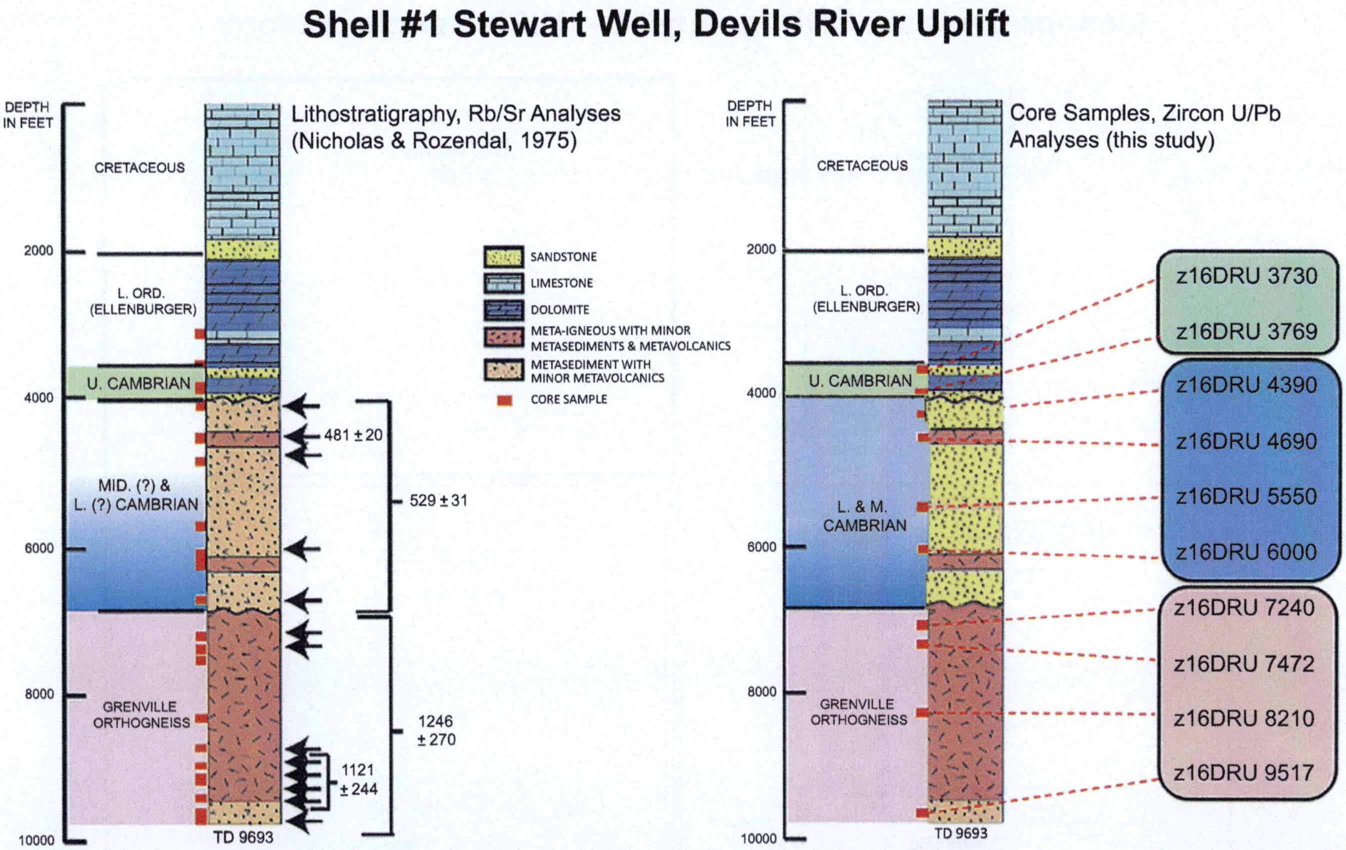

FIGURE 4.6 Shell No. 1 Stewart well, Devils River Uplift, cored lithologic section. Left—sampling sites and Rb/Sr ages (Nicholas and Rozendal, 1975). Right—sampling sites for LA-ICP-Ms analyses, zircon U/Pb (this study); see Figs 4.7–4.9 for data. Source: *Modified from Rodríguez, E., Stockli, D. F. and Dickerson, P. W., 2017, New zircon U/Pb geochronology from the Devils River Uplift – insights into the Neoproterozoic and early Paleozoic evolution of the southern margin of North America (abs): Geological Society of America, South-Central Section, Paper 1–2. PPT: https://gsa.confex.com/gsa/2017SC/webprogram/Paper289330.html; after Nicholas, R. L., and Rozendal, R. A., 1975, Subsurface positive elements within Ouachita foldbelt in Texas and their relation to Paleozoic cratonic margin: American Association of Petroleum Geologists Bulletin, 59, p. 193–216.*

Ages

Previous work

DRU cores long provided the sole evidence for colocated Neoproterozoic and early Cambrian magmatism in the region. Whole-rock Rb/Sr and K/Ar ages were determined for Mesoproterozoic through Cambrian strata in the Shell #1 Stewart well (R. E. Denison *in* Nicholas and Rozendal, 1975; Nicholas and Waddell, 1989), and among those data were Neoproterozoic to early Cambrian ages with direct bearing on Rodinia rift reconstructions (Figs. 4.2, 4.6). New zircon U/Pb analyses of cores from the same well expand on the earlier work; sampled intervals for both studies are indicated in Fig. 4.6. The similarity of ages obtained by the two methods is noteworthy and attests to the precision of the Rb/Sr determinations.

Devils River Uplift—new zircon U/Pb data

Zircons from four Mesoproterozoic felsic orthogneisses (depth of 1828–2954 m) yield uniform U-Pb crystallization ages of ~1230 Ma (Fig. 4.7). The gneisses are of equivalent age and composition to granitic orthogneisses (Comanche Creek, Red Mountain, and Inks Lake) in the Valley Spring Domain of the Llano Uplift, Central Texas (Reese et al., 2000).

Metaigneous basement is overlain by lower-middle Cambrian metasedimentary and metavolcanic rocks (depth of ~1219–1828 m; Fig. 4.8), with maximum depositional ages of ~533–545 Ma. Grenvillian ages predominate in the lower and upper metasediments, with lesser Cryogenian and Granite-Rhyolite Province components.

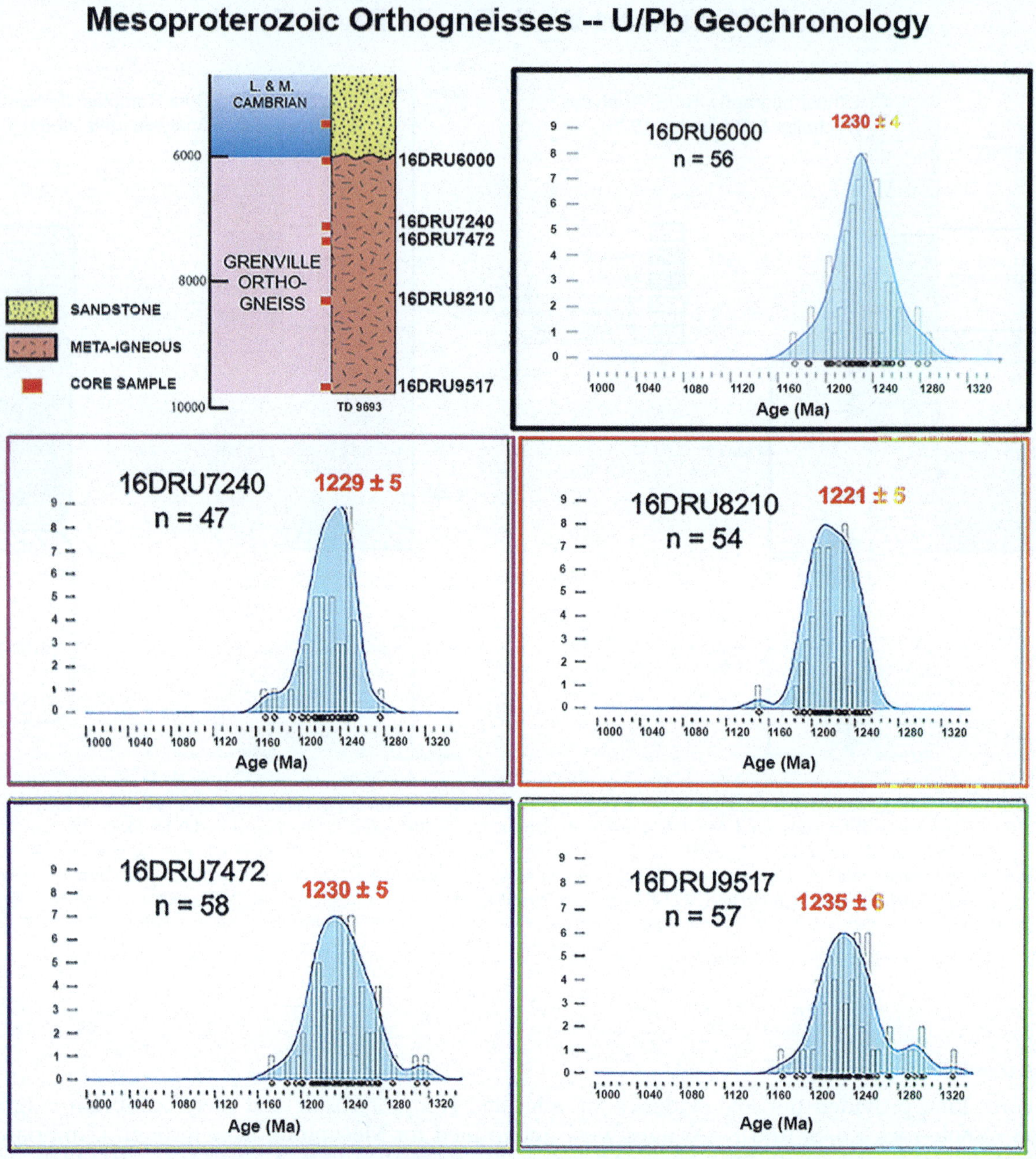

FIGURE 4.7 Mesoproterozoic (Grenvillian) orthogneisses—zircon U/Pb geochronology. Of equivalent age and composition to felsic orthogneisses in the Llano Uplift, Valley Spring Domain. Cored section and sampled intervals, Shell No. 1 Stewart well.

In contrast, the intervening metavolcanic/metasedimentary interval (at 1430 m) exhibits notable Cryogenian and early Cambrian peaks (780–700 Ma and 580–520 Ma, respectively), as well as a few Archean ages.

In unmetamorphosed upper Cambrian sandstones (Fig. 4.9), younger Grenvillian (1070–1030 Ma) detritus predominates, with less of the older Grenvillian (1275–1192 Ma), and there is substantial input from the Granite-Rhyolite Province (1435–1340 Ma). Rare early Cambrian and Cryogenian grains are present as well.

Devils River Uplift and Marathon/Solitario Basin equivalents

Cambrian and Ordovician sedimentary rocks in the Marathon/Solitario basin, about 75 km farther west along the margin, provide complementary data on rift-related volcanism. Locally sourced, unmetamorphosed

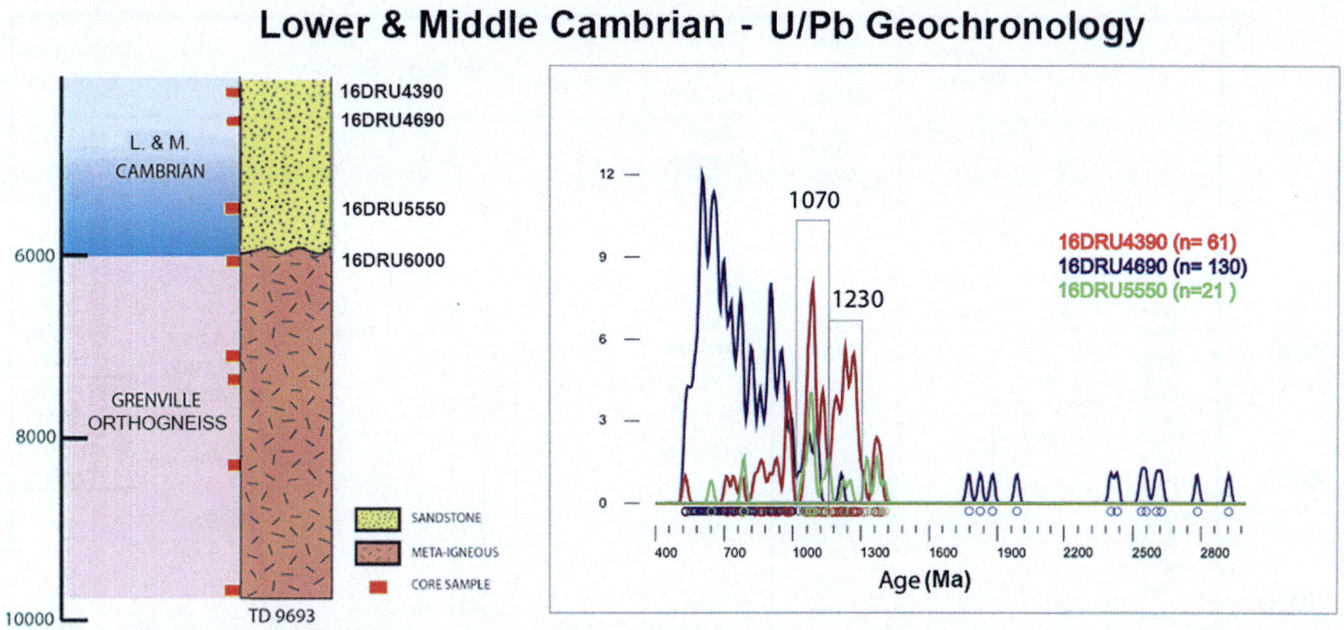

FIGURE 4.8 Lower to middle Cambrian metasediments/metaigneous rocks—zircon U/Pb geochronology. Cored section and sampled intervals, Shell No. 1 Stewart well.

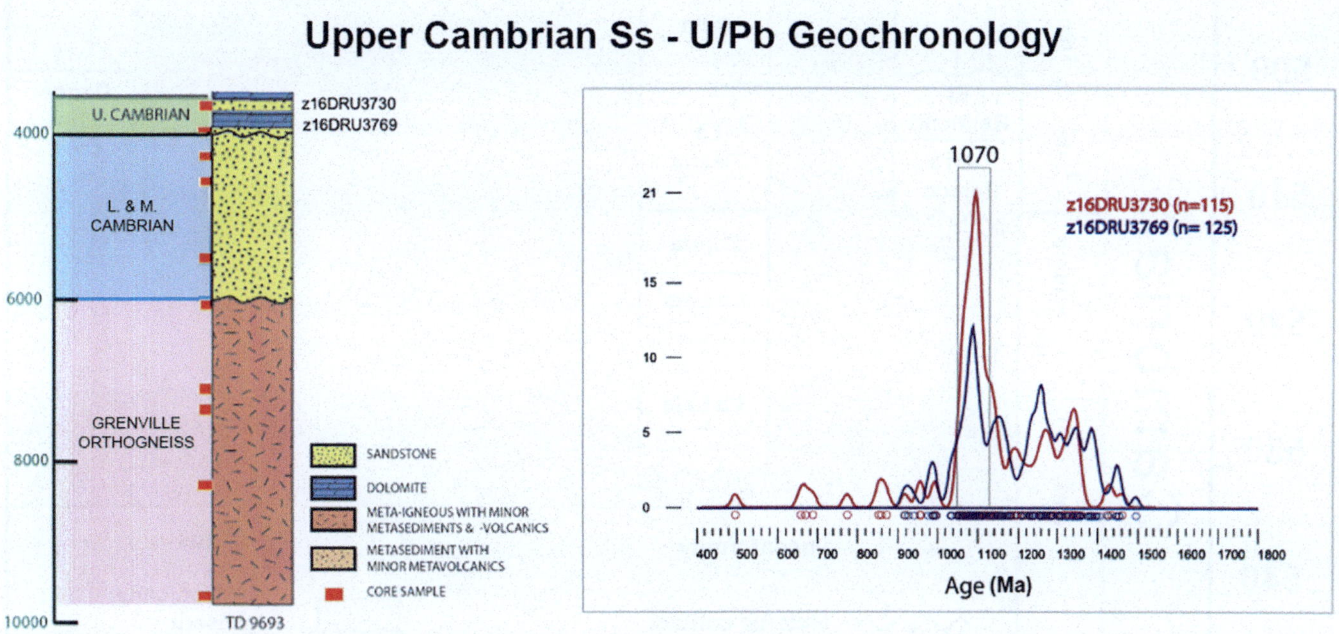

FIGURE 4.9 Upper Cambrian sandstones—zircon U/Pb geochronology. Cored section and sampled intervals, Shell No. 1 Stewart well.

Cryogenian (706 Ma, average crystallization age) volcanic boulders in shelf-derived Ordovician conglomerates (Marathon and Ft. Peña Formations; Fig. 4.10) exhibit intraplate rift geochemical character (geochronological and geochemical details are given in Hanson et al., 2016). The lack of metamorphism and of tectonic fabric precludes their derivation from underlying 1350- to 1000-Ma basement rocks (e.g., King and Flawn, 1953; Muehlberger et al., 1967; Flawn and Muehlberger, 1970). Sandstone and bentonitic shale of those same formations also contain early Cambrian zircons (580−520 Ma; Dickerson et al., 2014, 2022), albeit fewer than in DRU strata.

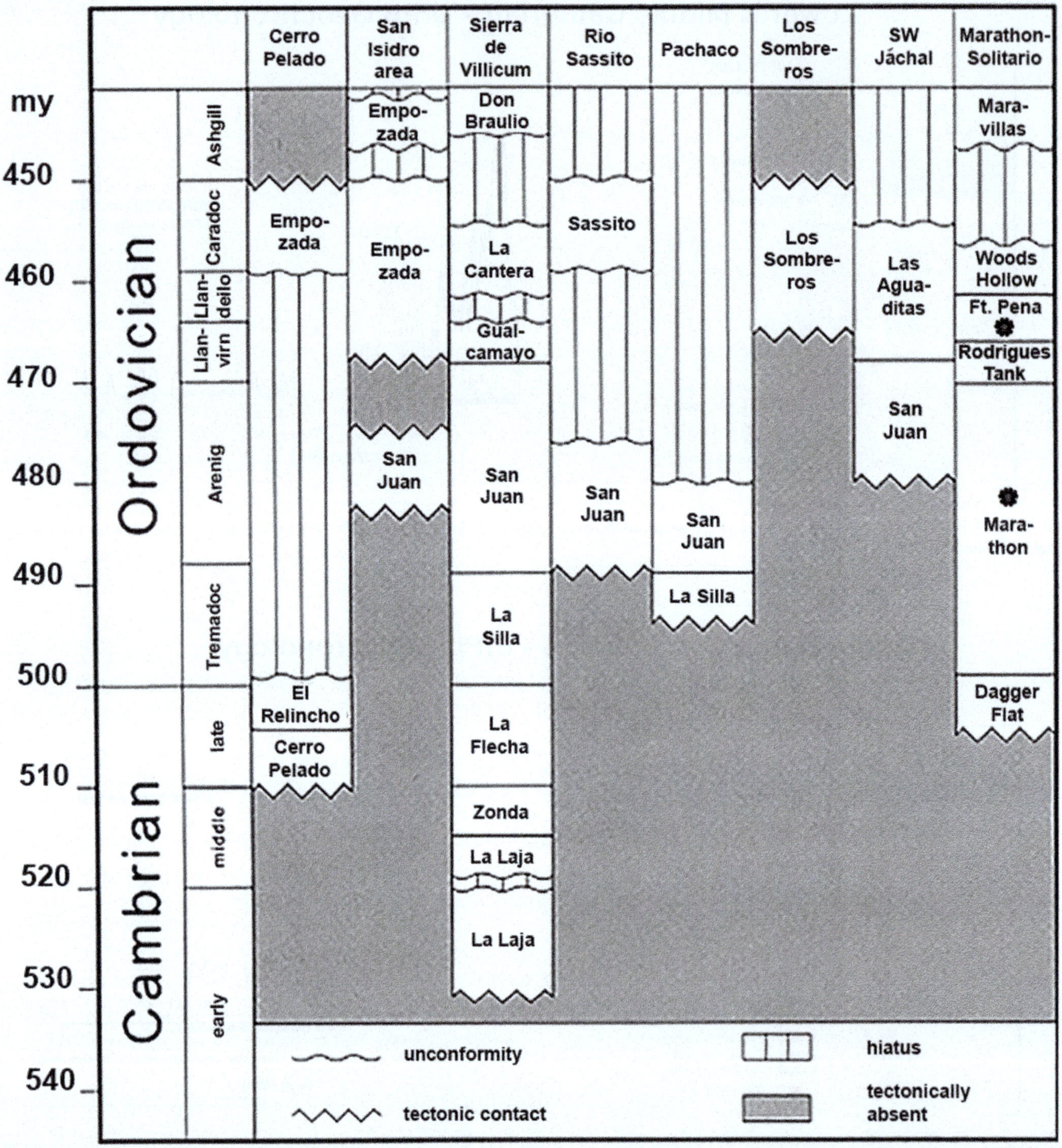

FIGURE 4.10 Correlation chart—Cuyania and Marathon/Solitario basin. Locations of Cuyania stratigraphic sections are shown in Fig. 4.12. In Marathon/Solitario basin, in Marathon and Ft. Peña Formations indicates Cryogenian rift volcanic boulder localities. Source: *Modified from Dickerson, P. W., and Keller, M., 1998, The Argentine Precordillera: its odyssey from the Laurentian Ouachita margin toward the Sierras Pampeanas of Gondwana, in R. J. Pankhurst and C. W. Rapela, ed., The Proto-Andean margin of Gondwana: The Geological Society (London) Special Publication 142, p. 85–105.*

Devils River Uplift—Cuyania correlatives

For all dated intervals in the Shell #1 Stewart well on the DRU, there are geochronological and compositional counterparts in the Cuyania terrane of western Argentina; notable correlatives are summarized here (see also tabulation in Dickerson et al., 2018). Our new ages (LA-ICP-Ms) and emerging zircon U/Pb data (SHRIMP) from both southern Laurentia and Cuyania further confirm their consanguinity and constrain the timing of departure of Cuyania from the Laurentian margin during Neoproterozoic to Cambrian rifting of Rodinia (Figs. 4.10, 4.11).

Mesoproterozoic—Granite-Rhyolite Province (1500–1350 Ma)

Devils River Uplift

Lower to middle Cambrian metavolcanics and upper Cambrian sandstone in the Shell #1 Stewart well contain magmatic zircons—clustered at 1500–1373 Ma and 1435–1340 Ma, respectively—which reflect derivation from the southern Granite-Rhyolite Province (GRP; Fig. 4.5; Rodríguez et al., 2017). In Cuyania, Neoproterozoic through Cambrian outcrops in the Precordillera and in the Western Sierras Pampeanas (WSP) have yielded suites of the same provenance (Fig. 4.11). At Sierra Chica de Zonda, eastern Precordillera, 1559- to 1316-Ma grains are common in a sandstone interbed within Laurentian carbonate rocks of the middle Cambrian La Laja Fm (Finney et al., 2005a,b; Finney, 2007). Analogous populations are known from the WSP at Sierra Pie de Palo and at Sierra de Maz.

Sierra Pie de Palo

GRP-derived detritus has been dated from two stratigraphic intervals—the Neoproterozoic Difunta Correa metasedimentary sequence and the Neoproterozoic to lower Cambrian Caucete Group. Magmatic zircons from the Difunta Correa are characterized by a strong peak at 1387 Ma for a paramphibolite and a notable peak at 1425 Ma for a metaconglomerate (Rapela et al., 2016, Fig. 4.3). Naipauer et al. (2010) emphasize the GRP component in metasiliciclastics of the Neoproterozoic to lower Cambrian Caucete Group: the El Quemado Fm (1434–1300 Ma) and the El Desecho Fm (1491–1326 Ma).

Sierra de Maz

Zircons from an equivalent source (1500–1303 Ma) are common in the metasedimentary strata of central and southeastern Sierra de Maz (Rapela et al., 2016, their Fig. 4.3). Underscoring the Laurentian origin of the WSP, those authors note the absence of Paleoproterozoic grains (2260–2020 Ma), which—had they been present—might have come from the Gondwanan Rio de la Plata craton. Data that preclude Gondwanan sources for the 1500- to 1350-Ma population are discussed in detail by Naipauer et al. (2010) and Rapela et al. (2016).

Mesoproterozoic—Grenville Orogen (1230–1000 Ma)

Devils River Uplift

Samples of felsic gneisses and metarhyolite from the lower half of the cored interval (3693 ft/1231 m) in the DRU well yielded a uniform U/Pb zircon age of 1230 ± 5 Ma. The gneisses are of equivalent age and composition to young granitic orthogneisses in the Grenvillian Valley Spring Domain of the Llano region, Central Texas (1239–1232 Ma; Reese et al., 2000). In Cuyania, detrital zircons dated from 1228 to 1240 Ma have been recovered from schists and metasandstones in the Neoproterozoic to lower Cambrian Caucete Group, Pie de Palo (Naipauer et al., 2010). The 1230-Ma population has not been reported elsewhere in the Laurentian WSP.

Younger Grenvillian grains, including volcanic zircons with ages from 1070 to 1030 Ma, predominate in the upper middle to upper Cambrian metamorphic and sedimentary rocks of the DRU. The same Grenvillian population, with youngest peak ages ranging from 1081 to 1038 Ma, is dominant in metasiliciclastics of the WSP—for example, the Neoproterozoic Difunta Correa sequence and the lower Cambrian La Paz Fm of Sierra Pie de Palo (Vujovich et al., 2004; Naipauer et al., 2010; Rapela et al., 2016). Farther north in the southern Puna, quartz-rich metasandstones of the Famabalasta Fm at Sierra de Chuquisaca, Antofalla, yielded a sizable suite of 1014–1068 Ma grains (Rapela et al., 2016).

Important for reconstructing the early Laurentian history of Cuyania are extensive younger Grenvillian ultramafic bodies of the Pecos Mafic Intrusive Complex, ~75 km north-northwest of DRU, and in Sierra de Maz in the WSP (Fig. 4.11). U/Pb data for zircons from gabbronorite in Sierra de Maz indicate emplacement at

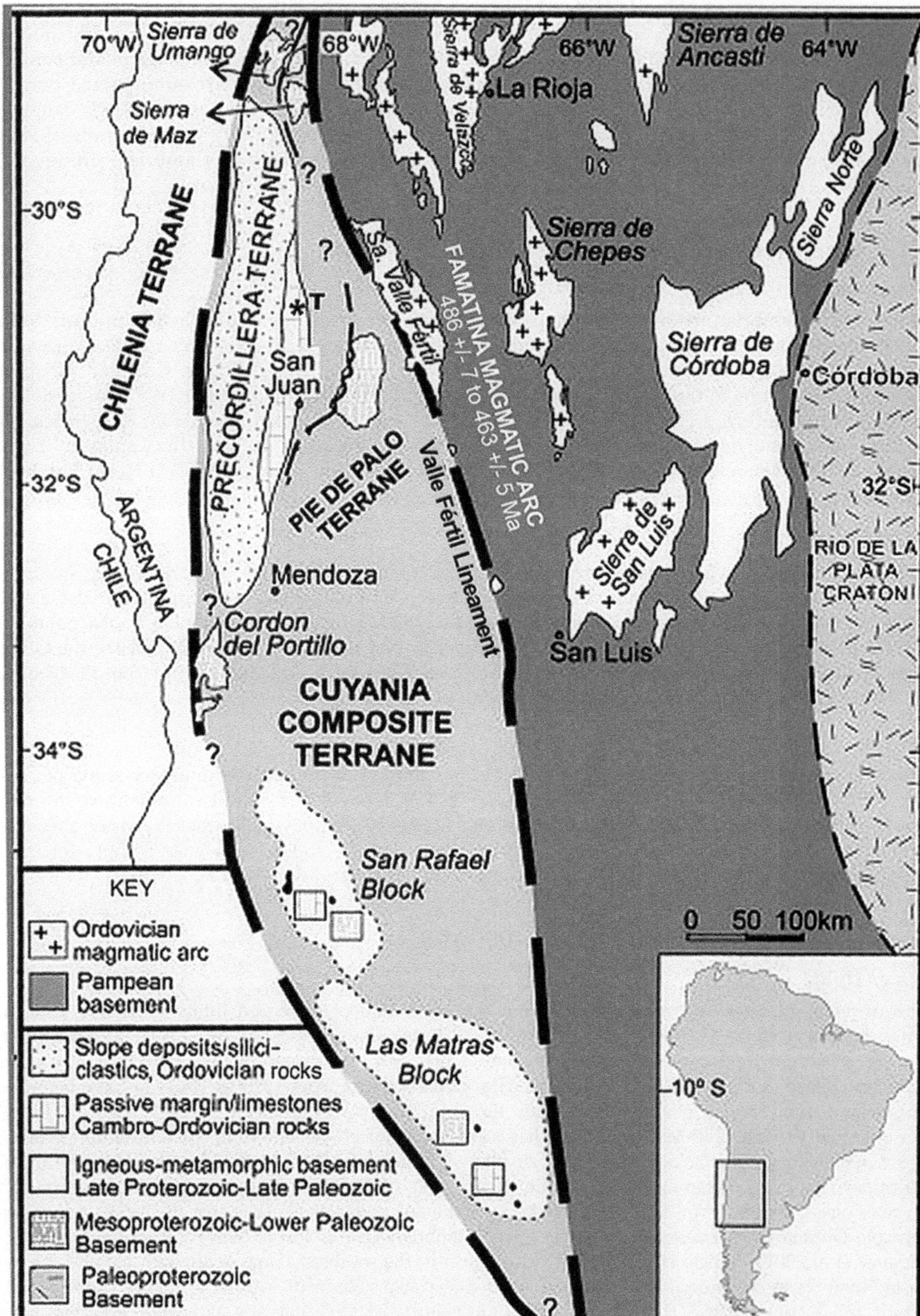
70°W
Sierra de Umango
68°W
Sierra de Velazco
La Rioja
66°W
Sierra de Ancasti
64°W
Sierra de Maz
-30°S
CHILENIA TERRANE
PRECORDILLERA TERRANE
Sa. Valle Fértil
FAMATINA MAGMATIC ARC
486 +/- 7 to 463 +/- 5 Ma
Sierra de Chepes
Sierra Norte
Sierra de Córdoba
Córdoba
San Juan
T
PIE DE PALO TERRANE
Valle Fértil Lineament
-32°S
32°S
ARGENTINA
CHILE
Sierra de San Luis
RIO DE LA PLATA CRATON
Mendoza
Cordon del Portillo
San Luis
CUYANIA COMPOSITE TERRANE
-34°S
San Rafael Block
KEY
Ordovician magmatic arc
Pampean basement
Slope deposits/silici-clastics, Ordovician rocks
Passive margin/limestones Cambro-Ordovician rocks
Igneous-metamorphic basement Late Proterozoic-Late Paleozoic
Mesoproterozoic-Lower Paleozoic Basement
Paleoproterozoic Basement
Las Matras Block
0 50 100km
-10° S

(*Continued*)

1070 ± 41 Ma (SHRIMP, zircon cores; Casquet et al., 2004, 2005). For gabbro and orthopyroxenite of the Pecos complex, an age range of 1072 to 1163 Ma was initially reported (TIMS, U/Pb, zircon, and titanite; Keller et al., 1989); further analyses are in progress. Based on magmatic chronology and geochemistry in the region, an age of 1080 ± 10 Ma appears most probable (C. G. Barnes, written communication, 2022). Major element and rare earth element data are markedly similar for the Pecos and Sierra de Maz bodies (Kargi and Barnes, 1995; Barnes et al., 1999; Casquet et al., 2004, 2005; Dickerson, 2012).

Neoproterozoic-Early Cambrian Rifting

Devils River Uplift

Magmatic zircons recovered from lower to middle Cambrian metasiliciclastics and upper Cambrian sandstones in DRU record a range of igneous activity on the southern Laurentian margin—Cryogenian (780–700 Ma) and early Cambrian (580–520 Ma).

Western Sierras Pampeanas

Cryogenian-early Cambrian magmatism in Cuyania is evidenced by plutons in the WSP. The Grenvillian basement of Sierra de Maz was intruded not only by 846–842 Ma and 774 ± 6 Ma A-type granites but also by a ~570 Ma carbonatite complex of Laurentian character (Baldo et al., 2006; Rapela et al., 2016 and sources therein). The crystallization age for the A-type granite protolith of the Quebrada Desecho orthogneiss in Sierra Pie de Palo is the same as that for the compositionally similar younger granite suite in Sierra de Maz (774 ± 6 Ma; Naipauer et al., 2010). Elsewhere in Cuyania, felsic tuff dated at 537.3 ± 0.9 Ma is found within turbidites of the Neoproterozoic Puncoviscana Fm in the Puna (Escayola et al., 2011).

Cryogenian-early Cambrian magmatic zircons are present in the metasediments of Sierra Pie de Palo in the WSP: the Neoproterozoic Difunta Correa sequence (587 ± 25 Ma; Rapela et al., 2016), as well as the El Quemado and El Desecho Fms of the Neoproterozoic to lower Cambrian Caucete Group (697–532 Ma; Naipauer et al., 2010). In contrast, at Sierra de Maz, where zircons derived from 846- to 774-Ma granitic intrusions might be expected, they have not been encountered in the metasedimentary strata (Rapela et al., 2016).

In summary, new zircon U/Pb determinations from the DRU in the subsurface of central southern Laurentia and data from correlative strata in outcrops of Cuyania affirm their coevolution.

Conclusions

Zircon U/Pb geochronological results from Mesoproterozoic to upper Cambrian metamorphic and sedimentary rocks in the DRU significantly extend the record of Cryogenian onset of intraplate extensional magmatism on the southern margin of Laurentia (Fig. 4.12). Abundant early Cambrian zircons in Cambrian strata are the evidence of episodic later-phase extension.

Temporal and compositional correlatives in the Cuyania rifted terrane of western Argentina and the DRU of West Texas refine the timing of the separation of Cuyania from the Marathon-Ouachita subbasin of western Iapetus.

The abundance of rift-vintage rocks and detritus documented in this investigation is relevant for later events as well. Uplifted and eroded during late Paleozoic plate convergence and assembly of Pangaea, the DRU became a proximal southern source for rift detritus that was deposited in foreland basins of the orogenic belt.

◀ FIGURE 4.11 The Laurentia-derived Cuyania composite terrane of western Argentina, including the Precordillera and Pie de Palo. Discussion in text of the Western Sierras Pampeanas, Sierra Pie de Palo (PdP), Puna, and Precordillera (Sierra Chica de Zonda). Source: *Modified from Dickerson, P. W., Hall, B. R., Stockli, D. F., Stockli, L. D., Hanson, R. E., Fanning, C. M., and et al., 2022, Pre-Pangean evolution of central southern Laurentia: insights from zircon U/Pb geochronology, Marathon-Solitario fold-thrust belt, west Texas, in Whitmeyer, S. J., Williams, M. L., Kellett, D. A., and Tikokff, B., eds., Laurentia: turning Points in the Evolution of a Continent: Geological Society of America, Memoir 220, chapter 20. https://doi.org/10.1130/2022(20) (in press); after Naipauer, M., Vujovich, G. I., Cingolani, C. A. and McClelland, W. C., 2010, Detrital zircon analysis from the Neoproterozoic–Cambrian sedimentary cover (Cuyania terrane), Sierra Pie de Palo, Argentina: Evidence of a rift and passive margin system?: Journal of South American Earth Sciences, 30, p. 84–96. https://doi.org/10.1016/j.jsames.2009.10.001.*

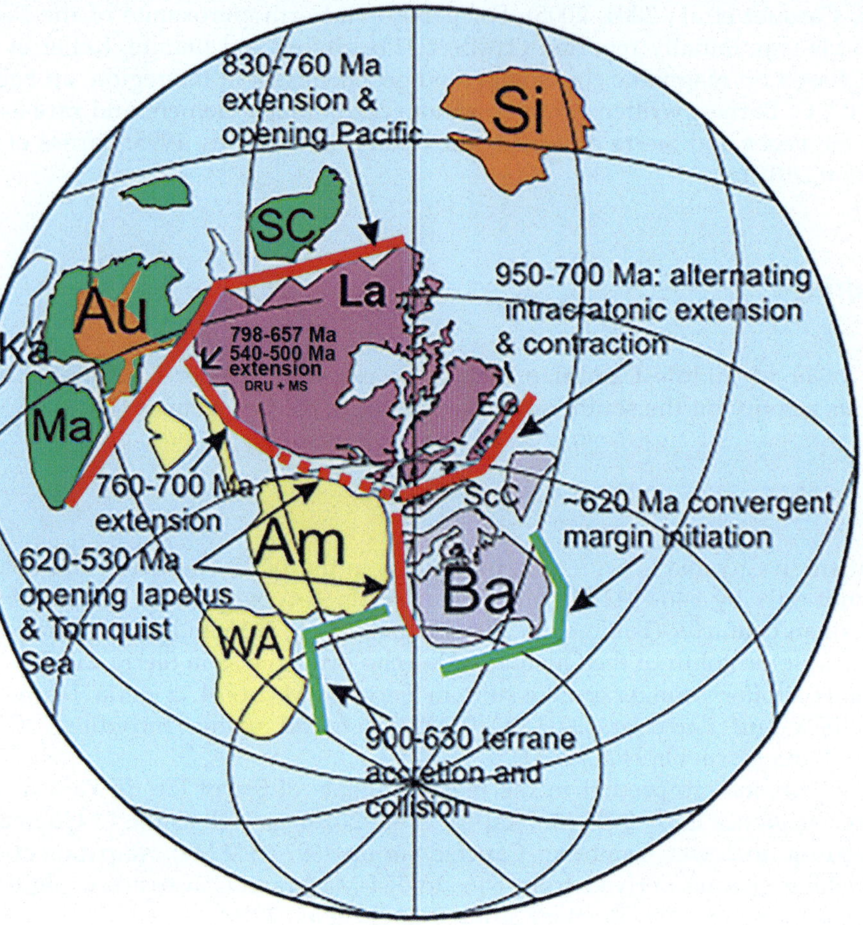

FIGURE 4.12 Rodinia rifting (red bands on figure). Cryogenian to early Cambrian rift magmatic phases, documented along the eastern edge of Laurentia, are increasingly recognized on the central southern margin. Zircon U/Pb ages (LA-ICP-Ms and SHRIMP) from the Devils River Uplift (DRU) and nearby Marathon/Solitario basin (Ms) are combined in the age ranges shown here. *Am*, Amazonia; *Au*, Australia; *Ba*, Baltica; *C*, Cuyania; *Ka*, Kalahari; *La*, Laurentia; *Ma*, Mawson; *SC*, South China; *Si*, Siberia; *WA*, West Africa. Source: *Modified from Dickerson, P. W., Hall, B. R., Stockli, D. F., Stockli, L. D., Hanson, R. E., Fanning, C. M., and et al., 2022, Pre-Pangean evolution of central southern Laurentia: insights from zircon U/Pb geochronology, Marathon-Solitario fold-thrust belt, west Texas, in Whitmeyer, S. J., Williams, M. L., Kellett, D. A., and Tikokff, B., eds., Laurentia: Turning Points in the Evolution of a Continent: Geological Society of America, Memoir 220, chapter 20. https://doi.org/10.1130/2022(20) (in press); after Cawood, P. A., Nemchin, A. A., Strachan, R., Prave, T. and Krabbendam, M. 2007, Sedimentary basin and detrital zircon record along East Laurentia and Baltica during assembly and breakup of Rodinia: Journal of the Geological Society, London, 164, p. 257–275.*

Acknowledgments

Essential for the execution of this study was the stewardship of the samples by the core repository of the Bureau of Economic Geology (University of Texas-Austin). The authors thank James Donnelly for retrieving and displaying the Shell well cores for investigation. Lisa D. Stockli of the UTChron Laboratory is gratefully acknowledged for contributing the tabulation of LA-ICP-Ms analytical methods. Our sincere appreciation to the editors—Nicola Scarselli, Jurgen Adam, and Domenico Chiarella—for the invitation to submit our work for this volume; in particular, the encouragement and patience of Nicola Scarselli is acknowledged. The interpretation presented here benefited from fruitful discussions with William A. Thomas, Timothy F. Lawton, and the late Albert W. Bally—all long intrigued by the litho- and chronostratigraphy of the Devils River Uplift.

References

Baldo, E., Casquet, C., Pankhurst, R.J., Galindo, C., Rapela, C.W., Fanning, C.M., et al., 2006. Neoproterozoic A-type magmatism in the Western Sierras Pampeanas (Argentina): evidence for Rodinia break-up along a proto- Iapetus rift? Terra Nova 18, 388–394.

Bankey, V., 2016. Texas magnetic and gravity maps and data: a website for distribution of data: U.S. Geological Survey, Data Series 232, 2 maps. [Page last modified: Monday, November 28, 2016] https://pubs.usgs.gov/ds/2006/232/.

Barnes, C.G., Shannon, W.M., Kargi, H., 1999. Diverse Mesoproterozoic basaltic magmatism in west Texas. Rocky Mountain Geology 34, 263–273. Available from: https://doi.org/10.2113/34.2.

Casquet, C., Pankhurst, R.J., Rapela, C.W., Galindo, C., Dahlquist, J., Baldo, E., et al., 2004. Grenvillian massif-type anorthosites in the Sierras Pampeanas. Journal of the Geological Society (London) 162, 9–12.

Casquet, C., Pankhurst, R.J., Rapela, C.W., Fanning, C.M., Galindo, C., Baldo, E., et al., 2005. The Maz suspect terrane: a new Proterozoic domain in the Western Sierras Pampeanas (abs.), *in* R. J. Pankhurst and G. D. Veiga, eds., Gondwana 12: Geological and biological heritage of Gondwana: Córdoba, Academia Nacional de Ciencias, Gondwana 12 Conference, Abstracts, p. 92.

Cawood, P.A., Nemchin, A.A., Strachan, R., Prave, T., Krabbendam, M., 2007. Sedimentary basin and detrital zircon record along East Laurentia and Baltica during assembly and breakup of Rodinia. Journal of the Geological Society, London 164, 257–275.

Denison, R.E., Burke, W.H., Jr., Hetherington, E.A. and Otto, J.B., 1971. Basement rock framework of parts of Texas, southern New Mexico and northern Mexico, The Geologic Framework of the Chihuahua Tectonic Belt: West Texas Geological Society, Publication 71–59, p. 3–14.

Dickerson, P.W., 2012. The circum-Laurentian carbonate bank, the western Ouachita-Cuyania basin, and the prodigal Llanoria landmass. In: Derby, J.R., Fritz, R.D., Longacre, S.A., Morgan, W.A., Sternbach, C.A. (Eds.), The Great American Carbonate Bank: The Geology and Economic Resources of the Cambrian–Ordovician Sauk Megasequence of Laurentia, 98. American Association of Petroleum Geologists Memoir, pp. 959–984.

Dickerson, P.W., Keller, M., 1998. The Argentine Precordillera: its odyssey from the Laurentian Ouachita margin toward the Sierras Pampeanas of Gondwana. In: Pankhurst, R.J., Rapela, C.W. (Eds.), The Proto-Andean margin of Gondwana: The Geological Society (London), 142. Special Publication, pp. 85–105.

Dickerson, P.W., Stockli, D.F., Hanson, R.E., and Roberts, J.M., 2014. Emerging data on Rodinia rifting from Marathon orogenic belt, southern Laurentia: Geological Society of America, Abstracts with Program, v. 46, no. 6, p. 371. PPT: https://gsa.confex.com/gsa/2014AM/webprogram/Paper250007.html.

Dickerson, P.W., Rodríguez, E., Stockli, D.F., Hanson, R.E., Roberts, J.M., Fanning, C.M., 2018. Mesoproterozoic to late Paleozoic supercontinent collisions and fragmentation – the record from south-central Laurentia and insights from zircon U/Pb data (abstract). Geological Society of America, South-Central Section Paper 9-1. PPT. Available from: https://gsa.confex.com/gsa/2018SC/meetingapp.cgi/Paper/310310.

Dickerson, P.W., Hall, B.R., Stockli, D.F., Stockli, L.D., Hanson, R.E., Fanning, C.M., et al., 2022. Pre-Pangean evolution of central southern Laurentia: insights from zircon U/Pb geochronology, Marathon-Solitario fold-thrust belt, west Texas, in Whitmeyer, S. J., Williams, M. L., Kellett, D. A., and Tikokff, B., eds., Laurentia: Turning Points in the Evolution of a Continent: Geological Society of America, Memoir 220, chapter 20. https://doi.org/10.1130/2022(20) (in press).

Escayola, M.P., van Staal, C.R., Davis, W.J., 2011. The age and tectonic setting of the Puncoviscana Formation in northwestern Argentina: an accretionary complex related to early Cambrian closure of the Puncoviscana Ocean and accretion of the Arequipa–Antofalla block. Journal of South American Earth Sciences 32, 438–459. Available from: https://doi.org/10.1016/j.jsames.2011.04.013.

Finney, S.C., 2007. The parautochthonous Gondwanan origin of the Cuyania (greater Precordillera) terrane of Argentina: A reevaluation of evidence used to support an allochthonous Laurentian origin. Geológica Acta 5 (2), 33.

Finney, S., Grove, M., Gehrels, G., Peralta, S. and Heredia, S., 2005a. Early Mesoproterozoic zircons from conglomerate clasts and sandstones in Cambrian and Ordovician siliciclastic strata of the Argentine Precordillera (abs.), in R. J. Pankhurst and G.D. Veiga, eds., Gondwana 12: Geological and biological heritage of Gondwana: Córdoba, Academia Nacional de Ciencias, Gondwana 12 Conference, Abstracts, p. 155.

Finney, S., Peralta, S., Gehrels, G. and Marsaglia, K., 2005b. U/Pb geochronology of detrital zircons from middle Cambrian sandstones, Argentine Precordillera and interpretations of sandstone provenance (abs.), in R. J. Pankhurst and G. D. Veiga, eds., Gondwana 12: Geological and biological heritage of Gondwana: Córdoba, Academia Nacional de Ciencias, Gondwana 12 Conference, Abstracts, p. 157.

Flawn, P.T., Goldstein, A., King, P.B., and Weaver, C.E., 1961. The Ouachita system: University of Texas Publication 6120, p. 401.

Flawn, P.T., Muehlberger, W.R., 1970. The Precambrian of the United States of America: south-central United States, The Precambrian, v. 4. Interscience, New York, pp. 72–144.

Hanson, R.E., Roberts, J.M., Dickerson, P.W., Fanning, C.M., 2016. Cryogenian intraplate magmatism along the buried southern Laurentian margin: Evidence from volcanic clasts in Ordovician strata, Marathon uplift, west Texas. Geology 44 (7), 539–542. Available from: https://doi.org/10.1130/G37889.1.

Jackson, S., Pearson, N., Griffin, W., Belousova, E., 2004. The application of laser ablation-inductively coupled plasma-mass spectrometry to in situ U-Pb zircon geochronology. Chemical Geology 211 (1–2), 47–69.

Kargi, H., Barnes, C.G., 1995. A Grenville-age layered intrusion in the subsurface of west Texas: Petrology, petrography, and possible tectonic setting. Canadian Journal of Earth Sciences 32, 2159–2166. Available from: https://doi.org/10.1139/e95-168.

Karlstrom, K.E., Amato, J.M., Williams, M.L., Heizler, M., Shaw, C.A., Read, A.S., et al., 2004. Proterozoic tectonic evolution of the New Mexico region: a synthesis. In: Mack, G.H., Giles, K.A. (Eds.), The geology of New Mexico: A geologic history. New Mexico Bureau of Geology, Socorro, New Mexico, pp. 1–34.

Keller, G.R., Peeples, W.J., 1985. Regional gravity and magnetic anomalies in West Texas. In: Dickerson, P.W., Muehlberger, W.R. (Eds.), Structure and tectonics of Trans-Pecos Texas, 85-81. West Texas Geological Society, Publication, pp. 101–105.

Keller, G.R., Hills, J.M., Baker, M.R., Wallin, E.T., 1989. Geophysical and geochronological constraints on the extent and age of mafic intrusions in the basement of west Texas and eastern New Mexico. Geology 17, 1049–1052. Available from: https://doi.org/10.1130/0091-7613(1989)017 < 1049:GAGCOT > 2.3.CO;2.

King, P.B., and Flawn, P.T., 1953, Geology and mineral deposits of the Precambrian rocks of the Van Horn area, Texas: University of Texas, Bureau of Economic Geology, Publication 5301, 218 p.

Mosher, S., 1998. Tectonic evolution of the southern Laurentian Grenville orogenic belt. Geological Society of America Bulletin 110, 1357–1375. Available from: https://doi.org/10.1130/0016-7606(1998)110 < 1357:TEOTSL > 2.3.CO;2.

Muehlberger, W.R., Denison, R.E., Lidiak, E.G., 1967. Basement rocks in the continental interior of the United States. American Association of Petroleum Geologists Bulletin 51, 2351–2380.

Naipauer, M., Vujovich, G.I., Cingolani, C.A., McClelland, W.C., 2010. Detrital zircon analysis from the Neoproterozoic–Cambrian sedimentary cover (Cuyania terrane), Sierra Pie de Palo, Argentina: Evidence of a rift and passive margin system? Journal of South American Earth Sciences 30, 84–96. Available from: https://doi.org/10.1016/j.jsames.2009.10.001.

Nicholas, R.L., Rozendal, R.A., 1975. Subsurface positive elements within Ouachita foldbelt in Texas and their relation to Paleozoic cratonic margin. American Association of Petroleum Geologists Bulletin 59, 193–216.

Nicholas, R.L., Waddell, D.E., 1989. The Ouachita system in the subsurface of Texas, Arkansas, and Louisiana. In: Hatcher, R.D., Jr, Thomas, W.A., Viele, G.W. (Eds.), The Geology of North America, Volume F-2. Geological Society of America, The Appalachian-Ouachita Orogen in the United States, pp. 661–672.

Paton, C., Hellstrom, J., Paul, B., Woodhead, J., Hergt, J., 2011. Iolite: freeware for the visualisation and processing of mass spectrometric data. Journal of Analytical Atomic Spectrometry 26 (12), 2508–2518.

Petrus, J., Kamber, B., 2012. VizualAge: a novel approach to laser ablation ICP-MS U-Pb geochronology data reduction. Geostandards and Geoanalytical Research 36 (3), 247–270.

Rapela, C.W., Verdecchia, S.O., Casquet, C., Pankhurst, R.J., Baldo, E.G., Galindo, C., et al., 2016. Identifying Laurentian and SW Gondwana sources in the Neoproterozoic to early Paleozoic metasedimentary rocks of the Sierras Pampeanas: Paleogeographic and tectonic implications. Gondwana Research 32, 193–212. Available from: https://doi.org/10.1016/j.gr.2015.02.010.

Reese, J.F., Mosher, S., Connelly, J., Roback, R., 2000. Mesoproterozoic chronostratigraphy of the southeastern Llano uplift, central Texas. Geological Society of America Bulletin 112 (2), 278–291. Available from: https://doi.org/10.1130/0016-7606(2000)112<278:MCOTSL>2.0.CO;2.

Rodríguez, E., Stockli, D.F. and Dickerson, P.W., 2017. New zircon U/Pb geochronology from the Devils River Uplift – insights into the Neoproterozoic and early Paleozoic evolution of the southern margin of North America (abs): Geological Society of America, South-Central Section, Paper 1–2. PPT: https://gsa.confex.com/gsa/2017SC/webprogram/Paper289330.html.

Slama, J., Kosler, J., Condon, D.L., Crowley, J.L., Gerdes, A., Hanchar, J.M., et al., 2008. Plesovice zircon - a new natural reference material for U-Pb and Hf isotopic microanalysis. Chemical Geology 241 (1), 1–35.

Thomas, W.A., 2014. A mechanism for tectonic inheritance at transform faults of the Iapetan margin of Laurentia: Geological Association of Canada. Geoscience Canada, Harold Williams Series 41, 321–344. Available from: https://doi.org/10.12789/geocanj.2014.41.048.

Vujovich, G.I., van Staal, C.R., Davis, W., 2004. Age constraints on the tectonic evolution and provenance of the Pie de Palo complex, Cuyania composite terrane, and the Famatinian orogeny in the Sierra Pie de Palo, San Juan, Argentina. Gondwana Research 7, 1041–1056. Available from: https://doi.org/10.1016/S1342-937X(05)71083-2.

Wiedenbeck, M., Alle, P., Corfu, F., Griffin, W., Meier, M., Oberli, F., et al., 1995. Three natural zircon standards for U-Th-Pb, Lu-Hf, trace element and REE analyses. Geostandards Newsletter 19 (1), 1–23.

CHAPTER

5

Observations, predictions, and implications of dynamic topography through space and time

Jonathan D. Paul

Department of Earth Sciences, Royal Holloway, University of London, Egham, United Kingdom

Introduction

Although an idea first proposed nearly 100 years ago (Pekeris, 1935), the maintenance of surface topography from a sublithospheric origin has only attracted substantial research interest since the late 1980s (e.g., Hager et al., 1985; Hager and Richards, 1989; Lithgow-Bertelloni and Silver, 1998; Conrad and Husson, 2009; Flament et al., 2013; Hoggard et al., 2016). Dynamic topography—variously non-isostatic topography, dynamically supported topography, or "dynamic topography," (with quotation marks)—essentially refers to broad-wavelength (10 s–1000 s km), relatively low-amplitude (typically on the order of 1–2 km, but consensus is limited, e.g., Molnar et al., 2015), and long-lived (Myr) topography arising from spatiotemporally varying patterns of subplate density anomalies (typically asthenospheric mantle convection). In this sense, it is "dynamic" as the density-driving mass anomalies are moving on geologically long timescales. Fig. 5.1 summarizes potential mechanisms that generate dynamic topography. Over the last decade, there has been growing recognition of the influence of dynamic topography on surface processes and geophysical observations including stratigraphic development of sedimentary basins, landscape evolution, oceanic/atmospheric circulation, long-term sea-level changes, and the long-wavelength gravitation field (Conrad and Gurnis, 2003; Poore et al., 2006; Spasojevic and Gurnis, 2012; Braun et al., 2013; Paul et al., 2014; Rovere et al., 2014; Cao et al., 2019; Hoggard et al., 2021).

Several permutations of the basic definition exist; dynamic topography has been described, variously, as:

- "[T]he lithosphere's surface deflection induced by sub-lithospheric mantle convection driven by buoyancy forces" (Hager and Richards, 1989);
- "... that topography due to flow within the mantle ..." (Flament et al., 2013);
- "[T]he deflection of a boundary between two different layers due to tractions or stresses normal to the plane of the boundary that are imposed by forces acting in either layer" (Eakin and Lithgow-Bertelloni, 2018);
- "... surface subsidence or uplift induced by vertical stress originating from sub-lithospheric mantle convection" (Chang and Liu, 2019).

These discrepancies, at first glance minor, have stimulated contentious and ongoing scientific debate, reflecting their origin in the different scientific backgrounds of workers interrogating dynamic support. The structure of this chapter reflects this: the Section "Observational estimates" considers "top-down" approaches of piecing together a database of geophysical and geological observables at the surface, which might yield useful insights into, and constraints on, mantle convection. The Section "Predictions" details modeling approaches that are "bottom-up" in the sense of starting with some form of density field for the mantle to predict dynamic topography and relative surface motions, sometimes back through geological time. These models can be calibrated, and predictions validated, against the geological record: both approaches are therefore closely interlinked and valuable for (1) understanding how "the whispers of the mantle" are manifest in the rock record, sea-level changes, oceanic circulation, and atmospheric dynamics; (2) constraining and better understanding the temporal evolution of mantle convection

Regional Geology and Tectonics: Volume 2
DOI: https://doi.org/10.1016/B978-0-444-64136-6.00006-3

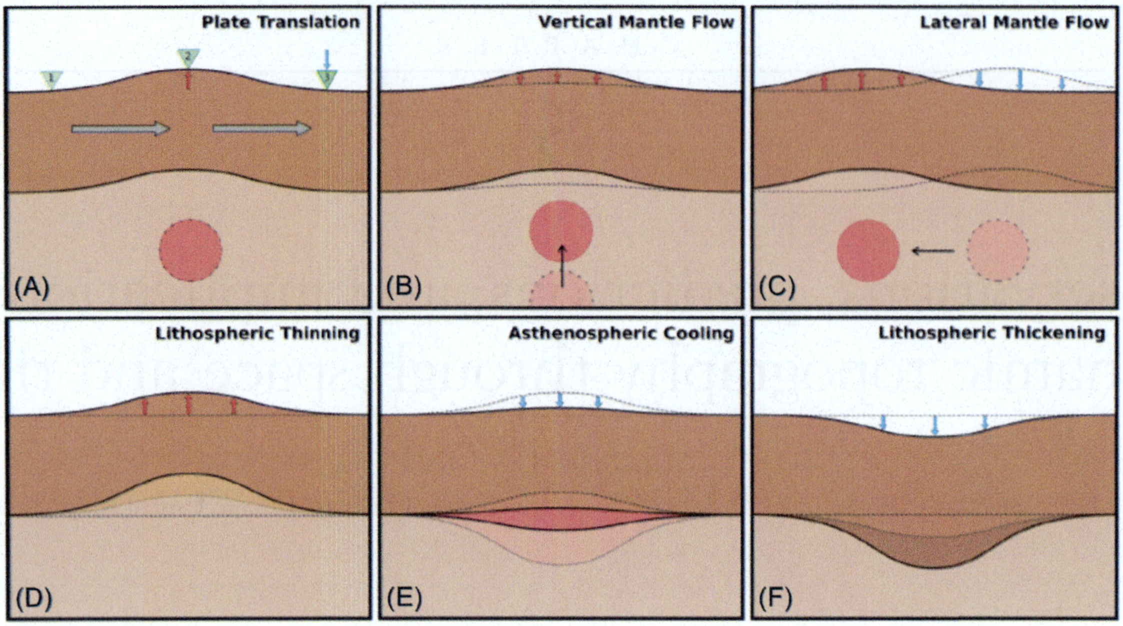

FIGURE 5.1 Mechanisms causing evolving ("dynamic") surface vertical motion. (A) Lateral translation of a plate across an underlying convective planform; (B) An evolving vertical flow field; (C) Lateral advection of convecting mantle; (D) Lithospheric thinning (e.g., by thermal erosion); (E) Progressive cooling of temperature anomalies; (F) Lithospheric thickening (e.g., during oceanic spreading or following horizontal extension). Source: *Modified from Hoggard M.J., Austermann J., Randel C., Stephenson S., 2021. Observational estimates of dynamic topography through space and time. In: Marquadt H., Ballmer M., Cottaar S., Konter J. (eds.), Mantle Convection and Surface Expressions. AGU Geophysical Monograph Series, Washington, DC, Chapter 15.*

(e.g., Hager et al., 1985; Gurnis et al., 2000; Flament, 2019). Eakin and Lithgow-Bertelloni (2018) note that, while our reliance on various geophysical observables like seismic tomography or gravity surveys to probe the deep mantle is hampered by their being present-day snapshots, the time-variant nature of dynamic topography holds great potential to unravel not just patterns of convection, but also the past history of our dynamic Earth.

Dynamic topography is often defined against a background of isostatically supported topography; the "tyranny of isostasy" has the potential to obscure the dynamic signal (for instance, many models predict a patch of negative dynamic topography centered on the Andes, where isostasy dominates; e.g., Ricard et al., 1993; Flament et al., 2013; Yang and Yang, 2021). The revolution in plate tectonics in the early 1960s is intimately associated with lateral plate motions in a way that dynamic topography is associated with vertical or epeirogenic motions at the Earth's surface (although the effect is theoretically possible across other surfaces, and has been observed at the core-mantle boundary: Deschamps and Li, 2019). Since tectonically active regions (notably plate boundaries) contain the largest deviations in surface topography, the quest for dynamic topography has tended to focus on tectonically more quiescent areas like intracratonic regions, where topographic variations due to mantle dynamics are more likely to be visible and not obscured/modified by shortening or other intense effects of plate tectonics.

Residual topography—calculated by removing the isostatic contribution (or sediments, ice, crust, and lithosphere) from observed topography—can be used as a constraint on dynamic topography induced by mantle flow (see the Section "Case study: Africa"; Crough, 1983; Flament et al., 2013). Residual topography highs are centered notably on southern Africa, the central Pacific, and Iceland. Dynamic topography is probably ubiquitous globally and can be of either sign: buoyant plumes of warm, rising asthenospheric material will exert an upward force on the surface, resulting in dynamic uplift; when dense and cold subducting slabs sink, they drag the mantle and surface downward (dynamic subsidence), a mechanism that can potentially enhance accommodation space in sedimentary basin development (see Section "Implications for basin evolution").

Spatial patterns of dynamic topography are easier to measure within the oceanic realm since isostatic signals can be removed with greater confidence (Winterbourne et al., 2009, 2014). Bathymetry is primarily determined by the cooling and thickening of the oceanic lithosphere as it moves away from spreading centers (e.g., McKenzie, 1967). However, ocean floor older than ~80 Ma appears to flatten compared to this simple physical model; proposed explanations include small-scale convection (e.g., Afonso et al., 2019) and dynamic support by mantle flow (Zhang et al., 2012). Winterbourne et al. (2019) exploited oceanic age-depth observations to determine positive and negative residual depth anomalies (i.e., deviations from an age-depth curve), which they attributed to thermal anomalies within the convecting mantle (Fig. 5.2). In the continental realm, by contrast, it is more challenging to establish a spatiotemporally rich grid of dynamic topography observations, due to (a) more pronounced variations in crustal thickness and density that are relatively poorly mapped; (b) complex small-scale changes in lithospheric temperature and magmatic processes; and (c) the lack of a single, unifying tectonic model for topographic development due to the continents' long and complex tectonic history. Instead, proxies must be used, such as the long-wavelength-filtered free-air gravity anomaly (a crude proxy for subplate mantle convection), and residual topography. Unfortunately, the use of residual topography results in continents being dynamically depressed by up to 3 km—the result of the nonnegligible isostatic contribution of the continental lithosphere—which is inconsistent with observations (Gurnis, 1993).

The remainder of the chapter is structured as follows. Section "Observational estimates" continues the exploration of the observational toolkit of recording "the whispers of the mantle" at the surface. Section "Predictions" explores modeling approaches, which often yield strikingly different patterns, magnitudes, and signs of dynamic topography and sometimes do not match observations, especially over southern Africa and southeast Asia (Fig. 5.3); Section "Case study: Africa" details a brief case study of evidence and predictions for dynamic topography across Africa, and excellent natural laboratory; Section "Implications for basin evolution" discusses implications of dynamic support on the architecture and accommodation of sedimentary basins; finally, Section "Summary and outlook" offers conclusions, outstanding challenges, and a future outlook.

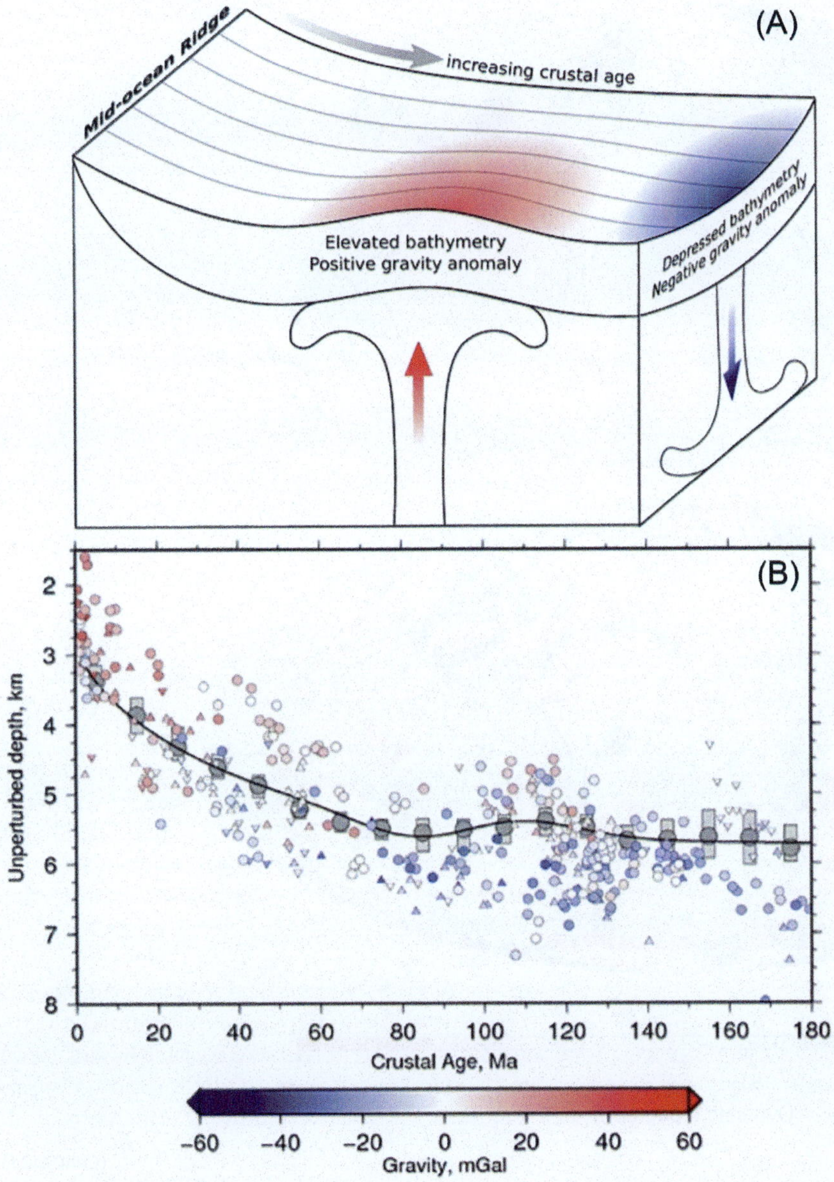

FIGURE 5.2 (A) Cartoon summarizing main controls of depth to oceanic basement as a function of plate age. General increase in plate thickness with age is controlled by the cooling and thickening of lithospheric mantle; while the negative and positive depth anomalies are caused by convective upwellings (red arrow) and downwellings (blue arrow), respectively. (B) Residual depth measurements as a function of plate age. Positive deviation due to thermal anomalies within the convecting mantle. Gray circles, error bars, and black line = age-depth relationship from Crosby et al. (2006); red and blue circles and triangles = 449 global measurements of water-loaded subsidence of oceanic crust (color indicates the amplitude of long-wavelength free-air gravity anomaly). Source: *Modified from Winterbourne J., White N., Crosby A., 2014. Accurate measurements of residual topography from the oceanic realm. Tectonics, 33, 982–1015.*

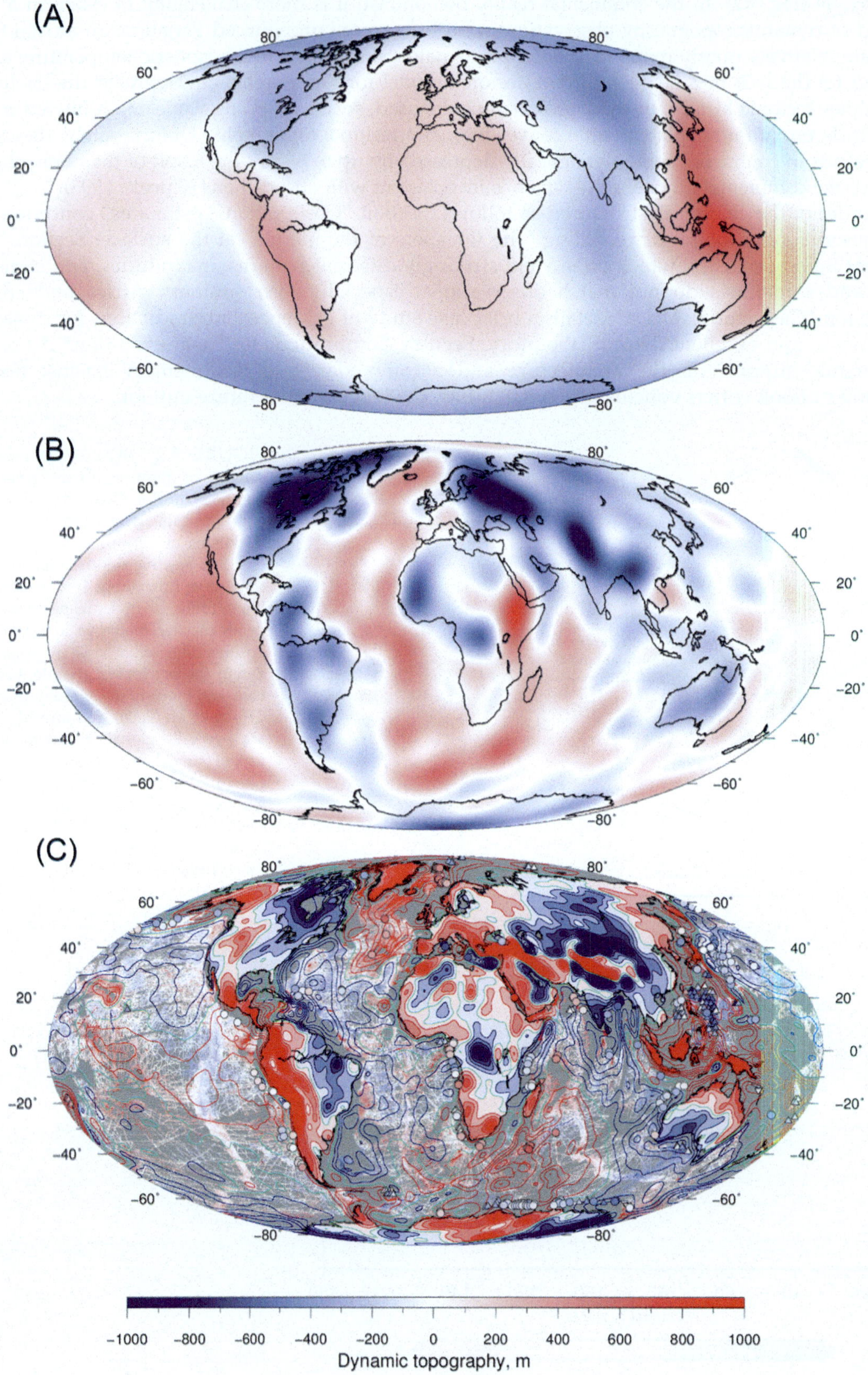
(A)
(B)
(C)
80°
60°
40°
20°
0°
-20°
-40°
-60°
-80°
−1000 −800 −600 −400 −200 0 200 400 600 800 1000
Dynamic topography, m

(*Continued*)

Observational estimates

Since convective circulation in the mantle cannot be directly observed, detailed surface observations of dynamic topography should provide useful physical constraints. Moreover, most predictive flow models of global dynamic topography are driven by convection in the sublithospheric mantle (Section "Predictions"). To test them, observational constraints on present-day dynamic topography are required. The principal problem is that it is difficult to discriminate between small dynamic topographic signals, which are probably driven by density changes beneath the lithospheric plate, and larger isostatic signals. Another issue arises from the difficulty in gridding or formalizing spatially sparse spot measurements of epeirogenic surface motion, which are often quite different in nature and carry variable uncertainty, such as uplifted marine fauna, immature river profiles, and palaeoaltimetric markers. Fig. 5.4 is a summary of surface responses to epeirogenic motion.

This study begins the discussion by interrogating geophysical datasets that can be exploited to infer dynamic support and/or surface epeirogeny. Satellite observations include the Global Positioning System (GPS) and the Gravity Recovery and Climate Experiment (GRACE) mission, which record direct measurements of ground motion and the temporal evolution of Earth's gravity field, respectively. However, GPS rates have yet to be used directly in models of dynamic topography, while the long-wavelength changes in geoid height predicted by these models remain elusively just below the detection threshold for GRACE (Ghelichkhan et al., 2018; Hoggard et al., 2021). The long-wavelength component of Earth's gravity field, in particular geoid anomalies, reveals a dominant spherical harmonic degree-2 pattern (and order 0, interpreted as Earth's dynamic oblateness or flattening), likely generated by mantle flow (Runcorn, 1965). Geoid anomaly highs of $\sim +100$ m are focused on the western Pacific Ocean and southern Africa. The governing variations in density, especially in the upper mantle, can be calculated by converting tomographic maps of seismic velocity using a variety of laboratory-based calibrations (these conversions are subject to significant uncertainty relating to the frequency content of seismic sources, and the content of minor phases such as water or melt in the mantle: Jackson and Faul, 2010; Yamauchi and Takei, 2016). Final density maps can be calibrated using observed gravity anomalies (Cowie et al., 2015). Another seismological technique elucidates the structure of the upper lithosphere. Observations of crustal thickness from joint inversion of teleseismic receiver functions can yield useful insights into whether long-wavelength topography is maintained by regional dynamic support, as has been suggested in northwest Scotland and Madagascar (Davis et al., 2012; Paul and Eakin, 2017).

The geomorphological record is rich in evidence of epeirogenic motion, which can potentially be linked to dynamic topography. Constraining dynamic uplift is perhaps more challenging than drawdown since erosion tends to destroy the geological record (cf. sediment deposition). The existence of surprisingly youthful flat or gently warped ancient erosion surfaces has been noted for some time across many regions, including western India, Brazil, southern Africa, and Madagascar (e.g., King, 1942). The topography of these surfaces or peneplains is "youthful" in the sense of deeply incised gorges and gullies, which often contrasts against much older basement bedrock (Fig. 5.5B; Paul, 2021). Initially, the conceptual model of their formation involved a steady, simultaneous erosion of all hillslopes in a process analogous to diffusion (e.g., Davis, 1899, 1830). While still controversial, it has been suggested that the elevation difference between successive surfaces represents a discrete vertical uplift event from which surface uplift rates may be calculated, only if the surfaces can be dated (Pelletier, 2010; Beauvais and Chardon, 2013; Paul, 2021). This is challenging; in practice, local constraints on the timing and magnitude of denudation—typically using one of a myriad of geochronologic techniques like apatite fission-track analysis—are more valuable for inferring changes in dynamic topography since erosion is isostatically related to the amplitude of tectonic or dynamic uplift (England and Molnar, 1990; Brodie and White, 1994; Jess et al., 2018).

◀ **FIGURE 5.3** Comparison between two models that predict the dynamic topography from variations in mantle viscosity structure, and an observational database. (A) Model based on density heterogeneity from slab subduction history. Dynamic topography is independent of the absolute value of mantle viscosity but is sensitive to the relative viscosity structure of the mantle (Lithgow-Bertelloni and Richards, 1998; Ricard et al., 1993). Predicted topography does not match observed residual topography over southern and eastern Africa. (B) Model based on density heterogeneity from seismic tomography (S40RTS: Ritsema et al., 2011). Density anomalies are assumed to be thermal in nature and can be simply related to seismic velocities by using a constant factor of $0.4\ \text{kg}\ \text{m}^{-3} \cdot \text{km}^{-1}\ \text{s}$. (C) Observations of dynamic topography. Offshore: circles = residual topographic anomalies calculated from well-constrained sedimentary and crustal thicknesses; triangles and inverted triangles = minimum and maximum estimates, respectively; sparse filigree = residual topography calculated from ship-track bathymetry and from National Geophysical Data Center (NDGC) sedimentary thickness archive (Divins, 2008; Winterbourne et al., 2009). Onshore: red, green, and blue contours = positive, zero, and negative values of dynamic topography plotted at 333 m (i.e., 10 mGal) intervals. These predictions are based upon long-wavelength (800–2000 km) free-air gravity anomalies, a crude proxy of mantle convection, assuming an admittance of $30\ \text{mGal} \cdot \text{km}^{-1}$.

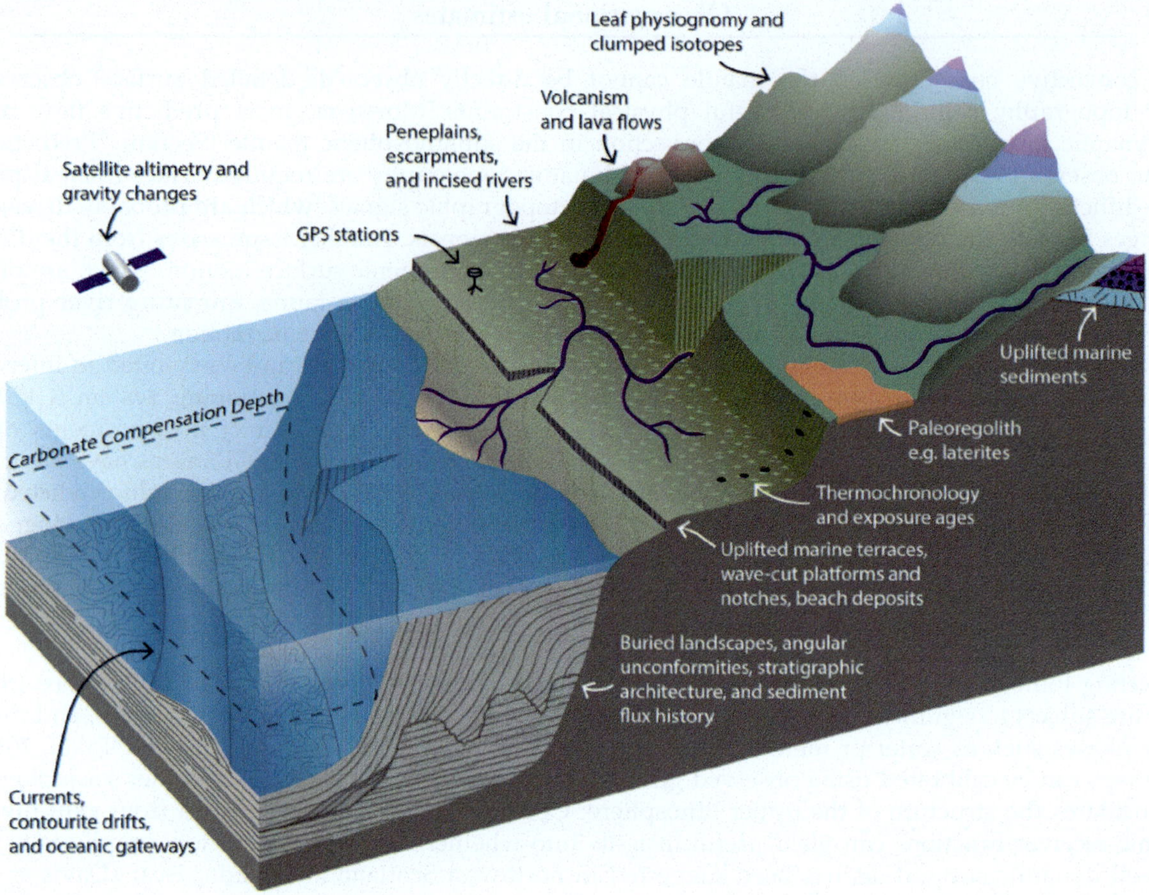

FIGURE 5.4 Cartoon summary of potential observations of surface response to evolving vertical motion. Such observables generally record relative changes between land surface elevation and sea level, except for GPS, which measures solid Earth deformation alone. *GPS*, Global Positioning System. Source: *From Hoggard M.J., Austermann J., Randel C., Stephenson S., 2021. Observational estimates of dynamic topography through space and time. In: Marquadt H., Ballmer M., Cottaar S., Konter J. (eds.), Mantle Convection and Surface Expressions. AGU Geophysical Monograph Series, Washington, DC, Chapter 15.*

Coastlines are particularly useful environments for analyzing vertical motions: Lyell (1830) and Darwin (1842) noted that morphological and biological features such as wave-cut platforms, sea cliffs, marine terraces, coral reefs, and beach deposits form in a very narrow palaeo-elevation band. Indeed, some of the most comprehensive constraints on geological-timescale continental vertical motion have come from observations of exposed marine sediments, provided that they could be (radiometrically) dated. Sahagian (1988) used shoreline deposits to reconstruct a two-dimensional surface representing the Cenomanian ($\sim$95 Ma) sea level across Africa, which is now gently warped by up to 3 km of topography, resulting from post-Cretaceous epeirogenic surface motions. Stephenson et al. (2019) dated a series of emergent coral reef fragments from Madagascar's northeast coastline to Marine Isotope Stage 5e (i.e., $\sim$130 ka), noting an important decrease in reef elevation of $\sim$6 m over a distance of 80 km (Fig. 5.5A). Since this spatial gradient is inconsistent with eustatic sea-level changes and glacio-isostatic adjustments, it is more likely that this region has experienced differential vertical motions resulting from changes in dynamic support over relatively short length and timescales.

Rivers link continental interiors to coasts and have been recognized as potentially sensitive "tectonic tape-recorders" whose longitudinal profiles are primarily shaped by patterns of surface uplift and then modified by erosion (Fig. 5.6; e.g., Whipple and Tucker, 1999; Roberts and White, 2010). Recent spectral analyses of African drainage patterns indicate that river elevation changes are dominated by power at wavelengths >100 km, suggesting that large coherent signals, such as regional uplift caused by dynamic topography, are potentially recoverable despite overprinting by shorter-wavelength complexities, such as variations in lithology, precipitation, and biota (Roberts et al., 2019).

FIGURE 5.5 Surface manifestations of dynamic support in Madagascar. (A–D) Outcrops of emergent marine deposits and coral reefs, northeast coast: (A) 52 m coral/dune rock cliff; (B) close-up of weathered and fresh consolidated dune rock at the top of the cliff; (C) Porites coral and other marine fauna detrital limestone cliff, halfway up slope to the summit; (D) close-up of old coral (Porites porites), infilled with reprecipitated calcite, perched 32 m above sea level. (E) Plan view of three peneplains, Cap Ste-Marie, southern coast: a coral-bearing beach rock bench (~3 m), an intermediate terrace (40–45 m), and an uppermost peneplain (~150 m). Two phases of dramatic river incision—in response to pulses in surface uplift—are apparent.

Drainage patterns can qualitatively and indirectly offer glances into the time-varying nature of dynamic support: in Africa, drainage tends to be radial around major topographic swells such as in Namibia: Section "Case study: Africa". Similarly, convex-upward or "immature" river profiles reflect the imposition of a pulse of uplift

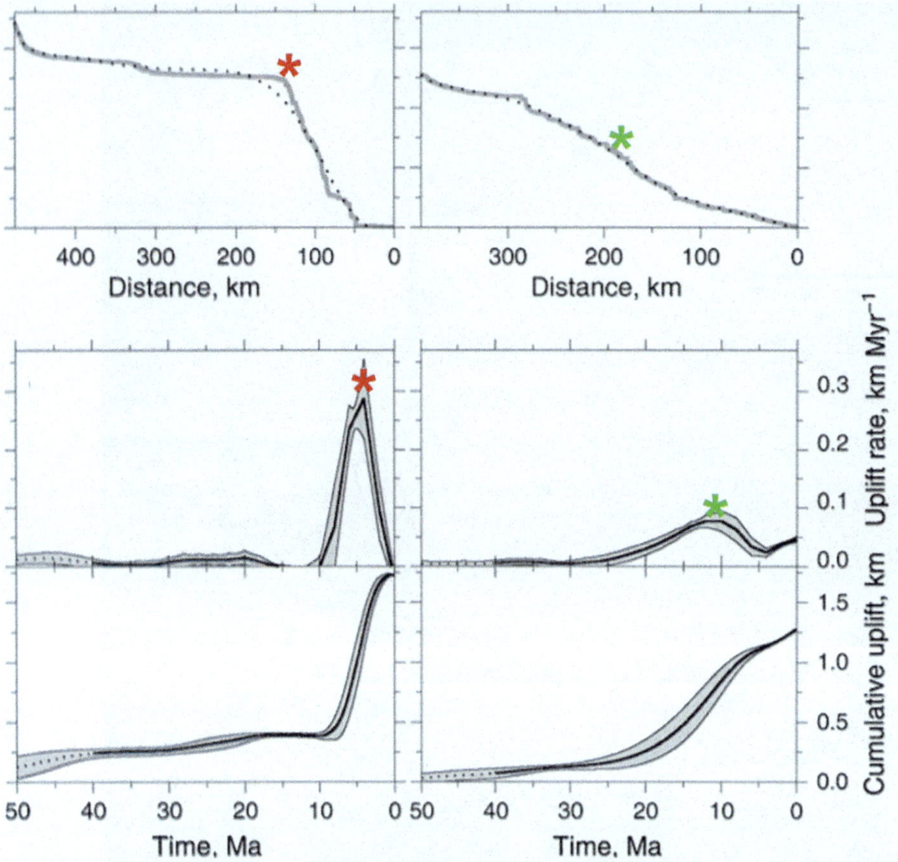

FIGURE 5.6 Two examples of rivers as "tectonic tape-recorders." Each river (gray line, top row) was inverted (modeled river: dotted line, top row) to retrieve an uplift rate history (middle row), which was then integrated to calculate cumulative uplift (bottom row). Asterisks link pulses in calculated uplift rate to headward-propagating convex-upward knickzones in river profiles. Source: *After Paul J.D., Roberts G.G., White N., 2014. The African landscape through space and time. Tectonics, 33(6), 898–935.*

at some point in the geological past, in the form of a knickzone that advects "backward" from the coast, assumed to be regional base level. Larger inventories of river profiles, appropriately parameterized, can be inverted to recover continental-scale uplift rate histories, the timing, and magnitude of which agreeing with independent geological evidence (e.g., Roberts et al., 2012a; Roberts et al., 2012b; Paul et al., 2014; O'Malley et al., 2021). These histories can be used to infer the evolution of dynamic topography in tectonically stable regions. Finally, histories of clastic sediment flux offshore can also be calculated and harnessed: discrete pulses of high sedimentary flux occur at times of increased uplift, while topographic quiescence suppresses the supply of clastic material to river deltas, promoting the growth of carbonates (Lavier et al., 2001; Walford et al., 2005; Paul, 2021). Sediment types and budgets, as well as diagnostic stratigraphic patterns offshore (on which more in Section "Case study: Africa"), can be used to infer evolving dynamic topography and to investigate long-term vertical motions of continental interiors (Walford et al., 2005; Beelen et al., 2019).

Predictions

Several techniques have been developed to model regional and global dynamic topography, including those that only consider the present-day (instantaneous models), time-dependent models, and either a forward or inverse matching of present-day mantle structure. Models either use or predict present-day quantities (such as the geoid, mantle seismic structure, surface and/or core-mantle boundary dynamic topography, the rate of change of surface topography, and plate motions) to infer mantle properties. One of the most important (and contentious) starting points is the density structure of the mantle, constrained by, for instance, seismic tomography models (Hager and Richards, 1989; Steinberger, 2007; Yang and Gurnis, 2016), or from time-dependent mantle flow models (e.g., Gurnis 1993; Moucha et al., 2008; Flament, 2019). Estimating the history of dynamic topography based on mantle convection models is difficult, because the past density structure of the mantle is not directly known. Moreover, plate reconstruction uncertainties, as well as those related to viscosity and compositional and thermal structures (e.g., the presence or lack of phase transitions), might affect the accurate prediction of dynamic topography at short wavelengths (Flament, 2019). These uncertainties, together with differences in starting models, initial datasets (e.g., seismic tomographic models), and model calibration, have resulted in suites of global predictions of dynamic topography that usually agree on the broadest-wavelength structure (e.g., global the degree-2 pattern), but often contain significant discrepancies in predicted magnitude (and sign: Fig. 5.7; Ricard et al., 1993; Steinberger, 2007; Conrad and Husson, 2009; Flament et al., 2013). On the other hand, the recent development of high-resolution (in time and space) plate kinematic models with continuous plate boundaries (e.g., Matthews et al., 2016) can be used to probe lower mantle structure (e.g., Young et al., 2019) as well as surface dynamic topography (Flament, 2019).

Predictions of global dynamic topography can be broadly divided into three flavors. First, analytical (Green's function) solutions were initially computed for viscous flow in either Cartesian or spherical domains with simply layered viscosity profiles. These solutions were the first to link mantle structure and flow to the geoid and surface topography (e.g., Ricard et al., 1984; Richards and Hager, 1984). Second, the present-day density structure of the mantle as constrained by global seismic tomography inversions informs instantaneous flow models, which strongly depend on the particular tomography model used, as well as whether all seismic velocity anomalies are included as density anomalies (e.g., Hager et al., 1985; Lithgow-Bertelloni and Silver, 1998). The dynamic topographic predictions of these models are then typically validated against the observed geoid and/or residual topography.

Finally, time-integrated and time-dependent flow models, most developed over the past 25 years, exploit multiple strategies that use data assimilation to model the evolution of mantle flow and of dynamic topography back through geological time. Ricard et al. (1993) developed a time-dependent analytical model of mantle flow in which subduction history (Lithgow-Bertelloni et al., 1993) was imposed over the past 200 Myr. In this model, slabs sink vertically at a constant rate and slow down as they enter the lower mantle. Gurnis et al. (1998) developed numerical regional mantle convection models with imposed plate kinematics to investigate the evolution of the dynamic topography of the Australian plate since Cretaceous times, with the location of subduction taken from plate reconstructions (Müller et al., 1997). They reproduced the regional Cretaceous flooding of Australia deduced from paleogeographic reconstructions.

Other time-dependent models of dynamic topography do not rely on the imposition of a subduction history, instead "backward advecting" tomographically derived mantle density structure (Steinberger and O'Connell,

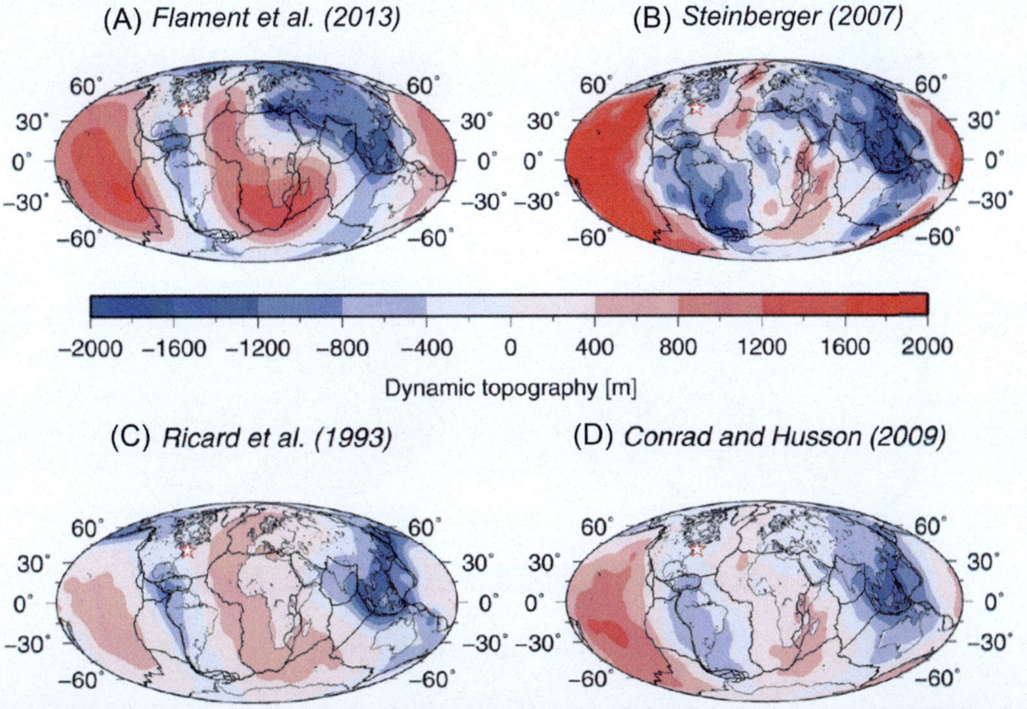

FIGURE 5.7 Comparison of four models of present-day dynamic topography (water-loaded in the oceans, air-loaded in the continents, and averaging to zero). (A) and (C) are based on time-dependent mantle flow models, while (B) and (D) exploit present-day instantaneous mantle flow models in which the density structure is based on a seismic tomographic model of the interior of the Earth. Source: *Modified from Flament N., Gurnis M., Müller R.D., 2013. A review of observations and models of dynamics topography. Lithosphere, 5(2), 189–210.*

1997). This method is relatively complex because of (1) the accumulation of dense material toward the surface, resulting in a decrease of information on the density structure of the mantle back in time; (2) the fact that thermal diffusion cannot be simply reversed (Ismail-Zadeh et al., 2004); and (3) the thermal boundary layer cannot explicitly be considered. Conrad and Gurnis (2003) used this technique to propose that the dynamic topographic high currently located under southern Africa migrated from eastern Africa over the past 36 Myr. Various adjoints of the governing equations have been introduced to tackle this method's weaknesses (while also greatly increasing computational cost), essentially by minimizing the cost function between seismic tomography and forward-predicted temperature and iteratively solving the forward and backward adjoint (e.g., Bunge et al., 2003; Liu et al., 2008). Yang and Yang (2021) recently modeled snapshots of the global dynamic topography field back to 410 Ma, using sophisticated 3D spherical models of mantle convection, and a new global plate motion history as a time-dependent surface mechanical boundary condition (Fig. 5.8 shows a combination of plate and dynamic topography reconstructions).

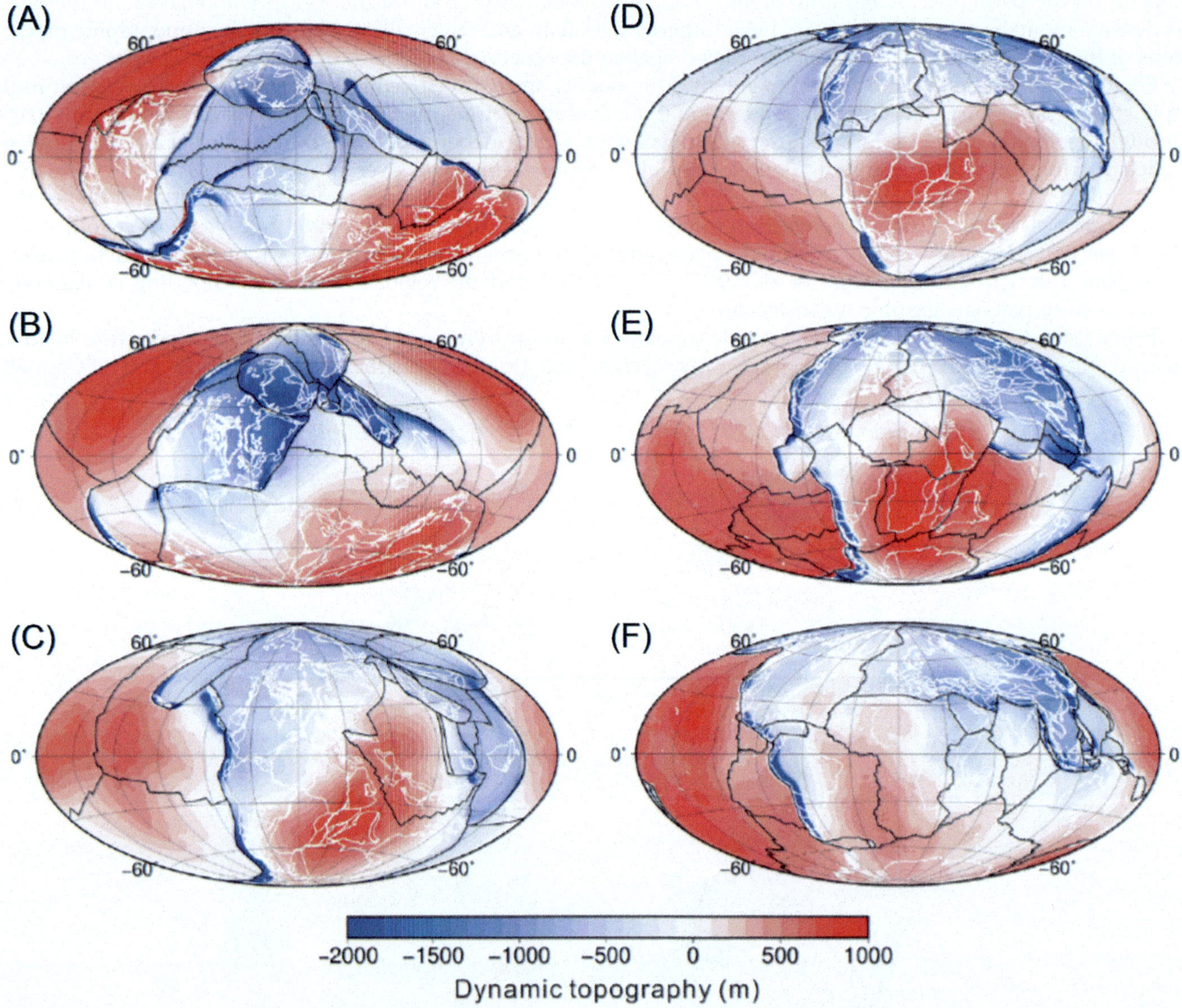

FIGURE 5.8 One estimate of the global dynamic topography field at (A) 405 Ma, (B) 337 Ma, (C) 262 Ma, (D) 190 Ma, (E) 104 Ma, and (F) present-day. Black lines = continental and oceanic plate boundaries; present-day coastlines are rotated according to plate reconstructions. Source: *Modified from Yang A., Yang T., 2021. Controls on the present-day dynamic topography predicted from mantle flow models since 410 Ma. Geophysical Journal International, 225, 1637–1652.*

Case study: Africa

The striking "basin-and-swell" physiography of Africa's landscape (Fig. 5.9A) and the likelihood that it is maintained by deep Earth processes dynamically have long been recognized (e.g., Holmes, 1944; Burke, 1996; Gurnis et al., 2000). It is now accepted that this physiography is a surficial manifestation of convective circulation within the sublithospheric mantle (e.g., Burke and Gunnell, 2008). Africa is a useful natural laboratory for studying regional epeirogeny and dynamic topography for several reasons. First, it is surrounded by passive margins and has been unaffected by regional shortening since Paleozoic times, aside from a short strip of plate convergence along the Mediterranean coast. Second, the African plate has moved slowly within a hotspot frame of reference since Early Oligocene times (Burke, 1996). Scotese et al. (1988) proposed that the African plate has moved northward by less than 10° since Chron 5 (~59 Ma). If so, the mechanical coupling between vertical stresses at the base of the lithosphere and vertical motion at the surface is not complicated by horizontal translation (Lithgow-Bertelloni and Richards, 1998).

The hypsometry of Africa is strikingly anomalous with ~50% of its topography standing at elevations of greater than 500 m. Long-wavelength elevation of subequatorial Africa comprises the most substantial contribution to this anomalous hypsometry. Nyblade and Robinson (1994) showed that this region is surrounded by anomalously shallow oceanic lithosphere, which suggests the existence of a convectively maintained "superswell" centered on the Kerguelen Plateau in the Southern Ocean. This "superswell" has a wavelength of $\sim 10^4$ km. More localized swells with wavelengths of $\sim 10^3$ km occur throughout Africa. These swells are clearly expressed by long-wavelength (800–2000 km) free-air gravity anomalies (Fig. 5.9B; Tapley et al., 2005). In North and East Africa, localized swells are usually magmatic (e.g., Hoggar, Tibesti, and Darfur: Cox, 1989; Fishwick and Bastow, 2011). The largest one occurs in the Afar region where the presence of a hot plume has been inferred (e.g., Ebinger and Sleep, 1998; Furman et al., 2006). In South and West Africa, three amagmatic swells straddle the continent-ocean boundary (Burke, 1996). Offshore, prominent unconformities in sets of prograding deltaic clinoforms have been used to suggest a stepped or pulsed history of uplift, reflective of a subplate origin (Paul et al., 2014; Fig. 5.10).

Larger upwellings and downwellings are associated with minor seismicity (Fig. 5.9C). The relationship between topography and crustal thickness variation suggests that regional uplift and subsidence are not accounted for by Airy crustal isostasy alone (Fig. 5.9D). Beneath elevated regions such as South Africa and the Afar region, crustal thicknesses are as low as 33 km and 20 km, respectively (Nair et al., 2006; Dugda et al., 2005). The northern margin of the Congo Basin has a crustal thickness of >45 km, whereas the adjacent Cameroon Line has crustal thicknesses of 25–33 km (Tokam et al., 2010). Likewise, crustal thickness beneath the central highlands of Madagascar is ~5 km thinner than at the southern coast (44 km vs 39 km: Paul and Eakin, 2017). Inferences based upon long-wavelength gravity anomalies, magmatism, and crustal thickness estimates are corroborated by the results of body wave and surface wave tomography.

In the lower mantle beneath southern Africa, body wave models show that there is a large slow velocity anomaly (Ni et al., 2002). This irregularly shaped feature has a footprint of 2000 x 4000 km on the core-mantle boundary and a height of 2000 km. It is thought to be predominantly a thermal anomaly, although the sharpness of its southwestern edge suggests that compositional variation plays a moderating role. Further north, a slow shear wave velocity anomaly occurs in the upper mantle beneath Afar and the Red Sea (e.g., Gurnis et al., 2000; Ritsema et al., 2011). Surface wave tomographic models can be used to estimate the lithospheric thickness beneath Africa (Fig. 5.9D; Priestley and McKenzie, 2013; Fishwick and Bastow, 2011). Three large cores of lithosphere, whose thickness exceeds 200 km, occur in subequatorial Africa and in northwest Africa. These cores are associated with negative long-wavelength free-air gravity anomalies, with positive anomalies occurring along their margins. Surface wave tomographic models show that slow velocity anomalies are associated with some, but not all, of these positive gravity anomalies.

These anomalies are consistent with the presence of warm material beneath the base of the lithosphere. A thickened lithospheric keel occurs beneath the Congo Basin and may coincide with convective downwelling and negative dynamic topography (Crosby et al., 2010).

Body wave tomographic models are often used as a basis for dynamic calculations that attempt to predict the spatial and temporal evolution of convectively maintained topography (e.g., Lithgow-Bertelloni and Richards, 1998; Forte et al., 2010). Some of these calculations have been tested using observed uplift rates from southern Africa (Gurnis et al., 2000; Conrad and Gurnis, 2003). Moucha and Forte (2011) simulate convective circulation backward in time from the present day to 30 Ma to predict the growth of dynamic topography. These different

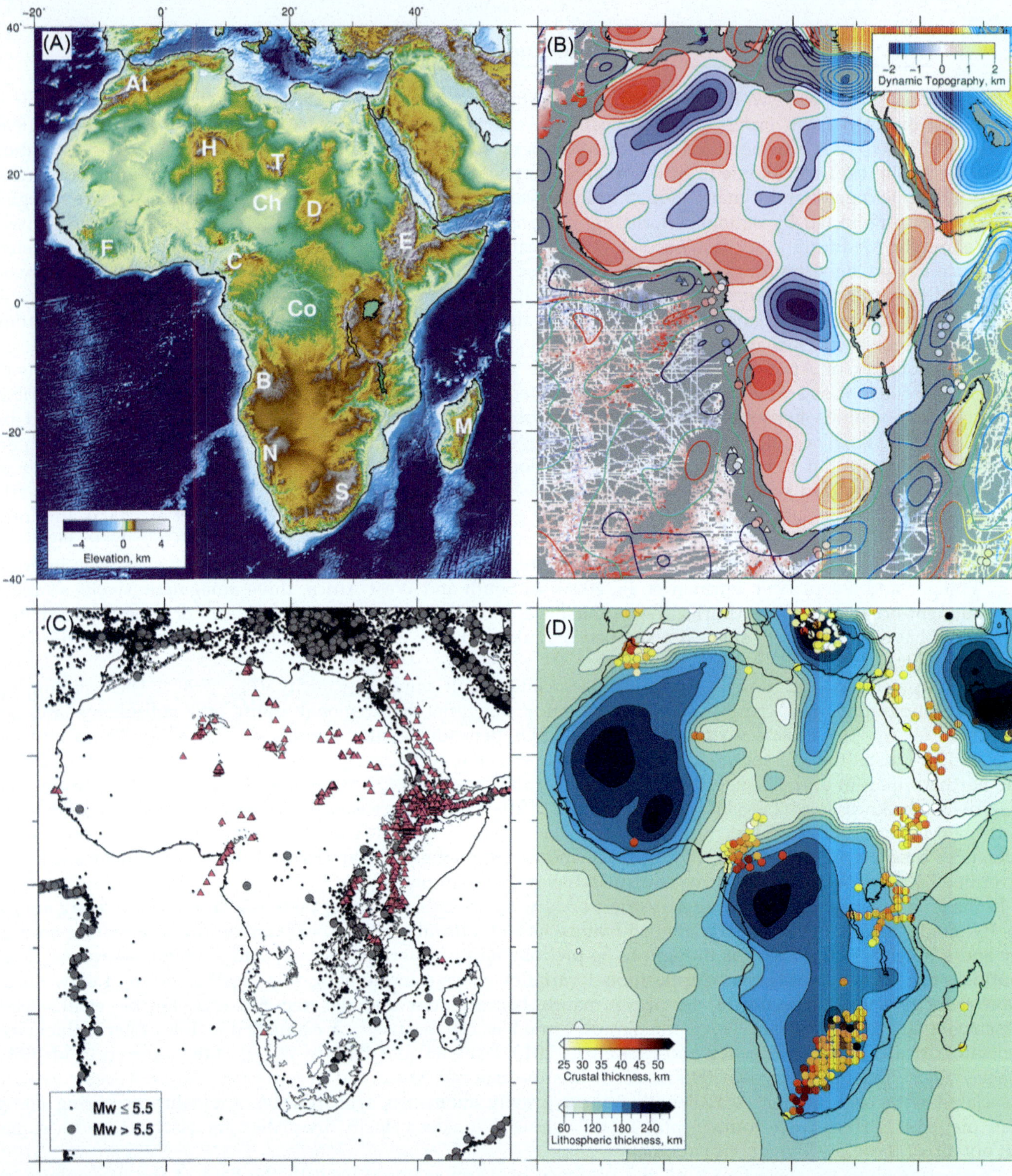

FIGURE 5.9 (A) Topographic map of Africa. *At*, Atlas Mountains; *H*, Hoggar; *T*, Tibesti; *Ch*, Chad Depression; *D*, Darfur; *E*, Ethiopian Highlands; *F*, Fouta Djallon; *C*, Cameroon Volcanic Line; *Co*, Congo Basin; *B*, Bié Dome; *N*, Namibian Dome; *S*, South African Plateau; *M*, Madagascar. (B) Proxy of dynamic topographic map of Africa calculated from a combination of long-wavelength (800–2000 km) free-air gravity anomalies, assuming an admittance of 30 mGal · km^{-1}, and residual topographic anomalies in the oceanic realm (Winterbourne et al., 2009). Symbols and colors as per Fig. 5.3. (C) Earthquake seismicity and volcanism. Open and solid circles = earthquakes of $M_w > 4$ between 1973 and 2013 from IRIS catalog (www.iris.edu). Pink triangles = Cenozoic volcanism (Thorpe and Smith, 1974; Woolley, 2001; Liégeois et al., 2005). Black line = 1 km topographic contour. (D) Lithospheric thickness of Africa was calculated from the surface wave tomographic model. Contour interval = 20 km. Colored circles = estimates of crustal thickness calculated from inversion of teleseismic receiver functions (Nair et al., 2006; Dugda et al., 2005; Tokam et al., 2010). Source: *Paul J.D., Roberts G.G., White N., 2014. The African landscape through space and time. Tectonics, 33(6), 898–935.*

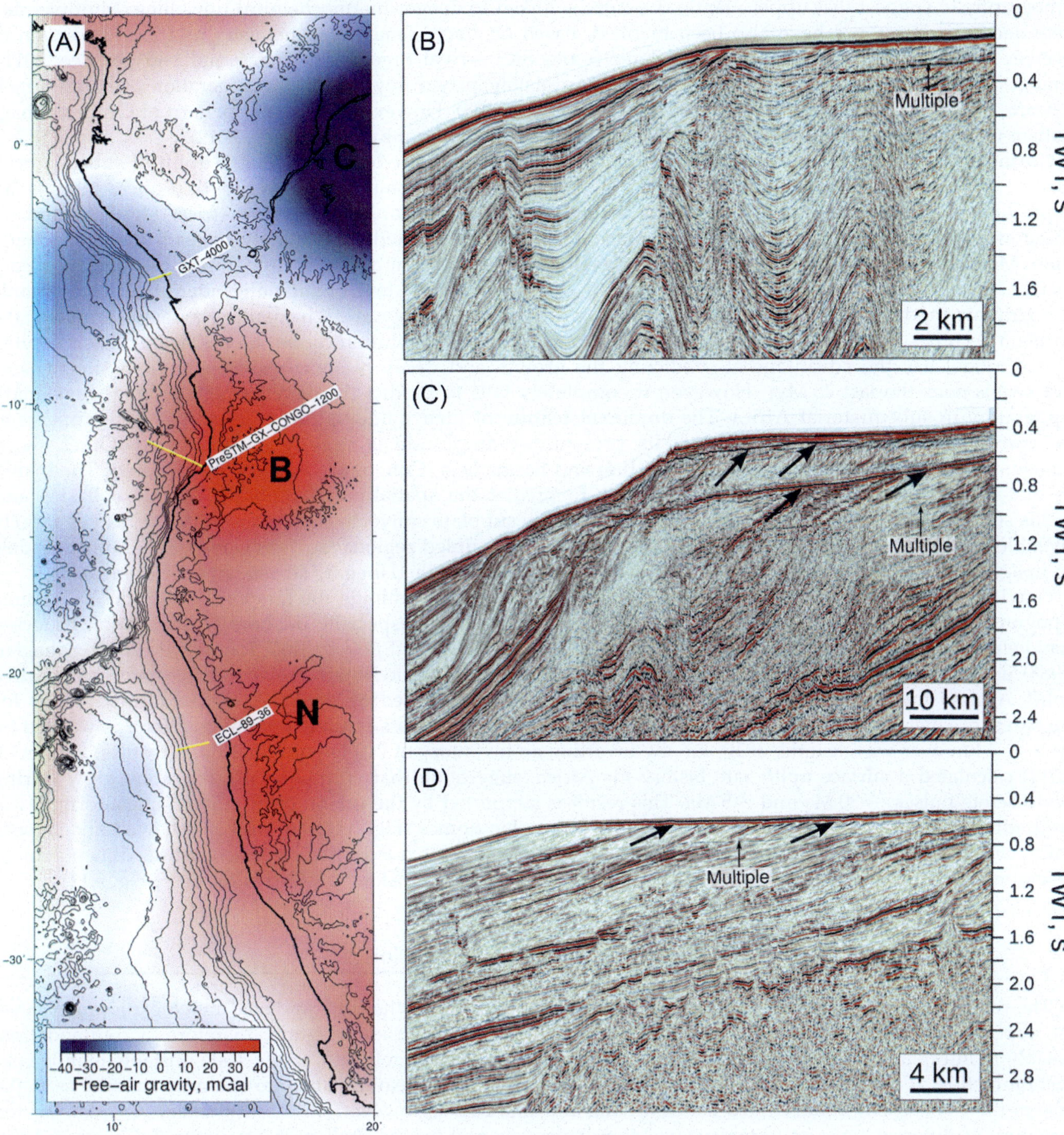

FIGURE 5.10 (A) Long-wavelength-filtered free-air gravity map of SW African coastline. Yellow lines = position of seismic lines. (B), (C), (D): 2D seismic reflection dip lines offshore Congo Basin (C), the Bié Dome (B), and the Namibian Dome (N), respectively. Arrows point up clinoforms to prominent unconformities in (C) and (D). Note the evidence of salt movement affecting deltaic stratigraphy in (B). Black lines = 500 m topographic/bathymetric contours.

approaches have been moderately successful in accounting for the growth of long-wavelength (>1000 km) topography, but they are less successful at predicting shorter-wavelength (<1000 km) basins and swells.

The spatial and temporal organization of convective circulation that gives rise to Africa's distinctive physiography remains poorly understood. While Ebinger and Sleep (1998) argue that lithospheric structure exerts a fundamental control on the spatial distribution of African magmatism, Hansen et al. (2012) use body wave tomography to suggest that upper mantle structure is linked to deeper thermochemical upwelling. Multiple, discrete deep-seated plumes have also been invoked, based on trace element geochemistry and joint inversion of receiver functions, gravity data, and teleseismic S and SKS arrival times (e.g., Chang and van der Lee, 2011). Changes in lithospheric thickness and architecture probably play a key role in the generation of basaltic melt beneath Africa (White and McKenzie, 1995; Priestley et al., 2008). However, the spatial pattern of Cenozoic magmatism may not necessarily reflect short-wavelength dynamic topography (Jones et al., 2012).

Gurnis et al. (2000) and Conrad and Gurnis (2003) showed that the long-wavelength uplift of subequatorial Africa could be supported by a gigantic lower mantle convective anomaly. However, both of these models fail to predict the existence and growth of shorter-wavelength swells that rim southern Africa (Fig. 5.9A and B). These swells may be supported by smaller convective plumes located immediately beneath the lithospheric plate. Moucha and Forte (2011) predicted the development of African dynamic topography by running a convective model backward in time to simulate the reverse advection of density anomalies identified from seismic tomographic models. By invoking the conservation of thermal energy at high mantle Rayleigh numbers, the problem becomes tractable, as diffusion is neglected (its temporal evolution cannot be predicted). Their sophisticated model has the advantage of predicting the growth and decay of short-wavelength ($\sim 10^3$ km) basins and swells over the last 25 Myr. However, it completely fails to predict the development of long-wavelength topography in subequatorial Africa. The dominant feature of their model is the Afar swell, which appears at the correct time and grows southward. Its present-day amplitude is slightly overpredicted. Other short-wavelength swells (e.g., Hoggar, Tibesti, and Bié) and basins (e.g., Congo) are also poorly captured. The inability of tomographically derived dynamic models to predict the spatial and temporal patterns of basins and swells suggests that these features could be supported by subplate convective anomalies that are, as yet, largely invisible in global tomographic models. Note that densely sampled regional surface wave tomographic models do image at least some of these features (e.g., the Cameroon Line, the Hoggar, Tibesti, and Bié domes).

Gravity- and seismology-based models are of limited use for constraining the temporal evolution of dynamic topography (e.g., Forte et al., 2010). However, stratigraphic records, especially offshore sedimentation histories, may help to constrain the development of basins and swells (White and Lovell, 1997; Jones et al., 2012). Sahagian (1988) used the elevation of Cretaceous shoreline deposits (Fig. 5.11) to obtain broad constraints on cumulative uplift. Around the edge of Africa, Winterbourne et al. (2009) calculated residual (i.e., dynamic) topography for the oldest oceanic lithosphere (Fig. 5.9B). Their results show that Africa's continent-ocean boundary is warped by up to ± 1 km over wavelengths of 10^3 km. By inverting an inventory of 710 longitudinal river profiles, Paul et al. (2014) calculated a surface uplift rate history for Africa, suggesting that its unique physiography grew during two discrete pulses, ~ 30 Ma and ~ 5 Ma. This result is supported by the timing of prominent unconformities in prograding clinoforms offshore two SW African topographic domes (Fig. 5.10), as well as renewed and intense volcanism across the continent, and a sudden contemporaneous switch from carbonate to clastic sedimentation at several major African deltas, suggestive of rapid onshore uplift and erosion (Walford et al., 2005; Fig. 5.11).

Implications for basin evolution

The relatively subtle amplitudes of dynamic drawdown have nevertheless stimulated much exploration interest, especially where it provides a contribution to total tectonic subsidence—and, therefore, accommodation, notably in intraplate settings (sag basins). Dynamic drawdown can be related to large-scale unstable mantle convection or flow (Farrington et al., 2010), or the subduction of cold oceanic slabs into the deep mantle, as is the case in the Great Basin in the United States (Chang and Liu, 2019). However, most sedimentary records bear the signature of different types of tectonic signal simultaneously and can be challenging to unravel. Chang and Liu (2019) used a forward landscape evolution model to demonstrate the way in which drainage networks and basin sedimentation are controlled by the type (isostatic or dynamic) and magnitude of tectonic subsidence (Fig. 5.12). Their results suggest that migratory patterns of dynamic drawdown produce migrating depocenters and subparallel tilted strata across an entire basin, relative to static dynamic drawdown or isostatic processes.

Several studies have detailed the manifestations of dynamic drawdown on sedimentary basin development across the globe, mostly in tectonically quiescent, intracratonic settings, but also in foreland basins, where the dynamic signal is harder to deconvolve from isostatic effects (Section "Introduction"). In the Brazilian Parnaiba Basin, backstripped and water-loaded subsidence curves calculated from boreholes distributed throughout the basin show that tectonic subsidence decreased exponentially over ~ 350 Ma, yet was punctuated by minor increases in subsidence that follow basin-wide erosional disconformities. These departures could be indicative of

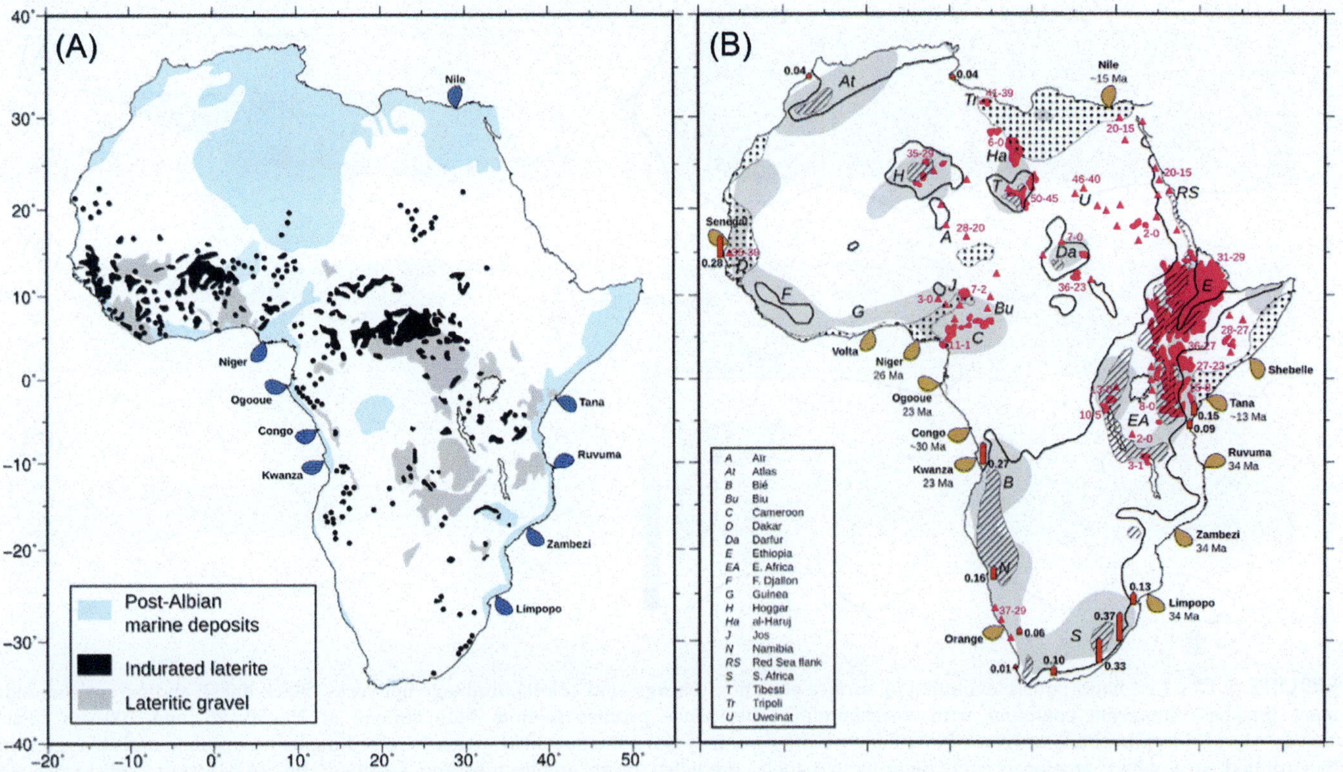

FIGURE 5.11 (A) Pre-Oligocene paleogeography of Africa (Paul et al., 2014). Blue shading = minimum extent of Cenomanian marine deposition (Sahagian, 1988); black shading = distribution of indurated laterite during Palaeogene times (Burke and Gunnell, 2008); gray shading = distribution of lateritic gravels; blue lobes at coastline = major deltas with Palaeogene carbonate reef deposits (Salman and Abdula, 1995; Séranne and Nzé Abeigne, 1999; Nyagah, 1995; Ismail et al., 2010). (B) Early Neogene paleogeography of Africa. Red triangles/shading = dated volcanic craters/flows (Thorpe and Smith, 1974; Ebinger et al., 1989; Reid et al., 1990; Woolley, 2001; Liégeois et al., 2005; Permenter and Oppenheimer, 2007; McDougall and Brown, 2009); warping of Cenomanian surface represented by stippled pattern < 0 km, black line = 1 km, and hatching > 2 km (Sahagian, 1988); gray shading = > 10 mGal long-wavelength free-air gravity anomaly; brown lobes at coastline = major deltas showing age of onset of clastic deposition (Sestini, 1989; Nyagah, 1995; Salman and Abdula, 1995; Séranne and Nzé Abeigne, 1999; Anka and Séranne, 2004; Reijers, 2011); numbered red arrows = radiometrically dated marine deposits with average uplift rate given in $mm \cdot yr^{-1}$ (Partridge, 1998; Partridge and Maud, 1987; Guiraud et al., 2010; Odada, 1996; Elmejdoub and Jedoui, 2009).

transient epeirogenic uplift events caused by changing patterns of localized dynamic topography (Rodriguez Tribaldos and White, 2018). Whole-mantle convection patterns have also been linked to dynamic drawdown of the central North China Craton (Ju et al., 2022), as well as Early Cretaceous inundation of Australia, which has important consequences for the lithology and distribution of first-order sedimentary sequences in the Great Artesian Basin inter alia (Braz et al., 2021).

Dynamic drawdown has likewise been invoked to account for the distribution of sediment records and migration of depocenters in western North America, notably in the Gulf of Mexico (GoM), and the Cordilleran foreland basin (CFB: Chang and Liu, 2019; Li and Aschoff, 2021). In the GoM, a continent-scale, temporally evolving dynamic subsidence signal is consistent with an eastward migratory wave of dynamic subsidence caused by the Cretaceous-aged shallow-to-flat subducting Farallon slab (Liu, 2015; Wang et al., 2016). This sinking slab has also strongly influenced the Cenozoic topographic evolution of eastern North America and surface processes including the recent relief growth of the Appalachians (Liu, 2015).

Careful, high-resolution reconstruction of regional stratigraphic frameworks using three-dimensional datasets is critical to constrain the influence of dynamic topography on basin development and depocenter migration. In this way, Li and Aschoff (2021) attempted to unravel the complex convolution of sedimentary systems, accommodation, and subsidence mechanisms through time and space, using the CFB as a case study. They developed a regional chronostratigraphic framework, constructing solid sediment isopachs and shoreline trends to argue that a wave of dynamic drawdown influenced accommodation creation in the CFB around 85–75 Ma, after which flexural effects dominated.

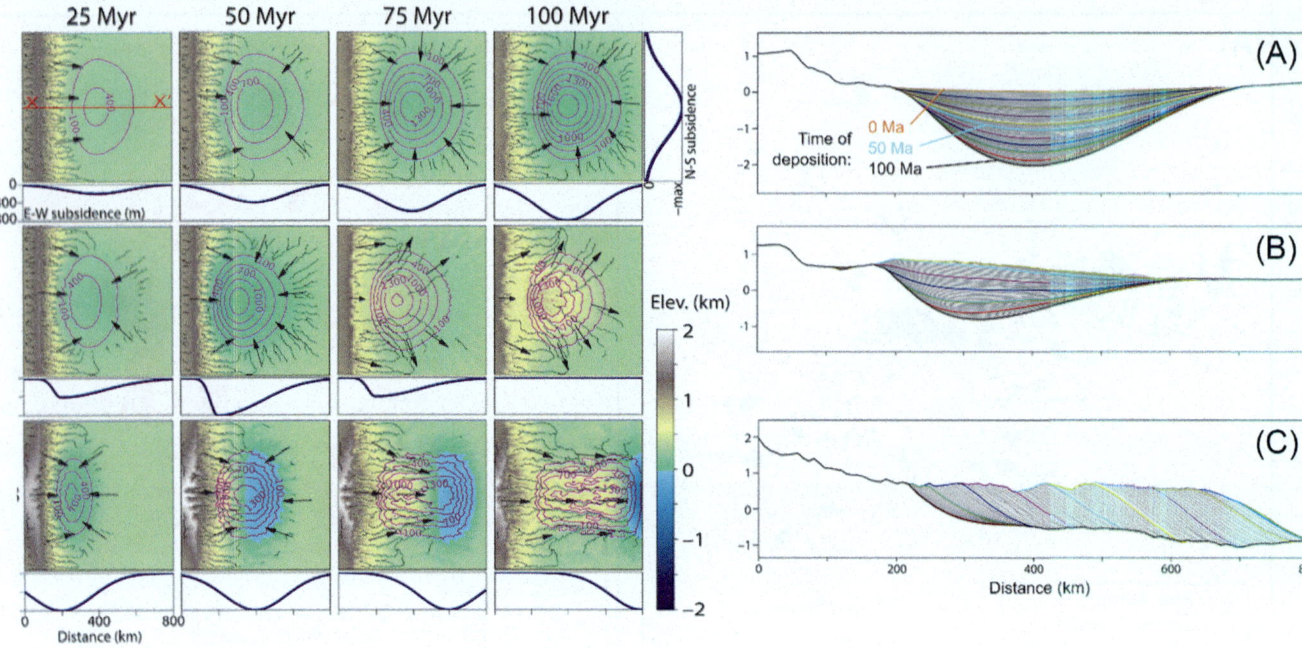

FIGURE 5.12 Left-hand panels: evolution of surface elevation (background colors), drainage networks (black lines), and cumulative sediment thickness (magenta contours) with corresponding subsidence profiles (bottom blue curves) at 25, 50, 75, and 100 Myr. Black arrows = local flow directions of rivers. Note that rivers converge toward the subsided area where sediments accumulate and diverge from the uplifted area where erosion occurs. Right-hand panels: predicted basin architecture and sediment fill. (A) Isostatic subsidence; (B) Stationary dynamic subsidence; (C) Migratory dynamic subsidence. Source: *Modified from Chang C., Liu L., 2019. Distinct responses of intraplate sedimentation to different subsidence mechanisms: insights from forward landscape evolution simulations. Journal of Geophysical Research: Earth Surface, 124(5), 1139–1159.*

Summary and outlook

There has been appreciable recent progress on the observational constraints on, and the modeling of, dynamic topography. In addition to constraining the nature of mantle convection and mantle rheology, studies of dynamic topography have led to a new interpretation of the stratigraphic record and of long-term sea-level change. Current global models and observations of dynamic topography are in broad agreement at very large scale (~10,000 km) but differ in some areas and at smaller scales (<5,000 km). Dynamic influences on surface topography help us better understand not only the shape of the Earth but also the evolution of the Earth system as a whole, including interactions among and between the landscape, life, and dynamics of the deep Earth interior. Several controversies nevertheless remain, which are the focus of ongoing and often contentious debate:-

1. Consensus on the definition of dynamic topography. This also speaks to developing a similar consensus on observed amplitudes, wavelengths, and timescales, and reconciling different approaches and languages between, for instance, physicists (developing numerical models) and field geologists (collecting surface observations). Regarding wavelengths, while many predictive models suggest that dynamic topography is dominated by a spherical harmonic degree-2 structure (~10,000 km), considerations of residual topography imply that dynamic topography can vary over shorter (~1,000 km) length scales (Hoggard et al., 2016). Deconvolving the two shallow- and deep-origin sources is the focus of ongoing research (e.g., Davies et al., 2019; Lamb et al., 2020).
2. Although the geological record contains theoretically abundant markers of evolving dynamic topography (see Section "Observational estimates"), it is challenging to extract, clean, and formalize the necessary measurements with associated uncertainties (e.g., Hoggard et al., 2021). Many observations are strictly qualitative and spatially sparse. Also, many observables are incomplete and fragmentary due to erosion and possibly relocation, introducing additional complexity into interpretation.
3. Significant uncertainties remain in time-dependent models of dynamic topography (see Section "Predictions"; Eakin and Lithgow-Bertelloni, 2018). Briefly, these include the fidelity of present-day 3D mantle rheological (mainly density) structures (seismic inversions have geographically and vertically variable resolution); the

viscosity structure of the lithosphere and convecting mantle (significant debate continues regarding the radial Earth viscosity profile); surface plate motion histories (numerical models cannot perfectly predict the time dependence of plate tectonics in a fully dynamic sense; Tackley, 2000; Stadler et al., 2010); and the validity of modeling tools used (for instance, the degree of refinement used on a spatially adaptive mesh, or the validity of approximating the Earth's surface with a free-slip boundary or as a free-surface formulation).

4. The degree of coupling between horizontal asthenospheric flow and plate motions. That convection simulations using mantle density anomalies derived from either seismic tomography or slab subduction histories can successfully reproduce the first-order characteristics of surface plate motions implies a degree of coupling between deep mantle flow and plates (Lithgow-Bertelloni and Richards, 1998; Conrad et al., 2013).

References

Afonso, J.C., Salajegheh, F., Szwillus, W., Ebbing, J., Gaina, C., 2019. A global reference model of the lithosphere and upper mantle from joint inversion and analysis of multiple data sets. Geophysical Journal International 217, 1602–1628.

Anka, Z., Séranne, M., 2004. Reconnaissance study of the ancient Zaire (Congo) deep-sea fan (ZaiAngo Project). Marine Geology 209, 223–244.

Beauvais, A., Chardon, D., 2013. Modes, tempo, and spatial variability of Cenozoic cratonic denudation: the West African example. Geochemistry, Geophysics, Geosystems 14 (5). Available from: https://doi.org/10.1002/ggge.20093.

Beelen, D., Jackson, C.A.-L., Patruno, S., Hodgson, D.M., Trabucho Alexandre, J.P., 2019. The effects of differential compaction on climothem geometries and shelf-edge trajectories. Geology 47 (11), 1011–1014.

Braun, J., Robert, X., Simon-Labric, T., 2013. Eroding dynamic topography. Geophysical Research Letters 40, 1494–1499.

Braz, C., Zahirovic, S., Salles, T., Flament, N., Harrington, L., Müller, R.D., 2021. Modelling the role of dynamic topography and eustasy in the evolution of the Great Artesian Basin. Basin Research 33 (6), 3378–3405.

Brodie, J., White, N.J., 1994. Sedimentary basin inversion caused by igneous underplating: northwest European continental shelf. Geology 22 (2), 147–150.

Bunge, H.-P., Hagelberg, C.R., Travis, B.J., 2003. Mantle circulation models with variational data assimilation: inferring past mantle flow and structure from plate motion histories and seismic tomography. Geophysical Journal International 152, 280–301.

Burke, K., 1996. The African plate. South African Journal of Geology 99, 339–410.

Burke, K., Gunnell, Y., 2008. The African Erosion Surface: a continental-scale synthesis of geomorphology, tectonics, and environmental change over the past 180 million years. Geological Society Memoir 201.

Cao, W., Flament, N., Zahirovic, S., Williams, S., Müller, R.D., 2019. The interplay of dynamic topography and eustasy on continental flooding in the late Paleozoic. Tectonophysics 761, 108–121.

Chang, C., Liu, L., 2019. Distinct responses of intraplate sedimentation to different subsidence mechanisms: insights from forward landscape evolution simulations. Journal of Geophysical Research: Earth Surface 124 (5), 1139–1159.

Chang, S.-J., van der Lee, S., 2011. Mantle plumes and associated flow beneath Arabia and East Africa. Earth and Planetary Science Letters 302, 448–454.

Conrad, C.P., Gurnis, M., 2003. Seismic tomography, surface uplift, and the breakup of Gondwanaland: integrating mantle convection backwards in time. Geochemistry, Geophysics, Geosystems 4 (3).

Conrad, C.P., Husson, L., 2009. Influence of dynamic topography on sea level and its rate of change. Lithosphere 1 (2), 110–120.

Conrad, C.P., Steinberger, B., Torsvik, T.H., 2013. Stability of active mantle upwelling revealed by net characteristics of plate tectonics. Nature 498, 479–482.

Cowie, L., Kusznir, N., Manatschal, G., 2015. Determining the COB location along the Iberian margin and Galicia Bank from gravity anomaly inversion, residual depth anomaly, and subsidence analysis. Geophysical Journal International 166, 533–573.

Cox, K., 1989. The role of mantle plumes in the development of continental drainage patterns. Nature 342, 873–877.

Crosby, A.G., McKenzie, D., Sclater, J.G., 2006. The relationship between depth, age, and gravity in the oceans. Geophysical Journal International 166, 553–573.

Crosby, A., McKenzie, D., Sclater, J., 2010. Structure and evolution of the intracratonic Congo Basin. Geochemistry, Geophysics, Geosystems 11, Q06010.

Crough, S.T., 1983. Hotspot swells. Annual Review of Earth and Planetary Sciences 11, 165–193.

Darwin, C.R., 1842. The Structure and Distribution of Coral Reefs. Being the First Part of the Geology of the Voyage of the Beagle, Under the Command of Capt. Fitzroy, R.N., During the Years 1832 to 1836. Smith Elder & Co, London.

Davies, D.R., Valentine, A.P., Kramer, S.C., Rawlinson, N., Hoggard, M.J., Eakin, C.M., et al., 2019. Earth's multi-scale topographic response to global mantle flow. Nature Geoscience 12, 845–850.

Davis, W.M., 1830. Rock floors in arid and humid climates: I. The Journal of Geology 38 (1), 1–27.

Davis, W.M., 1899. The geographical cycle. The Geographical Journal 14 (5), 481–504.

Davis, M.W., White, N.J., Priestley, K.F., Baptie, B.J., Tilmann, F.J., 2012. Crustal structure of the British Isles and its epeirogenic consequences. Geophysical Journal International 190, 705–725.

Deschamps, F., Li, Y., 2019. Core-mantle boundary dynamic topography: influence of postperovskite viscosity. Journal of Geophysical Research: Solid Earth 124 (8), 9247–9264.

Divins D., 2008. NGDC total sediment thickness of the world's oceans and marginal seas. Retrieved from: ngdc.noaa.gov/mgg/sedthick/sedthick.html.

Dugda, M., Nyblade, A., Julia, J., Langstone, C., Ammon, C., Simiyu, S., 2005. Crustal structure in Ethiopia and Kenya from receiver function analysis: implications for rift development in eastern Africa. Journal of Geophysical Research 110. Available from: https://doi.org/10.1029/2004jb003065.

Eakin, C., Lithgow-Bertelloni, C., 2018. An overview of dynamic topography: the influence of mantle circulation on surface topography and landscape. Mountains, Climate and Biodiversity 37.

Ebinger, C., Sleep, N., 1998. Cenozoic magmatism throughout east Africa resulting from impact of a single plume. Nature 395, 788–791.

Ebinger, C., Deino, A., Drake, R., Tesha, A., 1989. Chronology of volcanism and rift basin propagation: Rungwe Volcanic Province, East Africa. Journal of Geophysical Research 94, 15785–15803.

Elmejdoub, N., Jedoui, Y., 2009. Pleistocene raised marine deposits of the Cap Bon peninsula (NE Tunisia): records of sea-level highstands, climatic changes, and coastal uplift. Geomorphology 112, 179–189.

England, P., Molnar, P., 1990. Surface uplift, uplift of rocks, and exhumation of rocks. Geology 18, 1173–1177.

Farrington, R.J., Stegman, D.R., Moresi, L.N., Sandiford, M., May, D.A., 2010. Interactions of 3D mantle flow and continental lithosphere near passive margins. Tectonophysics 483, 20–28.

Fishwick, S., Bastow, I., 2011. Towards a better understanding of African topography, a review of passive-source seismic studies of the African crust and upper mantle. Geological Society of London, Special Publications 357, 343–371.

Flament, N., 2019. Present-day dynamic topography and lower-mantle structure from palaeogeographically constrained mantle flow models. Geophysical Journal International 216, 2158–2182.

Flament, N., Gurnis, M., Müller, R.D., 2013. A review of observations and models of dynamics topography. Lithosphere 5 (2), 189–210.

Forte, A., Quéré, S., Moucha, R., Simmons, N., Grand, S., Mitrovica, J., et al., 2010. Joint seismic-geodynamic-mineral physical modelling of African geodynamics: a reconciliation of deep-mantle convection with surface geophysical constraints. Earth and Planetary Science Letters 295, 329–341.

Furman, T., Bryce, J., Rooney, T., Hana, B., Yirgu, G., Ayalew, D., 2006. Heads and tails: 30 million years of the Afar Plume. Geological Society of London, Special Publications 259, 95–119.

Ghelichkhan, S., Murbock, M., Colli, L., Pail, R., Bunge, H.-P., 2018. On the observability of epeirogenic movement in current and future gravity missions. Gondwana Research 53, 273–284.

Guiraud, M., Buta-Neto, A., Quesne, D., 2010. Segmentation and differential post-rift uplift at the Angola margin as recorded by the transform-rifted Benguela and oblique-to-orthogonal rifted Kwanza basins. Marine and Petroleum Geology 27, 1040–1068.

Gurnis, M., 1993. Phanerozoic marine inundation of continents driven by dynamic topography above subducting slabs. Nature 364, 589–593.

Gurnis, M., Müller, R.D., Moresi, L., 1998. Cretaceous vertical motion of Australia and the Australian-Antarctic Discordance. Science (New York, N.Y.) 279, 1499–1504.

Gurnis, M., Mitrovica, J.X., Ritsema, J., van Heijst, H.-J., 2000. Constraining mantle density structure using geological evidence of surface uplift rates: the case of the African Superplume. Geochemistry, Geophysics, Geosystems 1 (7).

Hager, B.H., Richards, M.A., 1989. Long-wavelength variations in Earth's geoid: physical models and dynamical implications. Philosophical Transactions of the Royal Society A 328, 309–327.

Hager, B.H., Clayton, R.W., Richards, M.A., 1985. Lower mantle heterogeneity, dynamic topography, and the geoid. Nature 313, 541–545.

Hansen, S., Nyblade, A., Benoit, M., 2012. Mantle structure beneath Africa and Arabia from adaptively parameterised P-wave tomography: implications for the origin of Cenozoic Afro-Arabian tectonism. Earth and Planetary Science Letters 319, 23–34.

Hoggard, M.J., White, N.J., Al-Attar, D., 2016. Global dynamic topography observations reveal limited influence of large-scale mantle flow. Nature Geoscience 9 (6), 456–463.

Hoggard M.J., Austermann J., Randel C., Stephenson S., 2021. Observational estimates of dynamic topography through space and time. In: Marquadt H., Ballmer M., Cottaar S., Konter J. (eds.), Mantle Convection and Surface Expressions. *AGU Geophysical Monograph Series*, Washington, DC, Chapter 15.

Holmes, A., 1944. Principles of Physical Geology. Nelson, London and Edinburgh.

Ismail, A., Boukhary, M., Abdel Naby, A., 2010. Subsurface stratigraphy and micropaleontology of the Neogene rocks, Nile Delta, Egypt. Geologia Croatica 63 (1), 1–26.

Ismail-Zadeh, A., Schubert, G., Tsepelev, I., Korotkii, A., 2004. Inverse problem of thermal convection: numerical approach and application to mantle plume restoration. Physics of the Earth and Planetary Interiors 145 (1–4), 99–114.

Jackson, I., Faul, U.H., 2010. Grainsize-sensitive viscoelastic relaxation in olivine: towards a robust laboratory-based model for seismological application. Physics of the Earth and Planetary Interiors 183, 151–163.

Jess, S., Stephenson, R., Brown, R., 2018. Evolution of the central West Greenland margin and the Nuussuaq Basin: localised basin uplift along a stable continental margin proposed from thermochronological data. Basin Research 30 (6), 1230–1246.

Jones, S., Lovell, B., Crosby, A., 2012. Comparison and modern and geological observations of dynamic support from mantle convection. Journal of the Geological Society, London 169, 745–758.

Ju, Y., Wang, G., Li, S., Sun, Y., Suo, Y., et al., 2022. Geodynamic mechanism and classification of basins in the earth system. Gondwana Research 102, 200–228.

King, L.C., 1942. South African Scenery: A Textbook of Geomorphology. Oliver and Boyd, Edinburgh.

Lamb, S., Moore, J.D., Perez-Gussinye, M., Stern, T., 2020. Global whole lithosphere isostasy: implications for surface elevations, structure, strength, and densities of the continental lithosphere. Geochemistry, Geophysics, Geosystems 21, e2020GC009150.

Lavier, L., Steckler, M., Brigaud, F., 2001. Climatic and tectonic controls on the Cenozoic evolution of the West African continental margin. Marine Geology 178, 63–80.

Li, Z., Aschoff, J., 2021. Constraining the effects of dynamic topography on the development of Late Cretaceous Cordilleran foreland basin, western United States. GSA Bulletin 134 (1–2), 446–462.

Liégeois J.-P., Benhallou A., Azzouni-Sekkal A., Bonin B., 2005. The Hoggar swell and volcanism: Reactivation of the Precambrian Tuareg shield during Alpine convergence and West African Cenozoic volcanism. In: Foulger G.R., Natland J.H., Presnall D.C., Anderson D.L. (eds), Plates, Plumes and Paradigms, *Geological Society of America Special Paper*, 388, 379–400.

Lithgow-Bertelloni, C., Richards, M., 1998. The dynamics of Cenozoic and Mesozoic plate motions. Reviews of Geophysics 36 (1), 27–78.

Lithgow-Bertelloni, C., Silver, P., 1998. Dynamic topography, plate driving forces and the African superswell. Nature 395, 269–272.

Lithgow-Bertelloni, C., Richards, M.A., Ricard, Y., O'Connell, R.J., Engebretson, D.C., 1993. Toroidal-polodial partitioning of plate motions since 120 Ma. Geophysical Research Letters 20 (5), 375–378.

Liu, L., 2015. The ups and downs of North America: evaluating the role of mantle dynamic topography since the Mesozoic. Reviews of Geophysics 53 (3), 1022–1049.

Liu, L., Spasojevic, S., Gurnis, M., 2008. Reconstructing Farallon Plate subduction beneath North America back to the Late Cretaceous. Science (New York, N.Y.) 322, 934–938.

Lyell, C., 1830. Principles of Geology, Being an Attempt to Explain the Former Changes of the Earth's Surface, by Reference to Causes Now in Operation. John Murray, London.

Matthews, K.J., Maloney, K.T., Zahirovic, S., Williams, S.E., Seton, M., Müller, R.D., 2016. Global plate boundary evolution and kinematics since the Late Palaeozoic. Global and Planetary Change 146, 226–250.

McDougall, I., Brown, F., 2009. Timing of volcanism and evolution of the northern Kenya Rift. Geological Magazine 146 (1), 34–47.

McKenzie, D.P., 1967. Some remarks on heat flow and gravity anomalies. Journal of Geophysical Research 72 (24), 6261–6273.

Molnar, P., England, P.C., Jones, C.H., 2015. Mantle dynamics, isostasy, and the support of high terrain. Journal of Geophysical Research 120. Available from: https://doi.org/10.1002/2014JB011724.

Moucha, M., Forte, A.M., 2011. Changes in African topography driven by mantle convection. Nature Geoscience 4(10), 707–712.

Moucha, R., Forte, A.M., Mitrovica, J.X., Rowley, D.B., Quéré, S., Simmons, N.A., et al., 2008. Dynamic topography and long-term sea-level variations: there is no such thing as a stable continental platform. Earth and Planetary Science Letters 271, 101–108.

Müller, R.D., Roest, W.R., Royer, J.Y., Gahagan, L.M., Sclater, J.G., 1997. Digital isochrons of the world's ocean floor. Journal of Geophysical Research 102 (B2), 3211–3214.

Nair, S., Gao, S., Liu, K., Silber, P., 2006. Southern Africa crustal evolution and composition: constraints from receiver function studies. Journal of Geophysical Research 111. Available from: https://doi.org/10.1029/2005JB003802.

Ni, S., Tan, E., Gurnis, M., Helmberger, D., 2002. Sharp sides to the African superplume. Science (New York, N.Y.) 296, 1850–1852.

Nyagah, K., 1995. Stratigraphy, depositional history and environments of deposition of Cretaceous through Tertiary strata in the Lamu Basin, southeast Kenya, and implications for reservoirs for hydrocarbon exploration. Sedimentary Geology 96, 43–71.

Nyblade, A., Robinson, S., 1994. The African superswell. Geophysical Research Letters 21, 765–768.

Odada, E., 1996. Geological evolution of coastal Kenya as inferred from sedimentary sequences and marine terraces. Africa Geoscience Review 3 (3), 1–18.

O'Malley, C., White, N.J., Roberts, G.G., Stephenson, S.N., 2021. Large-scale tectonic forcing of the African landscape and its drainage patterns. Journal of Geophysical Research: Earth Surface .

Partridge, T., 1998. Of diamonds, dinosaurs and diastrophism: 150 Myr of landscape evolution in Southern Africa. South African Journal of Geology 101, 167–184.

Partridge, T., Maud, R., 1987. Geomorphic evolution of southern Africa since the Mesozoic. South African Journal of Geology 90, 167–184.

Paul, J.D., 2021. Controls on eroded rock volume, a proxy for river incision, in Africa. Geology 49 (4), 422–427.

Paul, J.D., Eakin, C.M., 2017. Mantle upwelling beneath Madagascar: evidence from receiver function analysis and shear wave splitting. Journal of Seismology 21 (4), 825–836.

Paul, J.D., Roberts, G.G., White, N., 2014. The African landscape through space and time. Tectonics 33 (6), 898–935.

Pelletier, J.D., 2010. How do pediments form? A numerical modelling investigation with comparison to pediments in southern Arizona, USA. Bulletin of the Geological Society of America 122, 1815–1829.

Pekeris, C.L., 1935. Thermal convection in the interior of the Earth. Geophysical Supplements to the Monthly Notices of the Royal Astronomical Society 3 (8), 343–367.

Permenter, J., Oppenheimer, C., 2007. Volcanoes of the Tibesti massif (Chad, northern Africa). Bulletin of Volcanology 69, 609–626.

Poore, H.R., Samworth, R., White, N.J., Jones, S.M., McCave, I.N., 2006. Neogene overflow of Northern Component Water at the Greenland-Scotland Ridge. Geochemistry, Geophysics, Geosystems 7 (6).

Priestley, K., McKenzie, D.P., 2013. The relationship between shear wave velocity, temperature, attenuation, and viscosity in the shallow part of the mantle. Earth and Planetary Science Letters 381, 78–91.

Priestley, K., McKenzie, D., Debayle, E., Pilidou, S., 2008. The African upper mantle and its relationship to tectonics and surface geology. Geophysical Journal International 175. Available from: https://doi.org/10.1111/j.1365-246X.2008.03951.x.

Reid, D., Cooper, A., Rex, D., Harmer, R., 1990. Timing of post-Karoo alkaline volcanism in southern Namibia. Geological Magazine 5, 427–433.

Reijers, T., 2011. Stratigraphy and sedimentology of the Niger Delta. Geolgos 17 (3), 133–162.

Ricard, Y., Fleitout, L., Froidevaux, C., 1984. Geoid heights and lithospheric stresses for a dynamic earth. Annales Geophysicae 2, 267–285.

Ricard, Y., Richards, C., Lithgow-Bertelloni, C., Le Stunff, Y., 1993. A geodynamic model of mantle density heterogeneity. Journal of Geophysical Research 98, 21895–21909.

Richards, M.A., Hager, B.H., 1984. Geoid anomalies in a dynamic earth. Journal of Geophysical Research 89 (B7), 5987–6002.

Ritsema, J., Deuss, A., van Heijst, H., Woodhouse, J., 2011. S40RTS: a degree-40 shear-velocity model for the mantle from new Rayleigh wave dispersion, teleseismic traveltime and normal-mode splitting function measurements. Geophysical Journal International 184 (3), 1223–1236.

Roberts, G.G., White, N.J., 2010. Estimating uplift rate histories from river profiles using African examples. Journal of Geophysical Research 115, B02406.

Roberts, G., Paul, J., White, N., Winterbourne, J., 2012a. Temporal and spatial evolution of dynamic support from river profiles: a framework for Madagascar. Geochemistry, Geophysics, Geosystems 13 (4), Q04004.

Roberts, G.G., White, N.J., Martin-Brandis, G.L., Crosby, A.G., 2012b. An uplift history of the Colorado Plateau and its surroundings from inverse modelling of longitudinal river profiles. Tectonics 31, TC4022.

Roberts, G.G., White, N., Lodhia, B.H., 2019. The generation and scaling of longitudinal river profiles. Journal of Geophysical Research: Earth Surface 124, 137–153.

Rodriguez Tribaldos, V., White, N., 2018. Implications of preliminary subsidence analyses for the Parnaiba cratonic basin. Geological Society of London, Special Publications 472 (1), 147–156.
Rovere, A., Raymo, M.E., Mitrovica, J.X., Hearty, P.J., O'Leary, M.J., Inglis, J.D., 2014. The mid-Pliocene sea-level conundrum: glacial isostasy, eustasy, and dynamic topography. Earth and Planetary Science Letters 387, 27–33.
Runcorn, S.K., 1965. Changes in the convection pattern in the Earth's mantle and continental drift: evidence for a cold origin of the Earth. Philosophical Transactions of the Royal Society A 258, 228–251.
Sahagian, D., 1988. Epeirogenic motions of Africa as inferred from Cretaceous shoreline deposits. Tectonics 7 (1), 125–138.
Salman, G., Abdula, I., 1995. Development of the Mozambique and Ruvuma sedimentary basins, offshore Mozambique. Sedimentary Geology 96, 7–41.
Scotese, C., Gahagan, L., Larson, R., 1988. Plate tectonic reconstructions of the Cretaceous and Cenozoic ocean basins. Tectonophysics 155, 27–48.
Séranne, M., Nzé Abeigne, C.-R., 1999. Oligocene to Holocene sediment drifts and bottom currents on the slope of Gabon continental margin (west Africa): consequences for sedimentation and southeast Atlantic upwelling. Sedimentary Geology 128, 179–199.
Sestini, G., 1989. Nile Delta: a review of depositional environments and geological history. Geological Society of London, Special Publications 41, 99–127.
Spasojevic, S., Gurnis, M., 2012. Sea level and vertical motion of continents from dynamic Earth models since the Late Cretaceous. AAPG Bulletin 96 (11), 2037–2064.
Stadler, G., Gurnis, M., Burstedde, C., Wilcox, L.C., Alisic, L., Ghattas, O., 2010. The dynamics of plate tectonics and mantle flow: from local to global scales. Science (New York, N.Y.) 329, 1033–1038.
Steinberger, B., 2007. Effects of latent heat release at phase boundaries on flow in the earth's mantle, phase boundary topography, and dynamic topography at the earth's surface. Physics of the Earth and Planetary Interiors 164, 2–20.
Steinberger, B., O'Connell, R.J., 1997. Changes of the earth's rotation axis owing to advection of mantle density heterogeneities. Nature 387 (6629), 169–173.
Stephenson, S.N., White, N.J., Li, T., Robinson, L.F., 2019. Disentangling interglacial sea level and global dynamic topography: analysis of Madagascar. Earth and Planetary Science Letters 519, 61–69.
Tackley, P.J., 2000. Self-consistent generation of tectonic plates in time-dependent, three-dimensional mantle convection simulations. 2: strain weakening and asthenosphere. Geochemistry, Geophysics, Geosystems 1 (8), GC000043.
Tapley, B., Ries, J., Bettadpur, S., Chambers, D., Cheng, M., et al., 2005. GGM02: an improved Earth gravity field model from GRACE. Journal of Geodesy 79, 467–478.
Thorpe, R., Smith, K., 1974. Distribution of Cenozoic volcanism in Africa. Earth and Planetary Science Letters 22 (1), 91–95.
Tokam, A.-P., Tabod, C., Nyblade, A., Juli, J., 2010. Structure of the crust beneath Cameroon, west Africa, from the join inversion of Rayleigh wave group velocities and receiver functions. Geophysical Journal International 183 (2). Available from: https://doi.org/10.1111/j.1365-246X.2010.04776.x.
Walford, H.L., White, N.J., Sydow, J.C., 2005. Solid sediment load history of the Zambezi delta. Earth and Planetary Science Letters 238, 49–63.
Wang, Y., Huang, J., Zhong, S., Chen, J., 2016. Heat flux and topography constraints on thermochemical structure below North China Craton regions and implications for evolution of cratonic lithosphere. Journal of Geophysical Research: Solid Earth 121, 3081–3098.
Whipple, K.X., Tucker, G.E., 1999. Dynamics of the stream-power river incision model: implications for height limits of mountain ranges, landscape response timescales, and research needs. Journal of Geophysical Research 107, B2–2039.
White, N., Lovell, B., 1997. Measuring the pulse of a plume with the sedimentary record. Nature 387, 888–891.
White, R., McKenzie, D., 1995. Mantle plumes and flood basalts. Journal of Geophysical Research 100, 17543–17585.
Winterbourne, J., Crosby, A., White, N., 2009. Depth, age and dynamic topography of oceanic lithosphere beneath the heavily sedimented Atlantic margin. Earth and Planetary Science Letters 287 (1–2), 137–151.
Winterbourne, J., White, N., Crosby, A., 2014. Accurate measurements of residual topography from the oceanic realm. Tectonics 33, 982–1015.
Woolley, A., 2001. Alkaline Rocks and Carbonatites of the World. Part 3. Geological Society of London, Africa. London, p. 372.
Yamauchi, H., Takei, Y., 2016. Polycrystal anelasticity at near-solidus temperatures. Journal of Geophysical Research: Solid Earth 121 (11), 7790–7820.
Yang, T., Gurnis, M., 2016. Dynamic topography, gravity, and the role of lateral viscosity variations from inversion of global mantle flow. Geophysical Journal International 207 (2), 1186–1202.
Yang, A., Yang, T., 2021. Controls on the present-day dynamic topography predicted from mantle flow models since 410 Ma. Geophysical Journal International 225, 1637–1652.
Young, A., Flament, N., Maloney, K., Williams, S., Matthews, K., Zahirovic, S., et al., 2019. Global kinematics of tectonic plates and subduction zones since the late Paleozoic Era. Geoscience Frontiers 10, 989–1013.
Zhang, N., Zhong, S., Flowers, R.M., 2012. Predicting and testing continental vertical motion histories since the Paleozoic. Earth and Planetary Science Letters 317–318, 426–435.

CHAPTER

6

The Songliao basin, China

Zhiqiang Feng[1] *and Stephan Alan Graham*[2]

[1]**Petroleum Exploration and Production Research Institute, SINOPEC, Beijing, China** [2]**Department of Earth and Planetary Sciences, Stanford University, Stanford, CA, United States**

Introduction

At 260,000 km^2, the NNE-SSW-elongate Songliao basin is the largest post-Paleozoic sedimentary basin in onshore northeastern China and is one of the world's most petroliferous basins. The basin is notable for the high quality of its nonmarine petroleum source rocks and its considerable hydrocarbon potential (Demaison, 1984; Yang, 1985; Feng et al., 2010a). In the early Paleozoic, what is now NE China was an ocean basin between the Siberian and Sino-Korean cratons. The Paleo-Asian Ocean closed by the end of the Paleozoic (Sengör et al., 1993; Johnson et al., 2008), trapping several smaller terranes, including the Erguna, Songnen, and Jiamusi terranes within a collisional suture collage, termed the Central Asian Orogenic Belt, which became the basement for the Mesozoic basins of southeast Mongolia and northeast China (Li et al., 2021) (Fig. 6.1). The structural weakness of this noncratonized basement of accreted terranes set the stage of localization of Mesozoic basins (e.g., Wang et al., 2016). Subduction of the Izanagi plate and subsequently the Kula plate beneath the Eurasia commenced by the end of the Jurassic. Extension associated with upwelling back-arc mantle likely driven by slab rollback gave rise to a series of Mesozoic rift basins, including the Songliao, the Sunwujiayin, and the Great Sanjiang basins (GSB) (e.g., Zhang et al., 2017a,b).

The Songliao basin, locally filled with about 10 km of Jurassic-Neogene deposits with Cretaceous rocks comprising the majority, overlies a basement complex composed of Paleozoic metamorphic and volcanic rocks intruded by Paleozoic granitic plutons (Tian and Han, 1993; Gao and Cai, 1997; Li et al., 2012). In this paper, we describe the structural geology and stratigraphic fill of the Songliao basin in terms of tectonostratigraphic units, which reflect the tectonic evolution of the basin, and briefly summarize how structure, stratigraphy, and basin evolution yielded a world-class petroliferous basin. In doing so, we build on and expand the synthesis of Feng et al. (2010a) and refer readers to that paper for additional details on the stratigraphy of the basin.

Geological setting

An E-W cross section of the Upper Jurassic–Cretaceous stratigraphic fill of the Songliao basin presents a steer's head geometry (e.g., Li, 1996; Gao and Cai, 1997; White and Mackenzie, 1988) (Fig. 6.2), reflecting its early history as a rift basin. Six principal structural units are recognized in Songliao basin: Northern Plunge, Northeastern Uplift, Central Downwarp, Southeastern Uplift, Southwestern Uplift, and Western Slope (Fig. 6.3).

The tectonic evolution of the Songliao basin has usually been described as reflecting four stages: mantle upwelling and doming, rifting, thermal subsidence, and basin inversion (Gao and Cai, 1997, Wang et al., 2016; Liu et al., 2017; Zhang et al., 2017a,b; Li et al., 2020). In this paper, we regard the inversion stage as representing a retroforeland phase comprising a foredeep that evolved into a wedgetop basin. (DeCelles and Giles, 1996). Fig. 6.4 summarizes the stratigraphy of the Songliao basin. Asthenospheric upwelling and subsequent cooling served as the primary driving mechanism of the late Mesozoic rifting and subsidence, as expressed in latest Jurassic extension and widespread volcanism. Upper Jurassic and Lower Cretaceous pre- and syn-rift deposits

Regional Geology and Tectonics: Volume 2
DOI: https://doi.org/10.1016/B978-0-444-64136-6.00003-8

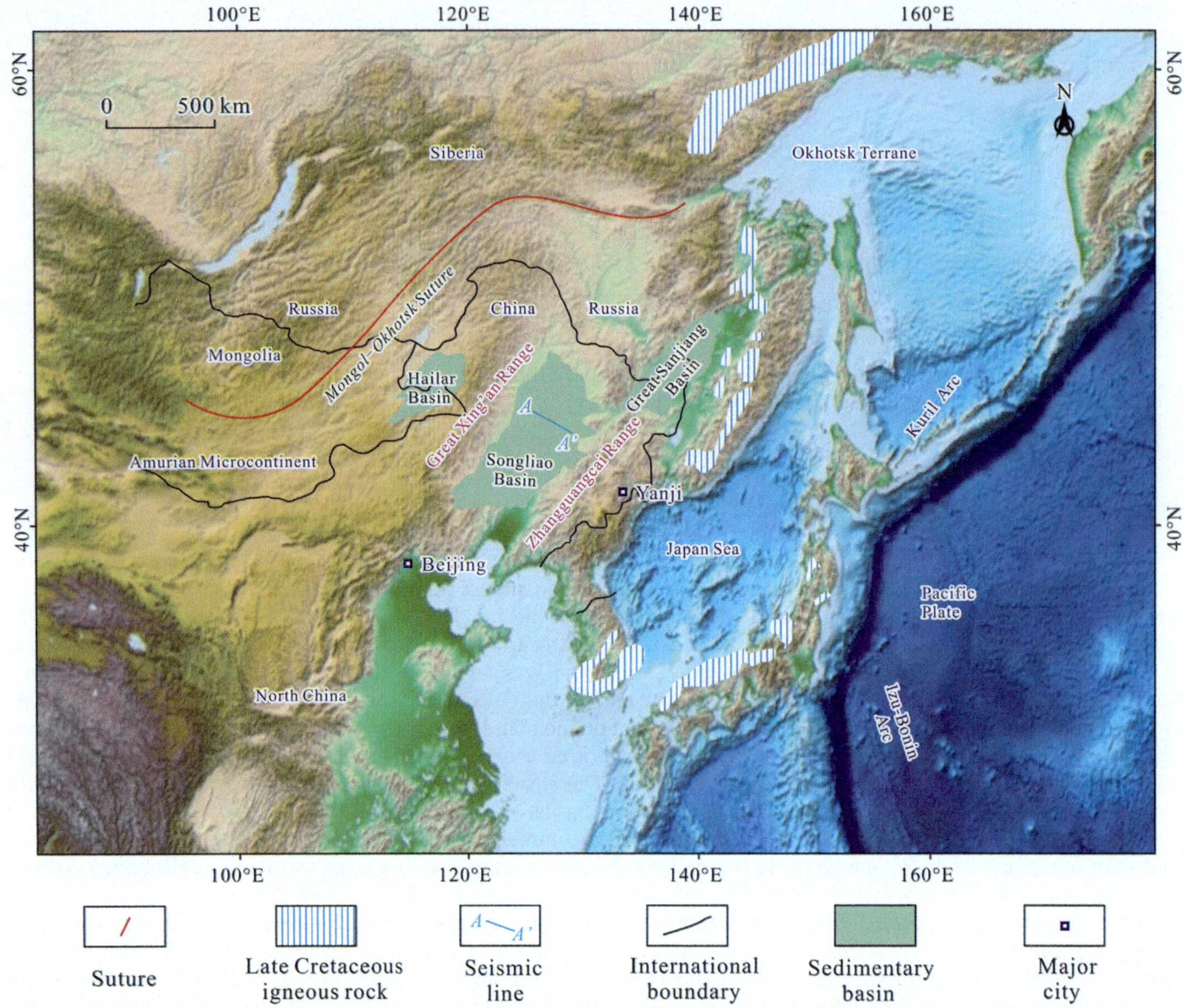

FIGURE 6.1 Regional map of eastern Asia showing the location of Songliao basin and major Mesozoic tectonic and basin features. A-A′ is seismic sections of Fig. 6.2.

are restricted to individual fault-block subbasins, where they range up to 7000 m thick in the Shiwu and Dehui Depressions (Xie et al., 2003; Yang et al., 2021). Postrift thermal subsidence permitted expansion to a large integrated basin, identifiable as the Songliao basin. Postrift Cretaceous strata, 3000–6000 m thick, unconformably overlie syn-rift strata and lap onto basement surrounding the isolated rift basins (Li, 1996; Li et al., 1997; Ren et al., 2002, Feng et al., 2010a). Geophysical data indicate that the depth to the Moho ranges from about 27 km parallel to the deepest part of the NNE-trending basin to more than 35 km beneath the surrounding mountains (Gao and Cai, 1997; Wang et al., 2016).

Pronounced structural inversion of the basin occurred in the latest Cretaceous with attendant erosional loss of considerable basin fill. Whereas many other authors have emphasized inversion, we (Feng and Graham, 2023) emphasize that the final evolutionary phase of Songliao basin history was as a foreland basin (Zhang et al., 2017a), as reflected in a foreland tectonostratigraphic unit with a preserved thickness of 2000–3000 m and characterized by significant migration of subsidence and deposition centers from the east to the west in the basin. The Songliao foredeep evolved into a foreland wedgetop basin in the latest Cretaceous, as shortening affected nearly all of the basin fill. Generally undeformed Paleocene and Eocene deposits, with a thickness of 0–510 m, are limited to the infill of erosional valleys developed at the end of the Cretaceous in the western basin.

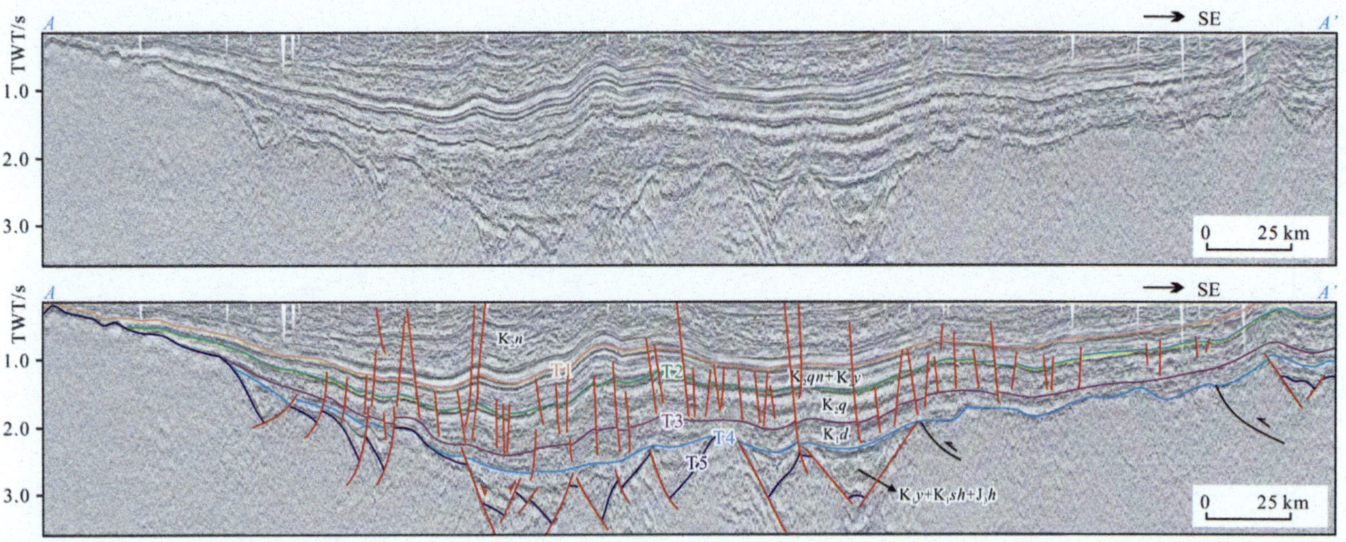

FIGURE 6.2 Regional seismic section and structural cross section across the central part of the Songliao basin. Lithologic and formation information is derived from geophysical logs and cores tied to seismic sections. Location is shown in Fig. 6.1 (A-A').

Prerift and syn-rift tectonostratigraphic units

Much of the basin fill remains in the subsurface, so subsurface data are essential for understanding the structure and stratigraphy of deeply buried prerift and syn-rift tectonostratigraphic units. Extensive gravity, magnetic, and 2-D and 3-D seismic surveys cover the basin. These data reveal the distribution of fault blocks and their structure.

The NNE-trending Sunwu-Shuangliao and Harbin-Siping faults partition the basin into western and eastern rifts separated by the central uplift; these three elements include 30 significant fault blocks (Fig. 6.5). Similar to other rifts (e.g., Rosendahl et al., 1986), structural motifs include half-grabens, grabens, and compound faulted basins. Extension and subsidence were greatest in the western rift, which later influenced the structural and stratigraphic architecture of the foreland phase of basin evolution.

Half-grabens dominate the western rift belt, and Changling (Fig. 6.6A), Gulong, and Lindian (Fig. 6.6C) are the three most prominent rift blocks in the western rift belt. Fault blocks in the south are larger than those in the north and contain alluvial, lacustrine, and volcanic strata. Andesitic and lesser felsic and basaltic volcanics attain thicknesses of 600 m. The eastern rift belt contains half-grabens and grabens. Dehui, Yushu, Shuangcheng (Fig. 6.6B), and Suihua are the principal fault blocks. Prerift and syn-rift alluvial and lacustrine deposits and volcanic rocks in the Shuangcheng fault block exceed 3350 m (Fig. 6.6D). Volcanic rocks range up to 1000 m thick (Yang et al., 2021).

The best documented of the nine fault blocks in the central uplift, Xujiaweizi half graben (Fig. 6.6D), trends NNE to NNW from south to north. Lowermost basin fill consists of volcanic strata overlain by lacustrine and alluvial facies. Felsic and andesitic volcanic rocks are widespread and range up to 2700 m thick. Surprisingly, these volcanic rocks form the main reservoir of the Qingshen gas field (Feng, 2008).

Late Cretaceous tectonic shortening, discussed below, modified the structure of the three rift belts. The eastern rift belt was differentially folded and uplifted (Fig. 6.2). The intensity of shortening deformation decreases to the west, *where an important depocenter persisted through the Cretaceous.*

The Upper Jurassic Huoshiling (J_3h), Early Cretaceous Shahezi (K_1sh), and Yingcheng (K_1y) formations comprise the prerift and syn-rift tectonostratigraphic units of the Songliao basin. Two additional unconformities corresponding to the $T4^2$ and $T4^1$ seismic markers divide the tectonostratigraphic unit into sequences I^1, I^2, and I^3 (Fig. 6.4). Seismic data demonstrate that the 500- to 1600-m-thick Huoshiling Formation is not confined to rift fault blocks, suggesting that deposition commenced before significant extension occurred. Huoshiling strata include extensive volcanic rocks interbedded with alluvial and fluvial facies.

The Lower Cretaceous Shahezi Formation is confined to fault blocks in the eastern and central rifts (Figs. 6.2, 6.5, and 6.6), where it attains thicknesses of 2500 m in the Xujiaweizi and Changling fault blocks. In contrast to underlying and overlying formations, the Shahezi lacks significant volcanic strata except for

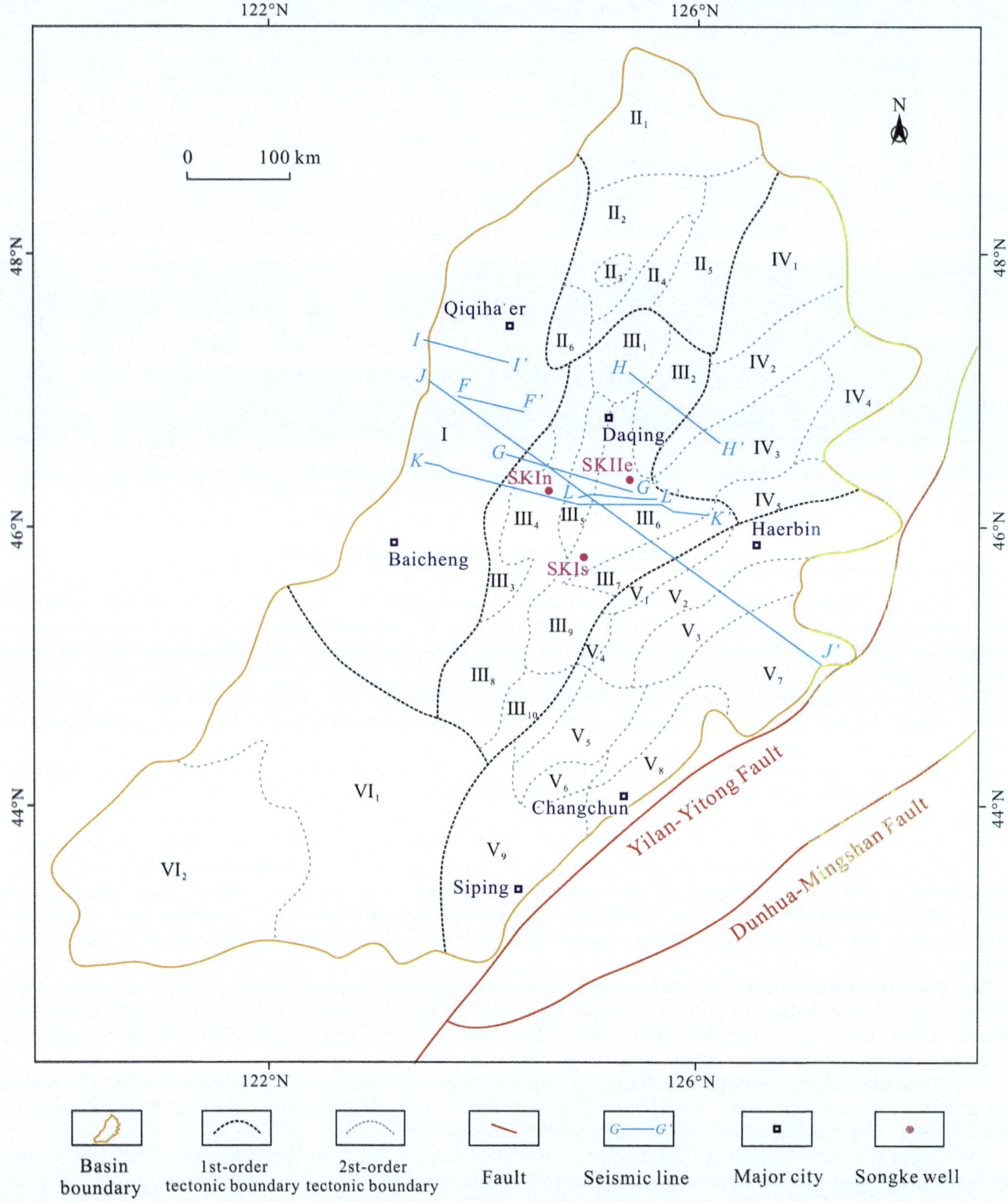

FIGURE 6.3 Structural subdivisions of the Songliao basin. I—the western slope zone; II—the northern plunge zone; II1—Nenjiang Terrace; II2—Yian Sag; II3—Sanxing Anticline; II4—Keshan-Yilong Anticlinal belt; II5—Qianyuan Anticlinal belt; II6—Wuyuer Sag; III—the central downwarp zone; III1—Heiyupao Sag; III2—Mingshui Terrace; III3—Longhupao-Honggang Terrace; III4—Qijia-Gulong Sag; III5—Daqing Anticline; III6—Sanzhao Sag; III7—Chaoyanggou Terrace; III8—Changling Sag; III9—Fuyu Uplift; III10—Shuangtuozi Terrace; IV—the northeastern uplift zone; IV1—Hailun Uplift; IV2—Suiling Anticline; IV3—Suihua Sag; IV4—Qing'an Uplift; IV5—Hulan Uplift; V—the southeastern uplift zone; V1—Changchunling Anticlinal belt; V2—Binxian-Wangfu Sag; V3—Qingshankou Uplift; V4—Denglongku Anticlinal belt; V5—Diaoyutai Uplift zone; V6—Yangdachengzi Anticlinal belt; V7—Yushu-Dehui Sag; V8—Jiutai Terrace; V9—Dehui-Lishu Sag; VI—the southwestern uplift zone; VI1—Gamatu Uplift; VI2—Kailu Sag. Source: *Modified from Feng, Z.Q., Jia, C.Z., Xie, X.N., Zhang, S., Feng, Z.H., Cross, T.A., 2010a. Tectonostratigraphic units and stratigraphic sequences of the nonmarine Songliao basin, northeast China. Basin Research, 22, 79–95.*

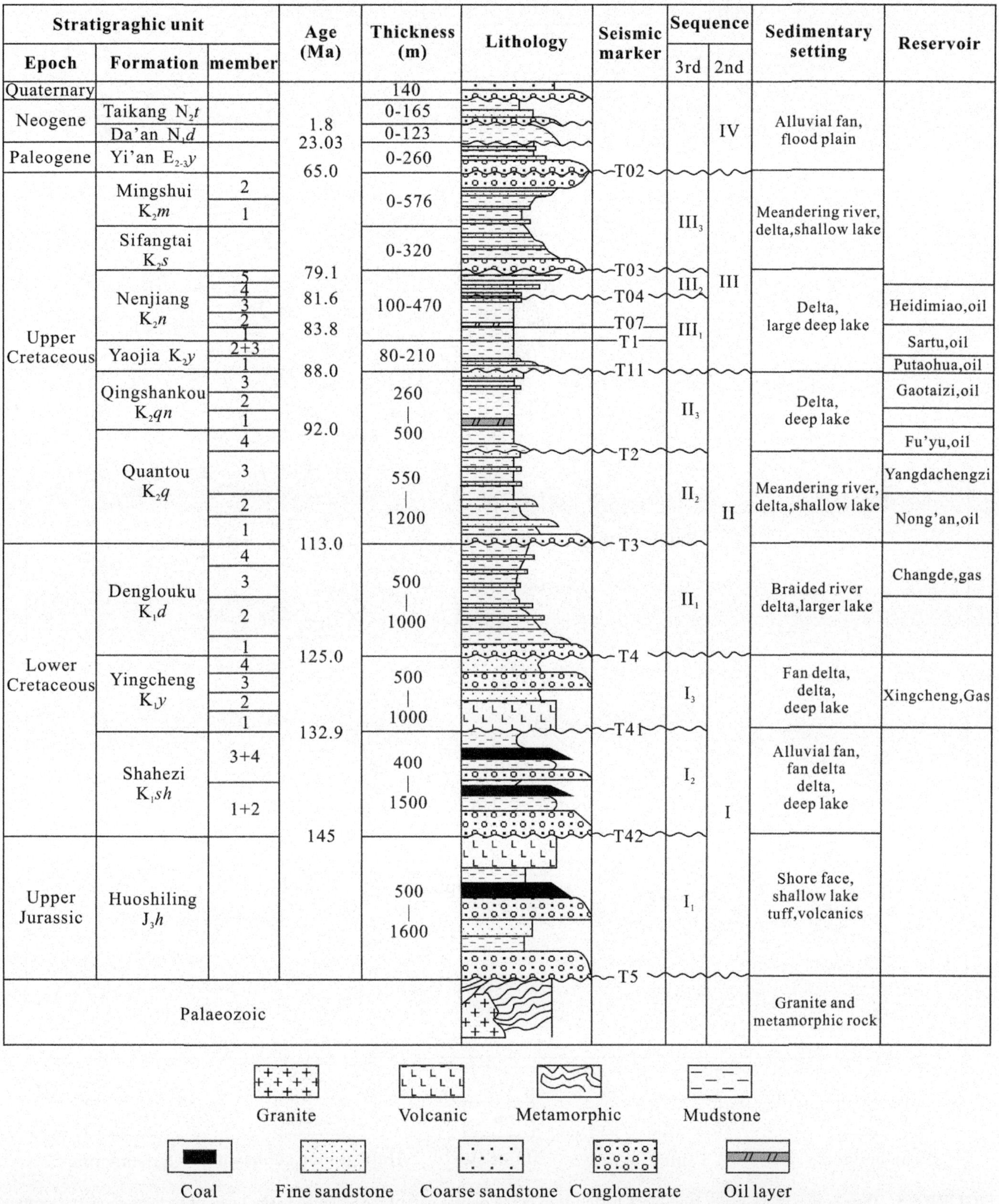

FIGURE 6.4 Generalized stratigraphic column of the Songliao basin. Source: *Modified from Feng, Z.Q., Jia, C.Z., Xie, X.N., Zhang, S., Feng, Z. H., Cross, T.A., 2010a. Tectonostratigraphic units and stratigraphic sequences of the nonmarine Songliao basin, northeast China. Basin Research, 22, 79–95.*

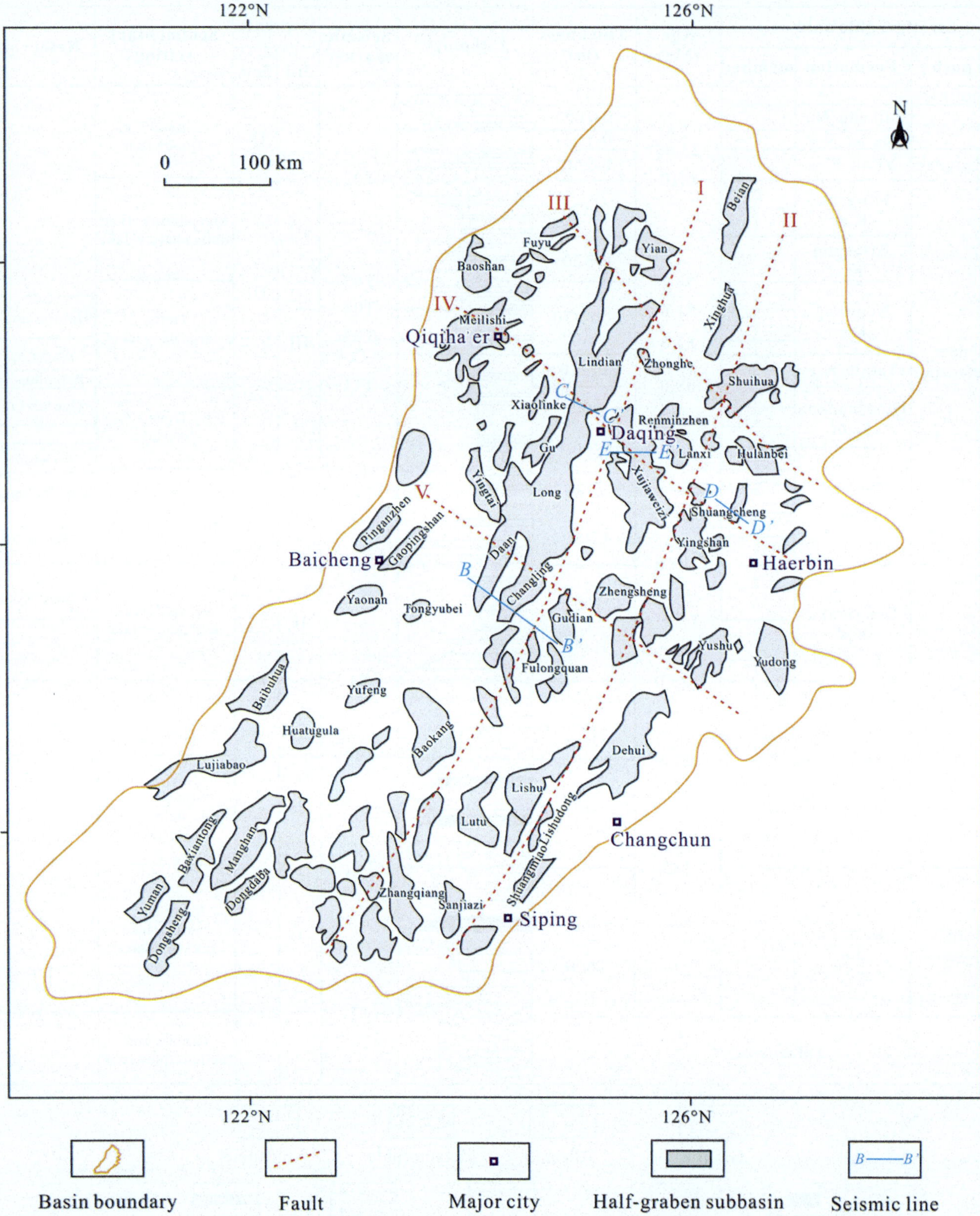

FIGURE 6.5 Distribution of the major fault-block basins deep in the central part of the Songliao basin. Lines of sections show the locations of seismic sections in Fig. 6.5. Roman numerals refer to the major faults. I: Sunwu-Shuangliao fault. II: Harbin-Siping fault. III: Suihua-Dedu fault. IV: Binzhou fault. V: Songhuajiang fault. Source: *Modified from Feng, Z.Q., Jia, C.Z., Xie, X.N., Zhang, S., Feng, Z.H., Cross, T.A., 2010a. Tectonostratigraphic units and stratigraphic sequences of the nonmarine Songliao basin, northeast China. Basin Research, 22, 79–95.*

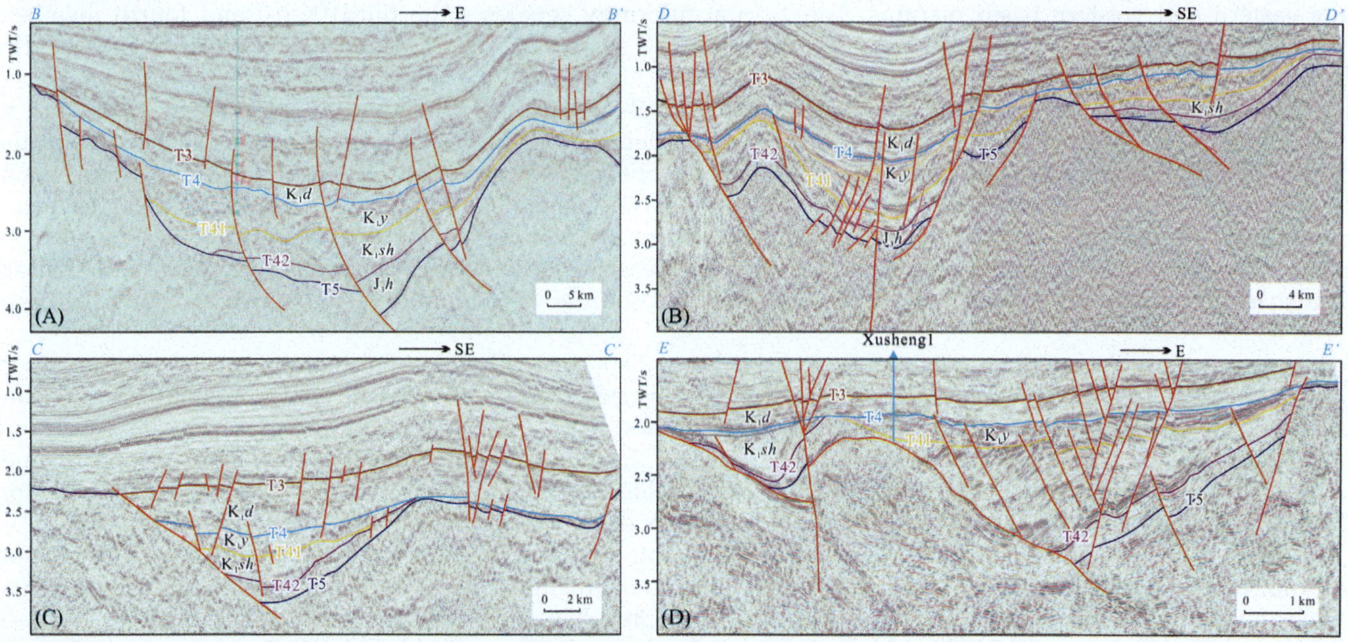

FIGURE 6.6 Seismic profiles through four deep fault-block basins in the Sangliao basin. Locations of profiles are shown in Fig. 6.5. (A) Changling fault block, (B) Shuangcheng fault block, (C) Lindian fault block, and (D) Xujiaweizi fault block. Source: *(A, B, and D) Modified from Feng, Z.Q., Jia, C.Z., Xie, X.N., Zhang, S., Feng, Z.H., Cross, T.A., 2010a. Tectonostratigraphic units and stratigraphic sequences of the nonmarine Songliao basin, northeast China. Basin Research, 22, 79–95; Feng, Z.Q., Dong, L., Tong, Y., Feng, Z.H., Zhang, S., Wu, G.K., et al., 2021. Impacts of the closure of eastern Mongolia-Okhotsk Ocean on formation and evolution of Songliao Basin (in Chinese). Oil & Gas Geology, 42, 251–264.*

tuffaceous strata at its base. As is typical in many rift basins (e.g., Lambiase, 1990), fluvial deposits low in the formation give way upward to lacustrine strata, as subsidence outpaced sediment supply and large lakes occupied the central rift belt. Subsurface data document stratal thickening toward graben-bounding growth faults. Lacustrine strata are overlain by upward-coarsening deltaic facies, reflecting slowing subsidence. Black lacustrine mudstone enriched in total organic carbon (TOC), averaging over 2%, comprises hundreds of meters of the second and third members of the Shahezi Formation. Coal beds occur in cumulative thicknesses of up to 105 m. These dark mudstones and coals are the key source rocks for the gas reservoirs in the deep Songliao basin (Feng et al., 2006).

The Lower Cretaceous Yingcheng Formation records a return to widespread volcanism in the basin (Fig. 6.4). Felsic volcanic rocks dominate in the lower part of the formation. Alluvial fan and fan-delta facies occur adjacent to basin-bounding faults, giving way laterally toward graben centers to shallow lacustrine mudstone, sandstone, and coal. The Yingcheng Formation attains 2900 m thickness in the Xujiaweizi fault block but elsewhere is 500–1000 m thick. The formation is capped by conglomerate that includes metamorphic and granitic clasts that reflect the unroofing of adjacent horst blocks (Feng et al., 2013; Wang et al., 2013).

Thermal subsidence tectonostratigraphic unit

Following cessation of most rift faulting, cooling subsidence gave rise to the thermal subsidence tectonostratigraphic unit consisting of the Denglouku (K_1d), Quantou (K_1q), and Qingshankou (K_2qn) formations (Fig. 6.4). The end of active rifting is recorded by a basal unconformity equivalent to seismic horizon T4. Two additional unconformities corresponding to the T3 and $T2^1$ seismic markers divide the tectonostratigraphic unit into sequences II_1, II_2, and II_3, comprising alluvial fan, fluvial, fluvial plain, shallow lacustrine, and delta facies. Thermal subsidence strata bury and extend beyond the rift blocks and lap onto the basin margins to form a large integrated basin.

The T3 unconformity truncates the Denglouku (K_1d) Formation and is blanketed by the Quantou Formation (K_1q). Denglouku strata occur locally in the basin, whereas Quantou strata cover the entire basin and lap onto the eastern and western basin margins. The T_2^1 unconformity separates the third (K_1q_3) and fourth members (K_1q_4) of the Quantou Formation. The $T2^1$ unconformity surface displays evidence of extensive and lengthy subaerial exposure, including red paleosols and calcretes.

The Denglouku Formation (K_1d), comprising sequence II_1, reflects the change from rift faulting subsidence to thermal cooling subsidence. The basin retained distinct depositional and subsidence centers inherited from the rifting period. Sequence II_1 is typically 500–1000 m thick but ranges upward to 2000 m in the Gulong depression. The sequence includes interbedded sandstone, mudstone, conglomerate, and a thick argillite unit, the latter being important as a detachment horizon during Late Cretaceous shortening in the basin. Deposition broadened from the rifting phase and expanded north to Mingshui and east to Harbin. Alluvial, lacustrine delta, and shallow lacustrine facies ring the basin because principal sediment sources were more or less concentric around the basin (Fig. 6.7A).

Sequence II_2, including the first, second, and third members of the Quantou Formation, consists of conglomerate, sandstone, and multicolored reddish mudstone deposited in relatively arid fluvial and floodplain environments. The Quantou Formation is 550–1650 m thick and laps onto the basin margins. Fluvial sandstone dominates the first and second members, whereas the third member notably contains more extensive fine-grained floodplain deposits, evincing a change from braided to meandering channels upward in the Quantou Formation. Third member shallow lacustrine environments occupied the central area of the basin (Fig. 6.7B).

Sequence II_3 comprises the fourth member of the Quantou Formation (K_1q_4) and the Qingshankou Formation (K_2qn). In aggregate, the two units are 260–635 m thick. The Quantou Formation includes dark fluvial and lacustrine mudstone, siltstone, and sandstone (Fig. 6.4). The Qingshankou Formation comprises 200–300 m of dark lacustrine mudstone interbedded with oil shale, sandstone, and siltstone. An upward deepening succession of deltaic and shallow lacustrine facies characterizes the base of the sequence. The Songliao lake reached its maximum depth and area of 87,000 km^2 (Fig. 6.7C) early in Qingshankou deposition. Up to 100 m of deep lacustrine black mudstone, comprising some of the most important petroleum source rocks in the Songliao basin, blankets the entire central basin. Large deltaic systems that comprise the second and third members of the Qingshankou Formation built southward and eastward into the basin. By upper Qingshankou deposition, the areal extent of lacustrine strata reduced by half to about 41,000 km^2 (Yang et al., 1985) (Fig. 6.7D).

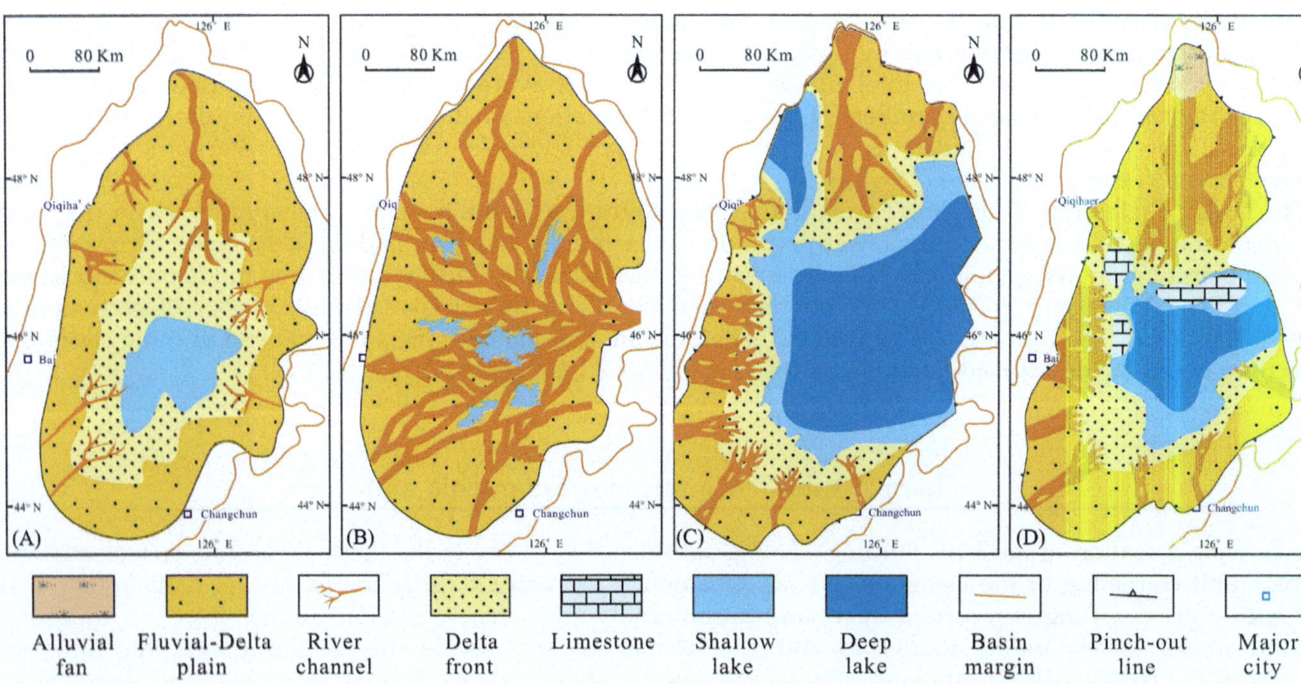

FIGURE 6.7 Depositional facies maps of (A) the Denglouku Formation, (B) third member of the Early Cretaceous Quantou Formation (K_{1q}), (C) the first (A), and (D) second and third members (B) of the Late Cretaceous Qingshankou Formation (K_{2qn}). Source: *(B–D) Modified from Feng, Z.Q., Jia, C.Z., Xie, X.N., Zhang, S., Feng, Z.H., Cross, T.A., 2010a. Tectonostratigraphic units and stratigraphic sequences of the nonmarine Songliao basin, northeast China. Basin Research, 22, 79–95.*

Foreland foredeep to wedgetop tectonostratigraphic unit

The foreland tectonostratigraphic unit, reflecting foredeep flexural subsidence and shortening of the basin comprises the Upper Cretaceous Yaojia (K_2y), Nenjiang (K_2n), Sifangtai (K_2s), and Mingshui (K_2m) formations (Figs. 6.2 and 6.4), and four major unconformities (seismic markers T11, T04, T03, and T02) divide the unit into stratigraphic sequences III_1, III_2, and III_3. The tectonostratigraphic unit records the basin's evolution from rift to a foreland setting as a result of Late Cretaceous regional shortening. Seismic marker T11 marks the base of the Yaojia Formation, at the Turonian-Coniacian boundary, about 88 Ma. Songke I (SK-I), scientific drilling in the Songliao basin that cored continuously through the Upper Cretacous strata (Feng et al., 2013; Wang et al., 2013) yielded the best chronostratigraphic framework to date based on magnetostratigraphy, radiometric ages, and biostratigraphy. The youngest sequences beneath the T11 unconformity are early Coniacian, ~88 ± 0.3 Ma (Wang et al., 2009), whereas the oldest sequences above the T11 unconformity are latest Coniacian, ~86.2 Ma (He et al., 2012; Deng et al., 2013). Seismic profiles clearly demonstrate stratal truncation (Fig. 6.8), and the T11 unconformity surface is characterized by evidence of extensive subaerial exposure, including calcareous nodules, caliche, mud cracks, red paleosols, and rhizoconcretions, as revealed by coring (Song et al., 2014).

T04, which separates Members 2–3 and 4–5 of the Nenjiang Formation, is an important unconformity marking a significant structural reorganization of the basin, as discussed below (Fig. 6.9). Seismic horizon T03 arises from a regional unconformity separating the Nenjiang and Sifangtai Formations, as demonstrated by stratal truncations and angularity. T02 caps the foreland subsidence tectonostratigraphic unit and separates Cretaceous and Cenozoic strata. This unconformity, which generally separates orogenic strata from postorogenic strata, records significant erosion of the basin after(?) the shortening deformation. The intensity of deformation and denudation increases

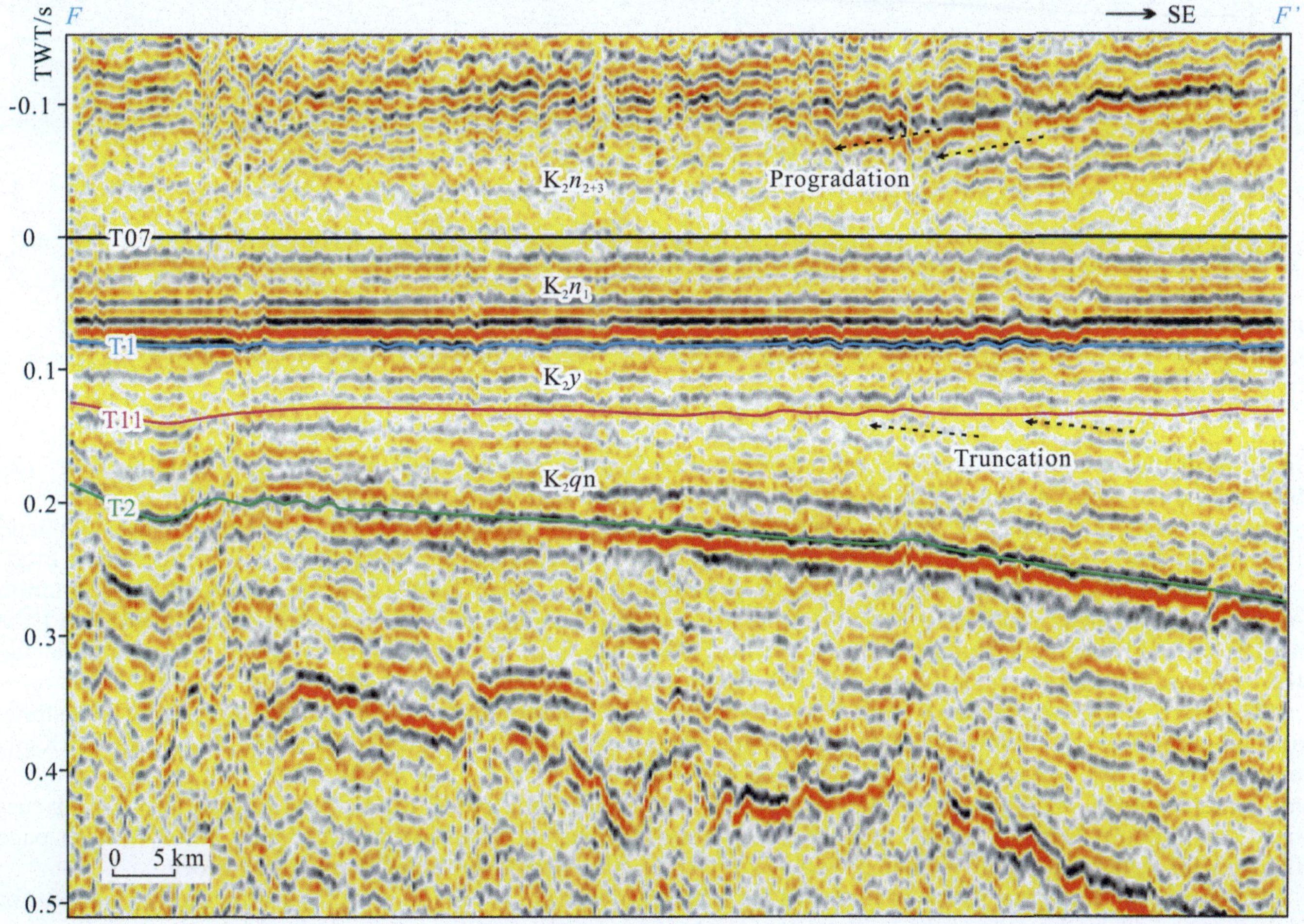

FIGURE 6.8 Seismic line illustrating stratal truncation by the T11 unconformity. Location of seismic line is shown in Fig. 6.3.

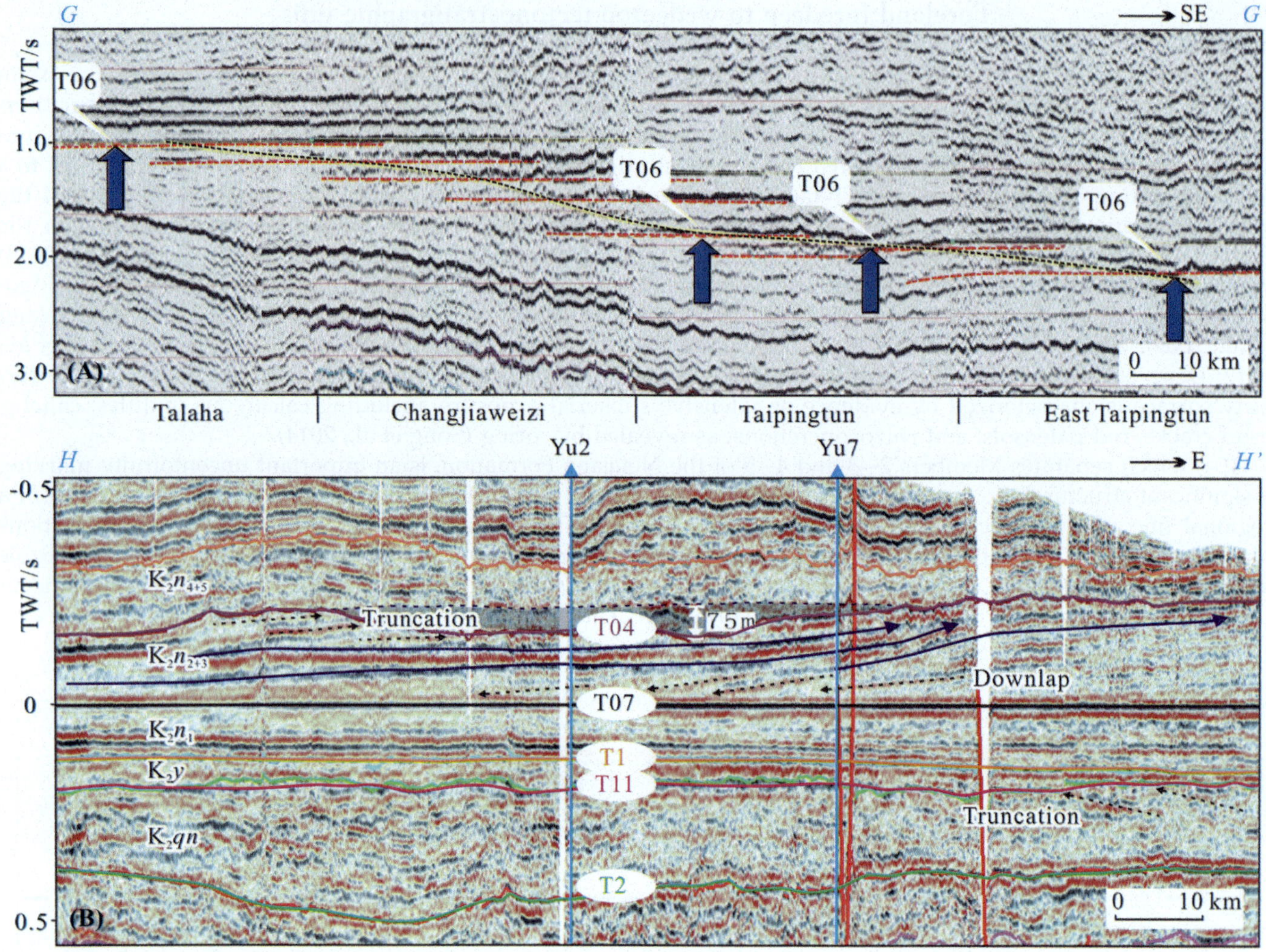

FIGURE 6.9 (A) Seismic profile G-G′ illustrates a facies change (seismic horizon T06) indicated by bright seismic amplitudes in the center of the Songliao basin. (B) Seismic profile (H-H′) showing stratal truncation by the T04 unconformity. The section is flattened on the T_{07} seismic horizon. Location of sections is shown in Fig. 6.3.

from northwest to southeast with T02 unconformity progressively cutting down to the Qingshankuo Fm (Fig. 6.2). Similarly, the western margin of the Mesozoic basin is truncated by the T02 unconformity (Fig. 6.10).

Sequence III_1 includes the Yaojia Formation (K_2y) and the first and second members of the Nenjiang Formation (K_2n_{1-2}), is 80–210 thick, and consists of lacustrine, fluvial, and deltaic facies of multicolored mudstone, siltstone, and sandstone. Fluvial sandstone and reddish floodplain mudstone of the basal Yaojia Formation overlie the T11 unconformity. Mud cracks, calcrete, and rhizoconcretions document subaerial exposure and the significance of the unconformity (Song et al., 2014). The Yaojia Formation constitutes a fluvial, deltaic, and lacustrine upward-fining succession (Figs. 6.4, 6.11A, and 6.11B).

The overlying lower Nenjiang Formation comprises deep lacustrine dark-colored mudstone, marl, limestone, and oil shale, with some interbedded siltstone and sandstone. Deep lacustrine facies of the first member (K_2n_1) occupied the basin center and quickly expanded to cover 2×10^5 km^2 (Chi et al., 2002). A sublacustrine turbidite fan system dominated the northern part of the lake (Fig. 6.11C; Feng et al., 2010b). Widespread and deep lacustrine conditions reflected a significant increase in basin subsidence where sediment supply failed to keep pace (Fig. 6.11D), thus forming the second period of large-scale lake expansion in Songliao basin history.

The second and third members of the Nenjiang Formation include voluminous westward-prograding wedges of deltaic to deep-lake deposits of the foreland foredeep, which for the first time were derived from beyond the eastern margin of the basin (Figs. 6.11E and 6.12). Reflectors above T07 downlap from east to west on seismic and borehole correlation sections, evincing sedimentary progradation from a sediment source area in the east. T06 marks a

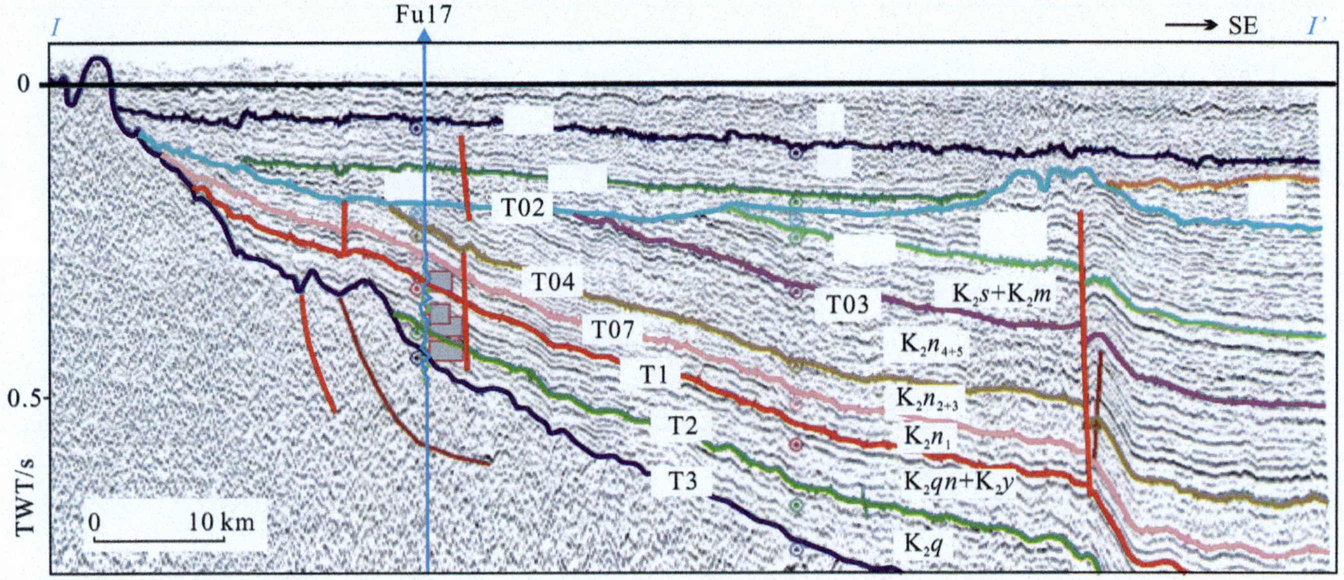

FIGURE 6.10 Seismic line showing stratal truncation of the western basin margin. Location of section is shown in Fig. 6.3.

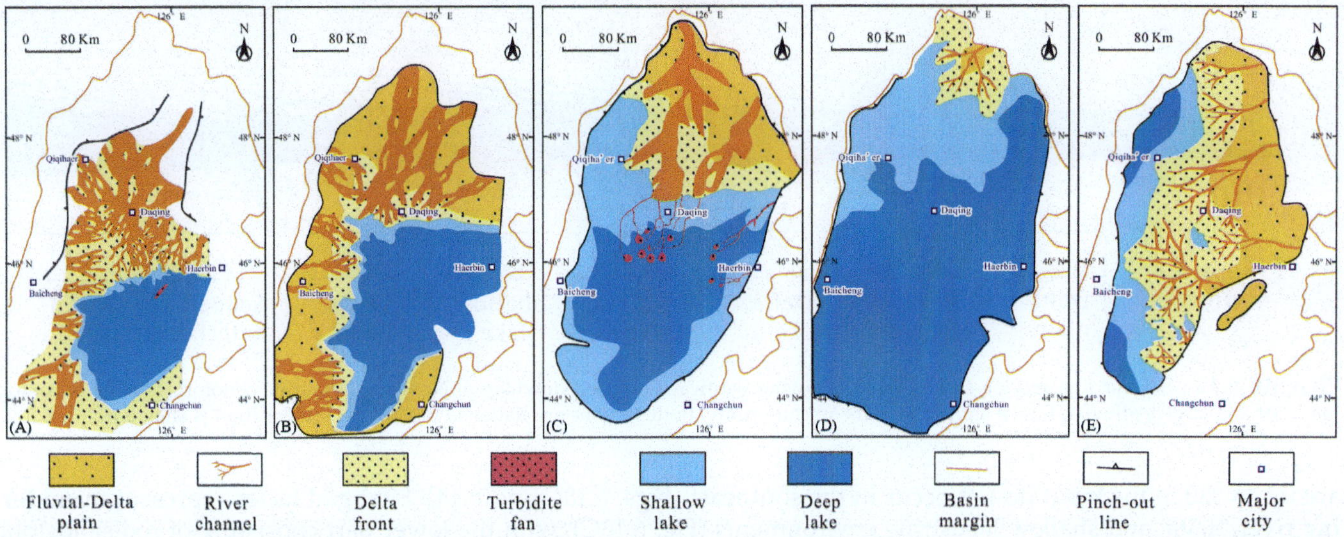

FIGURE 6.11 Depositional facies map of the (A) first and (B) second and third members of the Late Cretaceous Yaojia Formation (K_{2y}); (C) the first, (D) second, and (E) third members of the Late Cretaceous Nenjiang Formation (K_{2n}). Source: *(B–D) Modified from Feng, Z.Q., Jia, C.Z., Xie, X.N., Zhang, S., Feng, Z.H., Cross, T.A., 2010a. Tectonostratigraphic units and stratigraphic sequences of the nonmarine Songliao basin, northeast China. Basin Research, 22, 79–95.*

facies change indicated by bright seismic amplitudes associated with this change (Fig. 6.9A). The lower part of the Nenjiang Formation is a 200-m-thick transgressive fluvial-lacustrine sequence of deep lacustrine, organic-rich mudstone that progressively covered nearly the entire basin. The overlying balance of the formation consists of stacked, westward prograding, upward-coarsening sequences. The second and third members are truncated by the T04 unconformity, which signals the transition from foredeep to wedgetop deposition (Feng and Graham, 2023).

In Sequence III_2, which corresponds to the fourth and fifth members of the Nenjiang Formation, lacustrine facies are supplanted by delta plain and alluvial strata (Figs. 6.13A and 6.13B), so that the entirety of the Nenjiang Formation constitutes a grand upward shoaling and coarsening sequence. Following deposition of the third member, the lake decreased from 110,000 km^2 to 10,000 km^2 during deposition of the fourth and fifth members (Ye et al., 2002).

Sequence III_3 includes the Sifangtai (K_2s) and Mingshui (K_2m) formations (Fig. 6.4). Sediment sources for Sequence III_3 lay to the southeast. Sifangtai strata exist in the central and western parts of the basin, where they

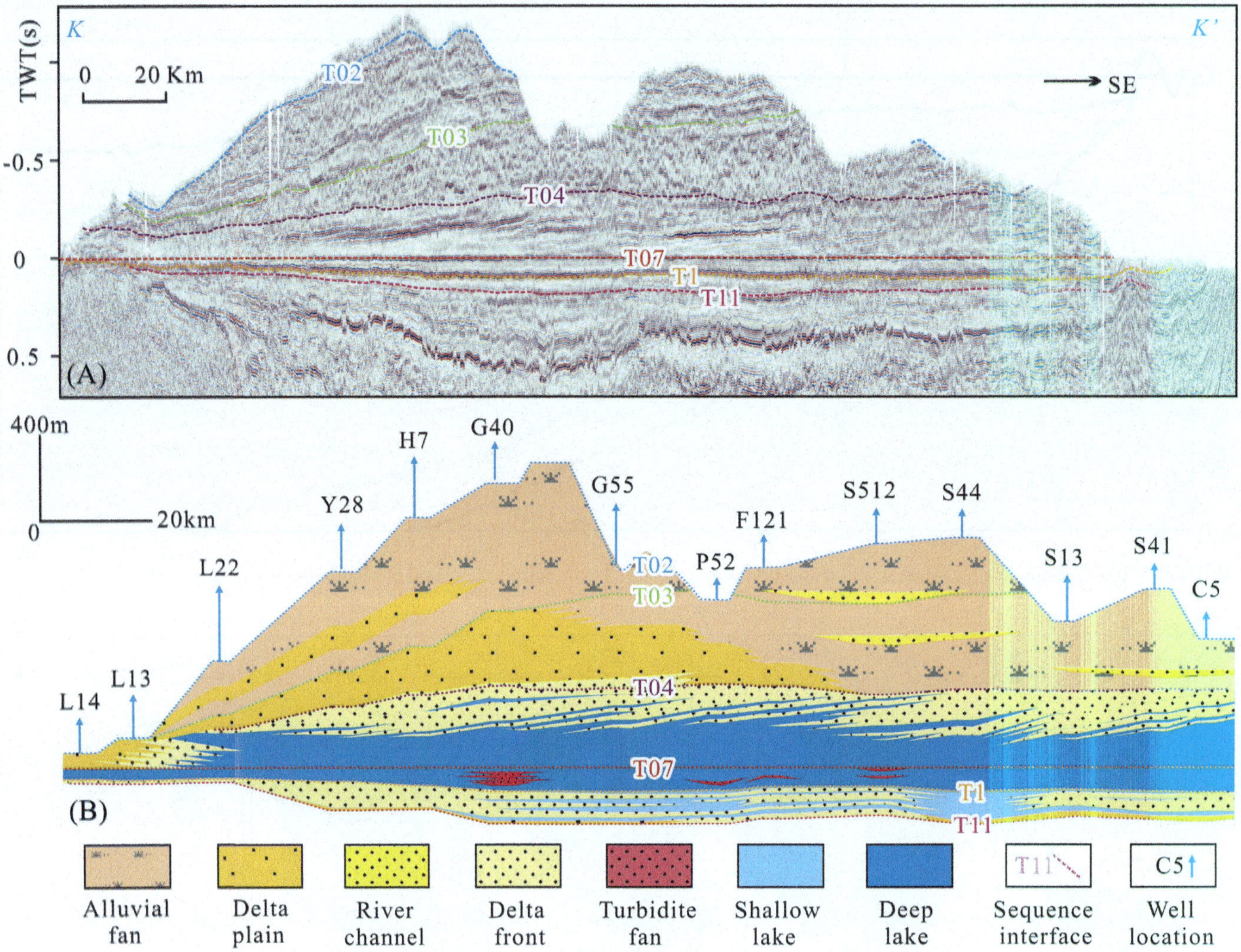

FIGURE 6.12 Seismic line and log section (K-K′) showing westward-directed downlap of Nenjiang lacustrine clinoforms below unconformity T04 and largely alluvial/fluvial strata above the unconformity. Both sections are flattened on T07. Section location is shown in Fig. 6.3.

are up to 320 m thick but do not occur in the southeast (Figs. 6.13C and 6.14). Sifangtai facies represent meandering river, delta, and shallow lacustrine environments (Fig. 6.13C), with the lower part consisting of red sandstone and shale interbedded with brown to gray sandstone and muddy siltstone. The middle part of the formation is overall finer grained, consisting of gray sandstone, siltstone, and muddy siltstone interbedded with red and tan shale. Red shale interbedded with gray-green mudstone dominates the upper Sifangtai Formation.

Interbedded shale and sandstone of the Mingshui Formation are limited to the central and western parts of the basin where it ranges up to 576 m (Fig. 6.14). Sediment derived mainly from the southeast, resulting in, from east to west, facies tracts of braided river, delta, and shallow lacustrine environments (Fig. 6.13D).

The original thickness of the Nenjiang-Sifangtai sequence is nowhere preserved due to truncation by the top Cretaceous unconformity. Based on projecting the preserved K-T boundary eastward toward the orogenic belt against the T11 datum, we estimate that the foreland tectonostratigraphic unit was originally about 3 km thick.

Tectonic drivers of Songliao tectonostratigraphic units

The Songliao basin evolved in stages reflecting plate interactions and geodynamic processes (Ma, 1987; Tian and Han, 1993; Gao and Cai, 1997; Li et al., 1997; Feng et al., 2010a, Zhang et al., 2015, 2017a,b; Liu et al., 2021; Zhao et al., 2022). Interactions between subducting Pacific Ocean basin plates and Eurasia and collisional closure

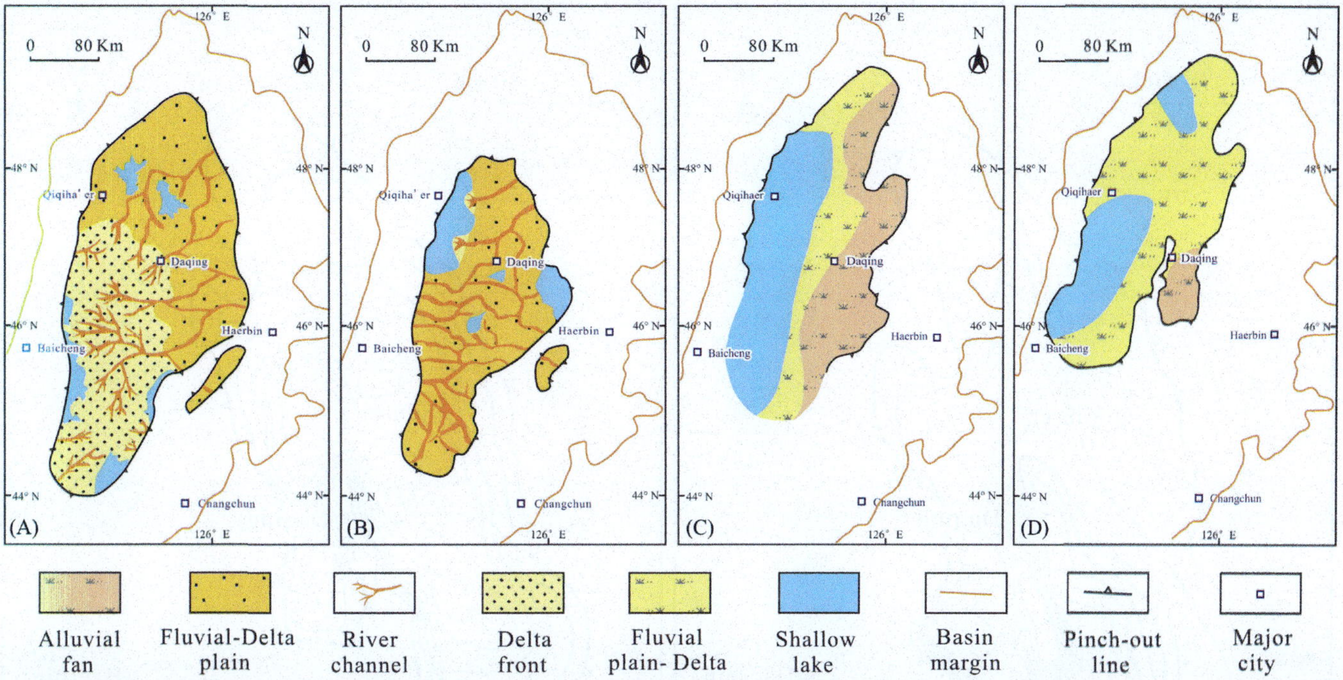

FIGURE 6.13 Depositional facies map of the (A) fourth (B) and fifth members of the Late Cretaceous Nenjiang Formation (K_{2n}); (C) the Sifengtai Formation (K_{2s}); (D) Mingshui Formation (K_{2m}).

of the Okhotsk Sea were the major tectonic drivers in the evolution of the Songliao basin (Watson et al., 1987; Li, 1996; Feng et al., 2010a, 2021; Wang et al., 2016; Zhang et al., 2017a,b; Wang et al., 2020). Prerift doming, rifting, and thermal cooling subsidence have long been recognized as drivers in the early phases of the evolution of Songliao basin, whereas late-stage shortening and its structural and stratigraphic manifestations have only recently been discussed, as discussed at greater length below.

As shown in Fig. 6.15, facies successions within the five Songliao tectonstratigraphic sequences reflect and record evolving tectonic and topographic controls. Syn-rift strata of the Huoshiling, Shahezi, and Yingcheng Formations are restricted to isolated fault blocks. Denglouku Formation strata covered the rift fault-block basins as thermal subsidence supplanted active rifting and strata symmetrically lapped onto the basin margins, setting the steer's head cross-sectional geometry of the basin (White and Mackenzie, 1988). During early thermal subsidence, sediment supply was dominantly from the Daxinganling orogenic belt to the north (Shao et al., 2005; Fan et al., 2008; Chamberlain et al., 2013), which was created by the collision of Siberia and the amalgamated North China–Mongolia block along the Mongol-Okhotsk suture (Graham et al., 2001; Ren et al., 2002; Feng et al., 2010a, 2021). During the final stage of basin evolution, the foreland stage, the depocenter migrated from the east to west, sediment supply was from the east, and strata in the southeastern basin were folded and truncated (Fig. 6.2).

Asthenospheric upwelling and lithospheric stretching perhaps may have been related to the south-directed subduction of Okhotsk oceanic crust from the Mongolia–China border region, although back-arc extension related to the rollback of subducting Pacific Ocean plates seems more likely (Feng et al., 2010a, 2021; Zhang et al., 2015; Wang et al., 2016; Zhao et al., 2022). In a variation of the latter, Liu et al. (2021) invoked slab retreat after flat slab subduction. Late Jurassic to Early Cretaceous rifting and attendant volcanism occurred across a broad area, yielding a set of NNE-oriented fault blocks in the Songliao basin (Fig. 6.1), as noted by Li et al. (1987). Two episodes of volcanism recorded in the Huoshiling Formation and the lower and middle members of the Yingcheng Formation occurred during rifting, signaling a close relationship between rift faulting and magmatism (Feng, 2008). Some authors have inferred that two minor episodes of structural inversion during the Songliao rifting stage were induced by the collisional closure of the Okhotsk seaway and subduction of the Kula Plate (Li et al., 1997; Zeng et al., 1997; Feng et al., 2021).

Cooling subsidence, reflected in the thermal subsidence tectonostratigraphic unit, was profound in the Songliao basin. Continentward in the China–Mongolia border region, the magnitude of rifting was comparatively modest and most rifts contain relatively thin thermal subsidence strata (Graham et al., 2001; Johnson, 2004), whereas

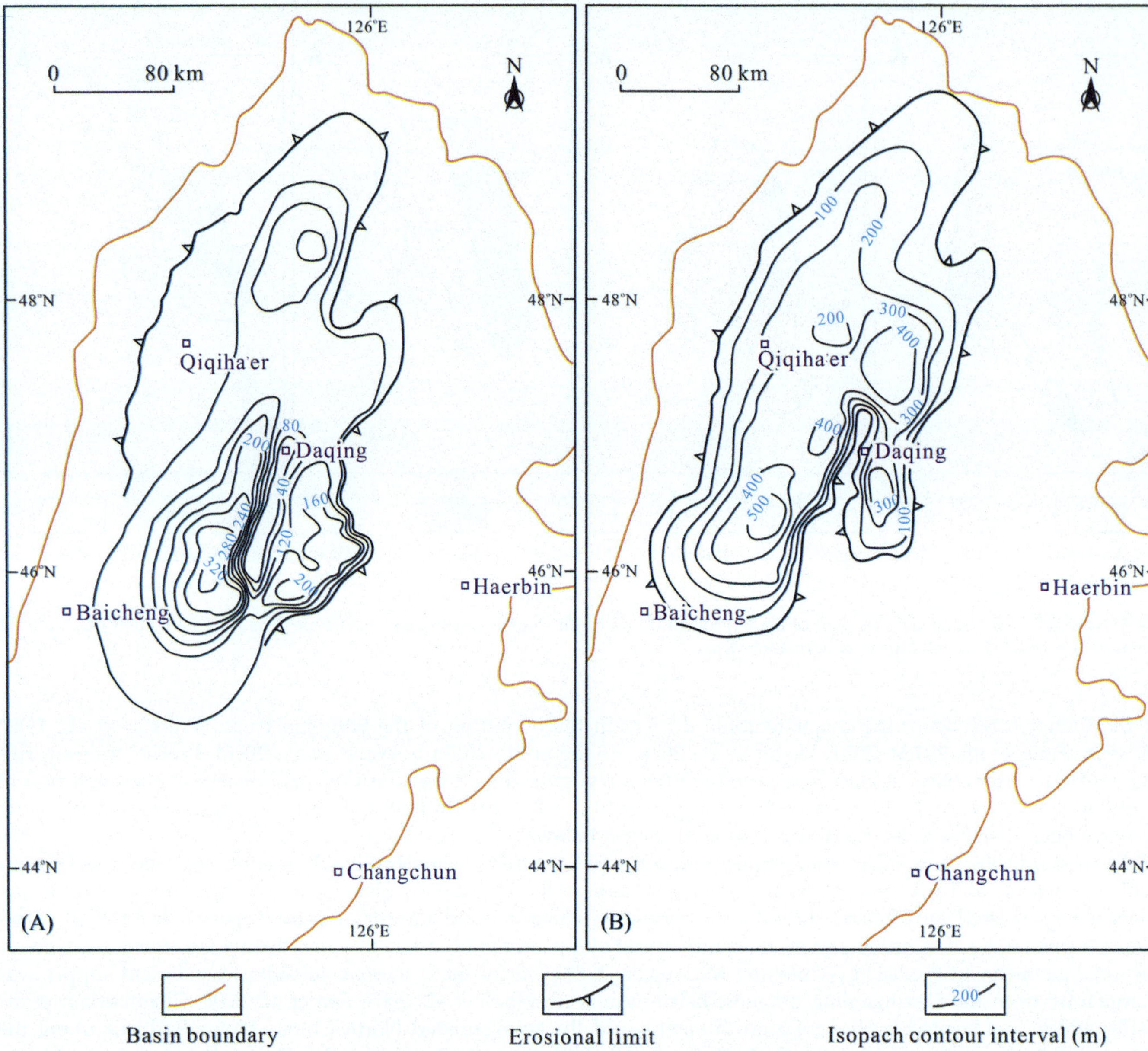

FIGURE 6.14 Isopach map of (A) the Late Cretaceous Sifangtai (K_{2s}) and (B) Mingshui Formations (K_{2m}). Source: *After Feng, Z.Q., Jia, C.Z., Xie, X.N., Zhang, S., Feng, Z.H., Cross, T.A., 2010a. Tectonostratigraphic units and stratigraphic sequences of the nonmarine Songliao basin, northeast China. Basin Research, 22, 79–95.*

coeval Cretaceous strata in Songliao are exceptionally thick. Asthenospheric upwelling and lithospheric heating in the Songliao basin region, reflected by Huoshiling and Yingcheng Formation deposition, likely enhanced subsequent cooling subsidence, while the distance from the Pacific margin likely resulted in lesser subsidence in southeast Mongolia.

Flexural subsidence associated with evolution to a foreland tectonic system characterized the final tectonic stages of evolution of the Songliao basin. In the later part of the Cretaceous, the regional stress field at the latitude of Songliao basin changed from extension to compression, driven by a change from Western Pacific style to Andean style plate margin interactions (Zhang et al., 2015). The transition from lithospheric cooling subsidence to dominantly compression-driven flexural subsidence began in Songliao about 88 Ma, signaled by the T11 unconformity between the Qingshankou and Yaojia Formations (Wang et al., 2013). Nearly all previous studies defined the last stage of Songliao basin evolution in terms of tectonic inversion, starting from the deposition of

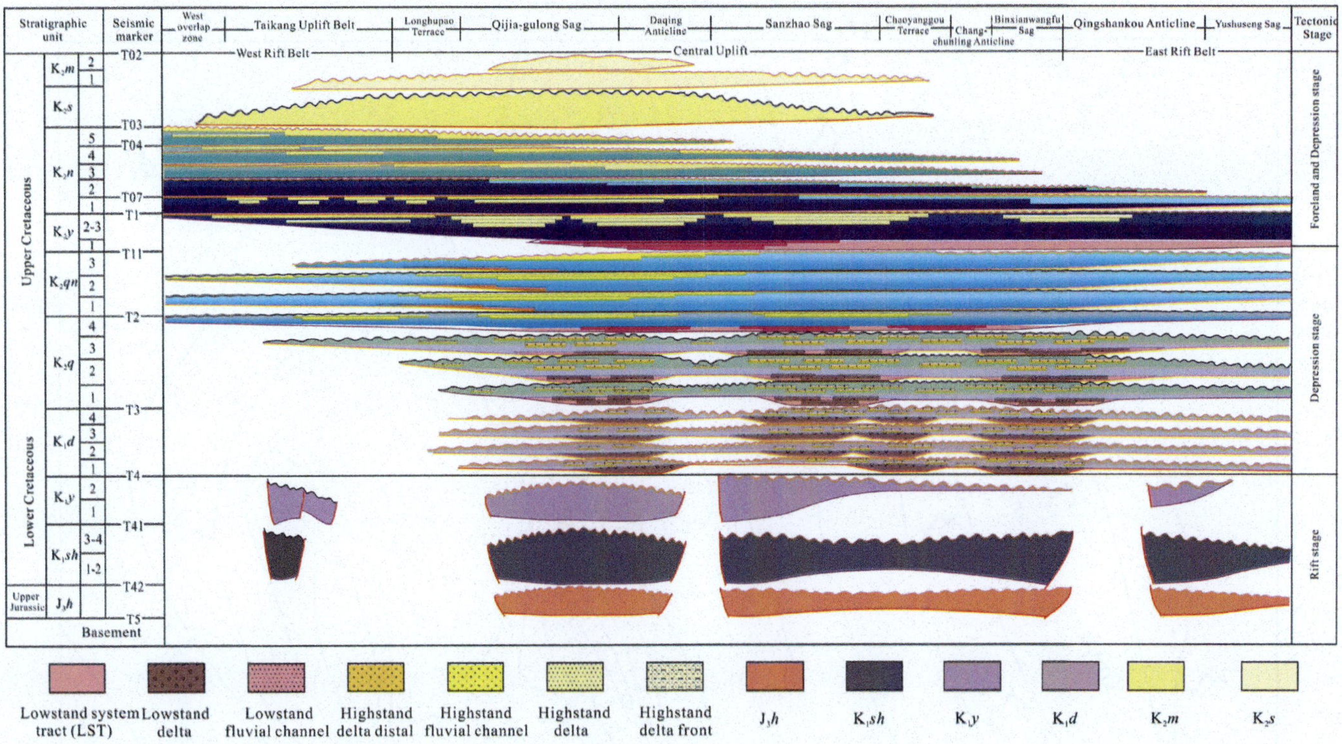

FIGURE 6.15 Tectonostratigraphic history of the Songliao basin as reflected in the east-west chronostratigraphy of the Songliao basin.

Yaojia Formation to the end of the Cretaceous. A number of aspects of regional geology are not well explained by simple inversion: (1) Why did several thousand meters of accommodation, as well as a maximum flooding surface, develop during the inversion stage? (2) The high maturity of petroleum source rocks relative to current burial depths, suggesting that erosion at the end of the Cretaceous may have been significantly underestimated and (3) exhumation of Zhangguancai orogenic belt at the end of the Cretaceous immediately east of the Songliao basin also may have been significantly underestimated, based on thermochronologically determined exhumation of nearly 5 km in the Zhangguangcai Range (Zhou et al., 2020).

Alternatively, Zhang et al. (2017a,b) proposed that during the latest Cretaceous, a foreland system spanned across nearly the entirety of northeast China, with Songliao basin as the foredeep, the Daxinganling orogenic belt as the forebulge, and the Hailar Basin as the backbulge depression (Fig. 6.1). We concur about the Songliao foredeep but view the Daxinganling Mountains and Hailar Basin as unlikely candidates for forebulge and backbulge because the Hailar Basin is over 1000 km from the orogenic front in the Zhangguangcai Range, which far exceeds expected distances in foreland systems, especially those associated with slab rollback (DeCelles, 2012) and the relative weak lithosphere of Paleozoic assemblies, which is related to narrower foreland system comparing to strong lithosphere of old cratons. Instead, we (Feng and Graham, 2023) contend that the sedimentary sequence and structural patterns of the uppermost Cretaceous fill of the Songliao basin are consistent with a foreland foredeep that largely evolved over time into a foreland wedgetop basin (DeCelles and Giles, 1996).

In our view, the Central Downwarp constituted the original foredeep, localized by rifting earlier in the Cretaceous and with the thinnest crust in the region (Fig. 6.16; Wang et al., 2020). The current shallowest depth of the Moho is about 27 km, and considering the thickness of sedimentary strata (more than 5 km), the thickness of the original crust was just 22 km. The fill of the Late Cretaceous foredeep is represented by the markedly westward progradational strata of the Nenjiang Formation, which filled the expansive lake that marked early foreland subsidence before coarse-clastic sedimentation overtook subsidence. Accordingly, provenance shifted in the basin from north to overwhelmingly east-southeast (contrast Figs. 6.11D and 6.11E). As noted earlier, organic maturation values for rift strata are greater than expected for the current burial depth, and we estimate that about 3 km of foredeep strata originally existed in the eastern part of the basin, based on a projection of the T02 horizon (K-T boundary) eastward to the basin boundary as measured against the T11 datum.

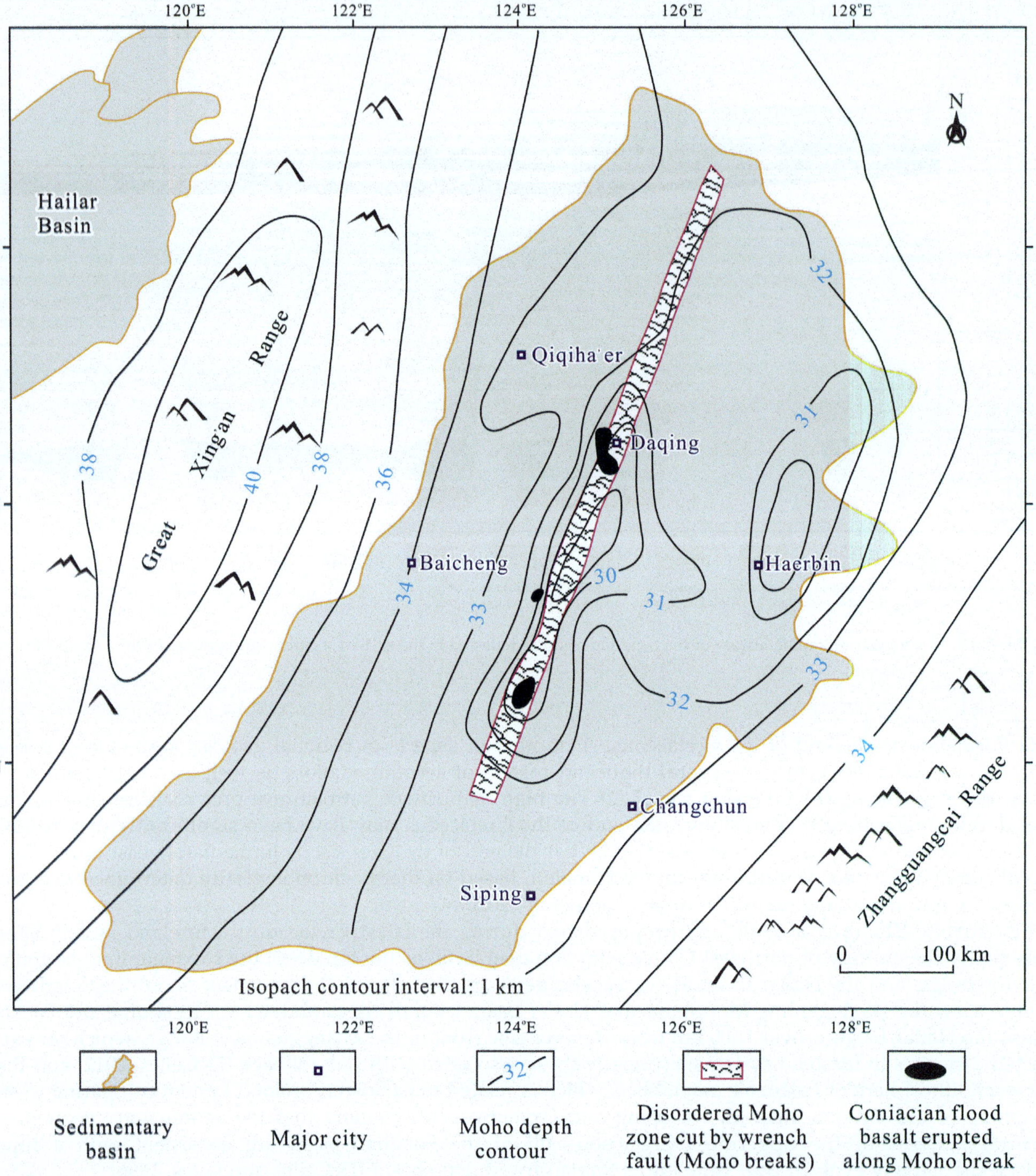

FIGURE 6.16 Map of the depth to the Moho (in kilometers) also showing axial strike-slip fault as associated basalt flows. Source: *Modified from Wang, P.J., Mattern, F., Didenko, N.A., Zhu, D.F., Singer, S., Sun, X.M., 2016. Tectonics and cycle system of the Cretaceous Songliao Basin: an inverted active continental margin basin. Earth-Science Reviews 159, 82–102. https://doi.org/10.1016/j.earscirev.2016.05.004.*

A paired, coeval thrust belt developed east of the basin in the Zhangguangcai Range, consistent with the general model of foreland systems (DeCelles and Giles, 1996). Numerous large thrust faults have been identified in the Zhangguancai orogenic belt, where thrusting and folding are the main expression of hinterland contraction. With time, shortening deformation propagated into the Songliao basin. The T04 horizon corresponds to a prominent unconformity in the eastern part of the basin and marks the evolution to wedgetop deposition. The T03 unconformity reflects regional uplift at about 72 Ma, as signaled by a rapid shift from deep lacustrine to fluvial and floodplain environments in the basin. Folding and faulting dominated the eastern part of the formerly expansive Songliao basin after T04 time, as most of the basin became incorporated into the foreland wedge and deposition in the latest Cretaceous occurred in a wedgetop basin setting (cf. DeCelles and Giles, 1996; Feng and Graham, 2023).

In general, contractile folding in the basin indicates E-W shortening, decreasing in amplitude and frequency to the west (Fig. 6.17). In the Southeastern Uplift Zone of the basin, nearly all the folds are characterized by the involvement of basements. Thrust faults, inverted from preexisting rift normal faults, are found in some of these anticlines, whereas thrust faults are not identified in other anticlines, which may indicate blind thrust faults developed in the basement. In the central part of the basin, the foredeep, preserved strata are thick and finer grained. Anticlines differ from the anticlines in the Southeastern Uplift Zone in that most are detached without obvious basement involvement (Fig. 6.17). Detachments are more likely between T3 and T4 in the shaley Donglouku Formation.

The growth of Daqing anticline, the most prominent fold in the basin, commenced after the T04 unconformity (Fig. 6.2; Feng and Graham, 2023). In past stratigraphic correlations, continuous and seismic events transitioning laterally from high to low amplitudes (T06) were incorrectly regarded as the top boundary of the second member of the Nenjiang Formation. However, T06 was in fact a lithologic interface instead of an isochronal interface located in the second and third members of the Nenjiang Formation. According to the new datum, T04 is regarded as the top boundary of the second and third members of the Nenjiang Formation (Fig. 6.9). Stratal thickening in synclines and thinning and truncation on anticlines above T04 demonstrate the syn-tectonic nature of upper Nenjiang and younger strata and support the notion that they constitute foreland wedgetop strata (Fig. 6.18).

Some of the preexisting rift faults appear to have been reactivated as transpressional features. Reactivated NNE-trending rift faults associated with strike-slip fault belts are common (Fig. 6.19). The large fault under the Daqing-Changyuan Anticline (Fig. 6.2), which trends NNE across nearly the whole Central Downwarp Zone, was commonly considered to be a normal fault reactivated with reverse slip that was responsible for the formation of the huge anticline. However, thrusting, normal faulting, and wrench characteristics are displayed in different sections across this fault, suggesting wrench tectonism (Christie-Blick and Biddle, 1985). Thus the NE-oriented *en echelon* subanticlines within the Daqing-Changyuan Anticline suggest transpressional stress regimes in this region (Fig. 6.19).

The forebulge and the backbulge depression of the Songliao foreland are not evident in preserved strata. It is possible that the West Slope Zone of the extant basin preserves the basinward slope of the forebulge with westward onlap and thinning of the foreland strata. Likely, forebulge and backbulge were developed to the northwest outside of the current basin limits and were eroded away at the end of Cretaceous during tectonic inversion, because deep-lake deposits of the second member of the Nenjiang basin are exposed at the west edge of the basin, indicating that the foredeep once extended to this region (Fig. 6.10).

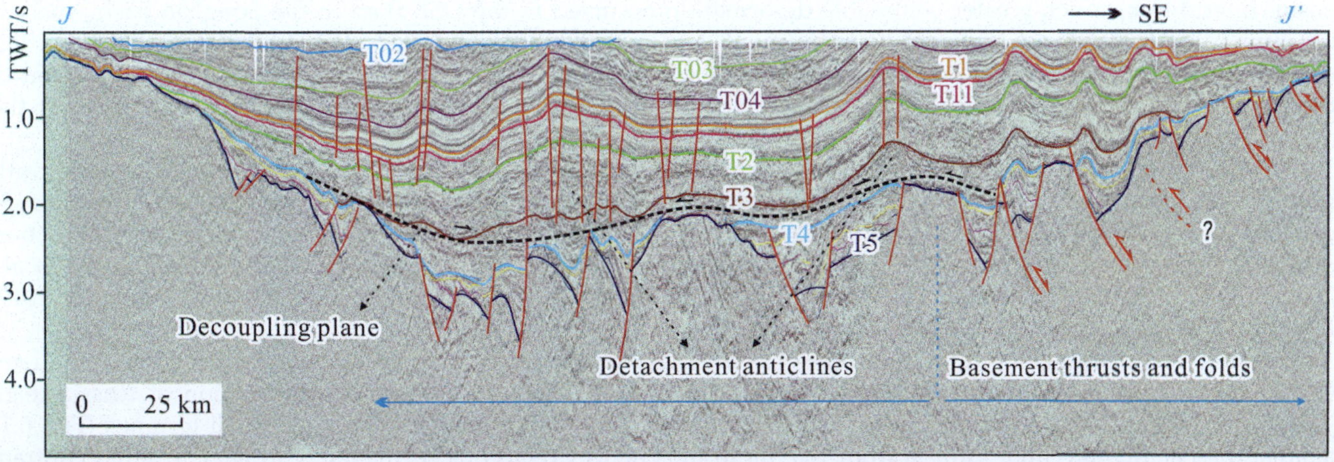

FIGURE 6.17 Interpreted seismic line J-J' showing major faults of the Songliao basin. Note the reversal of slip on basement-cutting rift normal faults in the eastern half of the line and detachment fault(s) in shaley strata below T3 in the central basin associated with latest Cretaceous evolution to a foreland basin. Location of seismic line is shown in Fig. 6.3. Source: *Modified from Feng, Z.Q., Graham, S.A., 2023. Songliao retroforeland basin, China, from Late Cretaceous foredeep to wedgetop. Geoscience Frontiers 14. https://doi.org/10.1016/j.gsf.2022.101527.*

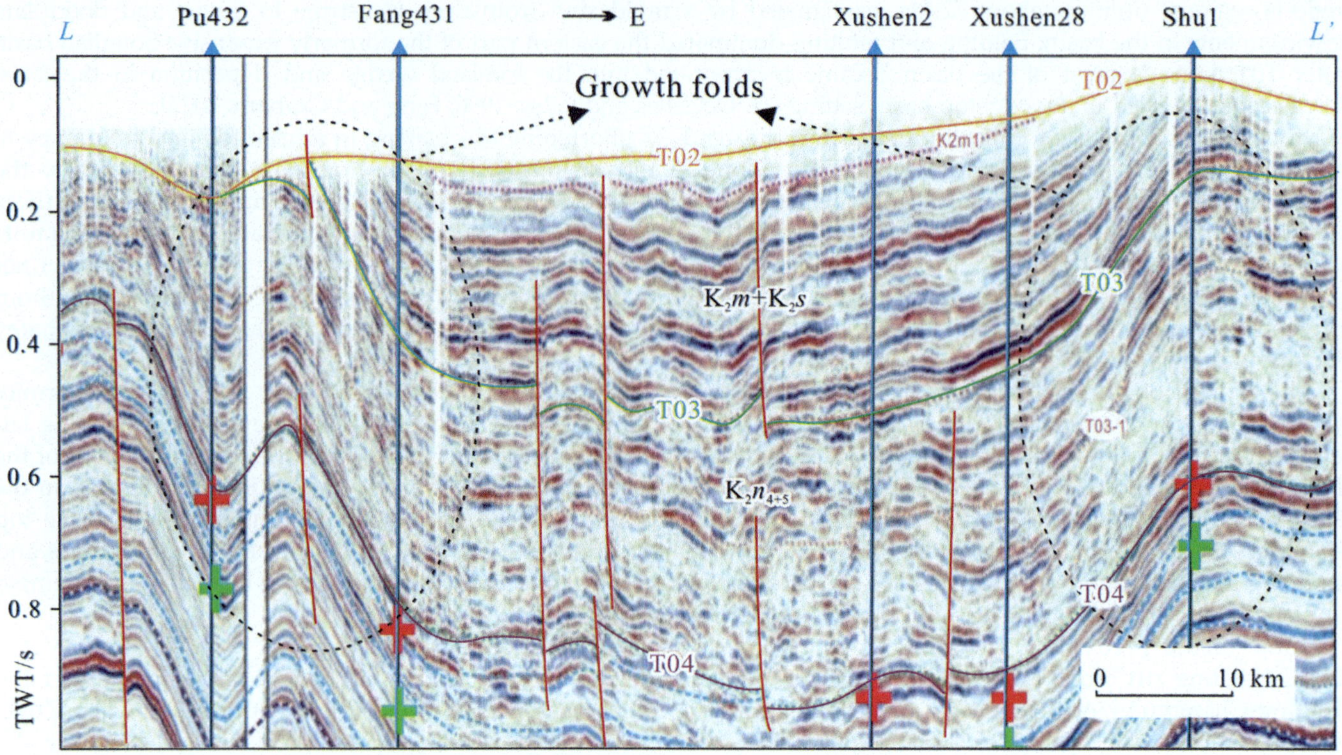

FIGURE 6.18 Seismic line L-L′ illustrating stratal thinning onto anticlines and thickening in synclines between T04 and T03 evincing structural growth as the basin evolved into a wedgetop basin. Location of seismic line is shown in Fig. 6.3.

Eastward toward the Pacific margin, a major unconformity developed in the GSB at the beginning of the Late Cretaceous (c.100 Ma) (Zhang et al., 2017b), whereas the T11 unconformity reflecting onset of foreland development in the Songliao basin dates from 88 Ma, a 12-Ma-lag cratonward, as is common in foreland systems elsewhere in the world (e.g., DeCelles and Giles, 1996; DeCelles, 2012). The later Cretacous of the GSB presents a foreland basin form as well. Similar to the Songliao basin, the foreland sequences overlap on top of the backpack rift-sag Early Cretacous sequences, whereas the differences are the 12 Ma lag of onsets of the foreland stages between them and the GSB involved to be thrusted belts at the end of Cretacous instead of wedgetop at Songliao basin. In addition, much greater contractile deformation occurred in the GSB than in the Songliao basin, including numerous large thrust faults, folded basement, and extensive erosion at the end of Cretaceous, which left comparatively little later Cretaceous strata preserved (Zhang et al., 2015, 2017b). As noted earlier, denudation of the intervening Zhangguancai orogenic belt likewise was extensive, with an exhumation of nearly 5 km at the end of Cretaceous (Zhou et al., 2020).

Many workers variously have argued that compression recorded by the flexural subsidence tectonostratigraphic unit and associated folding and faulting reflects Pacific Ocean plate motion reorganization following subduction of the Izanagi Plate beneath Eurasia and initial subduction of the Kula Plate, changes in subduction direction from oblique to orthogonal, and/or changes in subduction angle from high to low (Kravchinsky et al., 2002a; Li et al., 2012; Otofuji et al., 2003; Stepashko, 2006; Wang et al., 2016; Zhang et al., 2017a,b; Zhu et al., 2019; Li et al., 2020; Zhao et al., 2022).

Alternatively, docking of outboard terranes has been posited as a driver for contractile deformation of the Late Cretaceous hinterland of the Pacific margin of northeastern China and the peri-Pacific region of Russia (e.g., Yang, 2013; Zhang, et al., 2017a,b). Specifically, collision of the Okhotomorsk and other small terranes (Fig. 6.1) against the Pacific margin of Eurasia from 100 Ma to the end of Cretaceous seems likely to have fostered contractile deformation that swept across the GSB, Zhangguancai Range, and Songliao basin over a 35-Ma period from 100 to 65 Ma. Each collision formed a new orogenic belt at the site of the closed ocean basin, lead to the rejuvenation of nearby orogenic belts, and caused flexural bending of crust to form foreland basins on noncratonized, previously amalgamated terranes and on the cratonic margin.

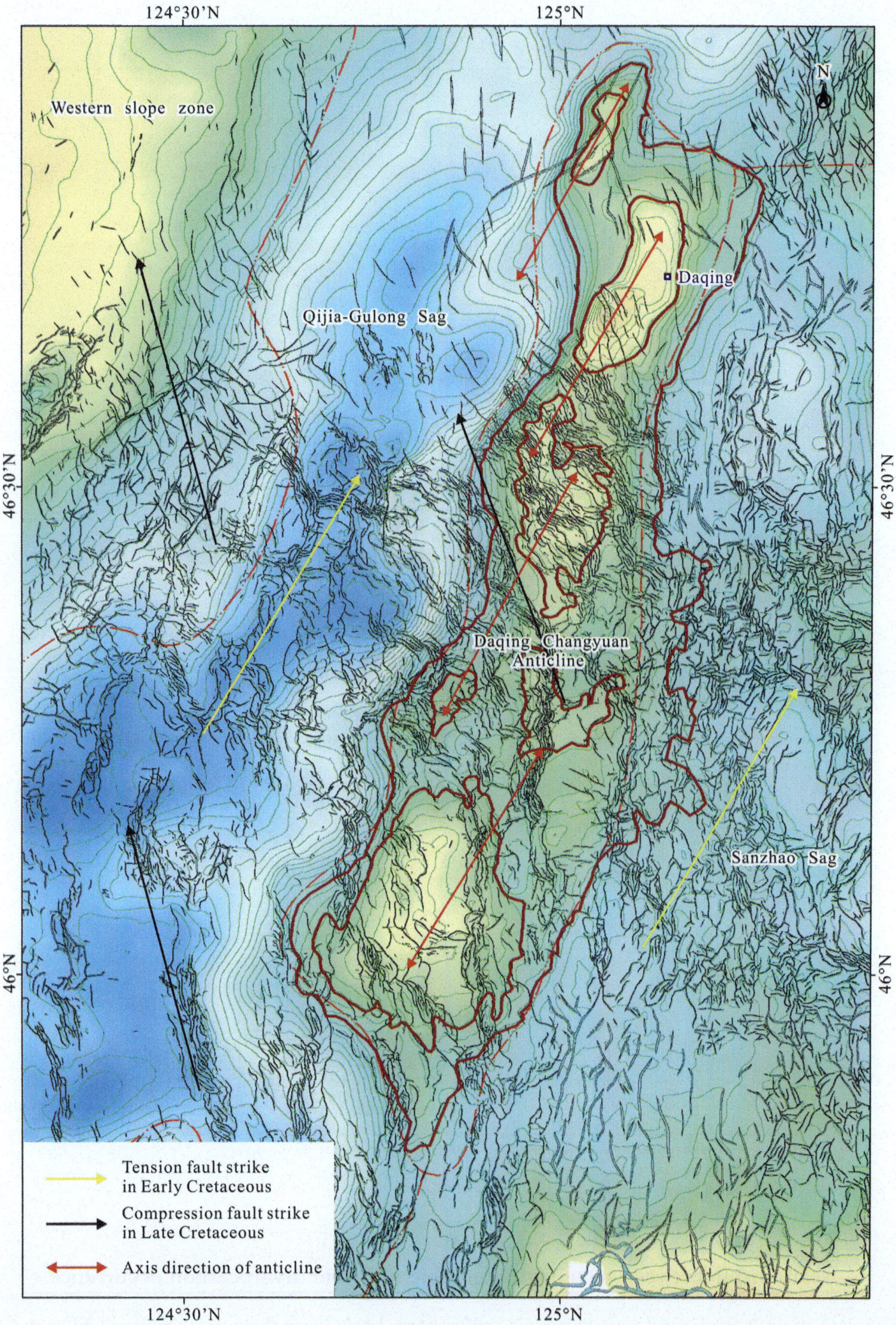

FIGURE 6.19 Wrench tectonics of Daqing anticline. Source: *Modified from Feng, Z.Q., Li, M., Guo, Y.L., Liu, G.X., 2022. Genetic analysis of typical strike-slip faults and related basins in China, Earth Science Frontiers, https://doi.org/10.13745/j.esf.sf.2022.8.6 (in Chinese with English abstract).*

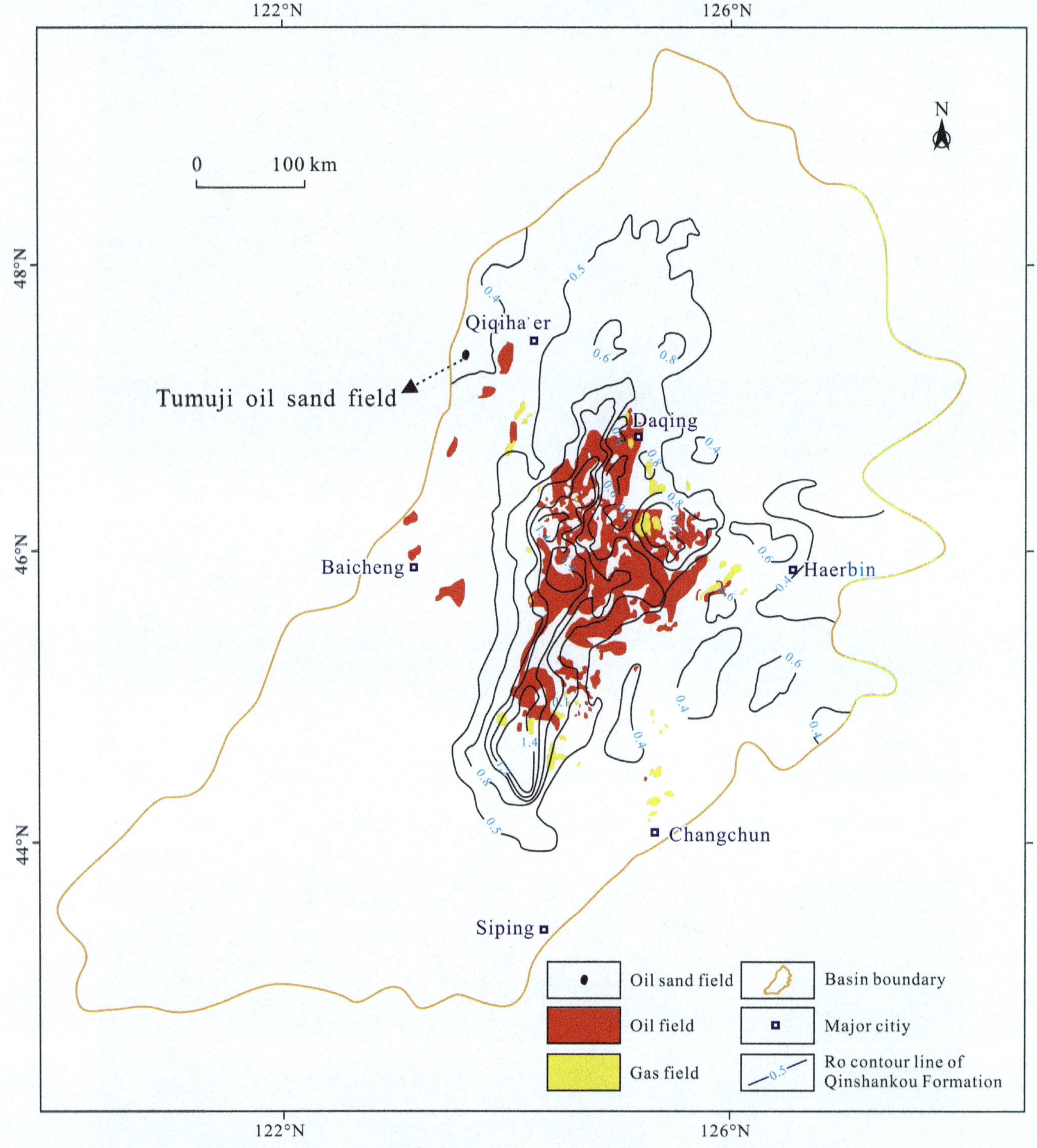

FIGURE 6.20 Map of oil and gas fields of Songliao basin.

Relationship between tectonostratigraphic units and hydrocarbon occurrence

The Songliao basin is the largest post-Paleozoic sedimentary basin in onshore northeastern China and is one of the world's most petroliferous basins (Fig. 6.20). Oil and gas production comes mainly from the Central Downwarp, specifically the Daqing anticline, the Qijia-Gulong, Sanzhao and Changling sags, and the

Chaoyanggou terrace. The basin is notable for the high quality of its nonmarine petroleum source rocks and its considerable hydrocarbon potential (Yang, 1985; Feng et al., 2010a, 2012a).

The voluminous syn-rift tectonostratigraphic unit contains excellent source rocks, reservoirs, and seals. Because of deep burial by the overlying thick sag and foreland sequences, the maturity of syn-rift source rocks is high and yields mainly gas fields. The thermal cooling megasequence comprises a full coarse-fine-coarse depositional cycle. During the early deposition of the Qingshankou Group, when cooling subsidence reached peak, exceptionally high-quality source rocks were deposited (Fig. 6.21). Large deltaic sequences prograding from the north of the basin lie directly atop the source rocks and formed world-class sandstone reservoirs (Figs. 6.4 and 6.7D). Significant subsidence and sediment accumulation during the foreland stage promoted maturation of the source rocks in the underlying thermal cooling tectonostratigraphic unit, and subsequent syn-sedimentary folding and basin inversion not only formed anticlinal traps but also focused oil migration. The thick blanket of foreland shale sequences that cover nearly the entire basin act as a regional seal for the underlying giant reservoirs.

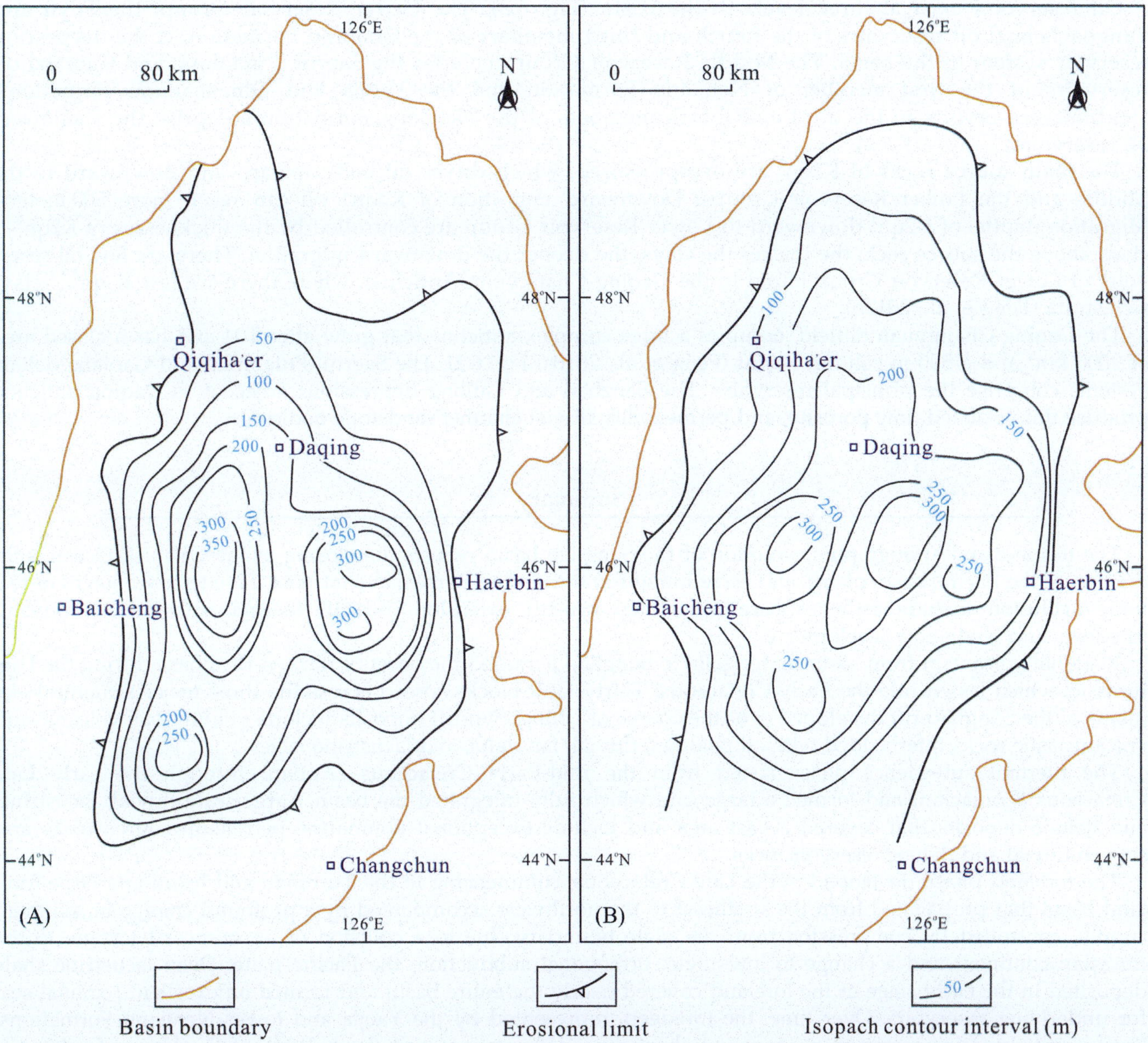

FIGURE 6.21 Isopach map of dark mudstones in the (A) Qingshankou Formation (K_{2qn}) and (B) the first and second members of the Nenjiang Formations (K_{2n}). Source: *After Feng, Z.Q., Jia, C.Z., Xie, X.N., Zhang, S., Feng, Z.H., Cross, T.A., 2010a. Tectonostratigraphic units and stratigraphic sequences of the nonmarine Songliao basin, northeast China. Basin Research, 22, 79–95.*

During the syn-rift and thermal cooling subsidence stages, gas-prone coaly source strata were deposited within the Lower Cretaceous Huoshiling (K_1h) along with reservoir-quality volcanic rocks and fluvial and fan-delta sandstone (Fig. 6.4). Both oil- and gas-prone source rocks occur in the Shahezi Formation, although depth of burial precludes oil preservation. Shahezi source rocks average 1.5%–2.0% TOC, average 1.76% Ro, cover 1000–3000 km^2, and attain thicknesses of up to 485 m in individual fault blocks. Conglomerate and volcanic reservoir rocks are interleaved with the source rocks. Volcanic rocks have porosities of 3%–18% and permeabilities of 0.01–1 mD, fostering large gas fields (Feng, 2008).

Oil-prone, organic-rich source rocks occur in the thermal subsidence and flexural tectonostratigraphic units. As summarized by Feng et al. (2010a), oil-prone lacustrine mudstone of the first member of the Qingshankou Formation covers 8.7×10^4 km^2 of the basin and is up to 135 m thick (Fig. 6.21A). TOC values of 3%–4% characterize profundal lacustrine mudstones in the basin center. Oil-prone lacustrine mudstone of 1.0%–3.0% TOC in the first and second members of the Nenjiang Formation covers 2×10^5 km^2 and up to 400 m thick (Fig. 6.21B). Maturation levels exceed 0.6% Ro in the central downwarp of the basin.

Oil reservoirs exist at three levels in the basin (Fig. 6.3). The Lower Reservoir Group, the Fuyu and Yangdachengzi oil reservoirs of the fourth and third members of the Quantou Formation, is the deepest oil reservoir system in the basin. The Middle Reservoir Group includes the Saertu, Putaohua, and Gaotaizi oil reservoirs of the first member of Nenjiang Formation and the Yaojia and Qingshankou formations. Heidimiao reservoirs in the third and fourth members of the Nenjiang Formation comprise the shallowest oil reservoirs.

The main source rocks of $K2qn^1$ are overpressured, which drives oil both upward and downward to the Middle and the Lower Reservoir Groups. Downward migration of K2qn1 oil can reach more 500 meters. Migration depths of $K2qn^1$ downward to Lower Reservoir Group are controlled by the thicknesses of $K2qn^{2+3}$ clays above the source rock; the thicker the clays, the deeper the downward migration. There are few oil reservoirs of Lower Reservoir Group found in the Daqing-Changyuan Anticline, where there are few K_2qn^{2+3} clay-rich strata (Feng et al., 2012b).

The Daqing-Changyuan oilfield, occupies a huge complex anticline that grew after T04 and has a closed area of 2000 km^2 and a 500-m column height (Feng et al., 2010a; Fig. 6.2). The Saertu, Putaohua, and Gaotaizi deltaic systems comprise the principal reservoirs. The Sanzhao and Gulong depressions adjacent to Daqing anticline produce thin, isolated, low porosity, and permeability fluvial channel sandstone bodies.

Summary

The tectonic and stratigraphic evolution of the Songliao basin reflects the closing of the Okhotsk Ocean, subduction of the Pacific basin plates, and accretion of terranes to the Euroasian continent. Five tectonic stages of tectonic evolution include asthenosphere upwelling, syn-rift extension, postrift thermal subsidence, foreland foredeep, and wedgetop development.

Regional volcanism from eastern Mongolia to Songliao heralded the onset of the syn-rift phase during the Late Jurassic, which lasted into the Early Cretaceous. Thirty fault blocks spread across the three major structural elements of the Songliao rift developed over the course of rifting. Syn-rift deposits include voluminous volcanic and volcaniclastic rocks interbedded with alluvial fan, fluvial, fan delta, and lacustrine facies.

The thermal subsidence phase lasted from the mid-Early Cretaceous (Aptian) through the early Late Cretaceous (Coniacian) and formed a large sag, which fully integrated the basin. Large-scale fluvial, lacustrine, and deltaic deposits that covered a vast area and include exceptional nonmarine petroleum source rocks and superb fluvial and deltaic reservoir rocks.

The foreland phase developed in the Late Cretaceous, commencing in the Turonian and forming a large foreland basin that propagated from the southeast to the northwest accompanied by a profound change in sediment supply, controlled by compression from the plate boundary due to a collision of an outboard terrane to the Eurasian continent and a change to low-angle, orthogonal subduction, the Pacific plate. Deep lacustrine shale deposited in the early stage of the foreland covered nearly the entire basin and formed an excellent regional seal for underlying reservoirs. Over time, the foredeep, represented by the Yaojia and lower Nenjiang Formations, became involved in westward propagating shortening deformation, and the balance of the Upper Cretaceous above the T04 horizon represents foreland wedgetop deposition.

Deeply buried syn-rift deposits generate gas from coal-bearing source rocks, which is housed in syn-rift sandstone, conglomerate, and volcanic rocks. Oil-bearing strata of the upper basin fill include voluminous high-quality lacustrine petroleum source rocks and fluvial and deltaic sandstone reservoirs among the thermal subsidence tectonostratigraphic units. Subsequent flexural subsidence promoted the maturation of source rocks of the thermal cooling phase, and later contraction not only formed anticline traps but also enhanced oil migration, leading to the formation of the basin's giant nonmarine oilfields.

Acknowledgments

The authors thank Li Dong and Gaokui Wu for assistance in manuscript preparation. This work was supported by the National Natural Science Foundation of China Grant 41790450. The authors thank Daqing Oilfield Ltd. for permission to publish the seismic lines in this paper.

References

Chamberlain, C.P., Wan, X., Graham, S.A., Carroll, A.R., Doebbert, A.C., Sageman, B.B., et al., 2013. Stable isotopic evidence for climate and basin evolution of the Late Cretaceous Songliao Basin, China. Palaeogeography, Palaeoclimatology, Palaeoecology 385, 106–124.

Chi, Y.L., Yun, J.B., Meng, Q.A., Yin, J.Y., Meng, G.T., 2002. Mantle framework, basin forming dynamic and hydrocarbon migration and accumulation in the Songliao Basin. Petroleum industry press (Beijing) 10–34.

Christie-Blick, N., Biddle, K.T., 1985. Deformation and basin formation along strike-slip faults. Society of Economic Paleontologists and Mineralogists (SEPM) Special Publication 37, 1–34.

DeCelles, P.G., 2012. Foreland basin systems revisited: variations in response to tectonic settings. In: Busby, C., Azor, A.P. (Eds.), Tectonics of Sedimentary Basins: Recent Advances. Blackwell Publishing Ltd, pp. 405–426.

DeCelles, P.G., Giles, K.N., 1996. Foreland basin systems. Basin Research 8, 105–123.

Demaison, G., 1984. The generative basin concept. American Association of Petroleum Geologists Memoir 35, 1–14.

Deng, C.L., He, H.Y., Pan, Y.X., Zhu, R.X., 2013. Chronology of the terrestrial Upper Cretaceous in the Songliao basin, northeast Asia. Palaeogeography, Palaeoclimtology, Palaeoecology 385, 44–54.

Fan, W.M., Guo, F., Gao, X.F., Li, C.W., 2008. Sr-Nd isotope mapping of Mesozoic igneous rocks in NE China: constraints on tectonic framework and crustal growth. Geochimica 37, 361–372.

Feng, Z.Q., Zhang, S., Xie, X.N., An, G.Z., Zhao, B., Hou, Y.P., 2006. Discovery of a large-scale lacus-trine subaqueous channel in the Nenjiang Formation of the Songliao Basin and its implication on petroleum geology (in Chinese). Acta Geologica Sinica 80, 1226–1232.

Feng, Z.Q., 2008. Volcanic rocks as prolific gas reservoir: a case study from the Qingshen gas field in the Songliao Basin, NE China. Marine and Petroleum Geology 25, 416–432.

Feng, Z.Q., Graham, S.A., 2023. Songliao retroforeland basin, China, from Late Cretaceous foredeep to wedgetop. Geoscience Frontiers 14. Available from: https://doi.org/10.1016/j.gsf.2022.101527.

Feng, Z.Q., Jia, C.Z., Xie, X.N., Zhang, S., Feng, Z.H., Cross, T.A., 2010a. Tectonostratigraphic units and stratigraphic sequences of the nonmarine Songliao basin, northeast China. Basin Research 22, 79–95.

Feng, Z.Q., Zhang, S., Cross, T.A., Feng, Z.H., Zhao, B., Fu, X.L., et al., 2010b. Lacustrine turbidite channels and fans in the Mesozoic Songliao basin, China. Basin Research 22, 96–107.

Feng, Z.Q., Zhang, S., Feng, Z.H., 2012a. Discovery of "enveloping surface of oil and gas overpressure migration" in the Songliao Basin and its bearings on hydrocarbon migration and accumulation mechanisms. Science China Earth Sciences 55 (12), 2005–2017.

Feng, Z.Q., Zhang, S., Fu, X.L., 2012b. Depositional evolution and accumulation response of Yaojia-Nenjiang Formation in Songliao Basin. Earth Science Frontiers 19, 78–88.

Feng, Z.Q., Wang, C.S., Graham, S.A., Koeberl, C., Dong, H.L., Yonjian Huang, Y.J., et al., 2013. Continental Scientific Drilling Project of Cretaceous Songliao Basin: Scientific objectives and drilling technology. Palaeogeography, Palaeoclimatology, Palaeoecology 385, 6–16. Available from: https://doi.org/10.1016/j.palaeo.2012.03.016.

Feng, Z.Q., Dong, L., Tong, Y., Feng, Z.H., Zhang, S., Wu, G.K., et al., 2021. Impacts of the closure of eastern Mongolia-Okhotsk Ocean on formation and evolution of Songliao Basin (in Chinese). Oil & Gas Geology 42, 251–264.

Feng, Z.Q., Li, M., Guo, Y.L., Liu, G.X., 2022. Genetic analysis of typical strike-slip faults and related basins in China (in Chinese). Earth Science Frontiers . Available from: https://doi.org/10.13745/j.esf.sf.2022.8.6.

Gao, R., Cai, X., 1997. Hydrocarbon Formation Conditions and Distribution Rules in the Songliao Basin. Petroleum Industry Press, Beijing (in Chinese).

Graham, S.A., Hendrix, M.S., Johnson, C.L., Badamgarav, D., Badarch, G., Amory, J., et al., 2001. Sedimentary record and tectonic implications of Mesozoic rifting in southeast Mongolia. Geological Society of America Bulletin 113, 1560–1579.

He, H., Deng, C., Wang, Y., Pan, Y., Zhu, R., 2012. Toward age determination of the termination of the Cretaceous Normal Superchron. Geochemistry, Geophysics, Geosystems 13, Q02002.

Johnson, C.L., 2004. Polyphase evolution of the East Gobi basin: sedimentary and structural records of Mesozoic-Cenozoic intraplate deformation in Mongolia. Basin Research 16, 79–99.

Johnson, C.L., Amory, J.A., Graham, S.A., Lamb, M.A., Badarch, G., Affolter, M., 2008. Accretionary tectonics and sedimentation during late Paleozoic arc collision, China-Mongolia border region. Special Paper 436. Geological Society of America, pp. 363–390.

Kravchinsky, V.A., Congn'E, J.P., Harbert, W., Kuzmin, M.I., 2002a. Evolution of the Mongol-Okhotsk Ocean with paleomagnetic data from the suture zone. Geophys. J. Int. 148, 34–57.

Lambiase, J.J., 1990. A model for tectonic control of lacustrine stratigraphic sequences in continental rift basins, American Association of Petroleum Geologists Memoir, 50. pp. 265–276.

Li, S.T., Yang, S.G., Wu, C.L., Huang, J.F., Chen, S.T., et al., 1987. Late Mesozoic rifting in northeast China and northeast Asia fault basin system. Science in China Series D – Earth Sciences 30, 185–195.

Li, D.S., 1996. Basic characteristics of oil and gas basins in China. Journal of Southeast Asian Earth Sciences 13, 299–304.
Li, S.T., Yang, S., Xie, X., 1997. Tectonic evolution of tertiary basins in Circum-Pacific Belt of China and their geodynamic setting. Journal of China University Geosciences 8, 4–10.
Li, S.Q., Chen, F.K., Siebel, W., Wu, J.D., Zhu, X.Y., Shan, X.L., et al., 2012. Late Mesozoic tectonic evolution of the Songliao basin, NE China. Evidence from detrital zircon ages and Sr–Nd isotopes. Gondwana Research 22, 943–955.
Li, Y., Xu, W.L., Zhu, R.X., Wang, F., Ge, W.C., Sorokin, A.A., 2020. Late Jurassic to early Early Cretaceous tectonic nature on the NE Asian continental margin: constraints from Mesozoic accretionary complexes. Earth-Science Reviews 200, 103042.
Li, Z.Q., Chen, J.L., Zou, H., Wang, C.S., Meng, Q.A., Liu, H.L., et al., 2021. Mesozoic-Cenozoic tectonic evolution and dynamics of the Songliao Basin, NE Asia: implications for the closure of the Paleo-Asian Ocean and Mongol-Okhotsk Ocean and subduction of the Paleo-Pacific Ocean. Earth-Science Reviews 218. Available from: https://doi.org/10.1016/j.earscirev.2020.103471.
Liu, L., Liu, L., Xu, Y.G., 2021. Mesozoic intraplate tectonism of East Asia due to flat subduction of a composite terrane slab. Earth-Science Reviews 214. Available from: https://doi.org/10.1016/j.earscirev.2021.103505.
Liu, S.F., Gurnis, M., Ma, P.F., Zhang, B., 2017. Reconstruction of northeast Asian deformation integrated with western Pacific plate subduction since 200 Ma. Earth-Science Reviews 175, 114–142. Available from: https://doi.org/10.1016/j.earscirev.2017.10.012.
Ma, X.Y., 1987. China Lithosphere Dynamics. Geology Press, Beijing.
Otofuji, Y., Matsuda, T., Enami, R., Uno, K., Nishihama, K., Halim, N., et al., 2003. Late Cretaceous palaeomagnetic results from Sikhote Alin, far eastern Russia: tectonic implications for the eastern margin of the Mongolia Block. Geophysical Journal International 152, 202–214.
Ren, J.Y., Kensaku, T., Li, S.T., Zhang, J.X., 2002. Late Mesozoic and Cenozoic rifting and its dynamic setting in Eastern China and adjacent areas. Tectonophysics 344, 175–205.
Rosendahl, B.R., Reynolds, D.J., Lorber, P.M., Burgess, C.F., McGill, J., Scott, D., et al., 1986. Structural expressions of rifting: lessons from Laake Tanganyika, Africa. Geological Society Special Publication 25, 29–43.
Sengör, A.M.C., Natal'in, B.A., Burtman, V.S., 1993. Evolution of the Altaid tectonic collage and Palaeozoic crustal growth in Eurasia. Nature v. 364, 299–307. Available from: https://doi.org/10.1038/364299a0.
Shao, J., Zhang, L.Q., Xiao, Q.H., Li, X.B., 2005. Rising of Da Hinggan Mts in Mesozoic: a possible mechanism of intracontinental orogeny. Acta Petrologica Sinica 21, 789–794.
Song, Y., Ren, J.Y., Stepashko, A.A., Li, J.G., 2014. Post-rift geodynamics of the Songliao Basin, NE China: origin and significance of T11 (Coniacian) unconformity. Tectonophysics 634, 1–18.
Stepashko, A.A., 2006. The Cretaceous dynamics of the pacific plate and stages of magmatic activity in northeasern Asia. Geotectonics 40, 225–235.
Tian, Z.Y., Han, P., 1993. Structure analysis and formation mechanism of Meso- Cenozoic Basin in East China. Petroleum Exploration and Development 20, 1–8.
Wang, P., Gao, Y., Ren, Y., Liu, W., 2009. $^{40}Ar/^{39}Ar$ age and geochemical features of mugearite from the Qingshankou Formation: significances for basin formation, hydrocarbon generation and petroleum accumulation of the Songliao Basin in Cretaceous. Acta Petrologica Sinica 25, 1178–1190.
Wang, C.S., Feng, Z.Q., Zhang, L.M., Huang, Y.J., Cao, K., Wang, P.J., et al., 2013. Cretaceous paleogeography and paleoclimate and the setting of SKI borehole sites in Songliao Basin, northeast China. Palaeogeography, Palaeoclimatology, Palaeoecology 385, 17–30. Available from: https://doi.org/10.1016/j.palaeo.2012.01.030.
Wang, P.J., Mattern, F., Didenko, N.A., Zhu, D.F., Singer, S., Sun, X.M., 2016. Tectonics and cycle system of the Cretaceous Songliao Basin: an inverted active continental margin basin. Earth-Science Reviews 159, 82–102. Available from: https://doi.org/10.1016/j.earscirev.2016.05.004.
Wang, J.M., Yang, B.J., Li, Z.L., Yu, C.L., Jiang, C.J., Li, P., et al., 2020. Crustal structure between the eastern margin of the Songliao Basin and its geological implication: deep seismic reflection. Chinese Journal of Geophysics (in Chinese) 63, 3478–3490. Available from: https://doi.org/10.6038/cjg2020M0131.
Watson, M.P., Hayward, A.B., Parkinson, D.N., Zhang, Z.M., 1987. Plate tectonic history, basin development and petroleum source rock deposition onshore China. Marine and Petroleum Geology 4, 205–225.
White, N., Mackenzie, D., 1988. Formation of the "steer's head" geometry of sedimentary basins by differential stretching of the crust and mantle. Geology 16, 250–253.
Xie, X.N., Jiao, J.J., Tang, Zhong, H., Zheng, C.M., 2003. Evolution of abnormally low pressure and its implications for the hydrocarbon system in the Southeast Uplift zone of Songliao Basin. China. American Association of Petroleum Geologists Bulletin 87, 99–119.
Yang, W.L., 1985. Daqing oil field, People's Republic of China; a giant field with of non-marine origin. American Association of Petroleum Geologists Bulletin 69, 1101–1111.
Yang, W.L., Gao, R.Q., Guo, Q.F., Liu, Y.G., 1985. Forming, migration and accumulation of the None-marine petroleum reservoir in the Songliao Basin. Heilongjiang Scientific and Technology Press 41–42.
Yang, Y.T., 2013. An unrecognized major collision of the Okhotomorsk Block with east Asia during the Late Cretaceous, constraints on the plate reorganization of the northwest Pacific. Earth-Science Reviews 126, 96–115.
Yang, X.B., Wang, H.Y., Li, Z.Y., Guan, C., Wang, X., 2021. Tectonic-sedimentary evolution of a continental rift basin: a case study of the Early Cretaceous Changling and Lishu fault depressions, southern Songliao Basin, China. Marine and Petroleum Geology 128105068. Available from: https://doi.org/10.1016/j.marpetgeo.2021.105068.
Zeng, Z.X., Li, S.T., Yang, S.G., Liu, L.L., Fu, Y.T., 1997. Three dimensional modeling of effect of India-Eurasia collision. Journal of China University Geosciences 8, 90–94.
Zhang, F.Q., Chen, H.L., Batt, G.E., Dilek, Y., An, M.A., Sun, M.D., 2015. Detrital zircon U-Pb geochronology and stratigraphy of the Cretaceous Sanjiang basin in NE China: Provenance record of an abrupt switch in the mode and nature of the NE Asian continental margin evolution. Tectonophysics 665, 58–78.
Zhang, F.Q., Dilek, Y., Chen, H.L., Yang, S.F., Meng, Q.A., 2017a. Late Cretaceous tectonic switch from a Western Pacific- to an Andean-type continental margin evolution in East Asia, and a foreland basin development in NE China. Terra Nova 29, 335–342. Available from: https://doi.org/10.1111/ter.12286.

Zhang, F.Q., Dilek, Y., Chen, H.L., Yang, S.F., Meng, Q.A., 2017b. Structural architecture and stratigraphic record of Late Mesozoic sedimentary basins in NE China: tectonic archives of the Late Cretaceous continental margin evolution in East Asia. Earth-Science Reviews 171, 598–620. Available from: https://doi.org/10.1016/j.earscirev.2017.05.015.

Zhao, X.Q., Li, C.C., Jiang, Y.X., Zhu, H.Y., Wang, F.D., Chai, P.R., 2022. Geological characteristics of the Mesozoic unconformities in eastern Heilongjiang, NE China: implications for the Mesozoic continental margin evolution of northeast Asia. Frontiers in Earth Science https://10.3389/feart.2022850324.

Zhou, J.P., Dunkl, I., Liu, Y.J., Li, W.M., von Eynatten, H., 2020. Late Cretaceous-Tertiary tectonic inversion of northeastern Asia continental margin: insight from the low temperature thermochronology in NE China. Gondwana Research . Available from: https://doi.org/10.1016/j.gr.2020.05.017.

Zhu, J.C., Feng, Y.L., Meng, Q.R., Li, S.H., Wu, G.L., Zhu, R.X., 2019. Decoding stratigraphic and structural evolution of the Songliao Basin: implications for late Mesozoic tectonics in NE China. Journal of Asian Earth Sciences. Available from: https://doi.org/10.1016/j.jseaes.2019.104138.

CHAPTER

7

Sequence stratigraphy of rift basins

Octavian Catuneanu

Department of Earth and Atmospheric Sciences, University of Alberta, Edmonton, AB, Canada

Introduction

Rifts are extensional basins that precede the opening of new oceans, formed as a result of crustal stretching before continental breakup. "Active" rifts originate commonly in triple-junction areas that develop over hotspots, in which three separate fracture zones diverge from the region of thermal uplift (e.g., the triple junction that includes the East African Rift Valley, the Red Sea, and the Gulf of Aden; Fig. 7.1). Rifting may also be "passive" (i.e., unrelated to mantle convection and hotspots), where the stretching of the lithosphere is driven by far-field tectonics (Miall, 2000; Allen and Allen, 2013; e.g., the Triassic Cuyana Basin in Argentina; Fig. 7.2). Both "active" (i.e., driven by mantle convection) and "passive" rifting lead to the brittle deformation of the crust in parallel with crustal thinning and the rise of the Moho discontinuity. At the same time, the mantle portion of the lithosphere also undergoes thinning, leading to the rise of the asthenosphere and a corresponding increase in heat flow into the lithosphere. These processes explain the common characteristics of rifts, which include the development of extensional faults, decompression melting, thermal metamorphism, volcanism, and regional thermal uplift.

A typical feature of rifts is the development of asymmetrical half-grabens bounded by normal faults on one side and uplifted flanks on the other, generally oriented along the rift axis (Fig. 7.3). Each rift, whether related to far-field "passive" rifting or "active" rifting within a triple-junction area, consists of a series of half-grabens that form together the relatively long and narrow rift zone (Fig. 7.3). Within a rift zone, individual half-grabens form discrete depocenters for sediment accumulation, most commonly in a range of 10^0 km wide and 10^1 km long (Figs. 7.2, 7.3). The orientation and polarity of the half-graben boundary faults often change along the rift axis, generating a structurally complex rift zone (Fig. 7.3). During the evolution of a rift, the initially isolated half-grabens become increasingly interconnected as individual fault segments expand and join into larger boundary faults. Volcanism is also typical of rifting, particularly along triple-junction rifts that lead to continental breakup. Most commonly, two of the rifts of a triple junction evolve to the point of seafloor spreading, while the third fails before reaching the point of continental breakup.

Depending on the location of rifts relative to the continental margins, rift basins may be flooded by the sea or may remain above the sea level; in the latter case, the rifts may host lacustrine, alluvial, fluvial, or eolian environments (Fig. 7.1). The place of preservation of rift sequences in the stratigraphic record depends on the evolution of rift basins and mainly on whether they lead to continental breakup. Half-grabens that belong to successful rifts (i.e., rifts that lead to the opening of new oceans and the formation of new spreading centers along mid-oceanic ridges) are found below the sedimentary fill of divergent continental margins (Figs. 7.4 and 7.5), whereas failed rifts remain preserved within the continental settings as continental rift basins. Irrespective of the rift type, half-grabens typically experience similar subsidence patterns, starting with an increase followed by a decrease in subsidence rates, which generate predictable changes in the ratio between accommodation and sedimentation during the evolution of the basin. This work presents the stratigraphic patterns that emerge in rift settings during the various stages of basin development, from initiation to maximum subsidence and final infill.

Regional Geology and Tectonics: Volume 2
DOI: https://doi.org/10.1016/B978-0-444-64136-6.00008-7

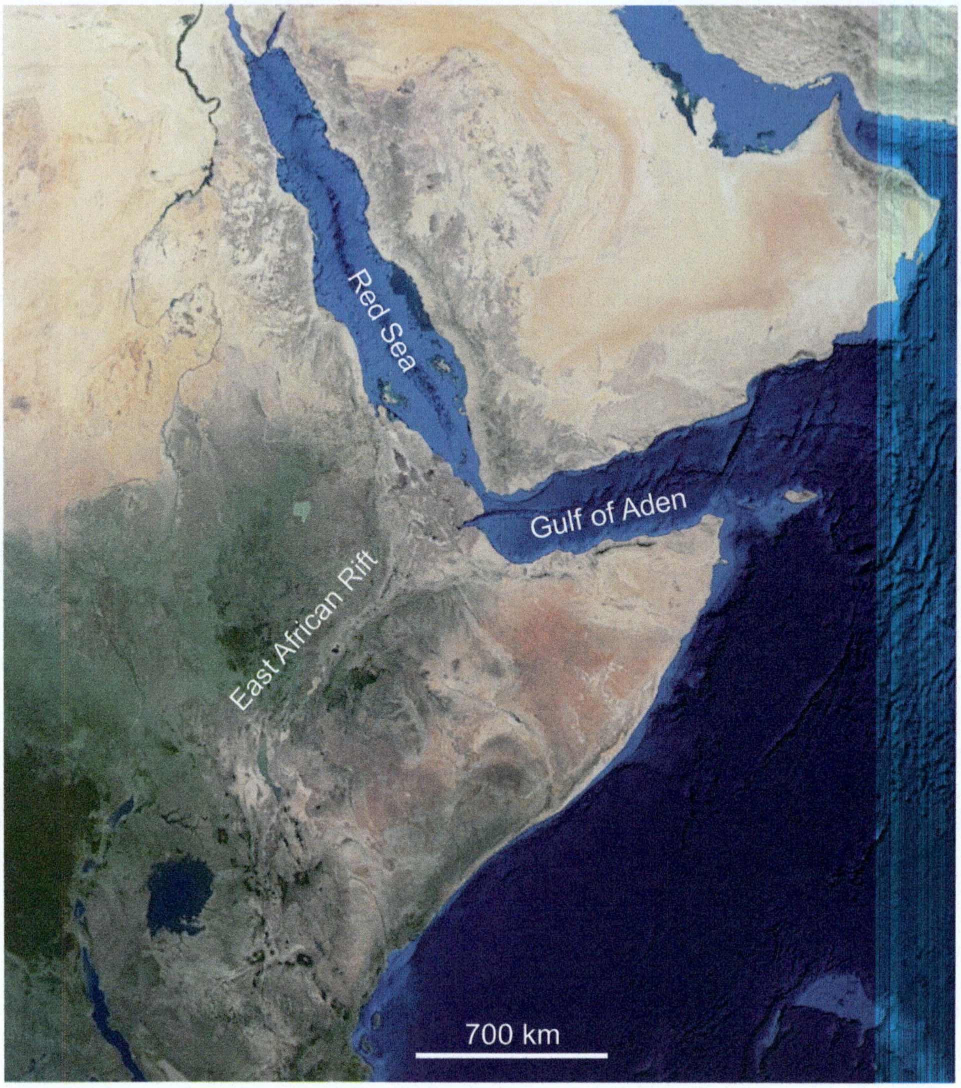

FIGURE 7.1 Triple junction related to hotspot activity (i.e., active rifting), connecting the Red Sea, the Gulf of Aden, and the East African Rift. *Courtesy: Google Earth*

Tectonism of rift basins

The evolution of rift basins follows a predictable pattern, herein referred to as the first-order rift cycle, whereby subsidence rates increase during the formation of the basin and decrease toward the end of rifting, commonly within a range of 10^{-3}–10^{0} m/kyr (dePolo and Anderson, 2000). Crustal stretching over hotspots, typically at horizontal rates of 10^{-2} m/year (Friedmann and Burbank, 1995), leads to extension and local mechanical subsidence across the area affected by thermal uplift. As a result, each arm of a triple junction includes multiple half-graben segments along the rift zone. Similar patterns are also recorded along the "passive" rifts generated by far-field tectonics. Whether rifts succeed or fail in producing continental breakup, subsidence typically starts with low rates in the early stage of rift initiation, continues with higher rates during the main stage of basin development, and ultimately ends with low rates in the late stage of rifting (Fig. 7.6). These stages result in changes in the balance between accommodation and sedimentation during the evolution of rifts, with an impact on the depositional environments that can occur within a basin at times of overfilled or underfilled accommodation (Fig. 7.7).

The three stages of basin development that define the first-order rift cycle (Figs. 7.6 and 7.7) generate total amounts of subsidence in a range of 10^{0} km, sometimes even exceeding 10 km (Bosence, 1998; Allen and Allen, 2013). However, this subsidence is differential, with higher rates adjacent to the boundary faults of the half-grabens, and

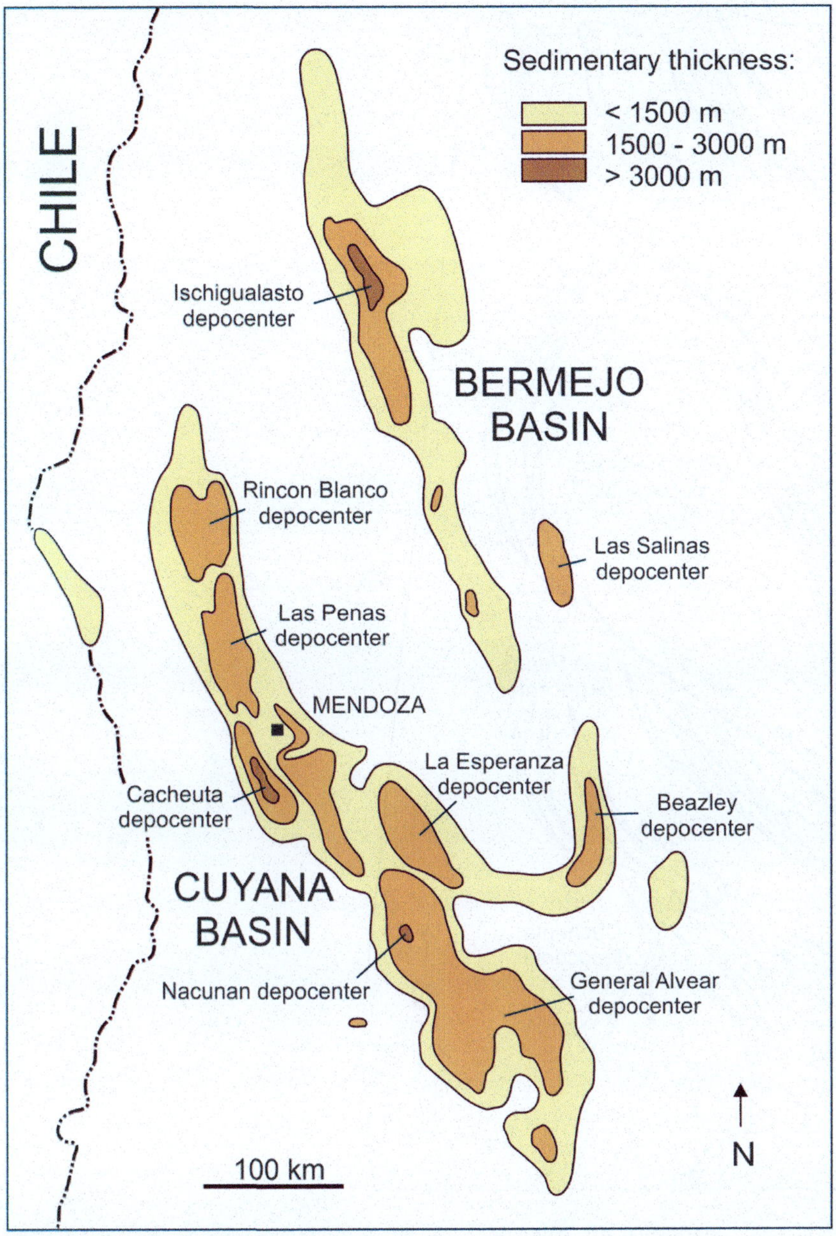

FIGURE 7.2 Triassic rift basins in west-central Argentina and their main subbasins. These continental rifts contain up to 3700 m of alluvial, fluvial, and lacustrine sediments interbedded with tuffs related to intraplate volcanism. Source: *Modified after Barredo, S.P., 2012. Geodynamic and tectonostratigraphic study of a continental rift: The Triassic Cuyana Basin, Argentina. In: E. Sharkov (Ed.) Tectonics – Recent Advances. IntechOpen, https://doi.org/10.5772/49958.*

gradually decreasing rates toward the flexural basin margins (Figs. 7.4 and 7.5). As a result, the basin fills are asymmetrical and wedge shaped, with the depocenters located closer to the boundary faults where greater amounts of accommodation are generated (Figs. 7.4 and 7.5). This asymmetry also implies that the balance between accommodation and sedimentation may vary at any given time across a basin in a dip direction, as well as in a strike direction due to variations in the rates of subsidence and sedimentation along the rift axis. Nevertheless, each stage of rift development records distinctive trends in terms of depositional environments and the degree in which the accommodation made available by extensional subsidence is consumed by sedimentation (Fig. 7.7).

As rifting is typically initiated in a continental setting, the early stage of basin development is prone to continental or shallow-water environments enabled by relatively low subsidence rates that are matched by sedimentation (Figs. 7.6 and 7.7). This depositional setting is maintained for as long as accommodation and sedimentation are in relative balance, resulting in overfilled or partially underfilled basins, which commonly host alluvial, fluvial, eolian, and lacustrine environments. In this early stage of evolution, the rift zone is a fragmented basin with multiple isolated depocenters dominated by a half-graben geometry controlled by single boundary faults. Fluvial drainage may be restricted to each half-graben, with no through drainage across the entire rift basin, and lakes may

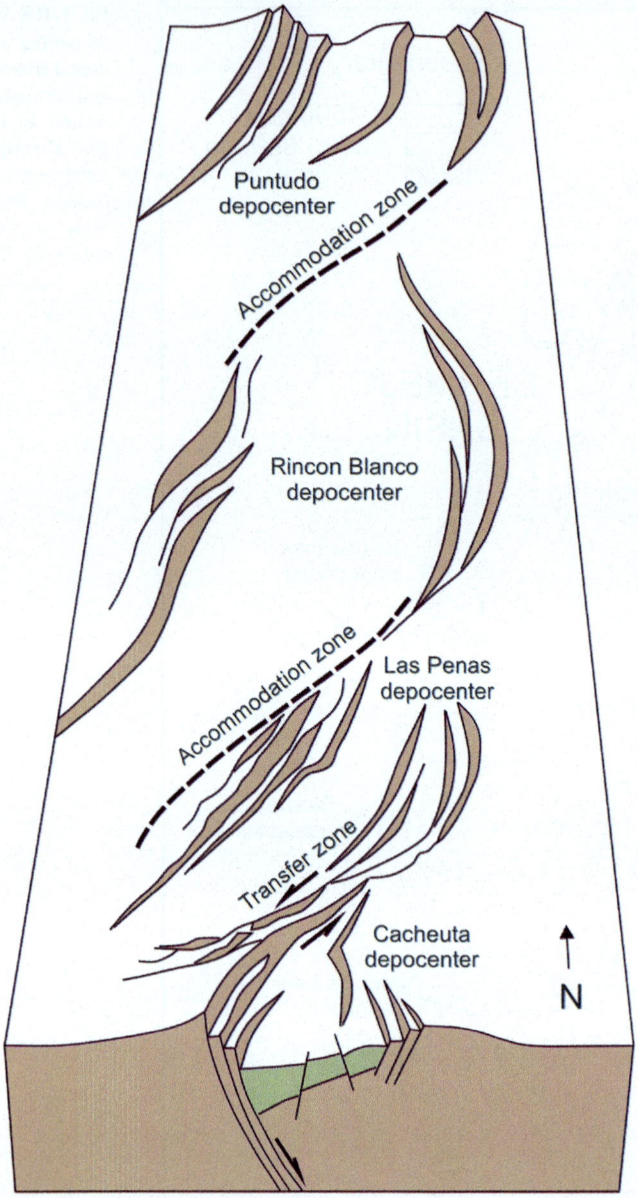

FIGURE 7.3 Main depocenters of the Cuyana Basin (Triassic continental rift, Argentina). The rift is fragmented into several half-grabens separated by intrabasinal highs (i.e., accommodation zones). The half-graben boundary faults trend obliquely relative to the main rift direction and may record opposite dip directions across the accommodation zones. Source: *Modified after Barredo, S.P., 2012. Geodynamic and tectonostratigraphic study of a continental rift: The Triassic Cuyana Basin, Argentina. In: E. Sharkov (Ed.) Tectonics – Recent Advances. IntechOpen, https://doi.org/10.5772/49958.*

form in the deeper and underfilled parts of the depocenters (e.g., as observed along the East African Rift; Fig. 7.1). During this stage, deposition may be either upstream-controlled, in the case of overfilled basins, or downstream-controlled, where lake-level changes influence shoreline trajectories and the related stratal stacking patterns.

The main stage of rift development is characterized by high subsidence rates, which typically outpace the rates of sedimentation across most of the basin. As a result, the rift becomes underfilled and dominated by deeper lacustrine or marine environments (Figs. 7.6 and 7.7). During this stage, the boundary faults of the individual half-grabens grow and coalesce into more regional extensional faults, which outline larger depocenters along the rift axis. Sediment supply to these lakes or interior seas may come from either side of the rift and may feed gravity flows directed to the depocenters and possibly along the rift axis if sufficient tilt supports the sediment transport in a strike direction. The resulting turbidites and debrites are commonly interbedded with fine-grained

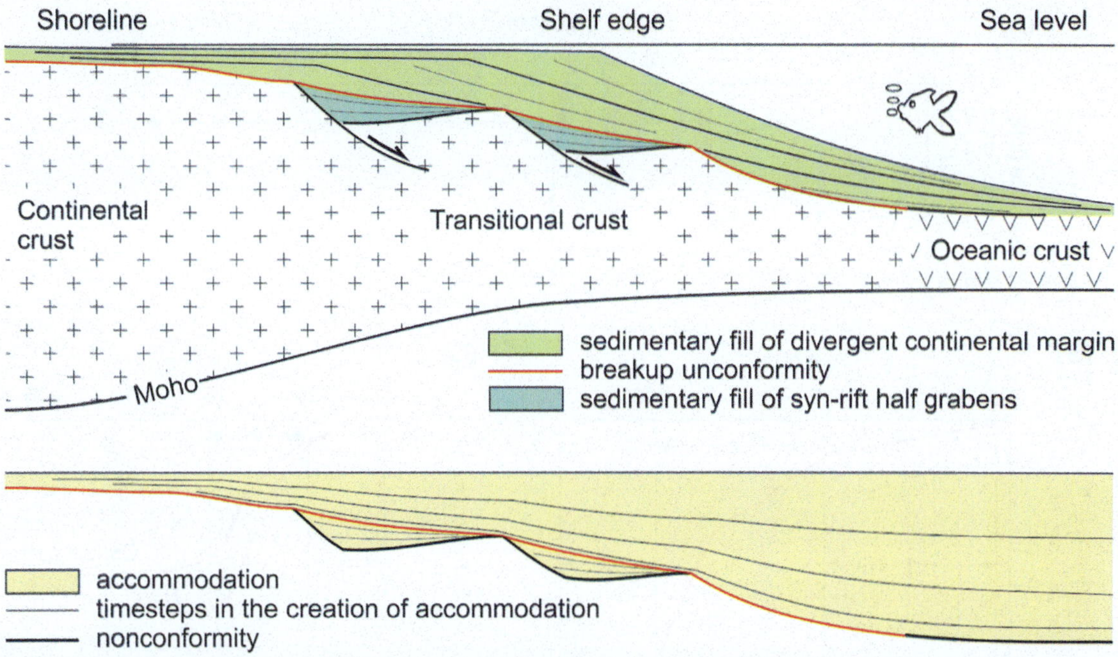

FIGURE 7.4 Subsidence patterns and stratigraphic architecture in a divergent continental margin setting. Below the breakup unconformity, subsidence rates in the rift half-grabens increase toward the boundary faults, typically in a landward direction. Above the breakup unconformity, subsidence rates increase in a basinward direction. Source: *From Catuneanu, O., 2022. Principles of Sequence Stratigraphy. Second Edition, Elsevier, Amsterdam.*

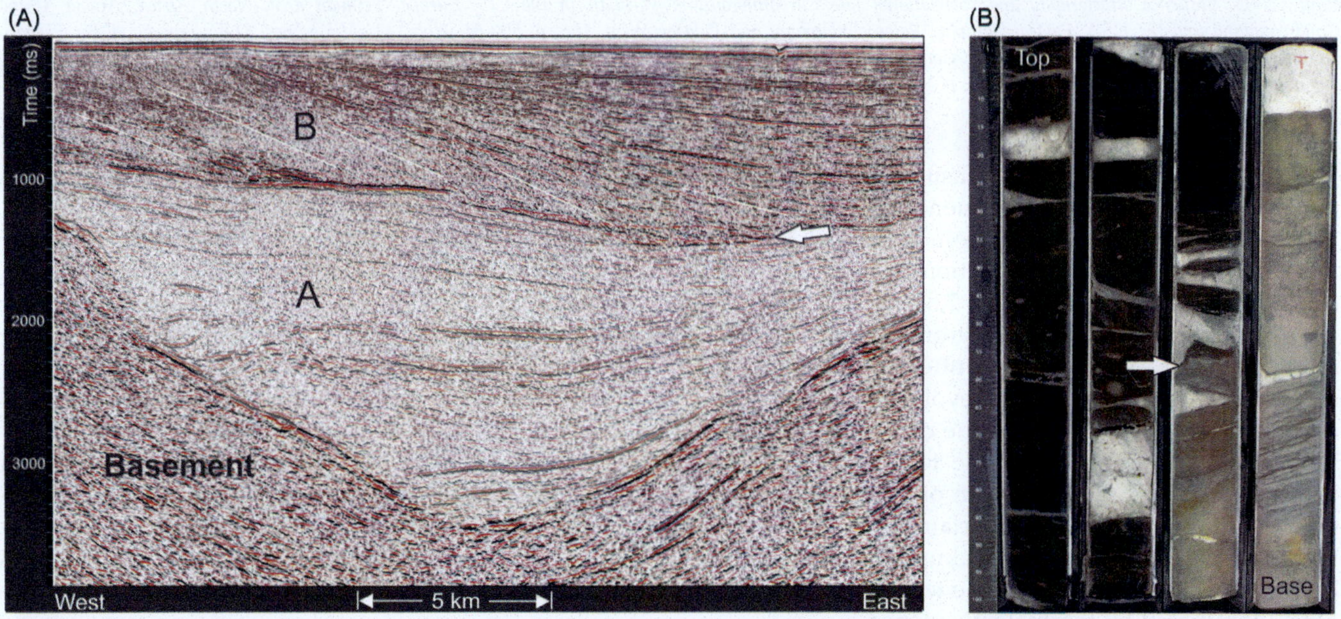

FIGURE 7.5 First-order sequence boundary (arrows) at the limit between a rift sequence (A) and the overlying "passive-margin" sequence (B). Core data indicate that the sequence boundary is represented by a wave-ravinement surface at the contact between fluvial deposits below and marine shales above (Krishna-Godavari Basin, offshore India). Source: *From Catuneanu, O., 2022. Principles of Sequence Stratigraphy. Second Edition, Elsevier, Amsterdam. Courtesy: Reliance Industries, India*

suspension sediment, which may generate economically significant petroleum source rocks. Such rift source rocks become important not only for the petroleum systems of rifts but also for the prospectivity of other juxtaposed basins (e.g., the overlying "passive-margin" in the case of rifts that lead to continental breakup; Figs. 7.4 and 7.5).

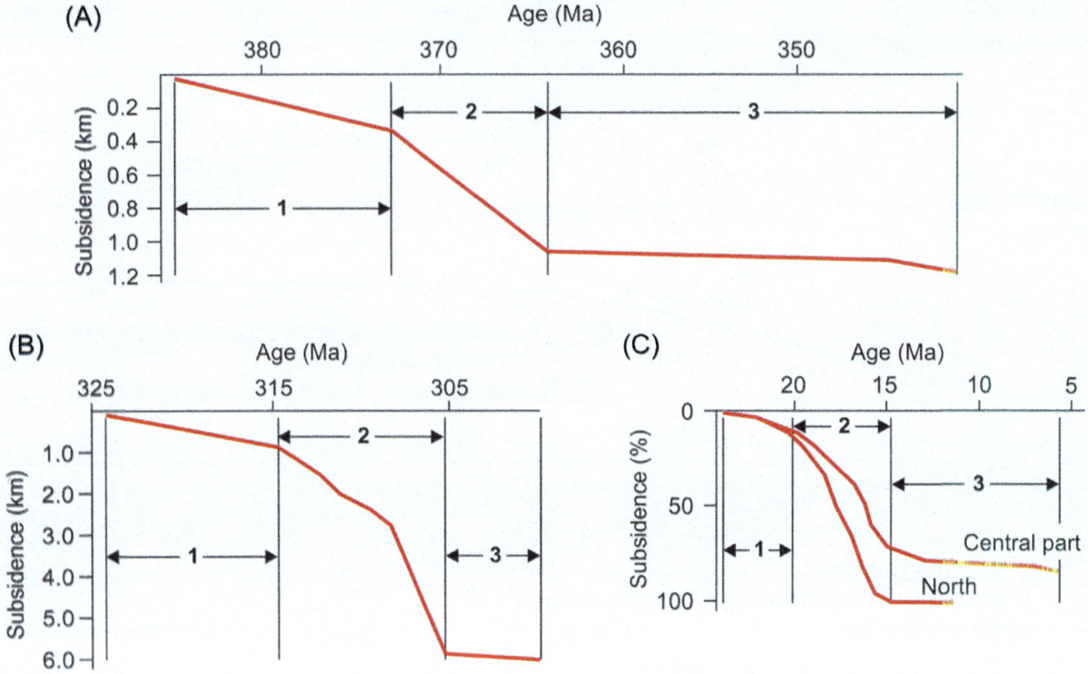

FIGURE 7.6 Subsidence curves in rift basins: (A) Dnieper–Donets Basin, Ukraine; (B) Asturias Basin, Spain; and (C) Suez rift, Egypt. Stages of basin development: 1—early stage of low subsidence rates; 2—main stage of high subsidence rates; and 3—late stage of low subsidence rates. Source: *Modified after Moretti, I., and Colletta, B., 1987. Spatial and temporal evolution of the Suez rift subsidence. Journal of Geodynamics, 7, p. 151–168; Van Wees, J.D., Stephenson, R.A., Stovba, S.M., and Shymanovskyi, V.A., 1996. Tectonic variation in the Dnieper-Donets Basin from automated modelling of backstripped subsidence curves. Tectonophysics, 268, p. 257–280; Izart, A., Stephenson, R.A., Vai, G.B., Vachard, D., Nindrfe, Y.L., Vaslet, D., et al., 2003. Sequence stratigraphy and correlationof late Carboniferousand Permian in the CIS, Europe, Tethyan area, North Africa, Arabia, China, Gondwanaland and the USA. Palaeogeography, Palaeoclimatology and Palaeoecology, 196, p. 59–84; Holz, M., Vilas-Boas, D.B., Troccoli, E.B., Carnairo Santana, V., and Vidigal-Souza, P.A., 2017. Conceptual models for sequence stratigraphy of continental rift successions. In: M. Montenari (Ed.) Stratigraphy and Timescales, Vol. 2, Academic Press, UK, p. 119–186.*

The underfilled nature of the basin at this time also implies that deposition is downstream-controlled, whereby the lake-/sea-level changes influence shoreline trajectories and the related stratal stacking patterns.

The late stage of rift evolution is defined by declining subsidence rates, which change the balance between accommodation and sedimentation in favor of the latter. As a result, rifts become overfilled and dominated by continental environments (Figs. 7.6 and 7.7). It is thus common to find fluvial or eolian sediments toward the top of rift sequences, below the first-order sequence boundaries (Kress et al., 2021; Fig. 7.5). Deposition may still be, at least in part, downstream-controlled during the transition from the main stage (underfilled basin) to the late stage (overfilled basin) of rift evolution and typically becomes upstream-controlled once all accommodation is consumed by sedimentation. The complete infill of rifts by the end of the first-order rift cycles explains the formation of unconformities at the top of rift sequences, which mark breaks in sedimentation before subsidence resumes following the formation of a new basin. For example, in the case of rifts that lead to the opening of new oceans, rift sequences are overlain by "passive-margin" sequences, with the intervening first-order sequence boundaries represented by breakup unconformities (Figs. 7.4 and 7.5). In this example, the first-order sequence boundaries mark a change in the tectonic setting from rifts dominated by extensional subsidence to "passive margins" dominated by thermal subsidence.

Stratigraphy of rift basins

The stratigraphic architecture of rift basins is to a large extent controlled by the subsidence trends recorded during the first-order rift cycles (Figs. 7.6 and 7.7). The first-order rift sequences are typically bounded by regional unconformities at the base and at the top, which mark changes in the tectonic setting and the dominant subsidence mechanisms. At the base, the "syn-rift" unconformity marks the onset of rifting and continues to

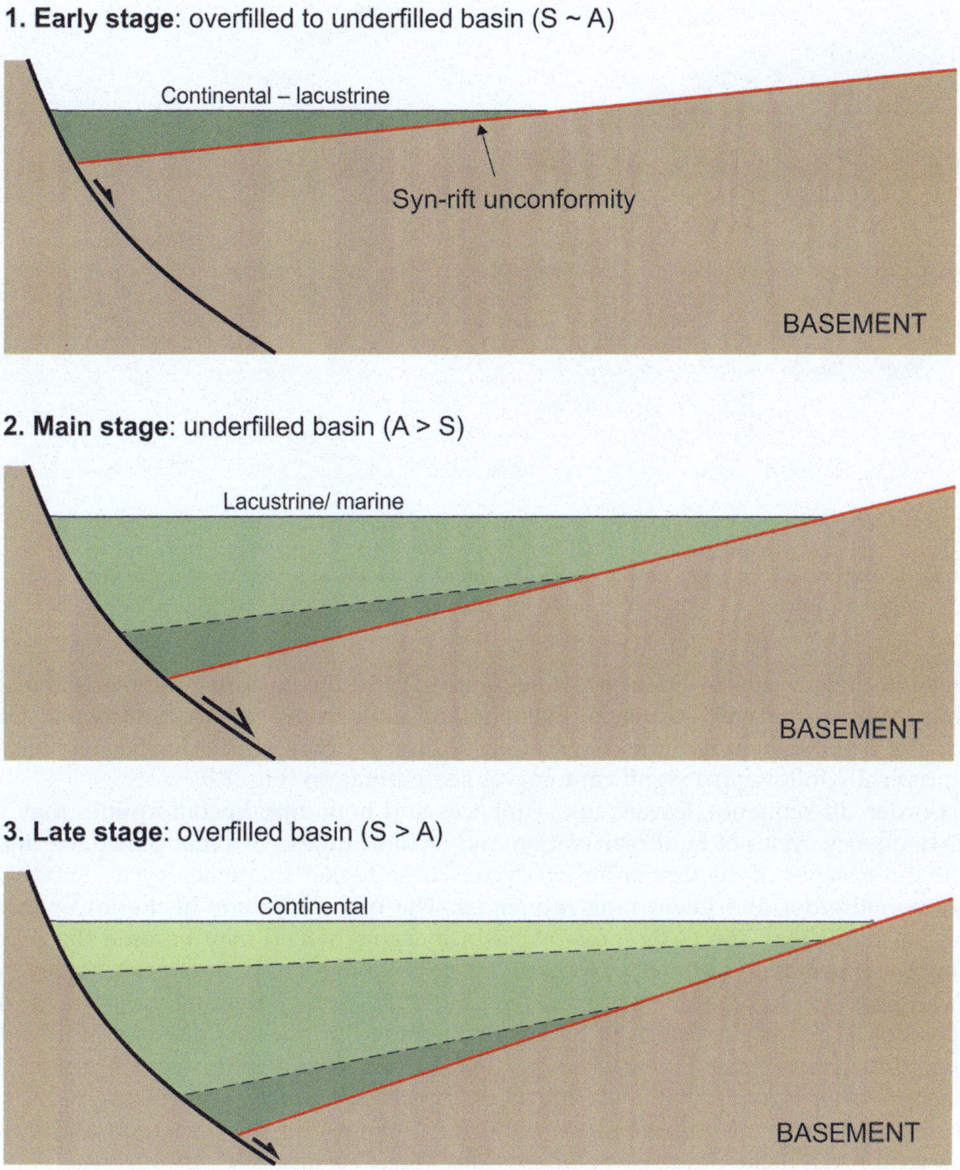

FIGURE 7.7 First-order rift cycle—stages of basin development: 1. low subsidence rates in relative balance with sedimentation; 2. high subsidence rates and basin deepening; and 3. low subsidence rates and basin infill (see Fig. 7.6 for examples of subsidence curves). Variations in the rates of subsidence and sedimentation across the basin may result in the formation of different but age-equivalent systems tracts along dip and strike directions (see text for details). *A*, Accommodation; *S*, sedimentation.

develop during rifting in response to the thermal uplift of the flexural basin margins (Bosence, 1998; Holz et al., 2017; Fig. 7.7). The fragmentation of the rift zone into multiple half-grabens, especially during the early stage of rift development, implies that the syn-rift unconformity may be a composite and diachronous surface, consisting of smaller fragments of local significance. At the regional rift scale, however, the syn-rift unconformity may be mapped as a throughgoing first-order surface at the base of the rift sequence, especially at the lower resolution scales afforded by seismic data.

The top of the first-order rift sequence is also marked by a regional unconformity that signifies the cessation of sedimentation following the complete infill of the rift basin at the end of the first-order rift cycle. This first-order sequence boundary can also be diachronous, reflecting variations in the rates of subsidence and sedimentation along the rift axis, which affect the timing of the end of rifting at any location. However, this unconformity is more likely to form a physically continuous surface across the rift basin, following the coalescence of half-grabens during the main stage of rift development. Where rifts succeed in producing continental breakup, the

FIGURE 7.8 First-order sequence boundary (arrow) at the limit between the Cuyana rift sequence and the overlying Neuquen backarc sequence (Triassic–Jurassic boundary, Mendoza, Argentina). (A) Upper Rio Blanco Formation: ephemeral fluvial sand sheets interbedded with tuffs and paleosols; (B) Barrancas Formation: alluvial fans and ephemeral fluvial systems. *Courtesy: Daniel Boggetti*

first-order sequence boundary at the top of a rift sequence is the breakup unconformity that separates the rift from the overlying "passive-margin" sequence (Figs. 7.4 and 7.5). In the case of continental failed rifts, the top sequence boundary of a rift sequence marks the shift to a different type of interior basin, such as a backarc or a retroarc foreland, typically following a significant gap in sedimentation (Fig. 7.8).

Within the first-order rift sequence, lower rank sequences and bounding unconformities may also form in relation to the higher frequency cycles of fault reactivation and tectonic quiescence that punctuate the rift development during each of the three stages of the first-order rift cycle. These higher frequency cycles subdivide the first-order rift sequence into second-order and lower rank sequences. The origin of some of the lower rank unconformities relates to the thermal uplift that affects the flexural basin margins, which may become the dominant control on accommodation and sedimentation at times of tectonic inactivity along the basin-margin faults. Periods of tectonic quiescence may also produce stratigraphic hiatuses, even in the absence of thermal uplift, particularly in overfilled basins where deposition gives way to sediment bypass once graded profiles are attained, until subsidence resumes and starts a new sedimentation cycle. Fluctuations in the lake/sea level during stages of underfilled accommodation may also generate lower rank depositional sequences, further complicating the stratigraphic architecture of rift basins. However, tectonism remains the dominant underlying control on the formation and evolution of rifts, as well as on the stratigraphic cyclicity observed at larger scales (Figs. 7.6 and 7.7).

Early stage of rift development: overfilled to underfilled basin

The early stage of rift development is conducive to deposition in continental to shallow-water environments due to the low subsidence rates that are in relative balance with the rates of sedimentation (Figs. 7.6 and 7.7; discussed above). Where basins are overfilled, continental sequences are upstream-controlled and consist of unconventional (i.e., shoreline-independent) systems tracts (e.g., high- and low-amalgamation systems tracts in fluvial systems; Fig. 7.9). Where subaqueous accommodation is made available (e.g., in lakes or shallow seas), all types of conventional (downstream-controlled) systems tracts may form in relation to shoreline trajectories, including lowstand systems tracts that require sedimentation in excess of accommodation at the shoreline following the onset of relative sea/lake-level rise (Fig. 7.10). The latter condition is afforded by the low rates of subsidence, which enable the manifestation of both lowstand and highstand normal regressions. Notably, highstand systems tracts are better developed than the lowstand systems tracts in rift basins, as the rates of subsidence related to the reactivation of basin-margin faults typically decline during each stage of relative sea/lake-level rise, providing more favorable conditions for the manifestation of highstand normal regressions. In the long term, the early stage of rift development starts with overfilled accommodation and gradually transitions toward the underfilled conditions that dominate the main stage of the first-order rift cycle.

FIGURE 7.9 Upstream-controlled depositional sequences in an overfilled rift basin (early stage of rift development; Triassic, middle Potrerillos Formation, Cuyana Basin, Argentina). (A) Amalgamated braided channels (high-amalgamation systems tract); (B) floodplain-dominated meandering-stream deposits (low-amalgamation systems tract); *SU*, subaerial unconformity.

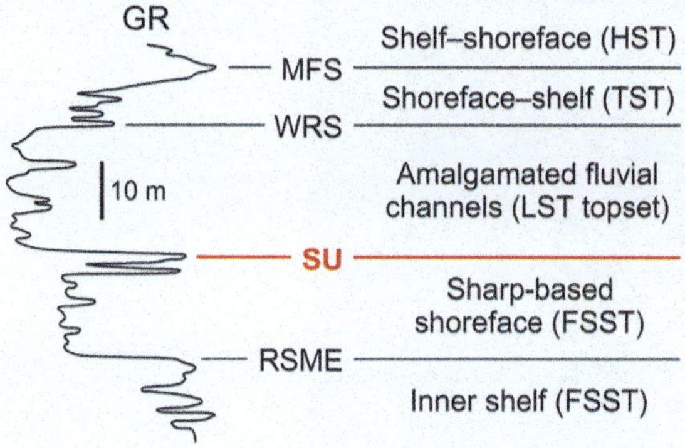

FIGURE 7.10 Stratigraphic architecture in an underfilled rift basin with relatively low rates of subsidence (Lower Cretaceous, Brooks-Mackenzie Basin, Arctic Canada). All downstream-controlled systems tracts may form where accommodation and sedimentation vary within overlapping ranges along the shoreline. *GR*, Gamma-ray log; *RSME*, regressive surface of marine erosion; *SU*, subaerial unconformity; *WRS*, wave-ravinement surface; *FSST*, falling-stage systems tract; *LST*, lowstand systems tract; *TST*, transgressive systems tract; *HST*, highstand systems tract. Source: *Modified after Millar, R.A., 2021. Sequence stratigraphy and underlying tectonism of the Northern Richardson Mountains and adjacent Mackenzie Delta related to the formation of the Arctic Ocean. M.Sc. Thesis, University of Alberta, 142 pp.*

Main stage of rift development: underfilled basin

The main stage of rift development is conducive to the formation of deeper water lacustrine or marine environments due to the high subsidence rates that outpace the rates of sedimentation in the depocenter(s) (Figs. 7.6 and 7.7; discussed above). As a result, rift basins are typically underfilled during this stage, with a stratigraphic architecture that can be described in terms of downstream-controlled stacking patterns related to shoreline trajectories. Bathymetric trends show a long-term deepening until rifting reaches its climax, which roughly corresponds to the timing of the first-order maximum flooding surface (i.e., the maximum expansion of the lake or sea; Fig. 7.11A). Following the rift climax, the bathymetry records a long-term shallowing trend as sedimentation starts to outpace accommodation toward the end of the main stage of rift development (Fig. 7.11B). This shallowing trend makes the transition to the late stage of the first-order rift cycle, in which the basin becomes overfilled. It should be noted, however, that the correlation between bathymetry and the physical stacking patterns (e.g., sediment grading) is only safely valid in areas adjacent to the shoreline. Toward the depocenters, the timing of sequence stratigraphic surfaces may diverge from bathymetric trends (discussion below; Naish and Kamp, 1997; Catuneanu et al., 1998; Vecsei and Duringer, 2003; Zecchin et al., 2021; Catuneanu, 2022).

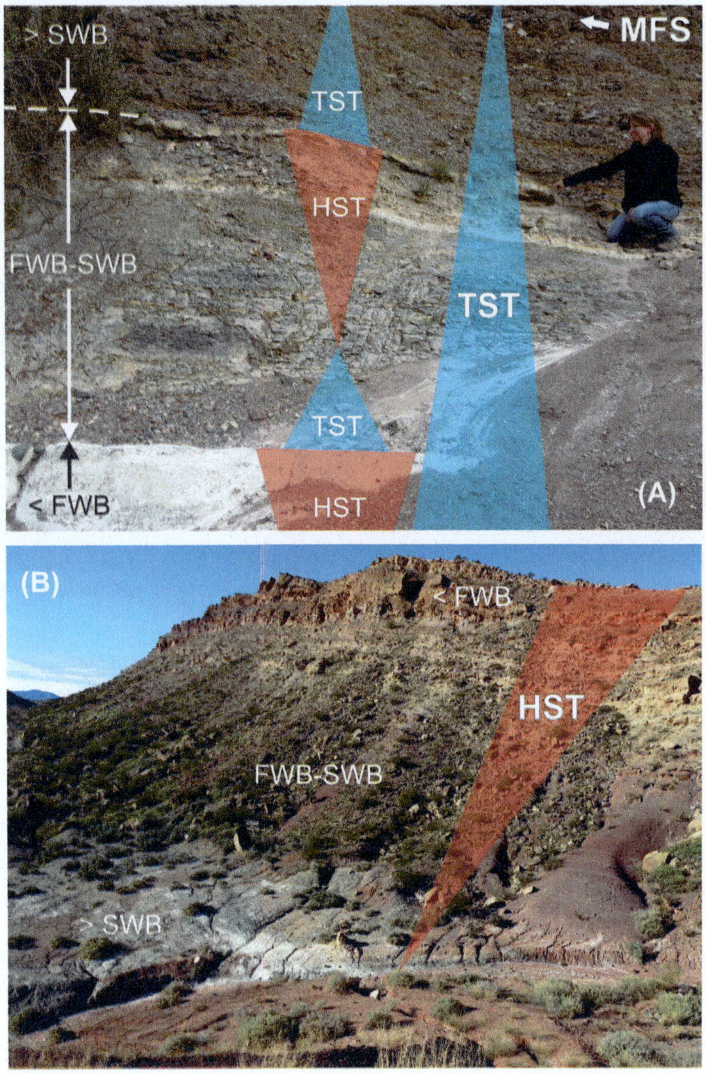

FIGURE 7.11 Bathymetric trends during the main stage of rift development (Triassic, Cuyana Basin, Argentina). (A) Deepening trend, as subsidence outpaces sedimentation in the long term (upper Potrerillos to Cacheuta); (B) shallowing trend, as sedimentation outpaces accommodation toward the end of the main stage of rift development (Cacheuta to lower Rio Blanco). The latter makes the transition to the late stage of the first-order rift cycle, in which the basin becomes overfilled. The first-order maximum flooding surface, which is defined by stratal stacking patterns (grading trends in this example), marks the maximum expansion of the lake at the climax of rifting. While a general correspondence exists between bathymetric trends (water deepening vs shallowing) and sediment grading (fining- vs coarsening-upward), the deepest water in the depocenter is typically recorded above the maximum flooding surface (e.g., as indicated by total organic carbon [TOC] or micropaleontology data; Naish and Kamp, 1997; Vecsei and Duringer, 2003; Catuneanu, 2006, 2022; Harris et al., 2018; Zecchin et al., 2021). *MFS*, Maximum flooding surface; *FWB*, fair-weather wave base; *SWB*, storm wave base.

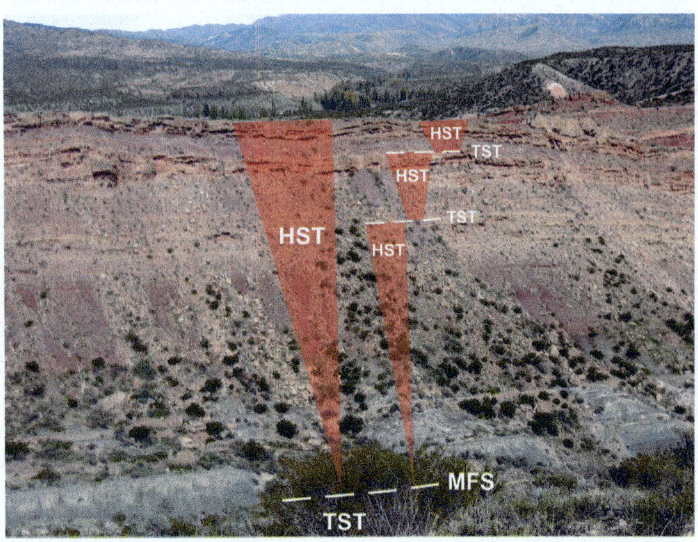

FIGURE 7.12 Long- and short-term prograding trends during the main stage of rift development, following the formation of the first-order maximum flooding surface (Triassic, transition from the Cacheuta to the lower Rio Blanco formations, Cuyana Basin, Argentina). Abrupt increases in accommodation associated with the reactivation of basin-margin faults inhibit the manifestation of lowstand normal regressions and result in the formation of thin transgressive systems tracts. As a result, sequences are dominated by highstand systems tracts. *MFS*, Maximum flooding surface; *TST*, transgressive systems tract; *HST*, highstand systems tract.

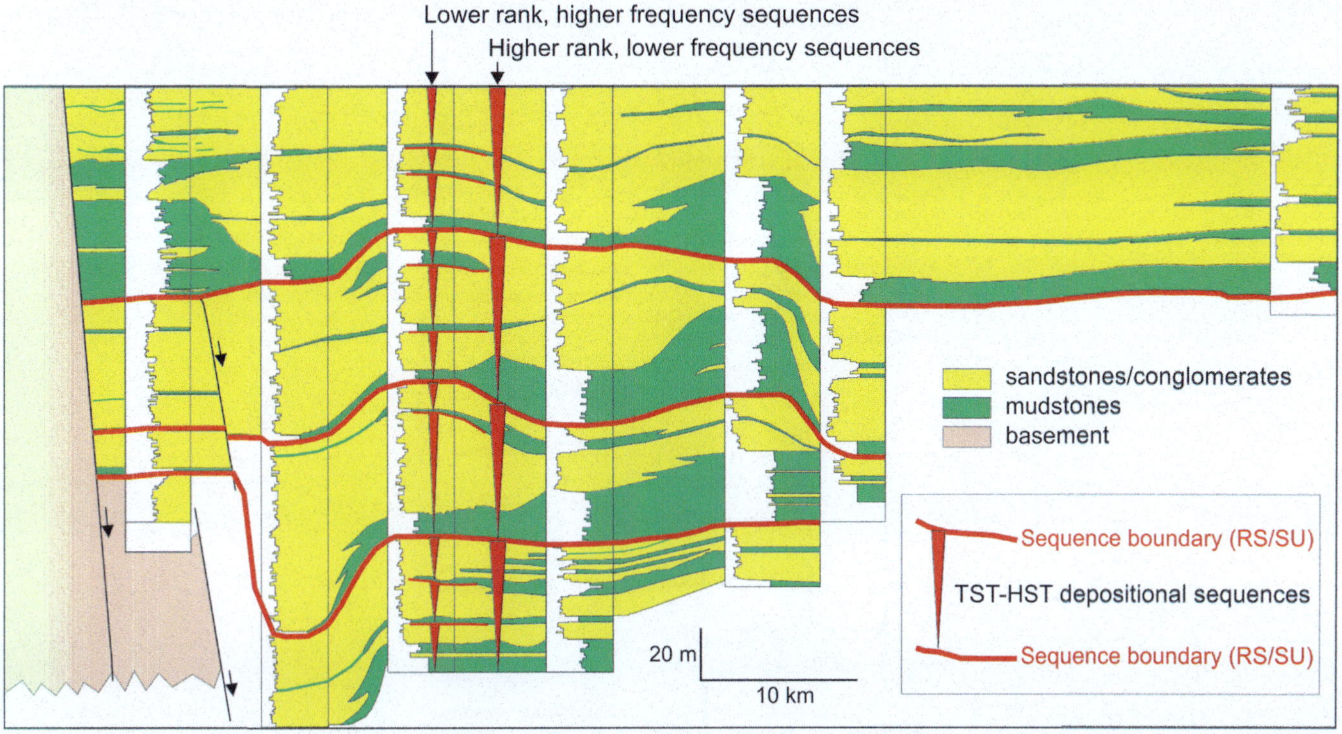

FIGURE 7.13 Stratigraphic architecture in an underfilled rift basin with relatively high rates of subsidence (Viking Graben, North Sea). Depositional sequences are initiated by the episodes of abrupt water deepening triggered by the reactivation of basin-margin faults. In this case, lowstand systems tracts are missing, depositional sequences are defined by transgressive and highstand systems tracts, and sequence boundaries are composite surfaces that originate as subaerial unconformities during stages of tectonic quiescence and thermal uplift of the flexural basin margins, subsequently reworked by transgressive ravinement surfaces. *TST*, Transgressive systems tract; *HST*, highstand systems tract; *RS*, transgressive ravinement surface; *SU*, subaerial unconformity. Source: *Modified after Frostick, L.E. and Steel, R.J., 1993. Sedimentation in divergent-plate basins. In: L.E. Frostick and R.J. Steel (Eds.) Tectonic Controls and Signatures in Sedimentary Successions. Special Publication of the International Association of Sedimentologists 20, p. 111–128.*

During both bathymetric trends of long-term water deepening and shallowing, shorelines may undergo higher frequency transgressions and regressions that reflect shorter term changes in the balance between accommodation and sedimentation along the shorelines (Fig. 7.11A). These lower rank cycles relate to the episodic reactivation of the basin-margin faults, which result in brief stages of high-rate subsidence followed by longer stages of tectonic quiescence that allow sedimentation to catch up with the newly generated accommodation. Therefore, these cycles are strongly asymmetrical, including thin transgressive deposits overlain by much thicker prograding clinoforms (Fig. 7.12). Sequence boundaries may be unconformable where the cessation of mechanical subsidence is accompanied by thermal uplift and exposure (Fig. 7.13). The high-rate subsidence events triggered by the reactivation of faults prevent the formation of lowstand systems tracts at the onset of relative sea/lake-level rise. As a result, the high-frequency sequences of the main stage of rift development are dominated by transgressive and highstand systems tracts (Figs. 7.11–7.13; Martins-Neto and Catuneanu, 2010).

Late stage of rift development: underfilled to overfilled basin

The late stage of rift development marks a change from underfilled to overfilled accommodation conditions, as subsidence rates decrease toward the end of the first-order rift cycle (Figs. 7.6 and 7.7; discussed above). This allows sedimentation to outpace accommodation and ultimately fill all the space made available by extensional subsidence. For this reason, rift basins typically end with sedimentation in continental environments, whether successful in leading to continental breakup (Fig. 7.5). For as long as accommodation is still available in lacustrine or marine environments, the stratigraphic architecture may be described in terms of downstream-controlled stacking patterns related to shoreline trajectories (Fig. 7.14). The low subsidence rates of this stage enable all types of shoreline trajectories, as accommodation and sedimentation vary within overlapping ranges. As a result, all conventional systems tracts may form, including the lowstand that requires sedimentation in excess of

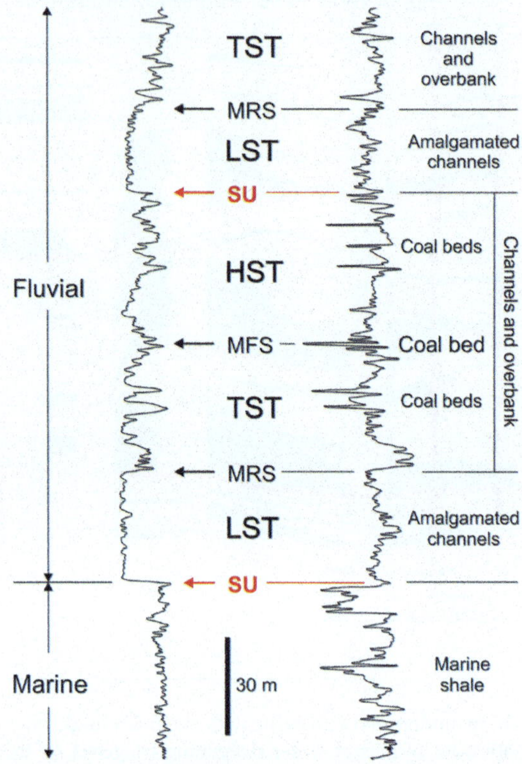

FIGURE 7.14 Stratigraphic architecture in a downstream-controlled fluvial succession (Jurassic, Sverdrup rift basin). The delineation of systems tracts relies on changes in fluvial style (degree of channel amalgamation), water table (development of coal beds), and the degree of marine influence (palynological marine index), as well as on the correlation with the coeval shallow-water systems in a downdip direction. *LST*, Lowstand systems tract; *TST*, transgressive systems tract; *HST*, highstand systems tract; *SU*, subaerial unconformity; *Mrs*, maximum regressive surface; *MFS*, maximum flooding surface. Source: *From Catuneanu, O., 2022. Principles of Sequence Stratigraphy. Second Edition, Elsevier, Amsterdam.*

accommodation at the shoreline at the onset of renewed subsidence (Fig. 7.14). Once all lacustrine or marine accommodation is consumed by sedimentation, the stratigraphic architecture becomes entirely upstream-controlled, with depositional sequences consisting of unconventional systems tracts (e.g., high- and low-amalgamation systems tracts in fluvial systems; Fig. 7.15).

Discussion: three-dimensional stratigraphic variability

The long-term stages in the evolution of rifts are commonly depicted with one-dimensional or two-dimensional diagrams, which capture the essence of the overall subsidence and sedimentation trends during the first-order rift cycles (Figs. 7.6 and 7.7). However, the rates of subsidence and sedimentation at any given time vary across a rift basin, leading to potentially out-of-phase bathymetric and/or depositional trends along dip and strike directions. This three-dimensional (3D) variability affords the coeval development of different types of systems tracts across a basin; for example, in relation to contrasts between the proximal and distal shorelines along the faulted versus the flexural basin margins, or along the same shoreline in a strike direction (Fig. 7.16). These variations affect the timing and the systems-tract composition of the lower rank sequences that form during each stage of rift development but may also offset the timing of the main stages of the first-order rift cycle along the rift axis. Therefore, different areas of a rift basin may record their own subsidence history (e.g., Fig. 7.6C) and the corresponding timing of the early, main, and late stages of the rift cycle.

The 3D tectonic and stratigraphic variability of rift basins is well documented with both field and modeling studies (Moretti and Colletta, 1987; Van Wees et al., 1996; Gawthorpe et al., 2003; Izart et al., 2003; Martins-Neto and Catuneanu, 2010; Csato et al., 2013, 2015; Holz et al., 2017; Catuneanu, 2019, 2022). Out-of-phase extensional

FIGURE 7.15 Upstream-controlled depositional sequences in an overfilled rift basin (late stage of rift development; Triassic, middle Rio Blanco Formation, Cuyana Basin, Argentina). (A) Amalgamated braided channels (high-amalgamation systems tract); (B) floodplain-dominated meandering-stream deposits (low-amalgamation systems tract); *SU*, Subaerial unconformity.

subsidence along the different half-graben boundary faults that coexist within the rift zone may lead to coeval subsidence and uplift between areas (individual depocenters or subbasins) dominated by mechanical subsidence or thermal uplift, respectively (e.g., Gawthorpe et al., 2003). As a result, forced regressions in areas subject to uplift and relative sea/lake-level fall can occur concurrently with transgressions or normal regressions in areas that are subject to subsidence and relative sea/lake-level rise. Variations in the rates of subsidence and sedimentation along strike may also generate coeval transgressions and normal regressions along the same shoreline, while similar variations along dip may result in coeval transgressions and normal regressions between the proximal and distal shorelines of the same basin (Fig. 7.16).

The concurrent formation of different types of systems tracts across a sedimentary basin is not unique to rifts, as it may occur in any basin that experiences variations in the rates of subsidence and sedimentation along dip and strike directions. Therefore, the 3D stratigraphic variability is the norm rather than the exception, in virtually all types of sedimentary basins (e.g., Helland-Hansen and Martinsen, 1996; Catuneanu et al., 1998, 2002; Schultz et al., 2020; Zecchin and Catuneanu, 2020; Catuneanu, 2022). Despite this tectonic and stratigraphic variability between and within sedimentary basins, the sequence stratigraphic methodology remains consistently based on the observation of stratal stacking patterns in a manner that is independent of tectonic setting, stratigraphic scales, and the types and resolution of the data available (Catuneanu, 2022). In downstream-controlled settings, systems tracts are defined by the stacking patterns generated by the interplay of accommodation and sedimentation at the shoreline, irrespective of their underlying controls (e.g., tectonism, sea/lake-level changes, climate, or autogenic processes; Fig. 7.17). In upstream-controlled settings, systems tracts are also defined by stratal stacking patterns, which account for the dominant depositional elements (Figs. 7.9 and 7.15).

Where rift basins are underfilled (i.e., with accommodation still available in lacustrine or marine environments), only the rates of accommodation and sedimentation *at the shoreline* matter to the formation of systems tracts, because it is the interplay of *these* rates that controls the type of shoreline trajectory and the depositional trends of progradation or retrogradation. For instance, a normal-regressive systems tract (lowstand or highstand) forms during the progradation and upstepping of the shoreline, which generates the diagnostic physical features (observable stacking patterns) that afford its identification as a stratal unit. At the same time, the syn-depositional bathymetric conditions may vary from water shallowing in the vicinity of the shoreline to water

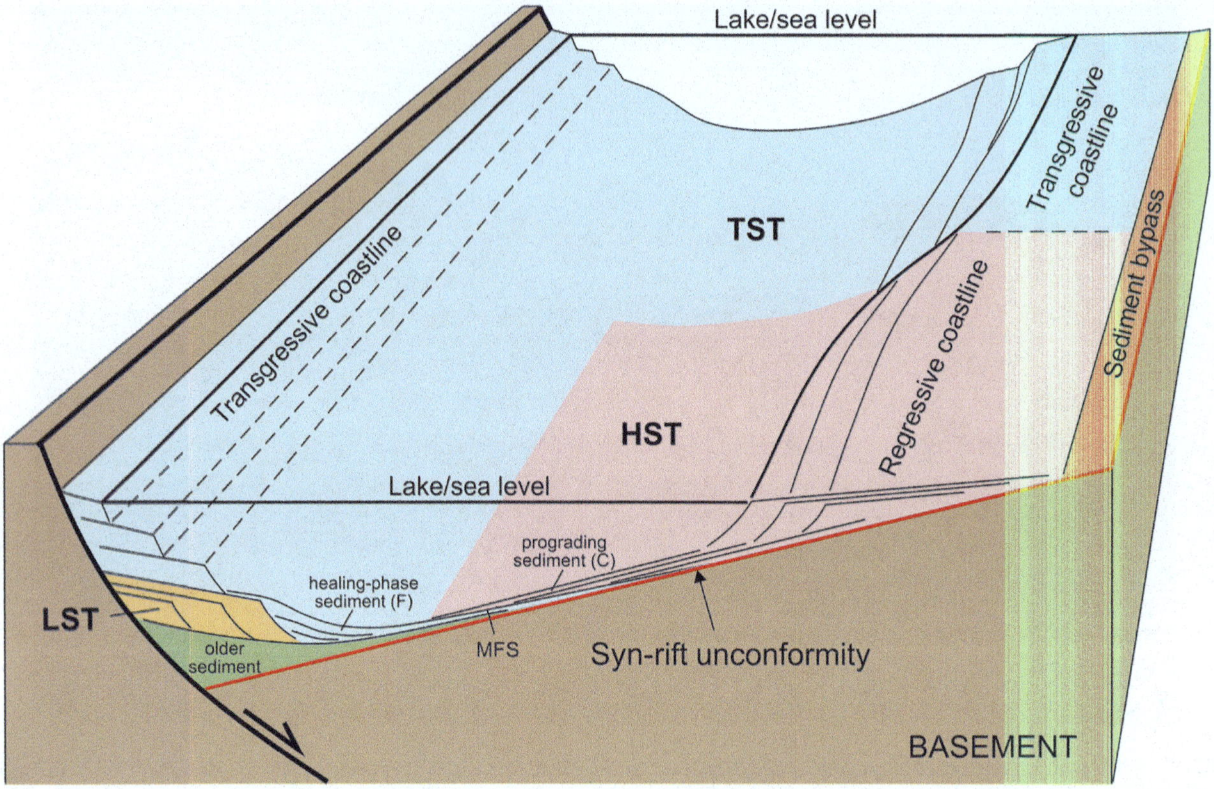

FIGURE 7.16 Stratigraphic variability in an underfilled rift basin: different systems tracts may form at the same time due to variations in the rates of accommodation and sedimentation along the shorelines. Systems tracts are defined by stratal stacking patterns that result from the interplay of accommodation and sedimentation *at the shoreline*, and not by offshore bathymetric trends. For example, the distal part of the HST may prograde into deepening water; as a result, the maximum flooding surface may form within an interval of water deepening (Naish and Kamp, 1997; Catuneanu et al., 1998; Vecsei and Duringer, 2003; Catuneanu, 2006, 2022; Harris et al., 2018; Zecchin et al., 2021). *LST*, Lowstand systems tract (sedimentation > accommodation at the shoreline); *TST*, transgressive systems tract (accommodation > sedimentation at the shoreline); *HST*, highstand systems tract (sedimentation > accommodation at the shoreline); *F*, fining upward; *C*, coarsening upward; *MFS*, maximum flooding surface.

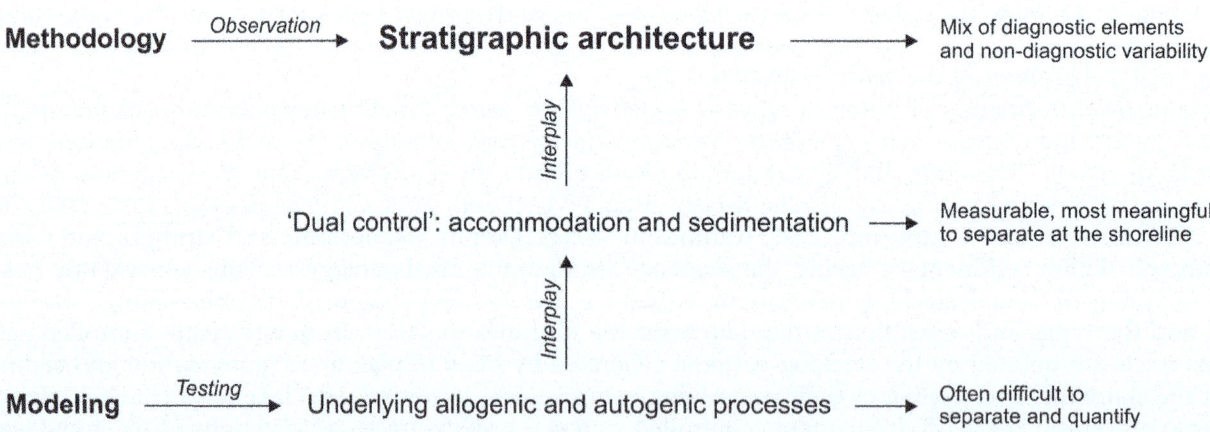

FIGURE 7.17 Methodology versus modeling in sequence stratigraphy. The interplay of accommodation and sedimentation explains the formation of stacking patterns in the stratigraphic record. The methodology involves the observation of the stratigraphic architecture in a manner that is independent of the interpretation of the underlying controls on sequence development. Source: *From Catuneanu, O., 2022. Principles of Sequence Stratigraphy. Second Edition, Elsevier, Amsterdam.*

deepening offshore. Nevertheless, an increase in water depth in the distal part of a prograding system, which reflects the *local* balance between accommodation and sedimentation, does not change the prograding nature of the stratal unit (Fig. 7.16). Systems tracts are defined by physical stacking patterns and not by offshore bathymetric trends or the inferred underlying controls (e.g., tectonism vs sea/lake-level changes). It is, in fact, most common that accommodation and sedimentation are controlled by the interplay of multiple allogenic and autogenic processes, whose relative contributions are difficult if not impossible to quantify. This underlines the difference between methodology and modeling in sequence stratigraphy, and the emphasis on objective criteria that afford a consistent application of the method despite any local variability (Fig. 7.17; Catuneanu, 2020, 2022).

Conclusions

Rifts are complex extensional basins that consist of multiple half-grabens bounded by normal faults on one side and uplifted flanks on the other, generally oriented along the rift axis. These half-grabens start as isolated depocenters for sediment accumulation but become increasingly interconnected during rifting as the individual fault segments expand and join into larger boundary faults. As a result, the lower boundary of a rift sequence tends to be fragmented into multiple segments, whereas the upper boundary may be mapped as a single physical surface. However, both rift-sequence boundaries are typically diachronous, as the timing of the onset and end of rifting usually varies along the rift zone. The rift-sequence boundaries are commonly unconformable and qualify as "first-order" in a hierarchical sense, as they mark changes in the tectonic setting (e.g., a shift from rift to continental margin where rifting leads to continental breakup, or from rift to other type of inland basin in the case of continental failed rifts; Figs. 7.5 and 7.8). Sedimentation in rift basins is influenced by the syn-depositional climate and the locus of rifting relative to the continental margins, but the underlying control that is common among all rifts is tectonism coupled with the thermal uplift that accompanies lithospheric thinning.

The evolution of rifts is described by the three stages of the first-order rift cycle (Figs. 7.6 and 7.7). Each stage of basin development is conducive to specific depositional and stratigraphic patterns, related to the underlying accommodation and sedimentation conditions. The early stage of the first-order rift cycle is characterized by low subsidence rates that are in relative balance with the rates of sedimentation, leading to deposition in continental to shallow-water settings. Where accommodation is overfilled, continental sequences are upstream-controlled and consist of unconventional systems tracts (e.g., high- and low-amalgamation systems tracts in fluvial systems; Fig. 7.9). Where accommodation is available in shallow-water environments, all types of downstream-controlled systems tracts may form in relation to shoreline trajectories, including lowstand systems tracts enabled by the low rates of subsidence (Fig. 7.10). The main stage of rift development is typified by deeper water lacustrine or marine environments due to the high subsidence rates that outpace the rates of sedimentation in the depocenter(s). As a result, the stratigraphic architecture is defined by downstream-controlled stacking patterns, typically excluding lowstand systems tracts due to the high subsidence rates at the onset of each episode of fault reactivation (Fig. 7.13). The late stage of rift development marks a change to overfilled accommodation conditions, as subsidence rates decrease toward the end of the first-order rift cycle. This allows sedimentation to outpace accommodation and ultimately fill all the space made available by extensional subsidence. As a result, rifts typically end with deposition in continental environments, with a stratigraphic architecture defined by upstream-controlled systems tracts (Fig. 7.15).

While the first-order rift cycle depicts the general trends in the evolution of rifts, variations in the rates of subsidence and sedimentation across a basin can lead to the coeval formation of different types of systems tracts along both dip and strike directions (Fig. 7.16). These variations affect the stratigraphic architecture at different scales, from the timing and the systems-tract composition of the lower rank sequences that form during each stage of rift development to the timing of the three stages of the first-order rift cycle along the rift axis. The 3D stratigraphic variability and the diachronous formation of sequences, systems tracts, and bounding surfaces are the norm in any type of sedimentary basin. For this reason, the sequence stratigraphic methodology must remain data-driven, based on the observation of stratal stacking patterns in a manner that is independent of the model and the interpretation of the underlying controls on sequence development (Fig. 7.17). This objective approach affords a standard application of the method across all geological settings, irrespective of stratigraphic scales and the types and resolution of the data available (Catuneanu, 2022).

Acknowledgment

The author thanks the editors, Jurgen Adam, Domenico Chiarella, and Nicola Scarselli, for their constructive comments and guidance during the completion of this work.

References

Allen, P.A., Allen, J.R., 2013. Basin Analysis: Principles and Application to Petroleum Play Assessment, Third edition Wiley-Blackwell, p. 632.

Barredo, S.P., 2012. Geodynamic and tectonostratigraphic study of a continental rift: The Triassic Cuyana Basin, Argentina. In: E. Sharkov (Ed.) Tectonics – Recent Advances. IntechOpen, https://doi.org/10.5772/49958.

Bosence, D.W.J., 1998. Stratigraphic and sedimentological models of rift basins. In: Purser, B.H., Bosence, D.W.J. (Eds.), Sedimentation and Tectonics of Rift Basins, Red Sea – Gulf of Aden. Chapman & Hall, London, p. 630.

Catuneanu, O., 2006. Principles of Sequence Stratigraphy. Elsevier, Amsterdam, p. 375.

Catuneanu, O., 2019. Model-independent sequence stratigraphy. Earth-Science Reviews v. 188, 312–388.

Catuneanu, O., 2020. Sequence stratigraphy in the context of the 'modeling revolution. Marine and Petroleum Geology 116 (104309), 1–19.

Catuneanu, O., 2022. Principles of Sequence Stratigraphy, Second Edition Elsevier, Amsterdam, p. 494.

Catuneanu, O., Willis, A.J., Miall, A.D., 1998. Temporal significance of sequence boundaries. Sedimentary Geology 121, 157–178.

Catuneanu, O., Hancox, P.J., Cairncross, B., Rubidge, B.S., 2002. Foredeep submarine fans and forebulge deltas: orogenic off-loading in the underfilled Karoo Basin. Journal of African Earth Sciences 35 (4), 489–502.

Csato, I., Granjeon, D., Catuneanu, O., Baum, G.R., 2013. A three-dimensional stratigraphic model for the Messinian crisis in the Pannonian Basin, eastern Hungary. Basin Research v. 25, 121–148.

Csato, I., Toth, S., Catuneanu, O., Granjeon, D., 2015. A sequence stratigraphic model for the Upper Miocene-Pliocene basin fill of the Pannonian Basin, eastern Hungary. Marine and Petroleum Geology 66, 117–134.

dePolo, C.M., Anderson, J.G., 2000. Estimating the slip rates of normal faults in the Great Basin, USA. Basin Research 12, 227–240.

Friedmann, S.J., Burbank, D.W., 1995. Rift basins and supradetachment basins: intracontinental extensional end-members. Basin Research 7, 109–127.

Frostick, L.E., Steel, R.J., 1993. Sedimentation in divergent-plate basins. In: Frostick, L.E., Steel, R.J. (Eds.), Tectonic Controls and Signatures in Sedimentary Successions, 20. Special Publication of the International Association of Sedimentologists, pp. 111–128.

Gawthorpe, R.L., Hardy, S., Ritchie, B., 2003. Numerical modeling of depositional sequences in half-graben rift basins. Sedimentology 50, 169–185.

Harris, N.B., McMillan, J.M., Knapp, L.V., Mastalerz, M., 2018. Organic matter accumulation in the Upper Devonian Duvernay Formation, Western Canada Sedimentary Basin, from sequence stratigraphic analysis and geochemical proxies. Sedimentary Geology 376, 185–203.

Helland-Hansen, W., Martinsen, O.J., 1996. Shoreline trajectories and sequences: description of variable depositional-dip scenarios. Journal of Sedimentary Research 66 (4), 670–688.

Holz, M., Vilas-Boas, D.B., Troccoli, E.B., Carnairo Santana, V., Vidigal-Souza, P.A., 2017. Conceptual models for sequence stratigraphy of continental rift successions. In: Montenari, M. (Ed.), Stratigraphy and Timescales, Vol. 2. Academic Press, UK, pp. 119–186.

Izart, A., Stephenson, R.A., Vai, G.B., Vachard, D., Nindrfe, Y.L., Vaslet, D., et al., 2003. Sequence stratigraphy and correlationof late Carboniferousand Permian in the CIS, Europe, Tethyan area, North Africa. Arabia, China, Gondwanaland and the USA. Palaeogeography, Palaeoclimatology and Palaeoecology 196, 59–84.

Kress, P.R., Catuneanu, O., Gerster, R., Bolatti, N., 2021. Tectonic and stratigraphic evolution of the Cretaceous western South Atlantic. Marine and Petroleum Geology 134, 105197.

Martins-Neto, M.A., Catuneanu, O., 2010. Rift sequence stratigraphy. Marine and Petroleum Geology v. 27, 247–253.

Miall, A.D., 2000. Principles of Sedimentary Basin Analysis., Third Edition Springer, p. 616.

Millar, R.A., 2021. Sequence stratigraphy and underlying tectonism of the Northern Richardson Mountains and adjacent Mackenzie Delta related to the formation of the Arctic Ocean. M.Sc. Thesis, University of Alberta, 142 pp.

Moretti, I., Colletta, B., 1987. Spatial and temporal evolution of the Suez rift subsidence. Journal of Geodynamics 7, 151–168.

Naish, T., Kamp, P.J.J., 1997. Foraminiferal depth palaeoecology of Late Pliocene shelf sequences and systems tracts, Wanganui Basin, New Zealand. Sedimentary Geology 110, 237–255.

Schultz, S.K., MacEachern, J.A., Catuneanu, O., Dashtgard, S.E., 2020. Coeval deposition of transgressive and normal regressive stratal packages in a structurally controlled area of the Viking Formation, central Alberta, Canada. Sedimentology 67, 2974–3002.

Van Wees, J.D., Stephenson, R.A., Stovba, S.M., Shymanovskyi, V.A., 1996. Tectonic variation in the Dnieper-Donets Basin from automated modelling of backstripped subsidence curves. Tectonophysics 268, 257–280.

Vecsei, A., Duringer, P., 2003. Sequence stratigraphy of Middle Triassic carbonates and terrigenous deposits (Muschelkalk and Lower Keuper) in the SW Germanic Basin: maximum flooding versus maximum depth in intracratonic basins. Sedimentary Geology 160, 81–105.

Zecchin, M., Catuneanu, O., 2020. High-resolution sequence stratigraphy of clastic shelves VII: 3D variability of stacking patterns. Marine and Petroleum Geology 121, 104582.

Zecchin, M., Caffau, M., Catuneanu, O., 2021. Recognizing maximum flooding surfaces in shallow-water deposits: an integrated sedimentological and micropaleontological approach (Crotone Basin, southern Italy). Marine and Petroleum Geology 133, 105225.

CHAPTER

8

Rift basin carbonate platforms

Dan Bosence

Department of Earth Sciences, Royal Holloway University of London, Egham, Surrey, United Kingdom

Introduction

Examination of ancient rift basins indicates that rifting commonly initiates in a continental environment. In this setting, lakes may develop within lows in the rifted topography that accumulate carbonate sediment. Subsequently, and as a consequence of crustal thinning, driven by fault-related extension, the basin subsides, base-level rises, and eventually may connect through to the world's oceans. The result of this is that carbonate platforms occur late in the evolution of rift basins. The processes behind this commonly observed pattern would appear to be that rift initiation, by whatever model, increases local relief and may also be linked to regional updoming. This process will lead to erosion and increased rates of clastic supply during the early stages of rifting, generating a basin-bounding or syn-rift unconformity overlain by an initial fluvial clastic infill. Subsequent stages of basin evolution lead to increased subsidence, lacustrine and marine flooding, the potential for carbonate accumulation, and the development of carbonate platforms. Carbonate platforms are large bodies of carbonate deposits with an essentially flat-top that develop in response to sediment building up to sea or lake level and with major facies differentiations in response to marginal buildups and slopes (rimmed platforms) or gently inclined margins (ramps) from shallow to deeper waters. As rift basins are narrow and with active faulting, they are also likely to have intermittent marine connections. Isolation from the world's oceans may also be driven by global sea-level lowering. In lacustrine settings, rifting may deviate fluvial systems and lead to isolation from freshwater supply, or climate change may result in increased aridity. All of these processes will result in an excess of evaporation over aqueous supply and the accumulation of evaporites. However, the timing of this is complex and will be determined by individual rift basin history, as it is controlled by prerift topography, subsidence, global/regional sea-level changes, and location of the rift with respect to drainage systems and adjacent marine basins. Examples of different fill successions come from the Lake Baikal rift basin with an entirely nonmarine siliciclastic fill, the South Atlantic marginal basins with an initial continental siliciclastic, then lacustrine carbonate followed by evaporites, and finally a marine carbonate fill, and parts of the Gulf of Aden, which have a largely marine carbonate fill above the syn-rift unconformity (Fig. 8.1). There do not appear to be any high-latitude (i.e., cold water) rift basins with carbonate deposits. Possibly the rates of carbonate production in such climates are too low to compete with the rates of clastic sediment supply commonly found in rift basin settings.

A pattern of basin fill from early syn-rift clastics to late syn-rift and postrift carbonates is commonly observed in passive or mature extensional margins today in low-latitude areas (Fig. 8.1). Well known are the extensive carbonate shelves of Mesozoic age of the Atlantic and the Tethyan margins of southern Europe, which have their foundations on rifted early Mesozoic clastics/volcaniclastics (Bernoulli and Jenkyns, 1974; Ziegler, 1990). Similarly, the syn- to postrift carbonate shelves of the Red Sea overlie earlier syn-rift siliciclastics (Orszag-Sperber et al., 1998) (Fig. 8.1). Both of these settings are examined in this chapter, and case studies have been chosen as representing particularly good examples of carbonate-dominated rifts; the Gulf of Suez–Red Sea–Gulf of Aden rifts are particularly well-exposed examples of Cenozoic marine carbonate stratigraphies in the rift margins and coastal plain areas. For Mesozoic lacustrine carbonate rifts, surface and subsurface examples of Early Cretaceous carbonate platforms are taken from the rifted southern Atlantic margins in Brazil (Campos and Santos basins), Angola (Kwanza Basin), and the late Jurassic northeastern Atlantic margin of the Wessex Basin, United Kingdom.

Regional Geology and Tectonics: Volume 2
DOI: https://doi.org/10.1016/B978-0-444-64136-6.00007-5

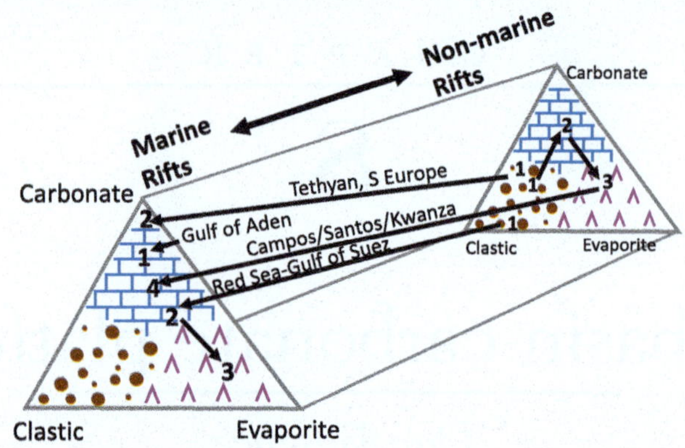

FIGURE 8.1 Diagram illustrating the range of lithologies present in marine and nonmarine rift basins. Time trends in the evolution of sedimentary environments are indicated by numbered black arrows and illustrate the commonly observed pattern of early rift basin siliciclastics to later carbonate and evaporite rift basin fills. Note that lithological successions are characterized by marine or nonmarine end-members of triangular block.

From these relatively well-studied regions, general patterns and processes within marine and lacustrine carbonate-dominated rifts are discussed, and models for the different types of carbonate platforms found within these rifts are presented.

Gulf of Suez–Red Sea–Gulf of Aden rift system

These rift basins show considerable heterogeneities along their 3500 km length. In the Red Sea and Gulf of Suez, the syn-rift phase of sedimentation started in the early Oligocene and was firstly dominated by clastic sediments, subsequently by carbonate sediments in the middle Miocene (Langhian-Serravalian), by evaporite deposits in the mid- to late Miocene and then, in marine areas, by carbonate accumulations. Marginal coastal areas have mixed carbonate-evaporite-siliciclastic deposits throughout the syn-rift. The Plio-Quaternary strata in the northern Red Sea–Gulf of Suez show features of syn-rift (local extensional faults and unconformities) but also features of postrift strata (basal onlap and downlap and a regional seaward dip of strata). Deposition, and subsequent deformation, of these strata is commonly affected by diapiric structures generated from the mid- to late Miocene salts. The southernmost Red Sea, by contrast, is a volcanic margin with eruption starting at 31 Ma (Bosworth, 2015).

The Gulf of Aden is a longer lived and more mature rift that has evolved into a young oceanic basin with a mixed carbonate-siliciclastic stratigraphy since early Oligocene times. Early Cenozoic marine (prerift) carbonates are unconformably overlain by syn-rift marine carbonates in southeastern Yemen (Brannan et al., 1997). Ocean floor spreading in the Gulf started at 18–19 Ma (Bosworth, 2015). Postrift sediments of late Miocene age lie unconformably on tilted syn-rift in offshore areas and these correspond to the emplacement of ocean floor. Having never been isolated from the Indian Ocean, the basin does not have thick evaporite deposits, or diapiric structures, affecting the syn- and postrift successions. Offshore and onshore records indicate rift development entirely in the marine, and largely carbonate-accumulating environment.

Within the syn- to postrift fills of the Gulf of Suez–Red Sea–Gulf of Aden, rift carbonate platforms develop in three main situations, where shallow-marine areas occur that can be isolated, either spatially or temporally (or both) from clastic supply; on top of rotated fault blocks, on top of salt diapirs, and on top of deltas.

Gulf of Suez–Red Sea fault-block platforms

Carbonate platforms that developed on subsiding and/or rotating fault blocks with a Proterozoic basement have been described in detail from mid-Miocene carbonate platforms in the northwest Red Sea and Gulf of Suez (Figs. 8.2 and 8.3; Burchette, 1988; James et al., 1988; Purser et al., 1998; Bosence et al.,1998b; Cross et al., 1998; Cross and Bosence, 2008). Early stages of rifting were largely dominated by clastic sedimentation with sediment transport mainly controlled by the evolving extensional fault systems and their oblique transfer zones (Purser et al., 1998; El-Azabi, 2021). During the syn-rift phase, hanging-wall subbasins generally persisted as sites of clastic sediment accumulation, whereas footwall areas, horsts, and proximal areas of transfer zones that were isolated from clastic supply accumulated subtropical and photozoan carbonates. These have proved difficult to date because of the

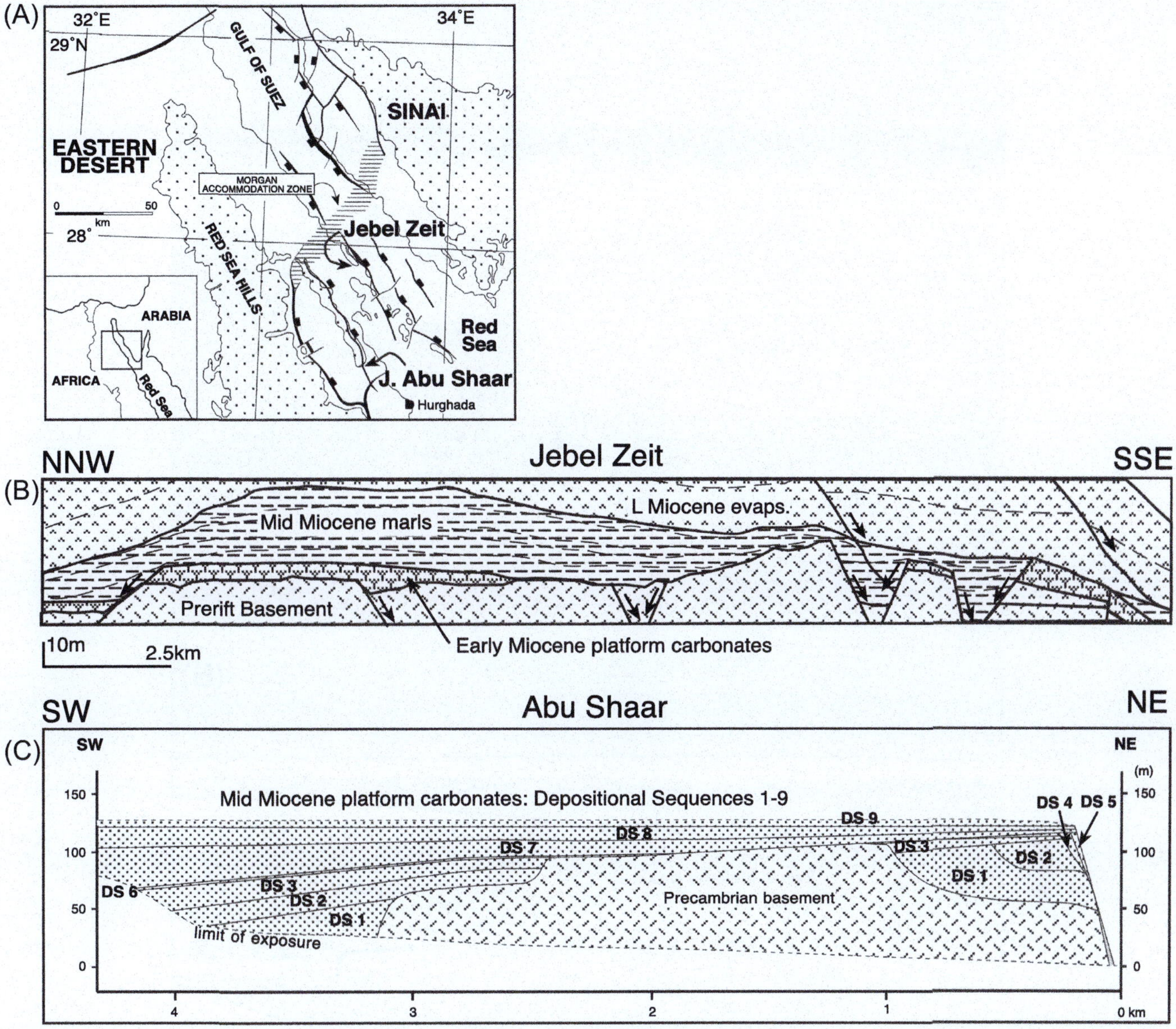

FIGURE 8.2 Setting and characteristics of Miocene carbonate platforms in the Gulf of Suez. (A) Gulf of Suez rift basin showing major faults and locations referred to in the chapter. (B) and (C) Cross sections of Jebel Zeit and Jebel Abu Shaar (respectively) showing platform carbonates on faulted basement highs. Syn-rift nature of Abu Shaar platform is indicated by the thickening of depositional sequences (DS1-DS9) to the southwest and thinned and fault-truncated sequences in footwall to the northeast. Source: *From Bosence, D.W.J. (2012) Carbonate dominated marine rifts. In: Roberts, D.G. & Bally, A.W. (Eds) Regional Geology and Tectonics: Phanerozoic Rift Systems and Sedimentary Basins: Volume 2: Phanerozoic Rift Systems and Sedimentary Basins. Elsevier. Chapter 5, 89–114.*

separation of the well-dated basinal marls and carbonates from the higher energy platform carbonates that show significant diagenetic alteration (Purser et al., 1996). Bosworth (2015) argues for a Serravalian age for the rift margin carbonate platforms that is consistent with Hughes' (2014) age of "latest Early Miocene to possibly earliest Middle Miocene" following extensive micropaleontological work and comparison with platform carbonates on the opposing margin in the Midyan area of Saudi Arabia. More recently, El-Azabi (2021) proposed a Langhian age for the Abu Shaar platform based on the correlation of sequence boundaries on the platform with those in the basinal facies of the Gulf of Suez, but no new paleontological ages are presented. The platforms are commonly fault-bounded, are rectilinear to trapezoidal in plan shape, and with maximum stratigraphic thicknesses of 130 m in outcrops in the Red Sea–Gulf of Suez margins (Fig. 8.3). Seismic scale sections through wadis illustrate Miocene clastics and carbonates onlapping onto the Proterozoic basement and thickening to the west into the Mellaha basin

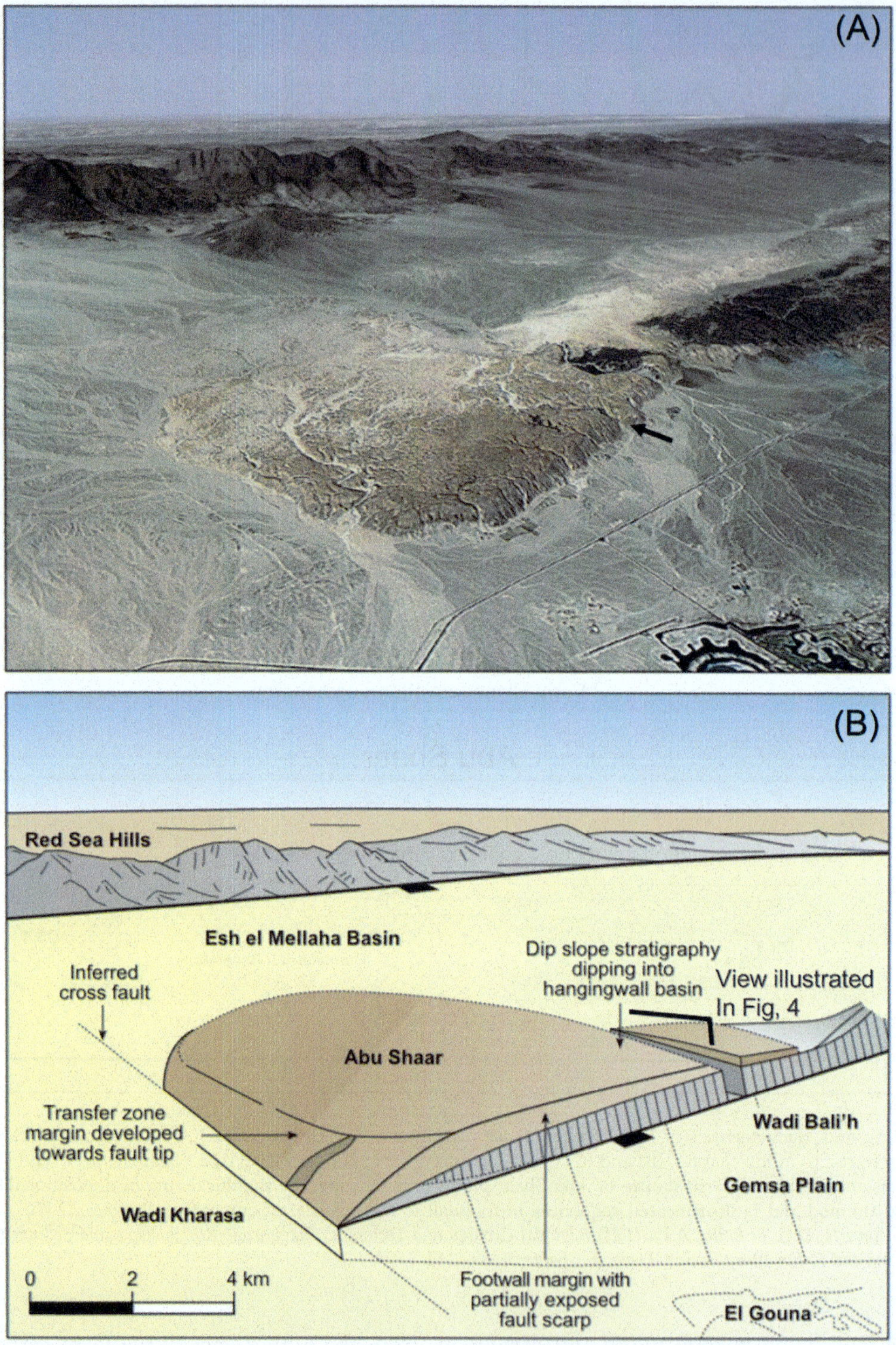

FIGURE 8.3 (A) Arial view of Jebel Abu Shaar (above) indicating light-colored carbonate lithologies, in center, overlying dark-colored Proterozoic basement to upper right. Wadi 3 detailed in Fig. 8.5 indicated with black arrow (author's photo). (B) Interpretation of major geological features of Jebel Abu Shaar and its setting in the rift margin. Source: *Modified from Cross, N. (1996) Sedimentology and sequence stratigraphy of a Miocene fault-block carbonate platform, Gulf of Suez, Egypt. PhD thesis Royal Holloway University of London. 271 pp.*

(Fig. 8.4). Within the platforms, rimmed-shelf margins, or fault-truncated margins, characterize the footwalls (Fig. 8.5) of fault blocks while hanging-wall dip-slopes are ramps and generally pass downdip into mixed clastics/carbonates in the hanging-wall subbasin (Fig. 8.4). Fault accommodation zones, such as Wadi Kharaza on the southern margin of Abu Shaar, have low relief and show progradational platform margin geometries (Cross and Bosence, 2008) or

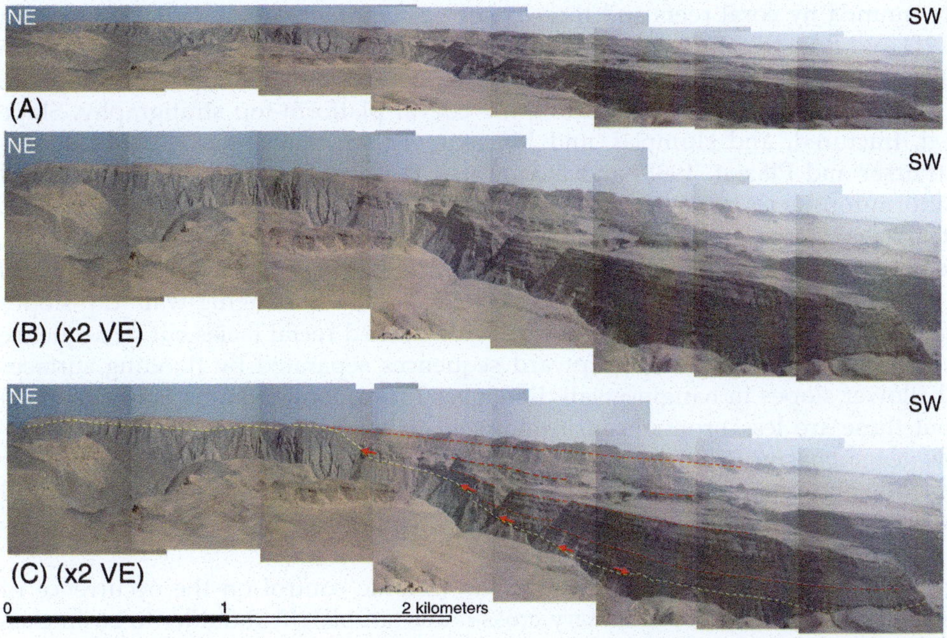

FIGURE 8.4 Northeast to southwest panoramic section through Miocene stratigraphy on the southern side of Wadi Bali'h, Jebel Abu Shaar (for view, see, Fig. 8.3B). (A) True to scale uninterpreted section, (B) with 2x vertical exaggeration uninterpreted, (C) with 2x vertical exaggeration and interpreted stratigraphy to illustrate syn-rift unconformity (dashed yellow line) above dark-colored basement, onlap (red arrows) onto footwall basement high and syn-rift stratigraphy (red dashed lines) showing stratal thickening into interbedded clastics and carbonates in the hanging-wall subbasin to the southwest. This southwesterly thickening is confirmed with measured sections (see also Fig. 8.2C) (author's photos).

FIGURE 8.5 (A) Panoramic view of the north face of Wadi 3 (indicated in Fig. 8.3) indicating main stratigraphic geometries and erosion surfaces of the footwall area of Abu Shaar. View to north with the coastal plain to Gulf of Suez to the right. Note that the present-day topography exhumes the Miocene carbonate platform morphology. (B) Summary diagram illustrating facies associations that accumulated in the footwall area of the Abu Shaar carbonate platform. Source: *After Cross, N. (1996) Sedimentology and sequence stratigraphy of a Miocene fault-block carbonate platform, Gulf of Suez, Egypt. PhD thesis Royal Holloway University of London. 271 pp.*

are dominated by through-flowing fan-deltas or submarine fans (Purser et al., 1998; Bosworth et al., 1998). These platforms may be attached to the rift margin, detached from the margin but influenced by marginal clastic supply or unattached and in an axial location. In the latter case (e.g., Gebel Zeit, Fig. 8.2B), the platform drowns and is overlain by deeper-water marls, and subsequently evaporites accumulate (Purser et al., 1998).

Platform margins are commonly rimmed by coral reefs and these occur preferentially on faulted margins and shallow (inner) ramp sites (Figs. 8.2D and 8.5). Locally, ooid shoals form the platform margin. Footwall platform slopes in these Egyptian examples are only exposed in their upper parts and comprise reef and platform-top bioclastic and intraclastic talus (Fig. 8.5), with a composition reflecting the coeval platform-top stratigraphy. Slope sediments are commonly cemented, fractured, and slumped, and Miocene erosion may remove 100 m to km-sized blocks leaving lunate scars (Purser and Plaziat, 1998) and a truncated footwall stratigraphy (Figs. 8.2 and 8.5; Cross and Bosence, 2008). Platform-top facies are bedded bioclastic and peloidal wackestones, packstones, and rudstones with scattered patch reefs. These are arranged in unconformity-bound depositional sequences, some of which can be demonstrated to be of tectonic origin (Cross and Bosence, 2008). Such sequences onlap and thin onto footwall sites where sequence boundaries converge and thicken down hanging-wall dip-slopes (Figs. 8.2C and 8.4). They consequently have wedge-shaped geometries with stacked reefal facies cut by sequence boundaries in footwall sites (Fig. 8.5) and coeval shallowing-upward sequences separated by flooding surfaces on hanging-wall dip-slopes. The shallower slopes in hanging-wall dip-slopes and accommodation zones result in more progradational geometries but these are less pronounced in footwall margins because of the steep, fault-related slopes and adjacent deepwater subbasins. The upper levels of these Middle Miocene-aged carbonates show a change in facies to more micritic and microbial facies in platform-top and slope environments indicating a more restricted environment that is interpreted to be a forerunner to the Serravalian to Tortonian evaporites formed during the isolation of the Gulf of Suez and Red Sea rift basins.

Further south, near Safaga, Egypt, Bosworth et al. (1998) describe the tectonic control on the occurrence of Serravalian carbonate and siliciclastic facies. Here, syn-sedimentary cross-faults supply rift shoulder to basin clastic corridors that cut through the shore-parallel carbonate platforms sited over fault blocks (Fig. 8.6A). Syn-sedimentary fault planes form steep submarine slopes that locally provide substrate for the overlying microbialite facies with well-developed stromatolites arranged in a hierarchy of domical morphologies (Fig. 8.6B).

Holocene examples of fault-block platforms include the abundant reef-rimmed platforms of the Red Sea and Gulf of Suez (e.g., Sanganeb Atoll, offshore platforms north of Jeddah, Shoab Ali) as reviewed by Dullo and Montaggioni (1998).

Red Sea—Gulf of Suez salt-diapir carbonate platforms

Salt diapirism can result in the formation of shallow-water sites that may be isolated from high rates of clastic supply, thus providing a suitable substrate for platform growth. The rise of diapirs relative to base-level results in seafloor highs that develop into domes, islands, and intervening minibasins. If these seafloor highs enter the zone of the photozoan carbonate factory (i.e., the photic zone), they may then develop carbonate platforms that are circular or ring-shaped, reflecting the morphology of the underlying top salt surface. Where salt movement is related to extensional faults in the Red Sea, then salt walls and more elongate platforms are developed with associated horsts and grabens. The stratigraphy and sedimentology of platforms developing over salt diapirs have been studied in outcrops within this rift basin along the northwestern Red Sea and Gulf of Suez (Orszag-Sperber, et al., 1998), Salif, NW Yemen (Bosence et al., 1998a; Davison et al., 1996; Heaton et al., 1996), the Farasan Islands offshore southwestern Saudi Arabia (Bantan, 1999), and from the opposing margin of the southern Red Sea in the Dahlak Islands, offshore Eritrea (Carbone, et al., 1998) (Fig. 8.7A).

These salt-cored archipelagos comprise low-relief, reef-fringed, late Cenozoic limestone islands. Morphologically, the islands and the surrounding submarine topography are the surface expression of carbonate platforms that, in plan view, have arcuate to subcircular bays, reentrants, and promontories. The platforms vary in horizontal section from about 1 km across at Salif (Bosence et al., 1998a; Davison et al., 1996) and the Gulf of Suez (Orszag-Sperber et al., 1998) to Dahlak Kebir, which is 140 km across (Carbone et al., 1998). The margins of the platforms are largely steep-sided and coral-reef rimmed (Fig. 8.7; Bantan 1999; Carbone et al., 1998). The internal lagoons are commonly seagrass meadows and fine-grained carbonates, or bioclastic sands and gravels in more exposed waters. Outcrops at Salif in northwest Yemen indicate colonization of the irregular dissolved upper surface of the salt diapir by reef corals and mollusks (Figs. 8.7D and 8.8A–C; Davison et al., 1996) indicating that the diapir has moved into the zone of shallow-water, photozoan carbonate production and that reef colonization followed shortly after salt dissolution. Carbonate stratigraphies are up to 450 m thick and most exposed successions commonly shallow upsection in both Farasan (Bantan, 1999) and the Salif area (Bosence, et al., 1998a). To achieve this maximum stratigraphic thickness on a substrate that is rising in relation to regional base level, the platforms must subside as well as rise and unconformities over circular diapirs are described from both Farasan and from Salif (Fig. 8.7D).

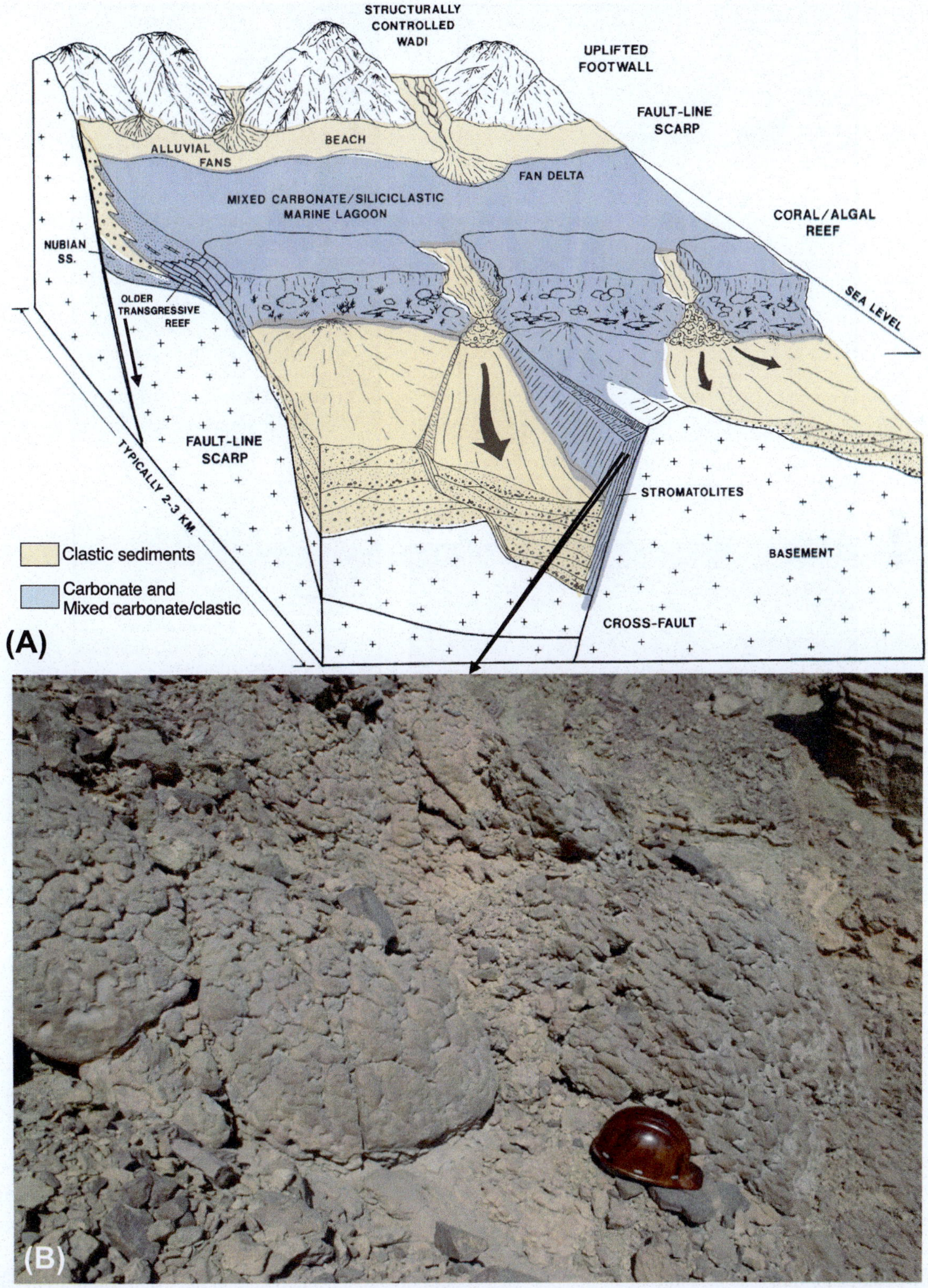

FIGURE 8.6 (A) Block diagram of reconstructed sedimentary environments from outcrops on Gebal Naqara, near Safaga, Egypt. Note rift parallel and rift normal or cross-faults largely controlling the occurrence of carbonate (blue gray) and siliciclastic (brown) facies. Photograph (B) is from arrowed location in (A) and shows an exhumed Miocene depositional slope constructed by stromatolites over a fault plane (scale from hard hat 27 cm across, author's photo). Source: *Modified from Bosworth,W., Crevello, P., Winn, R.D. & Steinmetz, J. (1998). Structure, sedimentation and basin dynamics during rifting of the Gulf of Suez and northwestern Red Sea. In: Purser, B. H & Bosence, D. W. J. (Eds.) Sedimentary and Tectonic Evolution of Rift Basins- The Red Sea-Gulf of Aden. Kleuwers, 77–96.*

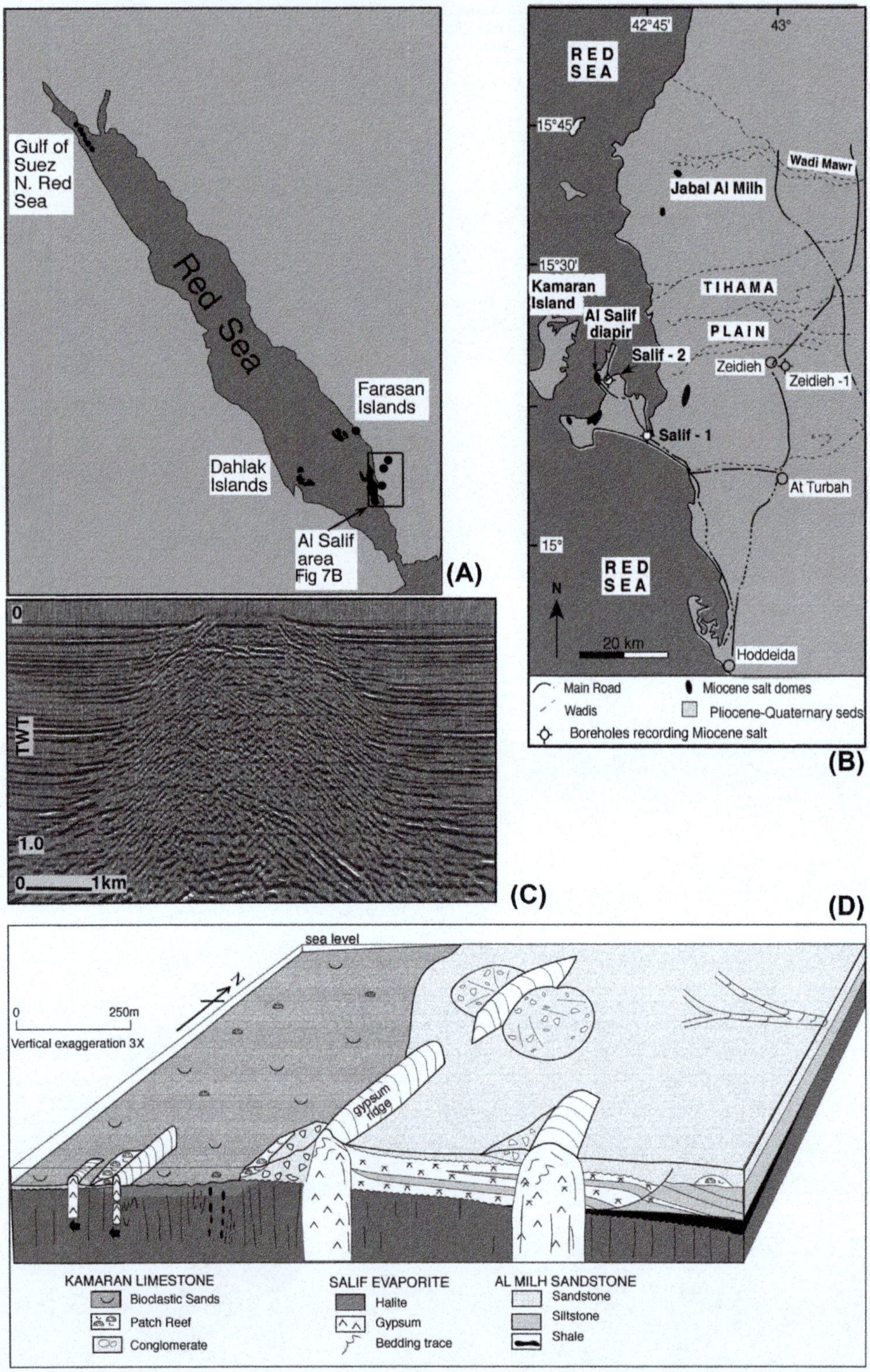

FIGURE 8.7 Salt-diapir carbonate platforms in the Red Sea–Gulf of Suez rift basin. (A) Locations of salt diapir platforms discussed in text. (B) Map of Salif area, Yemen, indicating Miocene salt diapirs and locations of wells. (C) Seismic profile from southern Gulf of Suez imaging Miocene salt diapir overlain by flat-topped carbonate platform. Note platform-top and slope morphologies from seafloor imaged in underlying reflectors (between 0.1 and 0.2 msec). (D) Block diagram to illustrate Quaternary facies and structures developed on top of salt diapir at Al Salif, Yemen. Note preferential erosion of upper surface of Miocene evaporites with prominent gypsum ridges. Source: *All from Bosence, D. W.J. (2012) Carbonate dominated marine rifts. In: Roberts, D.G. & Bally, A.W. (Eds) Regional Geology and Tectonics: Phanerozoic Rift Systems and Sedimentary Basins: Volume 2: Phanerozoic Rift Systems and Sedimentary Basins. Elsevier. Chapter 5, 89–114.*

Seismic sections illustrate multiple onlaps and surfaces adjacent to diapirs indicating the balance between diapir growth, subsidence, and platform production (Heaton et al., 1996; Bosence et al., 1998a). The stratigraphy of the platforms is complex with rapid lateral and vertical facies changes in these shallow-water, biogenic carbonates reflecting the complex morphology of the islands. Similarly depositional sequences have complex 3-D geometries with thickening and thinning and convergence of sequence boundaries over positive diapirs, eroded vertical strata on diapir walls, and their divergence into minibasins. Vertical movements are rapid (3–5 m/ky; in Bosence et al., 1998a). Extensional faulting (Fig. 8.7D), horst, and graben are recorded in stratigraphies developed over the areas of salt diapirs and many outcrops are blistered and brecciated (Fig. 8.8D).

The occurrence of this platform type within the Red Sea and Gulf of Suez rifts is a direct consequence of the isolation of these basins from the Mediterranean Sea and Indian Ocean in the mid- to late Miocene and the accumulation of thick evaporite deposits. This paleogeographical control is not present in the Gulf of Aden, which, with its wedge-like morphology opening into the Indian Ocean, was never isolated from marine waters.

Gulf of Aden–Red Sea delta-top platforms

In spite of the general acceptance that rates of siliciclastic supply restrict the occurrence of shelf carbonates (e.g., Hay et al., 1988; James and Jones 2016), deltas building out into shallow-marine areas are recurring sites of carbonate platform and reef development in the late Cenozoic of the Gulf of Suez, Red Sea, and Gulf of Aden (Roberts and Murray, 1983; Purser et al., 1998; Bosence et al., 1996). In this semiarid climate setting, river flow is rare, short-lived, and generally carries a coarse load, all factors that are considered to enable colonization by carbonate-producing communities.

The Oligo-Miocene (Shihr Group) on the northern margin of the Gulf of Aden west of Mukalla comprises rift-shoulder-derived clastics interfingering with shallow-marine carbonates in a complex three-dimensional pattern (Figs. 8.9 and 8.10; Bosence et al., 1996). Five depositional sequences have been mapped out that comprise point-sourced alluvial fans related to a basement-cutting, cross-fault with associated small reef-rimmed carbonate platforms. These sequences are bounded by presumed tectonically driven erosional sequence boundaries that are

FIGURE 8.8 Salt-diapir carbonate platforms from the southern Red Sea. (A) Salif salt mine, Yemen. Note upstanding gypsum rafts within halite that are not quarried (GR) and overlying, subhorizontal Quaternary sediments (Q), figure, circled, for scale. (B) Upper, eroded surface of vertically bedded, halite, unconformably overlain by gently dipping Quaternary carbonate sediment (encircled hammer, 33 cm, for scale). (C) Detail of unconformity surface in (B) illustrating dissolved halite surface overlain by *Stylophora* (coral), mollusk rudstone, suggesting sea-floor dissolution and redeposition in the shallo-marine photic zone (pen 10 mm diameter). (D) Irregularly blistered and brecciated shallow-marine limestones overlying Miocene salt in the Farasan Islands. (All author's photos).

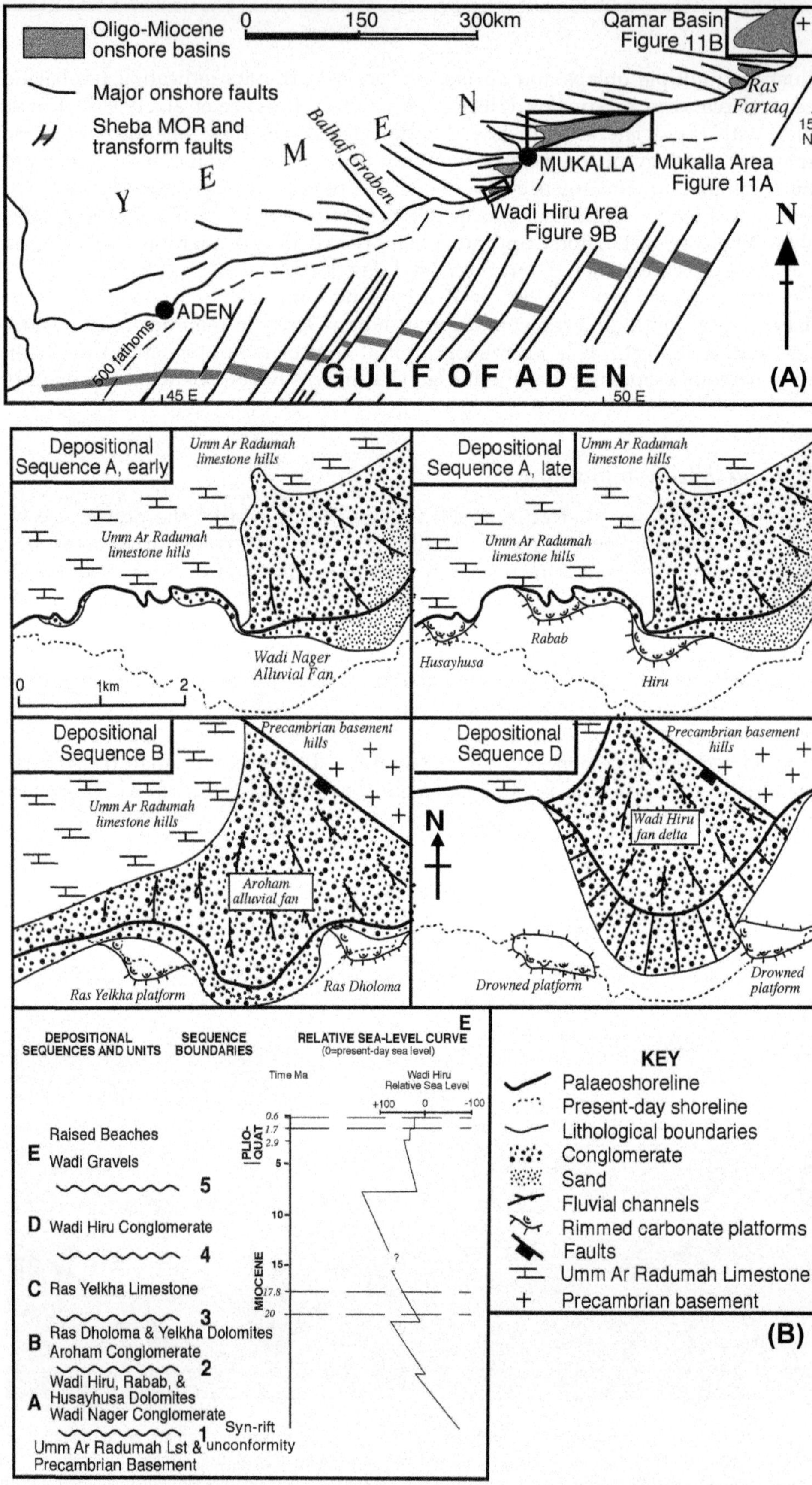

FIGURE 8.9 (A) Location map and geological setting for sites described on the northern margin of the Gulf of Aden rift. (B) Series of Miocene to Pliocene paleogeographic maps for four depositional sequences from the Wadi Hiru area, Yemen (located in Fig. 8.9A). The four depositional sequences (A–D) indicate the complex interplay beween rift margin clastic and carbonate sediments. Three fan deltas built out to the south and south-east (Wadi Najer, Aroham, and Wadi Hiru Conglomerates) and the first two of these developed small delta-top carbonate platforms on their seaward margins in late highstand periods. Small fringing reef-rimmed platforms also developed on the eroded shoreline of the Umm Ar Radhuma limestones in Sequence A. Source: *After Bosence, D.W.J. (2012) Carbonate dominated marine rifts. In: Roberts, D.G. & Bally, A.W. (Eds) Regional Geology and Tectonics: Phanerozoic Rift Systems and Sedimentary Basins: Volume 2: Phanerozoic Rift Systems and Sedimentary Basins. Elsevier. Chapter 5, 89–114.*

FIGURE 8.10 (A) View to the west of coastal outcrops of Oligo-Miocene, syn-rift Shihr Group at Wadi Hiru (located in Fig. 8.9A). Light-colored lithologies are dolomites and limestones and dark-colored lithologies in wadi sides are conglomerates. Precambrian basement forms the dark-colored mountains in the background (vehicle for scale on foreshore). (B) Outcrop detail from Wadi Hiru illustrating the complex interplay between clastic and carbonate sedimentation. Fluvial conglomerates (to the right) interfinger downslope, and to the left, with shallow-marine carbonates (now dolomite). The lateral facies transition from conglomerate to dolomite occurs over a distance of a few meters (author's photos).

overlain by lowstand, alluvial fan conglomerates, and sandstones. During transgression, carbonate platforms aggrade, interfingering with land-derived clastics, and then prograde during the highstand. Clastics and carbonates are then eroded by the following, fall in relative sea level (Fig. 8.9B). The platforms became established on the seaward margins of alluvial fans, fan-deltas, and adjacent paleo-seacliffs twice during the accumulation of this syn-rift succession (Fig. 8.9B; Bosence et al., 1996). In depositional sequence A, boulder-sized, fluvial conglomerates pass gradationally over a few meters and interfinger seaward into bioclastic rudstones, grainstones, and patch reefs (Fig. 8.10B). The carbonates subsequently aggrade and then prograde to form a 70-m-thick and 1.5-km-wide arcuate-shaped platform in plain view. Subsequent sea-level fall eroded the platform and a low-stand fan delta of sequence B was deposited (Fig. 8.9B).

Outcrops in the northwest margin of the Red Sea, south of Quseir at Sharm el Behari, also demonstrate the close association of clastic and carbonate facies in the syn-rift phase. Here, a long-lived (late Oligocene to Recent) fluvial system deposited continental and then marine, fluvio-deltaic sediments that formed the substrate for later platform growth (Purser et al., 1998). A mid-Miocene fan delta, with 30 m foresets, prograded southeastwards through a half-graben toward the NW Red Sea. These foresets comprise coarse conglomerates passing downslope to marls. Cor-algal patch reefs (<10 m thick) colonized the upper part of the delta cone, which subsequently formed the substrate for a larger (8 × 3 km) onlapping carbonate platform with up to 40 m of shallow peritidal marine facies. These are overlain by up to 30 m of microbialites and subaqueous clastics (Purser et al., 1998). These authors, along with Roberts and Murray (1983), working in the Gulf of Suez, consider that the intermittent, coarse clastic supply, from these semiarid drainage systems, was an important feature that allowed carbonates to accumulate in such settings.

In a review of modern reefs developing around fluvial sources in the Red Sea and Gulf of Aqaba, Dullo and Montaggioni (1998) described how the morphology of a delta and its distributary channels controlled the pattern of reef growth as corals preferentially colonize channel sides and delta margins. However, active channels prevented reef growth and instead provided downslope pathways for the export of reef and clastic debris to deeper-water sites. They considered that because of the intermittent nature of clastic supply, reefs, temporarily destroyed by flash floods, were able to recolonize or continue to grow elsewhere. In addition, these authors considered that periods of turbid flood waters are too brief to influence reef growth (Dullo and Montaggioni, 1998).

A common observation has been that along a shoreline, reefs occur in an arcuate band around a fan delta but that either side of the delta are reef-free beaches (Roberts and Murray, 1983; Dullo and Montaggioni, 1998). One explanation for this is that wave energy will be higher as wave orthogonals converge on the fan delta, thus winnowing away fine-grained sediment. Adjacent embayments will have lower energy regimes, where finer-grained sediment will accumulate, and be unfavorable for reef growth.

Gulf of Aden ramp margin clastics and carbonates

The extensive coastal plain to the northeast of Mukalla exposes a Cenozoic syn-rift stratigraphy that is faulted against Precambrian basement to the west and lies unconformably on earlier Cenozoic limestones to the east. Northerly dipping, east-west extensional faults delineate a series of hanging-wall subbasins infilled with Shihr Group sediments. The Shihr Group comprises two Oligo-Miocene syn-rift sequences separated by an unconformity that are overlain by a third, postrift succession of mid- to late Miocene age (Sequence 3) (Fig. 8.11A; Watchorn, et al., 1998). Sequence 1 (approx. 100 m thick) comprises a lower unit of clastic gravels, sands, marls, and calcretes that commonly fine upward and are interpreted to have formed in proximal alluvial fans with intervening areas with overbank deposits and paleosols (Watchorn et al., 1998). However, the greater part of Sequence 1 is formed of limestones and marls with local units of evaporites. Carbonate lithologies include bioclastic pack-grainstones, microbial laminites and oncoids, tufas, and local cor-algal framestones, some associated with conglomerates. Evaporite units comprise bedded successions of gypsum and marls with local lime mudstones. These units are interpreted to have formed in a low-relief coastal zone with shallow-marine sands and localized patch reefs. Emergent areas had evaporitic ephemeral playa lakes.

Sequence 2 forms the greater part of the onshore outcrops of the Shihr Group (Fig. 8.11A) and is interpreted to be about 1000–1500 m in thickness. A lower unit of mixed sandstone and conglomerate lithologies in cross-bedded and chanelled units indicate deposition in a braided river system with south-easterly directed channels. These lithologies are overlain by laminated red silts and marls with calcrete and gypsum interbeds interpreted to have formed in an ephemeral lake. The upper part of this sequence varies considerably through the area from shallow, reefal facies to coastal sabkha settings to thick units of gypsum-rich marls and localized conglomerates derived from the rift shoulder to the north.

This Oligo-Miocene succession of facies indicates an extensive low-relief, rift margin coastal area dominated by fine-grained marls and coastal evaporites, interspersed with small, reefal platforms (Watchorn et al., 1998). Rift shoulder uplift appears to have diverted most of the clastic supply inland to the Hadramaut drainage basin. Although this large area has not been studied in detail, there appears to be no clear link between facies distribution and the basin margin faults. The stratigraphy appears to be a passive fill of a previously faulted basin

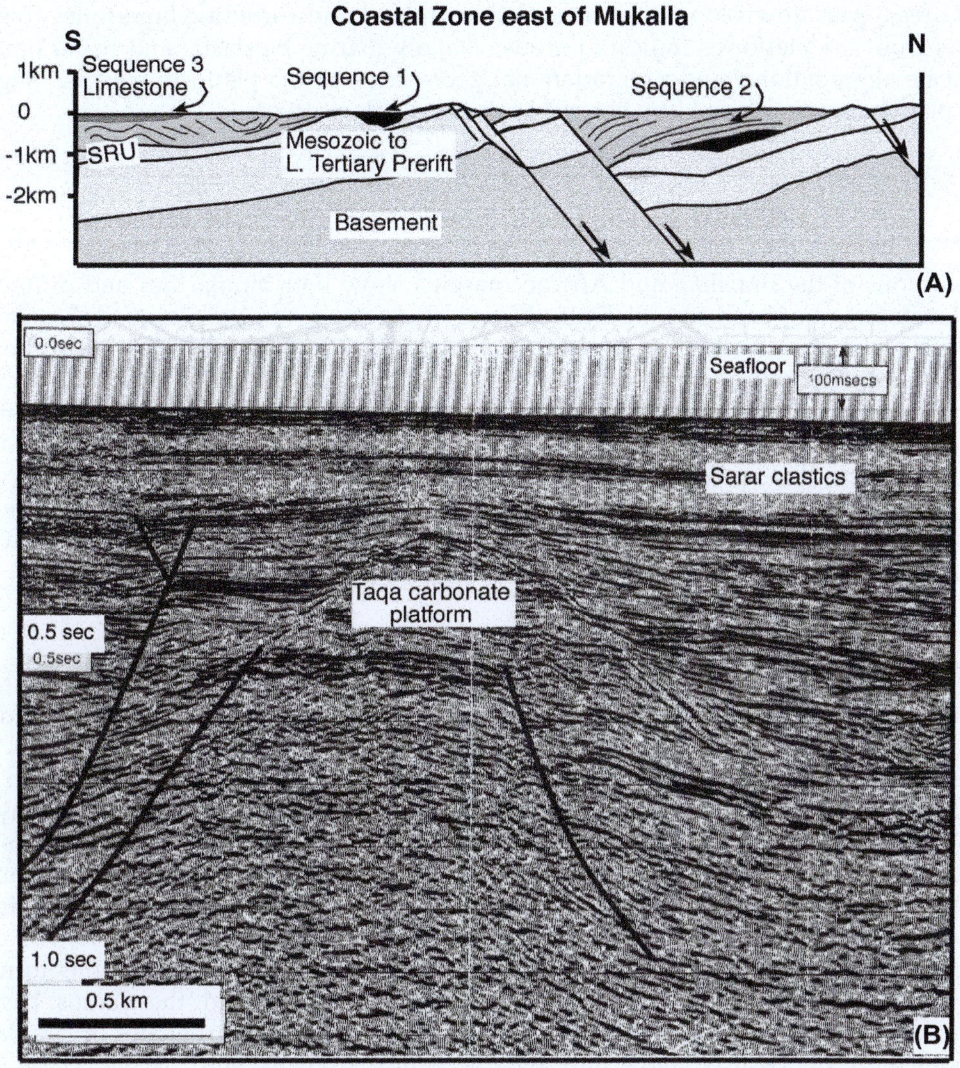

FIGURE 8.11 Syn-rift stratigraphy on the northern margin of the Gulf of Aden basin. (A) A generalized north to south section through the region to the east of Mukalla (for location, see Fig. 8.9A) illustrating preservation of folded and faulted Shihr Group strata preserved in hanging-wall basins. (B) Seismic profile from Qamar Basin, eastern Yemen (located in Fig. 8.9A) illustrating syn-rift Taqa carbonate platform sited extensional horst fault block with aggradational and progradational geometries. Source: *From Bosence, D.W.J. (2012) Carbonate dominated marine rifts. In: Roberts, D.G. & Bally, A.W. (Eds) Regional Geology and Tectonics: Phanerozoic Rift Systems and Sedimentary Basins: Volume 2: Phanerozoic Rift Systems and Sedimentary Basins. Elsevier. Chapter 5, 89–114; Brannan, J., Gerdes, K.D., Newth, I.R., (1997). Tectono-stratigraphic development of the Qamar basin, eastern Yemen. Marine & Petroleum Geology, 14, 701–730; Watchorn, F., Nichols. G.J. & Bosence, D.W.J. (1998). Rift related sedimentation and stratigraphy in southern Yemen. In: Purser, B.H. & Bosence, D.W.J. (Eds) Sedimentation and Tectonics of Rift Basins: Red Sea - Gulf of Aden, Kleuwers, 165-189.*

margin (Fig. 8.11A) and a low-relief, southerly sloping ramp margin is a likely model. Syn-rift, fault-block models (e.g., Leeder and Gawthorpe, 1987; Bosence 2005) like those described from the Gulf of Suez (above) do not apply to this section of the rift basin.

In the Qamar Basin (NE Gulf of Aden), the early Cenozoic prerift succession comprises shallow-marine limestones and shales (Hadramaut Group) with marginal evaporites of the Russ Fm (Brannan et al., 1997). These are cut by the syn-rift unconformity (middle Eocene), a major tectonostratigraphic break throughout the offshore area (Bott et al., 1992; Brannan et al., 1997). Offshore, seismic sections from the Qamar Basin indicate a pre- to syn-rift stratigraphy dominated by carbonates. The oblique setting of the rift basin at Qamar means that east-west extensional faults are essentially normal to main onshore-offshore facies belts and the paleo-coastline. Clastics from the Hadramaut drainage basin pass eastwards through hanging-wall basins within the carbonate province. Fault-block platforms such as the Taqa platform, oriented normal to the coastline, with up to 200 msec relief in footwall areas, pass downslope to presumed fine-grained and turbiditic limestones (Fig. 8.11B; Brannan et al., 1997). Cores from the platforms indicate porous, shallow-marine bioclastic and reefal limestones and seismic sections indicate aggradational and progradational geometries on the platform margins (Fig. 8.11B; Brannan et al., 1997). The platforms are onlapped by postrift shales and pelagic marls.

Lacustrine carbonates in South Atlantic rift margins

The Early Cretaceous of the Brazilian and African margins show both similarities and differences when compared with the development of basins in the Gulf of Suez–Red Sea area in the Cenozoic. Similarities include the local areas of prerift volcanics, the high-angle synthetic normal faults of the rift phase with the accumulation of syn-rift basal and marginal clastics followed by carbonates and a phase of thick evaporite accumulation in the postrift or sag phase of basin evolution. The subsequent postrift deposits being offshore marine in both cases. The major difference is that the syn-rift carbonates are of lacustrine origin in the South Atlantic as opposed to the marine in the Gulf of Suez–Red Sea–Gulf of Aden basins. The opening to marine conditions occurred in the late Aptian after the phase of lacustrine carbonate sedimentation and fed the chloride-rich evaporites (Davison et al., 2012; Ceraldi and Green, 2017). The lacustrine, presalt, and carbonate depositional system in these basins is very large, probably unique in the geological record, and has no comprehensive modern analog.

Campos and Santos basins lacustrine platforms

In the Campos and Santos and basins, offshore Brazil, the syn-rift is largely nonmarine, and accumulated in an Early Cretaceous rift basin setting (Fig. 8.12; Guardado, et al., 1990; Carminatti et al., 2009). Continental and lacustrine depositional environments persisted until marine flooding either from the south (Davison et al., 2012), or from the north (Pietzsch et al., 2018) in the late Albian. These petroleum-rich basins do not have an onshore expression, so that data comes from seismics and wells which, for commercial reasons, are sparse, and published figures and data are commonly unlocated. This lack of data has contributed to a number of controversies over the nature of the carbonate rocks, their interpretation from seismic data, and their environment of accumulation.

Conglomerates and sandstones of the early Barremian Atafona and Piçarros formations form the earliest rift sediments and fringe the western basin margins in both the Campos and the Santos basins, respectively (Fig. 8.12). These are overlain by unusually thick (100 s of meters) lacustrine carbonates of late Barremian to early Aptian age in both the Campos (Coqueiros Fm) and the Santos (Itapema Fm) basins. These are composed of bivalve-rich rudstones, grainstones, and packstones (or coquinas) that are part of the syn-rift stratigraphy of the basins (Winter et al., 2007; Thompson et al., 2015). They are classified as a bioclastic (or, BIO) facies association that existed in shallow lake areas with waters of variable (but nonmarine) salinity (Hunt et al., 2019; Hendry et al., 2021). Following a regional unconformity (pre-Alagoas or postrift unconformity), the overlying, Aptian aged, carbonates of the Macabu (Campos Basin), and Barra Velha (Santos Basin) formations were deposited in what are interpreted to have formed during the postrift or sag phase of basin evolution (Fig. 8.12; Moreira et al., 2007). This unit comprises carbonate Spherulites and Fibrous Fans (or "shrubs") of the SFF facies association (Hunt et al., 2019; Hendry et al., 2021). These unusual carbonate fabrics were initially interpreted to be of microbial origin (Terra et al., 2010) but subsequent work, based on cores from the Barra Velha (Santos Basin), has argued strongly for an abiotic origin with precipitation within shallow, hypersaline, and alkaline lake waters (Wright and Barnett 2015, Wright and Rodriguez, 2018). Microbial laminites and stromatolites are, however, described from the Macabu Formation from the Campos Basin (Muniz and Bosence, 2015).

In offshore areas of the Campos Basin syn-rift, fault-bound highs were isolated from siliciclastic supply and accumulated thick successions of coquinas (Fig. 8.13). A proximal to distal transect through this basin was studied by Muniz (2013) and Muniz and Bosence (2018) using seismic and well data (including cores, sidewall cores, and BHI data). The main conclusion from this study was that large bioclastic carbonate bodies hundreds of meters in thickness and tens of kilometers across were found to be sited over fault-block highs and adjacent subbasins (Fig. 8.13). These are similar in scale to marine carbonate platforms, have a range of different facies and facies associations, and demonstrate both aggradational and progradational geometries and are therefore regarded as lacustrine carbonate platforms. Facies range from coastal plains, fluvial conglomerates, and soils, through shallow lacustrine bivalve-rich packstones, grainstones, and rudstones, to deep lacustrine marls, shales,

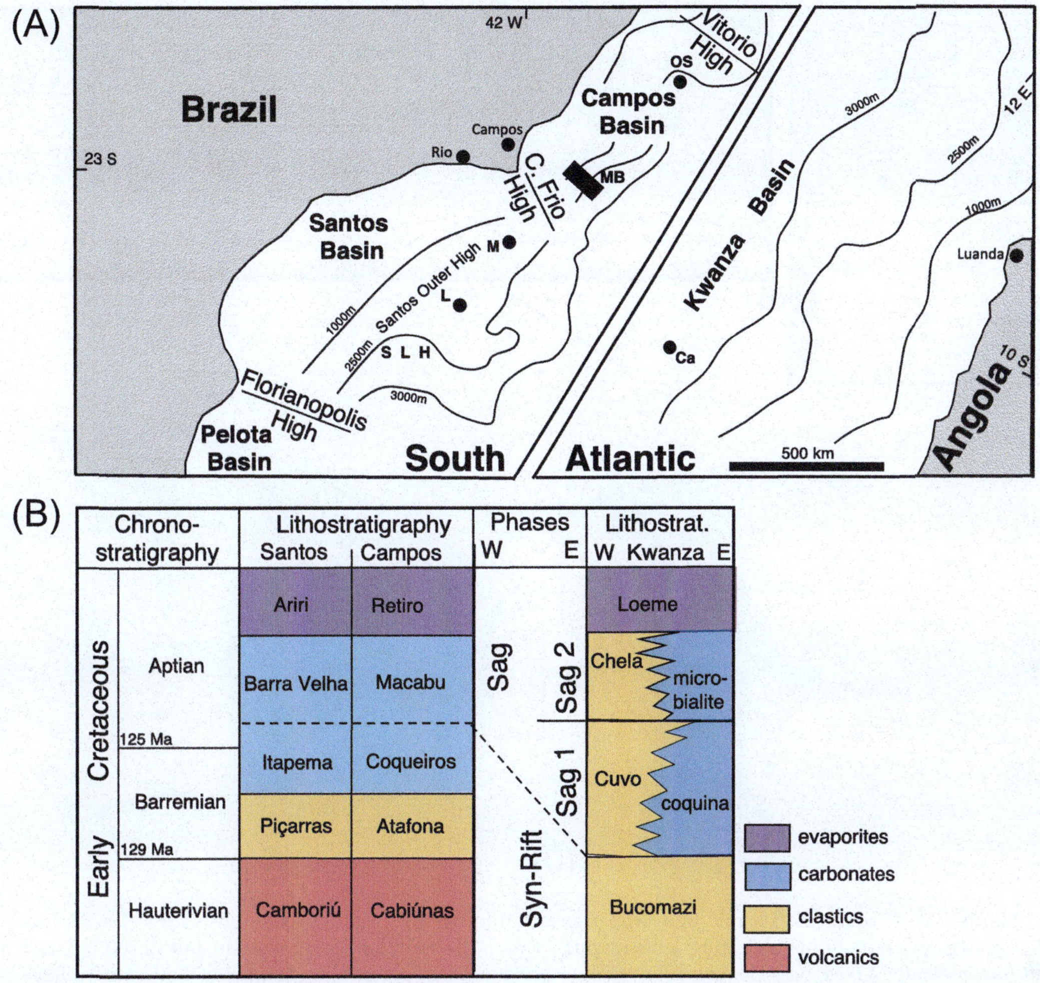

FIGURE 8.12 South Atlantic presalt basins. (A) Sketch map to indicate locations of Brazilian basins in western Atlantic and Kwanza Basin, Angola to the east (note; drawn to same scale but different orientations). Approximate locations/fields/case studies mentioned in text: SLH, Sugar Loaf High; L, Lula (or Tupi) field; M, Mero field (Barnett et al., 2020), MB, Muniz and Bosence (2018) transect; OS, Olivito and Souza (2020) case study; Ca, Cameia field (Hughes, 2014) (from many sources including Buckley et al. (2015), Spathopoulos (1996), and Cazier et al. (2014)). (B) Early Cretaceous stratigraphy of Santos, Campos, and Kwanza basins in the South Atlantic. Lithostratigraphy displayed as formations that are color coded by major lithology. The biostratigraphy and dating of these nonmarine basins are based on largely endemic ostracods, pollen and spores, and the ages and correlations of the formations between basins is tentative (after many sources including Barnett et al., 2020; Ceraldi and Green, 2017).

and some redeposited carbonates (Fig. 8.14). A core-based taphonomic study of the molluskan facies, their reworking, and preservation styles indicates grades of reworking from near in situ facies to highly abraded and transported facies (Muniz and Bosence, 2017).

A smaller-scale but more detailed study comes from the syn-rift of a northern area in the Campos Basin based on seismics, sidewall cores, and BHI, and some cored wells (Fig. 8.12A; Olivito and Souza, 2020). The lithologies are predominantly bivalve (locally enriched with gastropod) rudstones and grainstones and, less commonly, ostracod-peloid-ooid-grainstones, together with fine-grained siliciclastic facies. Typically, these reach minimum thicknesses of 120–180 meters in thickness. These authors present a similar model to that of Muniz and Bosence (2018) of fault-block highs characterized by high-energy, shallow-water mollusk-rich facies. Ostracod and peloid-rich mudstones and grainstones are interpreted to occur in sheltered nearshore sites behind a berm-foreshore area, where massive to cross-bedded and reworked molluskan rudstones to shoreface molluskan grain-rudstones and molluskan grain-rudstones. These shallow lacustrine facies pass basinward into more offshore siliciclastic mudstones and stevensitic peloidal sands.

Together these studies support a structural control on carbonate accumulation in the late syn-rift stage as fault-block carbonate platforms. An alternative model has been proposed by Goldberg et al. (2017), which interprets the thick bioclastic rudstones as being redeposited facies within hanging-wall subbasins. However, thick

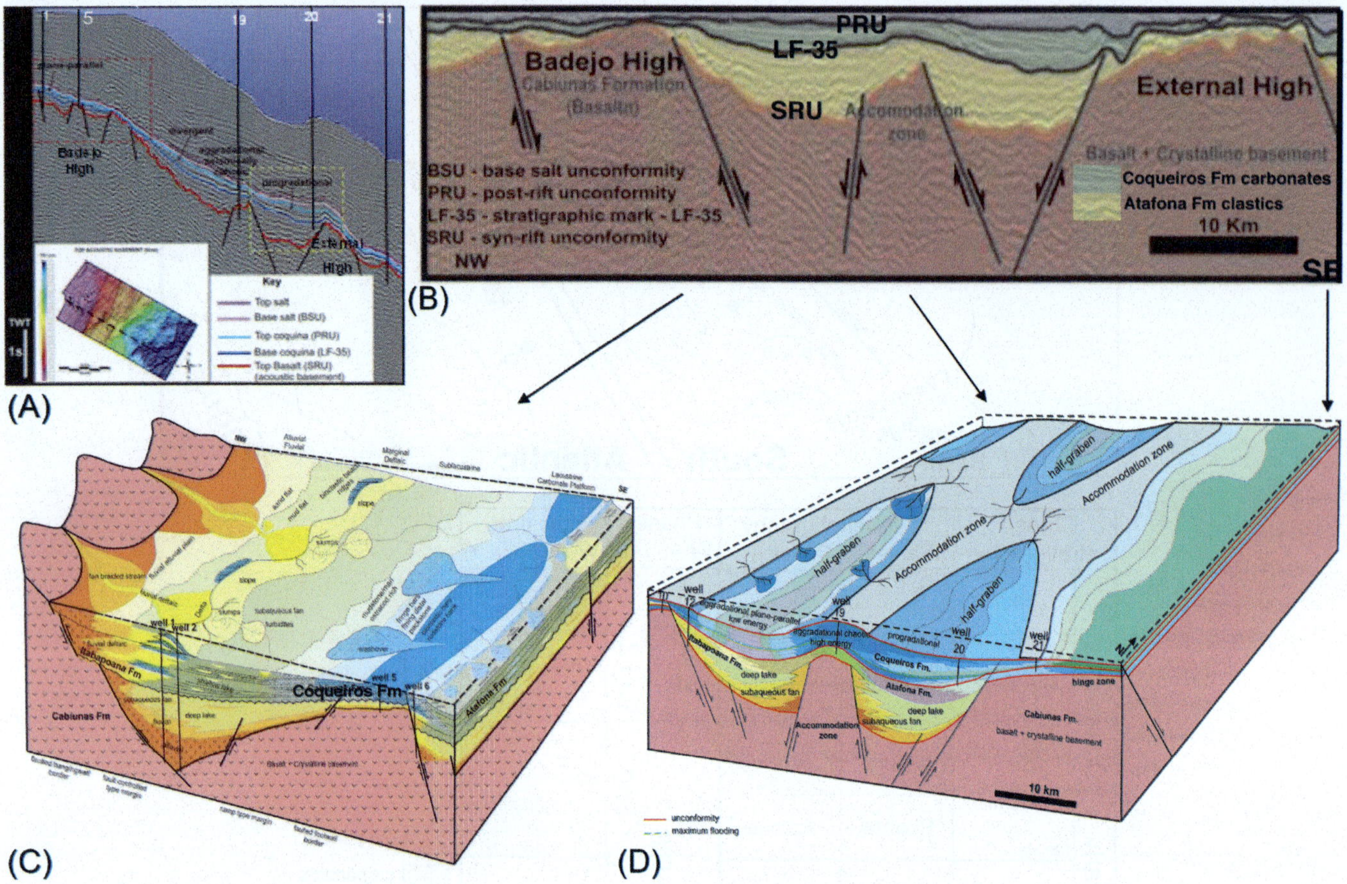

FIGURE 8.13 Northwest to southeast seismic and well transect through the southern Campos Basin (for location, see Fig. 8.12A). (A) Seismic section in two-way travel-time indicating units mapped within the southern Campos Basin and (inset) top acoustic basement surface (SRU, syn-rift unconformity; LF-35, base Coquieros Fm marker horizon; PRU, postrift [top Coqueiros Fm] marker; BSU, base salt horizon). (B) Seismic section flattened on base salt horizon illustrating main structural features along transect and surfaces mapped. Note the stratigraphic thickness variation of reflectors within the syn-rift Atafona and Coqueiros formations. (C) Tectonostratigraphic model based on seismic geometries and wells of rotated fault-block of Badejo High (located in Fig. 8.13A) illustrating mixed marginal siliciclastic facies and carbonate facies on isolated basinal high. (D) Tectonostratigraphic model based on seismic geometries and wells of horst fault-block in accommodation zone area (located in Fig. 8.13A). Note aggradational and progradational geometries of syn-rift Coqueiros Fm on horst block. Source: *Modified from Muniz, M. C. (2013) Tectono-Stratigraphic evolution of the Barremian-Aptian continental carbonates in the southern Campos Basin, Brazil. PhD thesis, Royal Holloway University of London. 324 pp.*

successions of facies with the characteristics of redeposited carbonates have not been described by these authors, whereas cross-bedding, texturally mature shell fragments, in situ preserved bivalves, and emergent facies described by Muniz and Bosence (2017, 2018) and Olivito and Souza (2020) all support a shallow lacustrine origin with carbonate production close to accumulation sites as described earlier.

A more detailed and localized study from the Itapema Formation of the Santos basin by Barnett et al. (2020) focused on a seismically imaged and drilled carbonate buildup in the Meru Field (Fig. 8.12). Here, bioclastic wackestones to packstones and grainstones to rudstones are shown through detailed investigation to have accumulated in progradational packages, several hundreds of meters in thickness. They argue convincingly that these packages of coarse, bivalve-rich sands, and gravels were transported in a shallow lacustrine site by wind-driven, shore-parallel, currents similar to the interpretation of the Coqueiros Formation in the Campos basin (Olivito and Souza, 2020).

While a syn-rift origin for the fault-block platforms of the Coqueiros Formation of the Campos Basin has been established (see above), the subsequent carbonates of the Macabu (Campos Basin) and Barra Velha (Santos Basin) show little evidence of syn-depositional tectonics and are classified as having formed in the sag, postrift phase of basin evolution (Fig. 8.15A; Winter et al., 2007; Moreira et al., 2007). Seismic images of layered and aggradational structures on fault-bound highs have been published from the Sugar Loaf High in the Santos Basin (e.g., Buckley et al., 2015) and faulted margins and suggest late syn-rift to sag phase for these structures. A large aggradational

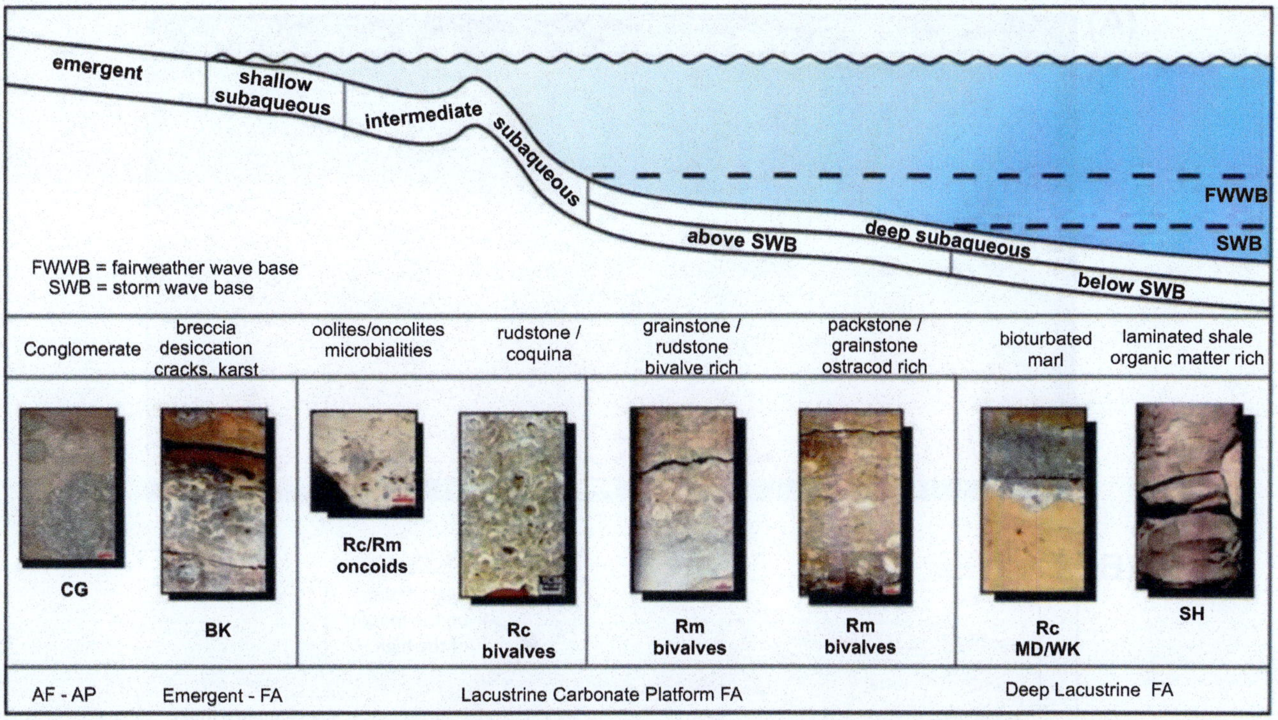

FIGURE 8.14 Facies model for proximal to distal lake margins of the Coqueiros Formation. Facies distribution interpreted with respect to fair- and storm-weather wave base below lake level surface. CG, Conglomerate; BK, Karst breccia; Rc, Rudstone clean; Rm, Rudstone muddy; MD/WK, Mudstone/wackestone; AF AP, Alluvial Fan and Emergent Plain; FA, Facies Association. Source: *Modied from Muniz, M. C. (2013) Tectono-Stratigraphic evolution of the Barremian-Aptian continental carbonates in the southern Campos Basin, Brazil. PhD thesis, Royal Holloway University of London. 324 pp.*

structure over a rotated fault block is imaged on the eastern margin of the Sugar Loaf High (Fig. 8.15B and C). This has an essentially horizontal upper surface and on its eastern margin aggradational seismic morphologies that pass downslope into inclined reflectors on its eastern edge. This feature has been interpreted as a carbonate platform and the seismic geometries and correlation indicate that it is part of the sag phase Barra Velha Formation. Following the criteria proposed by Burgess et al. (2013) for recognizing carbonate platforms from seismic data, Buckley et al. (2015) argue for a carbonate stratigraphy, as opposed to volcanic, siliciclastic, or evaporitic origin. Subsequently, Wright and Rodriguez (2018) argue against this interpretation and propose that the mound-like features with aggradational style reflectors may be either volcanic or inversion features, but do not present any evidence for either of these two interpretations. In addition, they infer from core-based studies that the lakes were essentially shallow-water and lacking the substantial bathymetry suggested by the seismic geometries. More recently, Minzoni et al. (2021) from an extensive seismic, well (including BHI), and core study of the Santos Outer High area argue strongly that the features on the Sugar Loaf High are carbonate platforms with aggradational and progradational geometries and that they had significant relief. The detailed nature of the carbonate factory generating these large platforms still needs to be addressed but the seismic geometries imply that there is a shallow-water or emergent carbonate factory that could build aggradational structures and slopes with hundreds of meters of relief in a lacustrine equivalent to the well-known marine photozoan carbonate platforms.

Kwanza Basin lacustrine carbonate platforms

Basins offshore from Angola have attracted exploration with targets in the coquina and microbialite facies that have been identified as hydrocarbon reservoirs. Preliminary findings have been published and the opposing Brazilian and African margins show broadly similar and correlatable phases of basin evolution with syn-rift and postrift or sag phases having been recognized (Fig. 8.12B; Saller et al., 2016; Ceraldi and Green, 2017). Syn-rift, continental fluvial and lacustrine siltstones and shales of the Bucomazi Formation accumulated in grabens during the Early Cretaceous (Hauterivian) (Spathopoulos, 1996; Ceraldi and Green, 2017). Subsequently, Barremian to

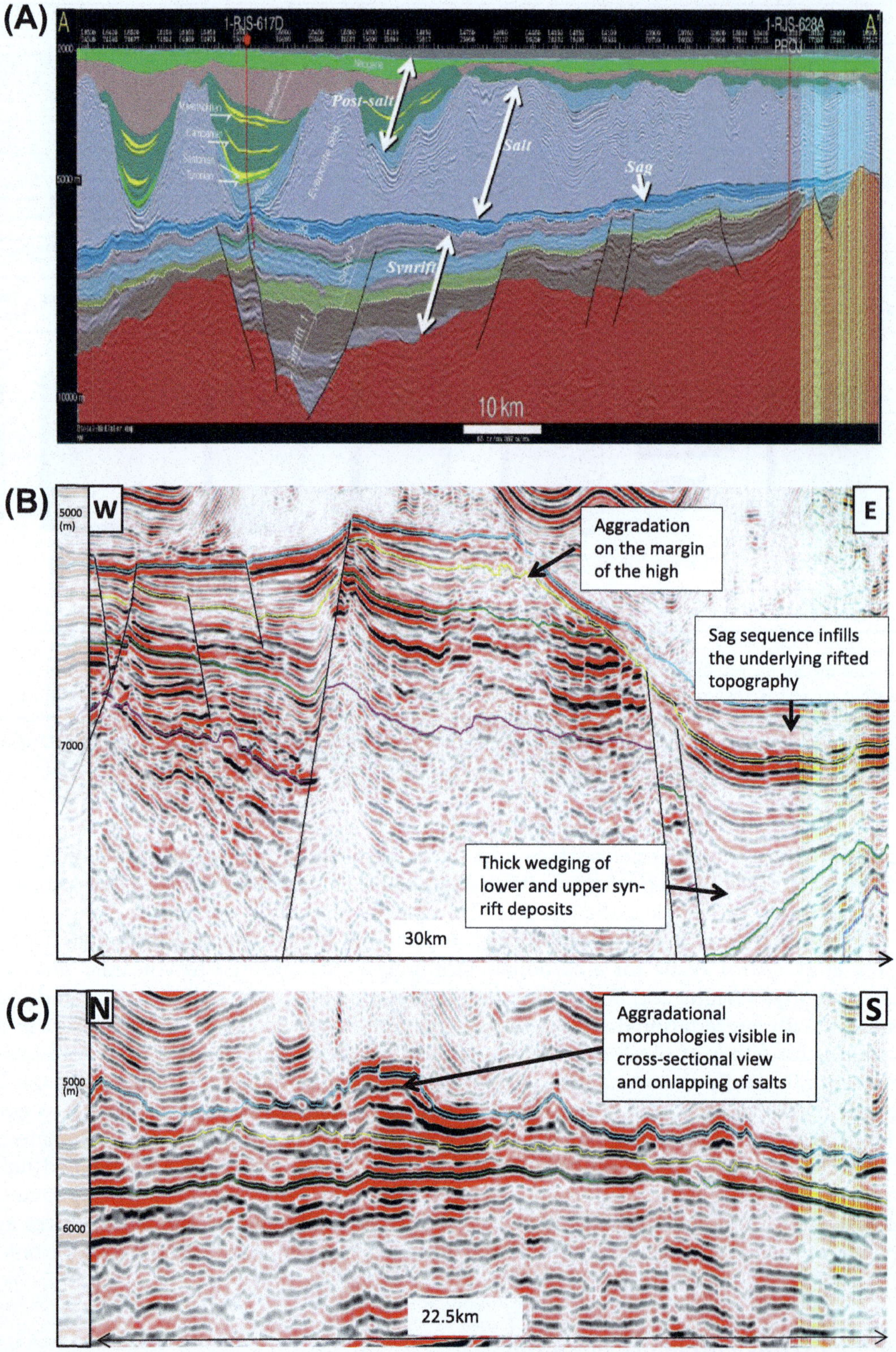

FIGURE 8.15 Interpreted seismic sections through Early Cretaceous stratigraphy, offshore Santos Basin, Brazil. (A) Section through Lula or Tupi field (right) and Pirati discovery (left) indicating downdip thickening of strata within the syn-rift and draping of postrift sag phase with nondeposition or erosion over fault-block high (right). For location, see Fig. 8.12A. (B) and (C) Seismic sections through basinward margin of Sugar Loaf High (location see Fig. 8.12A). (B) West to East dip-section showing aggradational geometries on eastern margin and eastern marginal slope. Key horizons: magenta, base syn-rift; green, intra syn-rift; yellow, base sag and blue- base salt. (C) Strike section through (B) showing aggradational morphologies above flat-lying reflectors on eastern margin. Aggradational features onlapped by base salt. *(A) Courtesy: Petrobras.* Source: *(B) and (C) after Buckley, J.P., Bosence, D.W.J. & Elders, C. (2015) Tectonic setting and stratigraphic architecture of an Early Cretaceous lacustrine carbonate platform, Sugar Loaf High, Santos Basin, Brazil. In: Microbial Carbonates in Space and Time: Implications for Global Exploration and Production. In: Bosence, D.W.J., Gibbons, K., Le Heron, D.P., Morgan, W.A., Pritchard, T.P. & Vining, B.A. (Eds), Geological Society London Special Publication 418, London, 175–191.*

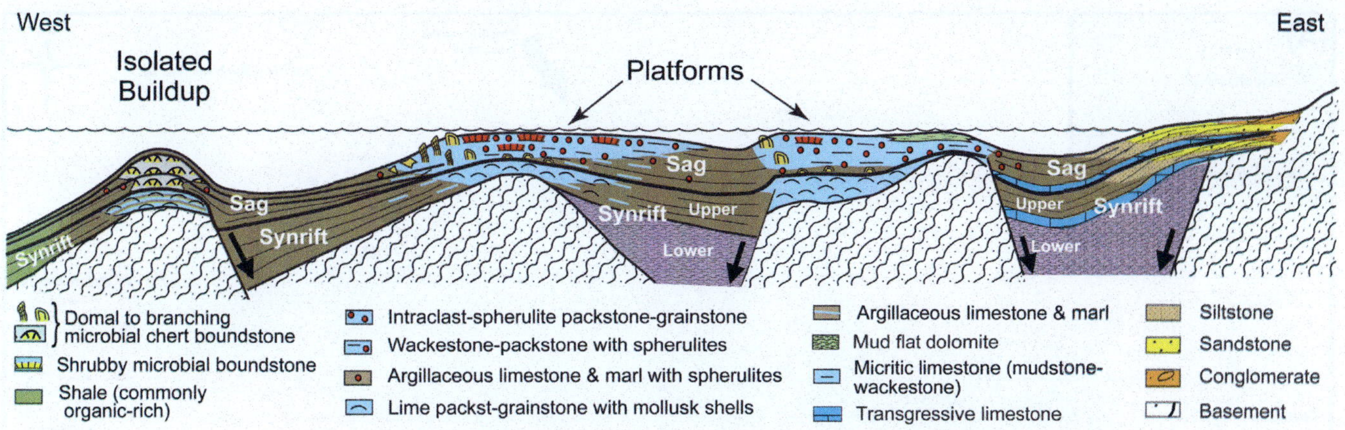

FIGURE 8.16 Schematic sedimentological and tectonostratigraphic model for syn-rift and sag units of the Kwanza Basin, Angola. Unlocated east to west profile. Source: *From Saller, A., Rushton, S., Buambua, L., Inman, K., McNeil, R. & J.A.D. Dickson (2016) Presalt stratigraphy and depositional systems in the Kwanza Basin, offshore Angola. American Association of Petroleum Geologists Bulletin, 100, 1135–1164, with permissions.*

early Aptian clastics and carbonates (Cuvo Formation and coquinas) accumulated that can be tentatively correlated with formations in offshore Brazilian Basins (Fig. 8.12B; Ceraldi and Green, 2017). Informally labeled as coquina on the Angolan margin, they comprise bioclastic carbonates with diverse mollusk faunas and ostracods, locally with stevensitic ooids, indicate brackish to saline lacustrine environments. These equate to the BIO facies association of the Coquieros and Itapema formations of the Brazilian margin (Hendry et al., 2021). Basinwards, these pass into organic-rich mudstones of the Cuvo Formation. Saller et al. (2016) classify these carbonates and shales of the Kwanza Basin as being part of the syn-rift with mollusk-rich limestones with packstone to grainstone textures accumulating on footwall highs on rotated fault-blocks and on horsts (Fig. 8.16). The tectonic setting of these carbonates is therefore very similar to that of the Coquieros Formation in the Campos Basin (cf. Figs. 8.13 and 8.16) and both examples can be interpreted as fault-block carbonate platforms. In the more extensive analysis by Ceraldi and Green (2017), the mollusk-rich coquinas are also interpreted to have accumulated on fault-block highs with molluskan mudstones and packstones in the hanging-wall subbasins (Ceraldi and Green, 2017, Fig. 8.8). However, they classify this interval as the early part (Sag 1) of the sag phase of basin evolution as it unconformably overlies the syn-rift clastics of the Bucomazi Formation with basal onlaps.

The succeeding sag phase deposits also have equivalents to the Brazilian margin with carbonate buildups on inherited fault-block highs and basinal shales. Then more inshore carbonates are shrubby microbial boundstones or the SFF (spherulite fibrous fan) facies association, while the offshore highs comprise silicified (clotted peloidal silica [CPS]) facies association (Saller et al., 2016; Hendry et al., 2021). More detailed descriptions of these platforms and their dimensions and geometries have yet to be published.

In the Kwanza Basin, the Barremian and early Aptian coquinas and Cuvo shales are unconformably overlain by the Aptian Chela Formation and microbialite unit that has equivalence with the postrift Macabu and Barra Velha formations in Brazil (Fig. 8.12). During the mid- to late Aptian subsidence-related rifting led to intermittent marine connections and the accumulation of halite and anhydrite evaporites of the Loeme Formation that is correlatable with the Ariri and Retiro salts of the sag phase of the Santos and Campos basins, respectively (Fig. 8.12).

Lacustrine carbonates in northeast Atlantic rift basin

Wessex Basin lacustrine carbonates

The Wessex Basin (Fig. 8.17) is an E-W-oriented extensional basin that was part of a much larger system of intracratonic Mesozoic basins associated with the opening of the northeast Atlantic (Underhill and Stoneley, 1998; Ziegler, 1990). The structural framework and stratigraphic evolution of the basin are known from extensive onshore and offshore seismic and well data and three main areas of outcrop: south Dorset, Vale of Wardour, and Boulonnais (Fig. 8.17A). Tectonostratigraphic phases of basin evolution have been documented by Underhill and Stoneley (1998) and the Permian to Lower Cretaceous mega-sequence comprises several extensional episodes between the Early Jurassic and Early Cretaceous times (Barton et al., 2011). Syn-depositional extensional faults are

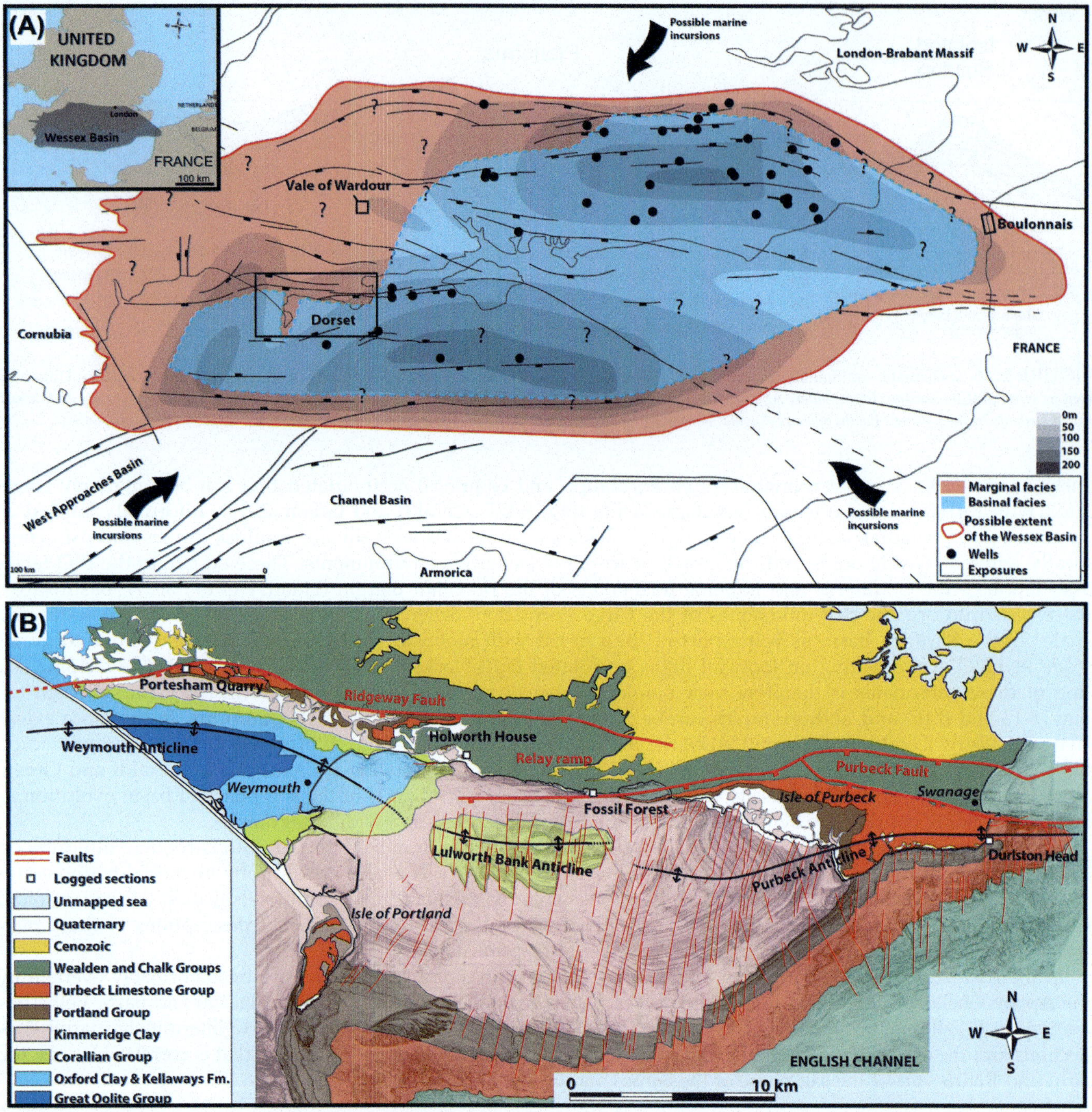

FIGURE 8.17 Wessex Basin, southern United Kingdom, and northern France. (A) Reconstructed area of Wessex Basin from subsurface and outcrop data indicating east-west extensional faults, marginal (microbialites, evaporites, and sandstones) and basinal (ostracod and molluskan shales, marls and evaporites) lacustrine facies. Outcrop-based study area of south Dorset (B) indicated by rectangle (after many sources including Lake (1985), Cope et al. (1980), West (1992), Underhill (2002) and examination of BGS cores). (B) Onshore and offshore geological map of south Dorset showing outcrop of Purbeck Limestone Group, inverted Mesozoic faults (Ridgeway and Purbeck faults) and central relay ramp area (after many sources). Source: *Bosence, D.W.J. & Gallois, A. (2021) How do thrombolites form? Multiphase construction of lacustrine microbialites, Purbeck Limestone Group, (Jurassic), Dorset, UK. Sedimentology, 69, 914–953.*

predominantly oriented east-west (Fig. 8.17) and many were inverted during Alpine compression (Underhill and Stoneley, 1998).

The lacustrine Purbeck Limestone Group (the PLG) was deposited during the final extensional episode of the Wessex Basin when the Ridgeway and the Purbeck faults created accommodation for lacustrine sedimentation in hanging-wall depocenters (Underhill, 2002; Gallois et al., 2018; Gallois et al., 2021). These two faults are linked by a relay ramp in the central part of the outcrop area (Fig. 8.17B). Syn-depositional fault displacements in the Jurassic are indicated by thickness increases in the hanging-wall strata relative to correlative footwall strata or unconformities (Hamblin et al., 1992; Hawkes et al., 1998; Stoneley, 1982; Underhill, 2002; Gallois et al., 2021).

Gallois et al. (2018) described and interpreted 10 facies grouped into 5 facies associations from the lower part (Mupe Member) of the PLG (Fig. 8.18): sublittoral lacustrine, mounded (microbial) marginal lacustrine, emergent (paleosoils) in a lower brackish-water phase, and a hypersaline lacustrine and bedded marginal lacustrine in an upper hypersaline-water phase (West, 1975).

Correlation of the beds and facies associations indicate two lacustrine subbasins separated by a paleohigh within the Purbeck lake (Fig. 8.18). Shallow-water microbial mounds occur along the northern, faulted margins of the hanging-wall blocks and to the south a hanging-wall dip-slope with extensive exposures on the Isle of Portland. In addition, a NE-SW facies trend is identified in the central part of the area coinciding with the tips of the Purbeck and Ridgeway faults. This is interpreted as a relay ramp comprising a shallow area that separated the two lake subbasins, to the east and to the west, where the sublittoral lacustrine and the deeper-water mounded marginal lacustrine FA occur (Fig. 8.18; Underhill, 2002; Gallois et al., 2018). This central, shallow zone is labeled as a "swell" by previous authors (West, 1975; Francis, 1982) separating more brackish and freshwaters bounded by the landmass to the west and more hypersaline waters in the basin to the east (Fig. 8.17A). Shallow-water relay ramps are considered to be favored sites for carbonate production and accumulation in lacustrine systems as they are the likely route for freshwater input into the lake and, in the Purbeck example, are enriched in dissolved calcium carbonate from the eroded limestones of the footwall hinterland. This increased saturation in calcium carbonate would have driven the calcification of the microbial filaments and microbial mound formation in the lower part of the PLG (Bosence and Gallois, 2022) as described for modern, freshwater microbialite mound formation in Laguna Bacalar, Mexico (Gischler et al., 2008).

Although earlier work suggested that the rifted topography would generate clastic sediment pathways from footwall areas to form clastic fringes to these subbasins (Underhill, 2002), subsequent facies analysis (Gallois et al., 2018) has not found these to be present and two factors appear to be important. First, the hinterland is dominated by carbonate rocks (Portland Limestone) would dissolve rather than erode by physical processes and second, the rift-related relief may have been too low so that surface fault scarps were not generated (Gallois et al., 2021).

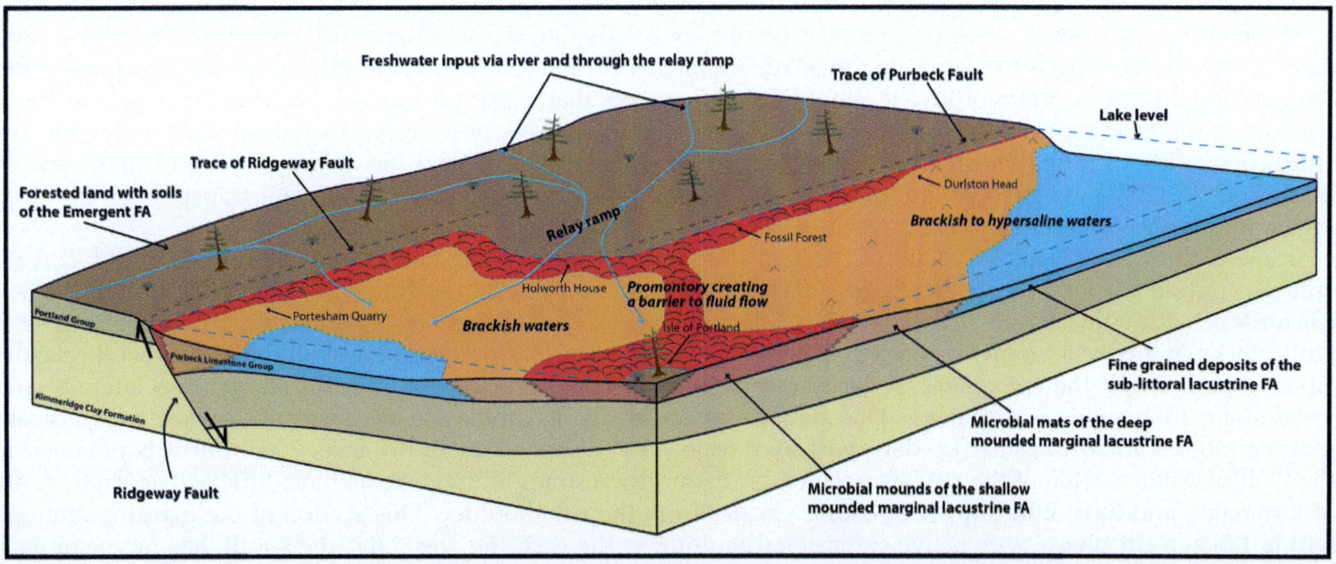

FIGURE 8.18 Tectono-sedimentary model for accumulation of the lower part of the Purbeck Limestone Group, southern Dorset. Two offset extensional faults (now inverted see Fig. 8.17B) provide accommodation in hanging-wall depocenters for two lacustrine subbasins separated by a relay ramp forming a shallow-water substrate for microbial carbonate production. Source: *From Gallois, A., Bosence, D.W.J. & Burgess, P. (2021) Influence of extensional faults and relay ramp on palaeo-topography and lacustrine carbonate facies: Purbeck Limestone Group (Upper Jurassic), Wessex Basin, UK. Basin Research, 33, 3159–3182.*

The carbonate sedimentation in the western part of the Wessex basin is therefore typified by accumulation in alkaline lakes in low-relief, hanging-wall subbasins. The footwall areas were presumed to be emergent and are represented by the pre-Aptian unconformity. In this low-relief setting and with the limitations of the coastal outcrops in the basinward (south-easterly) direction, the margin to these shallow lacustrine carbonates is not seen. Similarly, no steep margin is seen in the extensive seismic coverage in offshore areas (Gallois et al., 2021) and for these reasons the large-scale geometry of these carbonates is considered to be a ramp type of carbonate platform.

Synthesis and conclusions

This review of carbonate sedimentation in rift basins provides a number of case studies that illustrate the diversity of carbonate sediments and the platform types present in these tectonically active settings. In addition, the carbonate factories are variable within rifts with lacustrine, marine, and hypersaline aqueous environments. Subaerial carbonate accumulations have not been considered as these tend to form smaller-scale accumulations. The case studies exemplify carbonate sediment accumulation in lacustrine settings from the early phases of opening of the south and north Atlantic as well as from the younger, but marine-dominated, Gulf of Suez, Red Sea, and Gulf of Aden basins. In each case, the tectonic framework is an important driver on the location of the carbonate factories and the facies distribution. Typically, this is in the location of shallow aqueous sites that are somewhat, but not entirely, isolated from siliciclastic supply. This partial isolation can be both temporal or spatial. A common sedimentological pattern in rift basin evolution is that early syn-rift phases are dominated by clastic sediment, as topography and erosion rates are high, whereas later phases of the syn-rift can be dominated by carbonate sedimentation that then continues into the postrift or sag phase (Fig. 8.1). In addition, and as a consequence of rift-related extension and subsidence, early phases may be characterized by nonmarine or lacustrine carbonates and later phases, following marine flooding, by marine carbonates (Fig. 8.1). Spatially, environments with reduced clastic supply can originate through faulting and/or diapirism, which both generate syn-depositional topography where clastic pathways will follow the lows and carbonates will be produced on the highs. Such processes are well demonstrated in the Gulf of Suez–Red Sea basins described earlier.

Spatial variability in rift basin carbonate platforms

The concepts of prerift, syn-rift, and postrift phases (Badley et al., 1984) and rift initiation, rift climax, and postrift phases have been used to categorize rift basin fills (Prosser, 1993). While useful for understanding and categorizing rift basin stratigraphy, it should be understood that these phases are not synchronous and basin wide as portions of a rift basin may become quiescent due to changes in locations of crustal scale faults and lateral propagation of faults (Leeder and Gawthorpe, 1987). Three-dimensional forward modeling of platforms by Masiero et al. (2021) explores the important role of lateral fault propagation and sediment transport on developing stratigraphies on fault-blocks.

In the Gulf of Suez–Red Sea–Gulf of Aden basins, some of the best studied marine rifts in the world, considerable variation is found along its margins. Its northern parts represent a sedimentary basin whose margins are concordant with either basinward or landward dipping faults. Carbonates are concentrated on footwall highs to fault blocks in low-relief transfer zones and interfinger with clastics down hanging-wall dip-slopes while clastics either pass through the carbonates in fault accommodation zones, cut across fault blocks, or pass laterally and accumulate in hanging-wall basins. This pattern of sediment accumulation conforms to earlier tectonostratigraphic models for rift basins (Leeder and Gawthorpe, 1987; Bosence, 1998; Bosence, 2005) but it is not seen in the Gulf of Aden section of the rift basin. Here, on the northern margin, large areas have little active faulting, are of low relief, and have little supply of course clastics from the rift shoulder. This section of the margin, although within the syn-rift phase, with active extensional faulting to the east, the west, and the south, has more similarities to a passive margin fill than a rift basin. Strata appear to passively fill the accommodation space generated from regional subsidence of the margin. In addition, further to the east of the Gulf of Aden in this oblique rift the coastline is parallel to the direction of extension so that the major facies belts (deepwater carbonates, shelf carbonates and continental deposits) are normal to the extensional fault blocks. However, locally, the fault blocks control the occurrence of platform and slope carbonates, as fault-block platforms, with clastics funneled axially, through intervening hanging-wall basins (Fig. 8.11B). While the difference between the concordant and the oblique rift organization might be predicted from regional data, the different sections of the rift are in different stages of structural evolution.

Importance of base level in controlling rift stratigraphy

The position of base-level with respect to the evolving topography of rift basins is fundamental to the accumulation and preservation of carbonate stratigraphies. Base level will be controlled by a wide range of processes including previous geological history, structural evolution, sea-level changes, and climate. In this review, it is clear that following regional extension and crustal thinning, subsidence and base-level rise will occur and this is likely to be the driver for the evolution from continental to marine depositional systems. At the same time, early syn-rift or rift climax phases are characterized by active faulting and the generation of rift topography with high rates of erosion and sediment supply that will both decrease through to late syn-rift to postrift (or sag) phases. All these processes lead toward favoring carbonate sedimentation over siliciclastic sedimentation, so that this evolution in depositional systems is commonly observed in rift basins. Numerical forward modeling of carbonate platform development on rotating fault blocks indicates the importance of high rates of base-level rise (i.e., subsidence) in order to generate marine carbonates on footwall highs that otherwise are sites of uplift and erosion above base level (Bosence et al., 1998b). High rates of base-level rise are shown in this review with carbonate platforms of variable thickness developing over rotating fault-blocks in examples from Qamar, Gulf of Aden (Fig. 8.11B), on the Badejos High of the Campos Basin (Fig. 8.13), and the Sugar Loaf High in the Santos Basin (Fig. 8.15B). On the Abu Shaar platform (Gulf of Suez), the footwall stratigraphy is considerably thinned by the lack of accommodation and frequent erosional episodes on this footwall high (Fig. 8.4). The Wessex Basin provides an example of where the base level is low with respect to the rifted topography, so that the footwall is emergent and undergoing erosion and carbonate sedimentation is confined to the adjacent hanging-wall basin (Fig. 8.18).

Types of carbonate platforms in rift basins

Within carbonate-dominated rifts, there are certain sites that favor carbonate platform development if the basin is flooded by marine or lacustrine waters. During flooding, clastic systems will be pushed landwards so that deposition is focused in the coastal/palustrine zone or transported basinward into structural lows. Examination of carbonate successions in rift basins indicates four common sites where carbonate platforms develop (Fig. 8.19):

1) Half-graben or horst fault-blocks may form highs that are isolated from clastic supply, and in low-latitude, shallow-water settings, photozoan carbonates will be produced and accumulate. Carbonate platforms developed over rotating fault blocks and horsts during the syn-rift phase and have distinctive stratigraphy and sedimentology and have been labeled fault-block carbonate platforms (Bosence, 2005; Bosence et al., 1998b; Cross and Bosence, 2008).
2) If the rift basin has an early history of isolation as interior drainage basin or from marine connection, then evaporites accumulate. Following burial, later diapiric structures form submarine highs and if these are located within the shallow water, carbonate production zone will provide suitable sites for the development of salt-diapir carbonate platforms. Such is the case in the Red Sea and earlier in the history of the Atlantic margin.
3) During the marine flooding phase and in arid climate settings, deltas and fan deltas that protrude into more open-water sites are also, but counterintuitively, sites of reef and carbonate platform growth. Such platforms (labeled delta-top platforms by Bosence, 2005) are common in the Cenozoic Gulf of Suez–Red Sea–Gulf of Aden rift basins.
4) Ramp margins may develop in tectonically quiescent areas or phases of rift margins such as the central part of the northern Gulf of Aden rift east of Mukalla in the Oligo-Miocene. Similarly, ramp margins are present at a smaller scale on hanging-wall dip-slopes in both marine (Fig. 8.19; Burchette and Wright, 1992) and lacustrine settings (Platt and Wright, 1991) and as described above for the Wessex Basin hanging-wall subbasins.

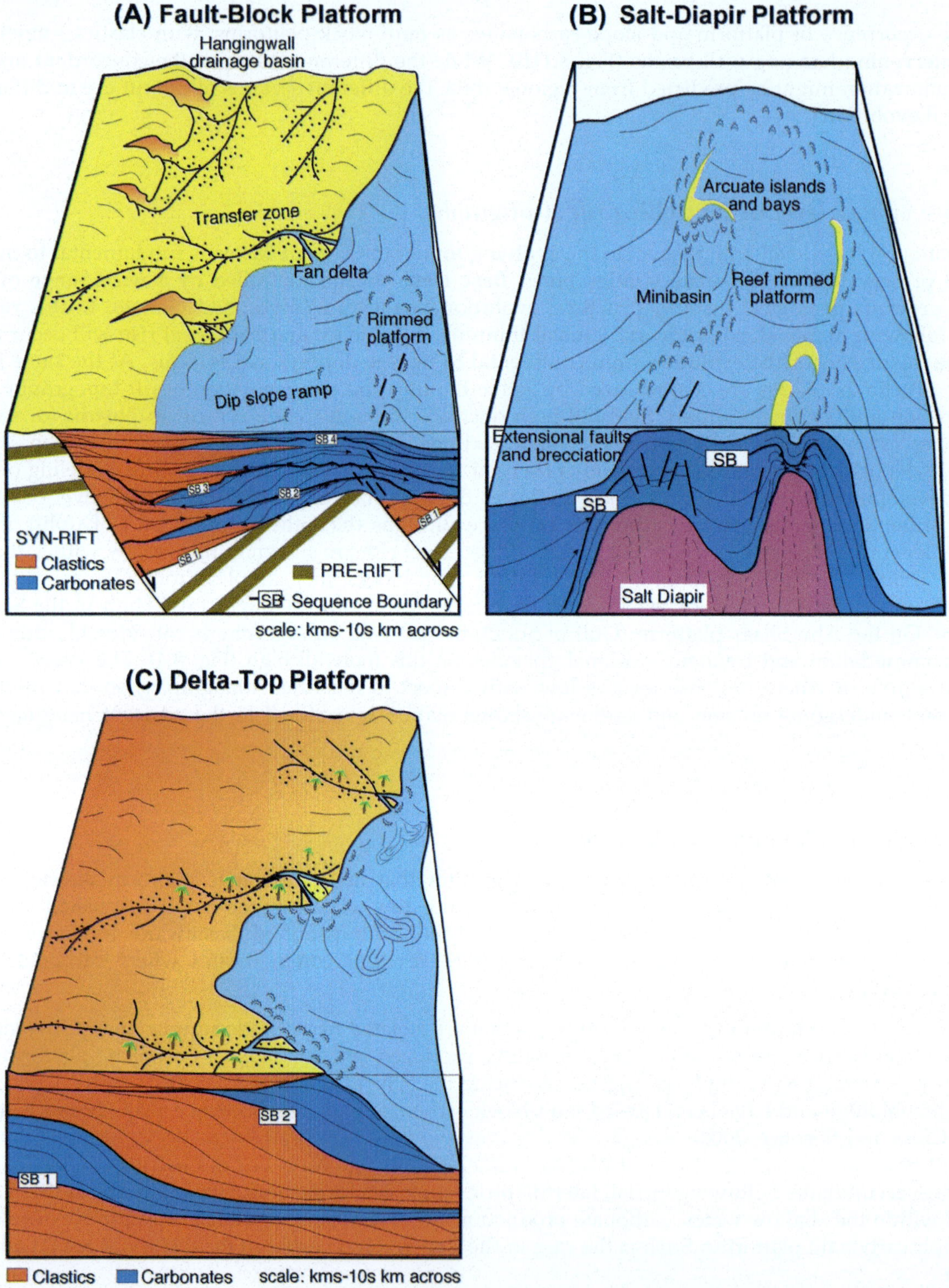

FIGURE 8.19 Facies and sequence stratigraphic models for carbonate platforms from rift basins. (A) Fault-block carbonate platform based on outcrop studies from marine carbonates in the Gulf of Suez and seismic sections from the Gulf of Aden. Note, footwall rimmed shelf and dipslope ramp with wedge-shaped depositional sequences. Clastic sedimentation is marginal and following offshore lows while carbonate sediments accumulate on highs. (B) Salt-diapir carbonate platform based on outcrop studies in Yemen and Saudi Arabia and seismic sections from the Gulf of Suez and southern Red Sea. Note arcuate shaped platform with minibasin reentrants and irregular-shaped depositional sequences generated by salt piercement and withdrawal. (C) Delta-top carbonate platform based on outcrop studies in the Red Sea, Gulfs of Suez, and Aqaba. Note predominant lowstand clastic deltas and transgressive and highstand carbonate platforms. Source: *From Bosence, D.W. J. (2012) Carbonate dominated marine rifts. In: Roberts, D.G. & Bally, A.W. (Eds) Regional Geology and Tectonics: Phanerozoic Rift Systems and Sedimentary Basins: Volume 2: Phanerozoic Rift Systems and Sedimentary Basins. Elsevier. Chapter 5, 89–114; Bosence, D.W.J. (2005). A genetic classification of carbonate platforms based on their basinal and tectonic settings in the Cenozoic. Sedimentary Geology, 175, 49–72.*

References

Badley, M.E., Egeberg, T.M., Nipen, O., 1984. Development of rift basins illustrated by the evolution of the Oseburg feature, Block 30/6, offshore Norway. Journal Geological Society of London 141, 639–649.

Bantan, R.A. (1999). *Geology and sedimentary environments of Farasan Bank (Saudi Arabia) southern Red Sea: A combined remote sensing and field study*. PhD thesis University of London. 297pp.

Barnett, A.J., Fu, L., Rapasi, T., Scotellaro, C., Guha, J., Cabolova, A., et al., 2020. Seismic characterisation and origin of clinoforms in lacustrine depositional environments: a case study from the Cretaceous of the South Atlantic. In: Hendry, J., Burgess, P., Hunt, D., Janson, X., Zampetti, V. (Eds.), Seismic Characterisation of Carbonate Platforms and Reservoirs, 509. Geological Society London Special Publication, London, pp. 127–145.

Barton, C., Woods, M., Bristow, C., Newell, A., Westhead, R., Evans, D., et al., 2011. Geology of South Dorset and South-East Devon and Its World Heritage Coast. British Geological Survey, Keyworth, Nottingham 161.

Bernoulli, D.L., Jenkyns, H.C., 1974. In: Dott, R.H., Shaver, R.W. (Eds.), Alpine, Mediterranean, and Central Atlantic Mesozoic facies in relation to the early evolution of the Tethys: Modern and Ancient Geosynclinal Sedimentation, 19. SEPM, Special Publication, pp. 129–160.

Bosence, D.W.J., 1998. Stratigraphic and sedimentological models of rift basins. In: Purser, B.H., Bosence, D.W.J. (Eds.), Sedimentary and Tectonic Evolution of Rift Basins- The Red Sea-Gulf of Aden. Kleuwers, pp. 9–25.

Bosence, D.W.J., 2005. A genetic classification of carbonate platforms based on their basinal and tectonic settings in the Cenozoic. Sedimentary Geology 175, 49–72.

Bosence, D.W.J., 2012. Carbonate dominated marine rifts. In: Roberts, D.G., Bally, A.W. (Eds.), Regional Geology and Tectonics: Phanerozoic Rift Systems and Sedimentary Basins: Volume 2: Phanerozoic Rift Systems and Sedimentary Basins. Elsevier, pp. 89–114. , Chapter 5.

Bosence, D.W.J., Al-Aawah, M.H., Davison, I., Rosen, B.R., Vita Finzi, C., Whitaker, E., 1998a. Salt domes and their control on basin margin sedimentation: a case study from the Tihama Plain, Yemen. In: Purser, B.H., Bosence, D.W.J. (Eds.), Sedimentary and Tectonic Evolution of Rift Basins- The Red Sea-Gulf of Aden. Kleuwers, pp. 448–464.

Bosence, D.W.J., Cross, N.E., Hardy, S., 1998b. Architecture and depositional sequences of Tertiary fault-block carbonate platforms; an analysis from outcrop and computer modelling. Marine and Petroleum Geology 15, 203–221.

Bosence, D.W.J., Gallois, A., 2022. How do thrombolites form? Multiphase construction of lacustrine microbialites, Purbeck Limestone Group, (Jurassic), Dorset, UK. Sedimentology 69, 914–953.

Bosence, D.W.J., Nichols, G.J., Al-Subbary, A.-K., Al-Thour, K.A., Reeder, M., 1996. Synrift continental to marine depositional sequences, Tertiary, Gulf of Aden, Yemen. Journal of Sedimentary Research 66, 766–777.

Bosworth, W., 2015. Geological evolution of the Red Sea: historical background, review and synthesis. In: Rasul, N.M.A., Stewart, I.C.F. (Eds.), The Red Sea. *Springer*, pp. 45–78.

Bosworth, W., Crevello, P., Winn, R.D., Steinmetz, J., 1998. Structure, sedimentation and basin dynamics during rifting of the Gulf of Suez and northwestern Red Sea. In: Purser, B.H., Bosence, D.W.J. (Eds.), *Sedimentary and Tectonic Evolution of Rift Basins- The Red Sea-Gulf of Aden*. Kleuwers. pp. 77–96.

Bott, W.F., Smith, B.A., Oakes, G., Sikander, A.H., Ibrahim, A.I., 1992. The tectonic framework and regional hydrocarbon prospectivity of the Gulf of Aden. Journal of Petroleum Geology 15, 211–243.

Brannan, J., Gerdes, K.D., Newth, I.R., 1997. Tectono-stratigraphic development of the Qamar basin, eastern Yemen. Marine & Petroleum Geology 14, 701–730.

Buckley, J.P., Bosence, D.W.J., Elders, C., 2015. Tectonic setting and stratigraphic architecture of an Early Cretaceous lacustrine carbonate platform, Sugar Loaf High, Santos Basin, Brazil. In: Bosence, D.W.J., Gibbons, K., Le Heron, D.P., Morgan, W.A., Pritchard, T.P., Vining, B.A. (Eds.), Microbial Carbonates in Space and Time: Implications for Global Exploration and Production, 418. Geological Society London Special Publication, London, pp. 175–191.

Burchette, T.P., Wright, V.P., 1992. Carbonate ramp depositional systems. Sedimentary Geology 79, 3–57.

Burchette, T.P., 1988. Tectonic control on carbonate platform facies distribution and sequence development: Miocene, Gulf of Suez. Sedimentary Geology 59, 179–204.

Burgess, P.M., Winefield, P., Minzoni, M., Elders, C., 2013. Methods for identification of isolated carbonate buildups from seismic reflection data. American Association Petroleum Geologists Bulletin 97, 1071–1098.

Carbone, F., Matteucci, R., Angelucci, A., 1998. Present day sedimentation on the carbonate platform of the Dahlak Islands, Eritrea. In: Purser, B.H., Bosence, D.W.J. (Eds.), Sedimentary and Tectonic Evolution of Rift Basins- The Red Sea-Gulf of Aden. Kleuwers, pp. 523–536.

Carminatti, M., Dias, J.L. & Wolff, B. (2009) From turbidites to carbonates: Breaking paradigms in deep waters. Offshore Technology Conference, Houston, USA, May 2009. Extended abstract OTC 20124, pp 1–7.

Cazier, E.C., Bargas, C., Buambua, L., Cardoso, S., Ferreira, S., Inmanl, K., et al., 2014. Petroleum geology of cameia field, deepwater pre-salt Kwanza Basin, Angola, West Africa. American Association of Petroleum Geologists, Search and Discovery Article #A.20275.

Ceraldi, T.S., Green, D., 2017. Evolution of the South Atlantic lacustrine deposits in response to Early Creatceous rifting, subsidence and lake hydrology. In: Ceraldi, T.S., Hodgkinson, R.A., Backe, G. (Eds.), Petroleum Geoscience of the West Africa Margin, 438. Geological Society London Special Publication, pp. 77–98.

Cope, J.C.W., Duff, K.L., Parsons, C.F., Torrens, H.S., Wimbledon, W.A., Wright, J.K., 1980. A Correlation of the Jurassic Rocks in the British Isles: Part Two: Middle and Upper Jurassic. Blackwell Scientific Publications, p. 109.

Cross, N., Bosence, D.W.J., 2008. 3D tectono-sedimentary models for marine, rift-margin, carbonate systems. In: Lukasik, J., Simo, T. (Eds.), Controls on Carbonate Platform and Reef Development. *Society Economic Paleontologists and Mineralogists*, 89. *Special Publication*, pp. 83–106.

Cross, N. (1996) Sedimentology and sequence stratigraphy of a Miocene fault-block carbonate platform, Gulf of Suez, Egypt. PhD thesis Royal Holloway University of London. 271 pp.

Cross, N.E., Purser, B.H., Bosence, D.W.J., 1998. The tectono-sedimentary evolution of a rift margin carbonate platform: Abu Shaar, Gulf of Suez, Egypt. In: Purser, B.H., Bosence, D.W.J. (Eds.), Sedimentary and Tectonic Evolution of Rift Basins- The Red Sea - Gulf of Aden. Kleuwers, pp. 271–295.

Davison, I., Anderson, L., Nuttall, P., 2012. Salt deposition, loading and gravity drainage in the Campos and Santos basins. In: Alsop, G.I., Archer, S.G., Hartley, A.J., Grant, N.T., Hodgkinson, R. (Eds.), Salt Tectonics Sediments and Prospectivity, 363. Special Publication Geological Society of London, pp. 159–173.

Davison, I., Bosence, D.W.J., Alsop, I., Al-Aawah, M.H., 1996. Deformation and sedimentation around active Miocene salt diapirs on the Tihama Plain, northwest Yemen. In: Alsop, I., Blundell, D.J., Davison, I. (Eds.), Salt Tectonics, 100. Special Publication Geological Society of London, pp. 23–39.

Dullo, W.-C., Montaggioni, L., 1998. Modern Red Sea coral reefs: a review of their morphologies and zonation. In: Purser, B.H., Bosence, D.W. J. (Eds.), *Sedimentary and Tectonic Evolution* of Rift Basins- The Red *Sea* - Gulf *of Aden*. Kleuwers, pp. 583–594.

El-Azabi, M., 2021. Depositional models and sequence stratigraphy of the middle Miocene clastic/carbonate deposits in the Esh El-Mellaha range, southwest Gulf of Suez rift basin, Egypt. Sedimentary Geology 418. Available from: https://doi.org/10.1016/j.sedgeo.2021.105902.

Francis, J.E., 1982. The Fossil Forest of the basal Purbeck Formation (Upper Jurassic) of Dorset, southern England. Palaeobotanical and palaeoenvironmental investigations (PhD thesis). University of Southampton 317.

Gallois, A., Bosence, D.W.J. & Burgess, P. (2018) Brackish to hypersaline facies and facies transition in lacustrine carbonates: Purbeck Limestone Group, Upper Jurassic-Lower Cretaceous, Wessex Basin, Dorset, U.K. *Facies*, 64, 1–39.

Gallois, A., Bosence, D.W.J. & Burgess, P. (2021) Influence of extensional faults and relay ramp on palaeo-topography and lacustrine carbonate facies: Purbeck Limestone Group (Upper Jurassic), Wessex Basin, UK. *Basin Research*, 33, 3159–3182.

Gischler, E., Gibson, M.A., Oschmann, W., 2008. Giant Holocene fresh-water microbialites, Laguna Bacalar, Quintana Roo, Mexico. Sedimentology 55, 1293–1309.

Goldberg, K., Kuchle, J., Scherer, C., Alvarenga, R., Ene, P.L., Armelente, G., et al., 2017. Re-sedimented deposits in the rift section of the Campos Basin. Marine and Petroleum Geology 80, 412–431.

Guardado, L.R., Gamboa, L.A.P., Lucchesi, C.F., 1990. Petroleum geology of the Campos Basin, Brazil, a model for a producing Atlantic type basin. In: Edwards, J.D., Santogrossi, P.A. (Eds.), *Divergent/Passive Margin Basins*. American Association Petroleum Geologists Memoir, 48. pp. 3–80.

Hamblin, R.J.O., Crosby, A., Balson, P.S., Jones, S.M., Chadwick, R.A., Penn, I.E., et al., 1992. *The Geology of the English Channel*. British Geological Survey. United Kingdom Offshore Regional Report. HMSO, London, p. 106.

Hawkes, P.W., Fraser, A.J., Einchomb, C.C.G., 1998. The tectono-stratigraphic development and exploration history of the Weald and Wessex Basins, Southern England, UK. In: Underhill, J.R. (Ed.), The Development, Evolution and Petroleum Geology of the Wessex Basin, 133. Special Publications Geological Society of London, pp. 39–65.

Hay, W.W., Rosol, M.J., Sloan, J.L., 1988. Plate tectonic control of global patterns of detrital and carbonate sedimentation. In: Doyle, L.J., Roberts, H.H. (Eds.), Carbonate-Clastic Transitions. Elsevier, pp. 1–34.

Heaton, R.C., Jackson, M.P.A., Bamahmoud, M., et al., 1996. Superposed Neogene extension, contraction and salt canopy emplacement in the Yemneni Red Sea. In: Jackson, M.P.A., Roberts, D.G., S. Snelson, S. (Eds.), Salt Tectonics A Global Perspective, 65. American Association Petroleum Geologists Memoir, pp. 333–352.

Hendry, J., Burgess, P., Hunt, D., Janson, X. and Zampetti, V. (2021) Seismic characterization of carbonate platforms and reservoirs: an introduction and review. In: Hendry, J., Burgess, P., Hunt, D., Janson, X. and Zampetti, V. (Eds). *Seismic Characterization of Carbonate Platforms and Reservoirs*. Geological Society, London, Special Publications, 509. Available from: https://doi.org/10.1144/SP509-2021-51.

Hughes, G.W., 2014. Micropaleontology and palaeoenvironments of the Miocene Wadi Waqb carbonate of the northern Daudi Arabian Red Sea. GeoArabia 19, 59–108.

Hunt, D., de Luca, P.H.V., Cartesio, L.E.G., Lapponi, F., Masiero, I., Waldum, I., et al. (2019) A very different Barremian-Aptian lacustrine pre-salt facies association; biotic self-organisation in BMC-33, Outer Basin, Campos Basin, Brazil. First EAGE workshop on pre-salt reservoirs: from exploration to production, 5–9 December 2019, Rio de Janeiro, Brazil. 4pp.

James, N.P., Jones, B., 2016. Origin of Carbonate Sedimentary Rocks. John Wiley & Sons, Chichester, p. 446.

James, N.P., Coniglio, M., Aissaoui, D.M., Purser, B.H., 1988. Facies and geological history of an exposed Miocene rift margin carbonate platform, Gulf of Suez, Egypt. American Association Petroleum Geologists Bulletin 72, 555–572.

Lake, S.D. (1985) The structure and evolution of the Wessex Basin. PhD thesis, University of Durham, 390 pp.

Leeder, M., Gawthorpe, R.L., 1987. Sedimentary models for extensional tilt-block/half-graben basins. In: Coward, M.P., Dewey, J.F., Hancock, P.L. (Eds.), *Continental Extensional Tectonics*. Geological Society London, 28. Special Publication, pp. 139–152.

Masiero, I., Burgess, P., Hollis, C., Manifold, L., Gawthorpe, R., Lecomte, I., et al. (2021) Syn-rift carbonate platforms in space and time: testing and refining conceptual models using stratigraphic and seismic numerical forward modelling. In: Hendry, J., Burgess, P., Hunt, D., Janson, X. and Zampetti, V. (Eds). *Seismic Characterization of Carbonate Platforms and Reservoirs*. Geological Society, London, Special Publications, 509. Available from: https://doi.org/10.1144/SP509-2019-217.

Minzoni, M., Cantelli, A., Thorton, J., Wignall, B., 2021. Seismic scale geometries and sequence-stratigraphic architecture of Early Cretaceous sun-rift carbonate systems, Presalt Section, Brazil. In: Hendry, J., Burgess, P., Hunt, D., Janson, X., Zampetti, V. (Eds.), Seismic Characterization of Carbonate Platforms and Reservoirs, 509. Special Publications, LondonGeological Society. Available from: https://doi.org/10.1144/SP509-2019-78.

Moreira, J.L.P., Madiera, C.V., Gil, J.A., Machado, M.A.P., 2007. Bacia de Santos. Boletin Geosciencias Petrobras 15, 531–549.

Muniz, M.C. (2013) Tectono-Stratigraphic evolution of the Barremian-Aptian continental carbonates in the southern Campos Basin, Brazil. PhD thesis, Royal Holloway University of London. 324 pp.

Muniz, M.C. & Bosence, D.W.J. (2015). Pre-salt microbialites from the Campos Basin (offshore Brazil): image log facies, facies model and cyclicity in lacustrine carbonates. In: *Microbial Carbonates in Space and Time: Implications for Global Exploration and Production*. In: Bosence, D. W.J., Gibbons, K., Le Heron, D.P., Morgan, W.A., Pritchard, T.P. & Vining, B.A. (Eds), 418, 221–242, Geological Society London Special Publication.

Muniz, M.C. & Bosence, D.W.J. (2017) Reservoir layering controlled by the taphonomy and taphofacies of coquinas from the early Aptian Campos Basin, Brazil. AAPG, International Conference and Exhibition, October 15–18, London, England, Extended Abstract 2720168, 1–9.

Muniz, M.C., Bosence, D.W.J., 2018. Lacustrine carbonate platforms; facies, and Tectono-sedimentary models for the pre-salt Lagoa Feia Group (Lower Cretaceous), Campos Basin, Brazil. American Association Petroleum Geologists Bulletin 102, 2569–2597.

Olivito, J.P.R., Souza, F.J., 2020. Depositional Model of early Cretaceous lacustrine carbonate reservoirs of the Coqueiros Formation- northern Campos Basin, southeastern Brazil. Marine and Petroleum Geology 111, 414–439.

Orszag-Sperber, F., Harwood, G., Kendall, A.C., Purser, B.H., 1998. A review of the evaporites of the Red Sea – Gulf of Suez rift. In: Purser, B. H., Bosence, D.W.J. (Eds.), Sedimentary and Tectonic Evolution of Rift Basins- The Red Sea - Gulf of Aden. Kleuwers, pp. 409–426.

Pietzsch, R., Oliviera, D.M., Tedeschi, L.R., Neto, J.V.Q., Figueirdo, M.F., Vazquez, J.C., et al., 2018. Palaeohydrology of the Lower Cretaceous pre-salt lacustrine system from rift to post-rift phase, Santos Basin, Brazil. Palaeogeography, Palaeoclimatology and Palaeoecology 507, 60–80.

Platt, N.H., Wright, V.P., 1991. Lacustrine carbonates: facies models, facies distributions and hydrocarbon aspects. In: Anadón, P., Cabrera, L. L., Kelts, K. (Eds.), Lacustrine Facies Analysis, 13. International Association of Sedimentologists Special Publication, pp. 57–74.

Prosser, S., 1993. Rift-related linked depositional systems and their seismic expression. In: Williams, G.D., Dobb, A. (Eds.), Tectonics and Seismic Sequence Stratigraphy, 71. Geological Society Special Publication, pp. 35–66.

Purser, B.H., Plaziat, J.-C., 1998. Miocene periplatform slope sedimentation in the northwestern Red Sea rift, Egypt. In: Purser, B.H., Bosence, D.W.J. (Eds.), Sedimentation and Tectonics of Rift Basins: Red Sea - Gulf of Aden. Kleuwers, pp. 320–346.

Purser, B.H., Barrier, P., Montenat, C., Orszag-Sperber, F., Ott d'Estevou, P., Plaziat, J.-C., et al., 1998. Carbonate and siliciclastic sedimentation in an active tectonic setting: Miocene of the Northwest Red Sea rift Egypt. In: Purser, B.H., Bosence, D.W.J. (Eds.), Sedimentation and Tectonics of Rift Basins: Red Sea - Gulf of Aden. Kleuwers, pp. 239–270.

Purser, B.H., Plaziat, J.-C., Rosen, B.R., 1996. Miocene reefs of the northwestern Red Sea. In: Franseen, E.K., Esteban, M., Ward, W.C., Rouchy, J.M. (Eds.), Models for Carbonate Stratigraphy from Miocene Reef Complexes of Mediterranean Regions, 5. SEPM Concepts in Sedimentology and Paleontology, pp. 347–365.

Roberts, H.H., Murray, S.P., 1983. Gulfs of the northern Red Sea: depositional settings of abrupt siliciclastic-carbonate transitions. In: Doyle, L. J., Roberts, H.H. (Eds.), Carbonate-Clastic Transitions. Elsevier, pp. 99–142.

Saller, A., Rushton, S., Buambua, L., Inman, K., McNeil, R., Dickson, J.A.D., 2016. Presalt stratigraphy and depositional systems in the Kwanza Basin, offshore Angola. American Association of Petroleum Geologists Bulletin 100, 1135–1164.

Spathopoulos, F., 1996. An insight on salt tectonics in the Angola Basin, South Atlantic. In: Alsop, G.I., Blundell, D.J., Davison, I. (Eds.), 100. Salt Tectonics, Geological Society London. Special Publication, pp. 153–174.

Stoneley, R., 1982. The structural development of the Wessex Basin. Journal of the Geological Society 139, 543–554.

Terra, G.J.S., Spadini, A.R., et al., 2010. Classifição de rochas carbonáticas aplicavel à bacias sedimentaires. Boletin Geosciencias Petrobras 18, 9–29.

Thompson, D.L., Stillwell, J.D., Hall, M., 2015. Lacustrine carbonate reservoirs from Early Cretaceous rift lakes of Western Gondwana: pre-salt coquinas of Brazil and West Africa. Gondwana Research 28, 26–51.

Underhill, J.R., Stoneley, R., 1998. Introduction to the development, evolution and petroleum geology of the Wessex Basin. In: Underhill, J.R. (Ed.), The Development, Evolution and Petroleum Geology of the Wessex Basin, 133. Special Publication Geological Society of London, pp. 1–18.

Underhill, J.R., 2002. Evidence for structural controls on the deposition of the Late Jurassic-Early Cretaceous Purbeck Limnestone Group, Dorset, southern England. In: Milner, A.R., Batten, D.J. (Eds.), Life and Environments in Purbeck Times. Palaeontogical Association, London, pp. 21–40.

Watchorn, F., Nichols, G.J., Bosence, D.W.J., 1998. Rift related sedimentation and stratigraphy in southern Yemen. In: Purser, B.H., Bosence, D. W.J. (Eds.), Sedimentation and Tectonics of Rift Basins: Red Sea - Gulf of Aden. Kleuwers, pp. 165–189.

West, I.M., 1992. Contribution on Purbeck Group. In: Cope, J.C.W., Ingham, J.K., Rawson, P.F. (Eds.), Atlas of Palaeogeography and Lithofacies. Geological Society of London, pp. 124–126.

West, I.M., 1975. Evaporites and associated sedimentsof the basal Purbeck Formation (Upper Jurassic) of Dorset. Proceedings of the Geologists' Asoociation 86, 205–225.

Winter, W.R., Jahnert, R.J., França, A.B., 2007. Bacia de Campos. Boletin Geosciencias Petrobras 15, 511–529.

Wright, P., Rodriguez, K., 2018. Reinterpreting the South Atlantic Pre-Salt "Microbialite" reservoirs: petrographic, isotopic and seismic evidence for a shallow evaporitic lake depositional model. First Break 36, 71–77.

Wright, V.P., Barnett, A.J., 2015. An abiotic model for the development of textures in some South Atlantic early Cretaceous lacustrine carbonates. In: Bosence, D.W.J., Gibbons, K., Le Heron, D.P., Morgan, W.A., Pritchard, T.P., Vining, B.A. (Eds.), Microbial Carbonates in Space and Time: Implications for Global Exploration and Production, 418. Geological Society London Special Publication, pp. 209–219.

Ziegler, P.A. (1990). *Geological Atlas of Western and Central Europe*. Shell Internationale Petroleum Maatschappij B.V. 239 pp.

CHAPTER

9

Dniepr-Donets Basin

Sergiy Stovba[1] *and Randell Stephenson*[2]

[1]Institute of Geophysics of National Academy of Sciences of Ukraine, Kyiv, Ukraine [2]School of Geosciences, University of Aberdeen, Aberdeen, Scotland

Introduction

The Dniepr-Donets Basin (DDB) is located in the southeastern part of the East European Craton (EEC) along a NW-SE trending axis between the present-day Ukrainian Shield and Voronezh Massif (Fig. 9.1). It is part of the

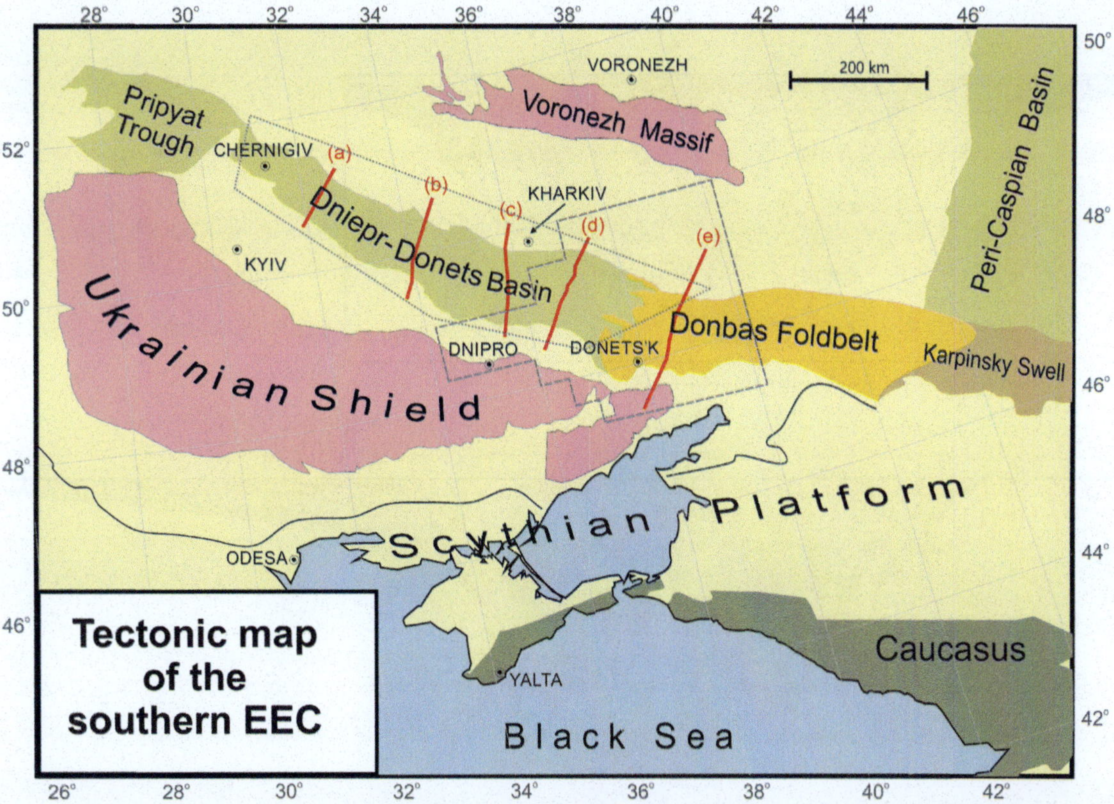

FIGURE 9.1 Regional setting of the DDB (including the Donbas Foldbelt), the darker colors corresponding to the lateral extent of the active Late Devonian rift; postrift sediments overlie the flanks of the rift up to the Ukrainian Shield and Voronezh Massif, which expose the Archean-Paleoproterozoic crystalline basement of the East European Craton (EEC) to the south and the north, respectively. The Scythian Platform overlies the southern margin of the EEC north of the younger Alpine-Tethyan belt (including the Caucasus region). Red lines show the locations of DDB cross-sections in Fig. 9.2 (panels A–E). The gray dashed polygon indicates the location of the map in Fig. 9.4 and the gray dotted polygon indicates the approximate location of the maps in Figs. 9.5 and 9.7. *DDB*, Dniepr-Donets Basin.

Regional Geology and Tectonics: Volume 2
DOI: https://doi.org/10.1016/B978-0-444-64136-6.00004-X

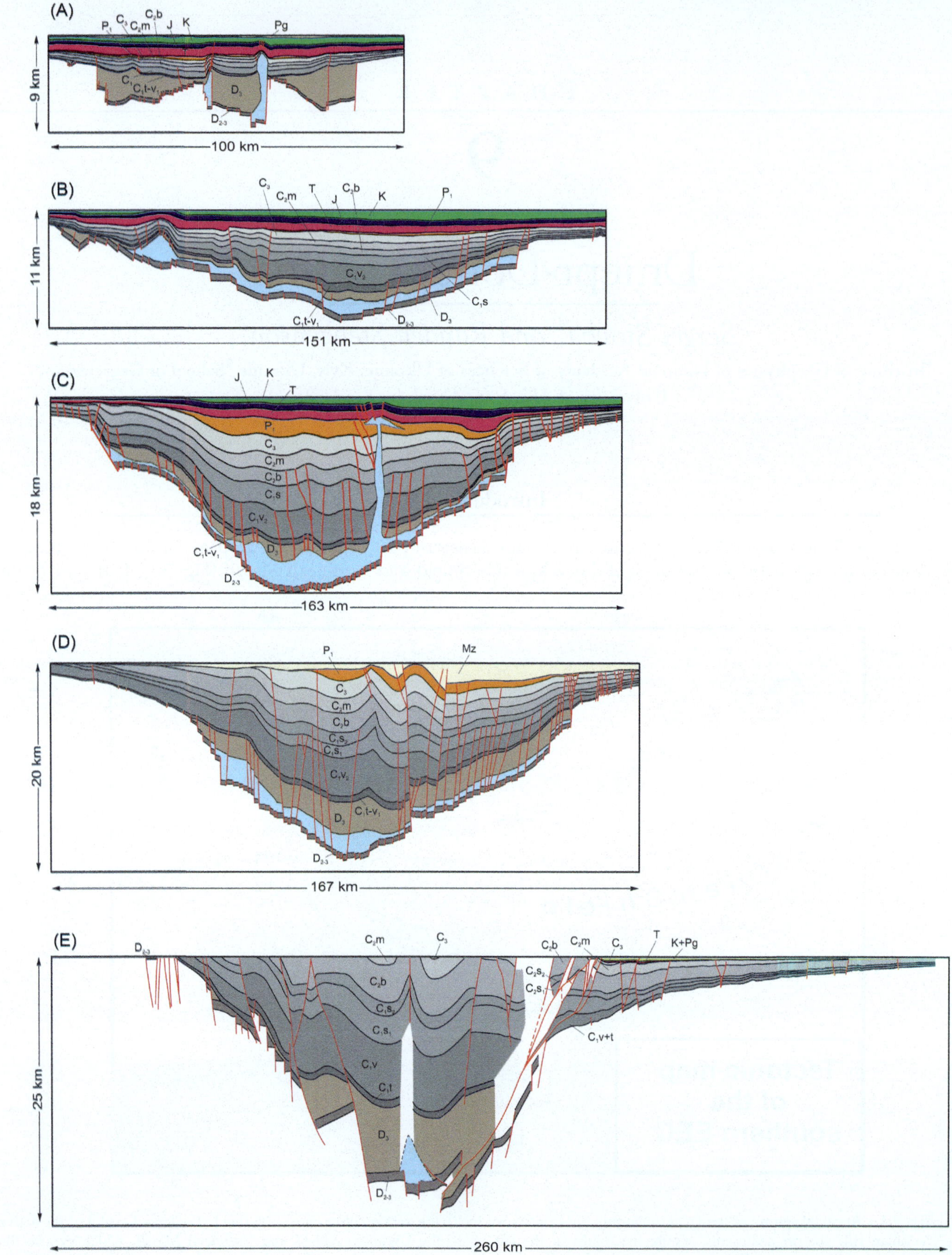

FIGURE 9.2 Structural cross-sections through the DDB, based on the depth-converted versions of interpreted regional seismic reflection profiles (panels A–D) and the DOBRE deep seismic reflection profile (panel E; Maystrenko et al., 2003; Stovba et al., 2005). For locations, see Fig. 9.1. Blue areas without stratigraphic labeling represent salt bodies. The Late Devonian rift (brown layers) has a generally symmetric structure separated from

(Continued)

same rift basin system as the shallower Pripyat Trough to the northwest (mainly in Belarus) and the inverted Donbas Basin (Donbas Foldbelt [DF]) and its southeastward extension to the southeast (straddling the Ukraine–Russia border). The DDB is the most important hydrocarbon province of Ukraine, accounting for some 90% of the Ukraine's oil and gas production. The most recent published estimate, made at the beginning of 2019, of ultimate recoverable reserves in established accumulations (produced and remaining reserves) is 278×10^6 tons of oil, 114×10^6 tons of gas-condensate, and 2414×10^9 m^3 gas (State Balance of Mineral Reserves of Ukraine, 2019). The annual production in 2019 came to some 1.2×10^6 tons of oil, 0.75×10^6 tons of gas-condensate, and 19×10^9 m^3 of gas (National Report of Ukraine, 2019) but this meets only a fraction of the total hydrocarbon requirements of Ukraine.

The DDB is one of the deepest basins of Europe (up to 19 km of sediment fill; Stovba et al., 1996). The Upper Devonian syn-rift sequence, which was deposited in a shallow marine environment and contains significant quantities of salt, is overlain by a thick postrift sequence of Carboniferous age and younger. Structural reactivation, including inversion and salt tectonics, has played an important role in forming the present-day basin architecture. Major postrift structural reactivations took place at the beginning of the late Visean and in the middle of Serpukhovian, in the Early Permian, and between the Mesozoic and Cenozoic (Stovba et al., 1996; Stovba and Stephenson, 1999). Salt movements occurred during the whole history of basin formation starting with the Late Devonian rift stage (Stovba and Stephenson, 2003). Four example cross-sections—drawn from a dense set of interpreted, depth-converted regional seismic profiles acquired in the 1980s and 1990s (Stovba et al., 1996)- plus one deep seismic profile acquired in 2000–01 (Maystrenko et al., 2003; Stovba et al., 2005) crossing the basin, located in Fig. 9.1, are shown in Fig. 9.2.

A number of tectonic subsidence models of the DDB have been published, including 2-D forward models, along profiles in the Dniepr (Kusznir et al., 1996; Starostenko et al., 1999) and Donets (Stovba et al., 2003) segments, and 1- and 2-D inverse models throughout the DDB (van Wees et al., 1996) and in the Dniepr segment (Poplavskii et al., 2001). Although these studies employed a variety of modeling approaches, with significant differences in how they were parameterized, a common result of all of them was that conventional concepts of rift basin formation (involving syn-rift crustal thinning and postrift thermal reequilibration) sufficed to explain the regional basin architecture of the DDB. Some key inferences of these models included the probable role of additional mantle heating ("active" rifting, implying possible mantle plume involvement; van Wees et al., 1996; Stovba et al., 2003), regional uplift before and during rifting (also consistent with mantle plume effects; Kusznir et al., 1996), the importance of postrift tectonic reactivations (van Wees et al., 1996), and the high degree of crustal thinning, almost to continental breakup, in the southeasternmost segment of the basin (Donets segment adjacent to DF; van Wees et al., 1996; Stovba et al., 2003; cf. Stephenson et al., 2021).

Tectono-stratigraphic successions of the Dniepr-Donets Basin

The sedimentary succession of the DDB can be readily subdivided into pre-, syn-, and postrift series, corresponding to prelate Frasnian (D_{2-3}), late Frasnian-Famennian (D_3), and post-Devonian units, respectively (Fig. 9.3). Rifting may have begun slightly earlier in the southeast, propagating northeastwards (cf. Stephenson et al., 2001). The postrift succession is well developed, displaying evidence of multiple extensional reactivations as well as compressional tectonic events. A lack of stratigraphy of suitable age in the DF (due to the uplift, deformation, and erosion affecting it) and, therefore, an absence of diagnostic structural relationships led to uncertainty regarding the timing and nature of postrift tectonic events controlling its development (cf. Fig. 9.4). However, by comparing the exposed and drilled geology of the DF with seismic images from the adjacent Donets segment of the DDB, Stovba and Stephenson (1999) demonstrated that the main preinversion events affecting the DF do not significantly differ from the DDB. This has been confirmed by subsequent structural

◀ the southern and northern rift shoulders by regional, large amplitude boundary faults. *D_{2-3}*, Middle Devonian-Upper Devonian; *D_3*, Upper Devonian; *C_1*, Lower Carboniferous (t, Tournaisian; v, Visean; v_1, lower Visean; v_2, upper Visean; s, Serpukhovian; s_1, lower Serpukhovian; s_2, upper Serpukhovian); *C_2*, middle Carboniferous (b, Bashkirian; m, Moscovian); *C_3*, Upper Carboniferous (Kasimovian and Gzelian); *P_1*, Lower Permian; *Mz*, Mesozoic; *T*, Triassic; *J*, Jurassic; *K*, Cretaceous; *K_2*, upper Cretaceous; *Pg*, Paleogene. *DDB*, Dniepr-Donets Basin. Source: *From al. (2006). Late Palaeozoic intra- and pericratonic basins on the East European Craton and its margins. In: Gee, D.G., Stephenson, R.A. (Eds.), European Lithosphere Dynamics, Geological Society of London, Memoir 32, 463-479; Stovba, S.M, Stephenson, R.A. & Kivshik, M. (1996). Structural features and evolution of the Dniepr-Donets Basin, Ukraine, from regional seismic reflection profiles, Tectonophysics, 268, 127-147; Stovba, S.M & Stephenson, R.A. (1999). The Donbas Foldbelt: its relationships with the uninverted Donets segment of the Dniepr-Donets Basin, Ukraine, Tectonophysics, 313, 59-83.*

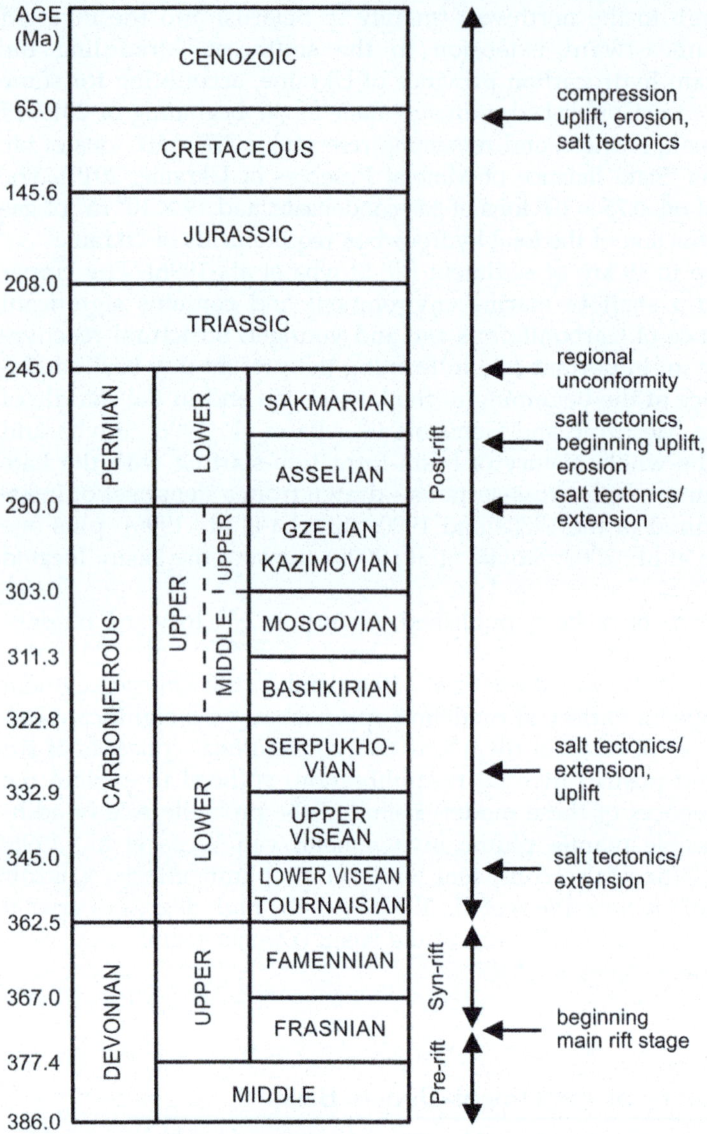

FIGURE 9.3 Summary chrono-stratigraphic column for the Dniepr-Donets Basin showing the main tectonic phases discussed in text. The subdivision of Upper Carboniferous into separate Middle and Upper units corresponds to common usage in Russian and Ukrainian literature. Source: *From Stephenson, R.A., Stovba, S.M. & Starostenko, V.I. (2001). Pripyat-Dniepr-Donets Basin: implications for rift geodynamics and northern Peri-Tethyan tectonic history. Mémoires du Muséum National d'Histoire Naturelle, 186, 369-406.*

studies (Saintot et al., 2003a,b) and the deep seismic DOBREflection profile (e.g., Stovba et al., 2005; cf. Maystrenko et al., 2003; Stephenson et al., 2006; 2021).

Prerift succession

The oldest sediments in the DDB are Eifelian to Middle Frasnian, so-called "undersalt," prerift sediments. These were deposited in platformal terrestrial and shallow marine environments and comprise sandstones, siltstones, clays, and carbonates. These prerift Devonian sediments correlate with equivalent Devonian sequences of the East European Platform (Eisenverg, 1988). They are characterized by homogeneous lithofacies, an average thickness of 300–400 m and a series of stratigraphic gaps, the most significant being between the Eifelian and Givetian and between the Givetian and Frasnian. Thickness variations of the prerift succession are independent of the modern basement relief. The sequence is observed only locally on the rift shoulders. The prerift sediments were probably deposited over a much wider area but were truncated during syn-rift uplift of the rift flanks. Similarly, they are locally absent atop intrabasinal structural highs developed during rifting.

There is no evidence for the presence of a coaxial but narrower Riphean-aged graben underlying the DDB as reported in much of the older literature (e.g., Chekunov et al., 1992). This was based on deep seismic sounding (DSS) velocity models but is not observed on seismic reflection profiles recorded up to 12 s TWT (Stovba et al., 1996). No strata

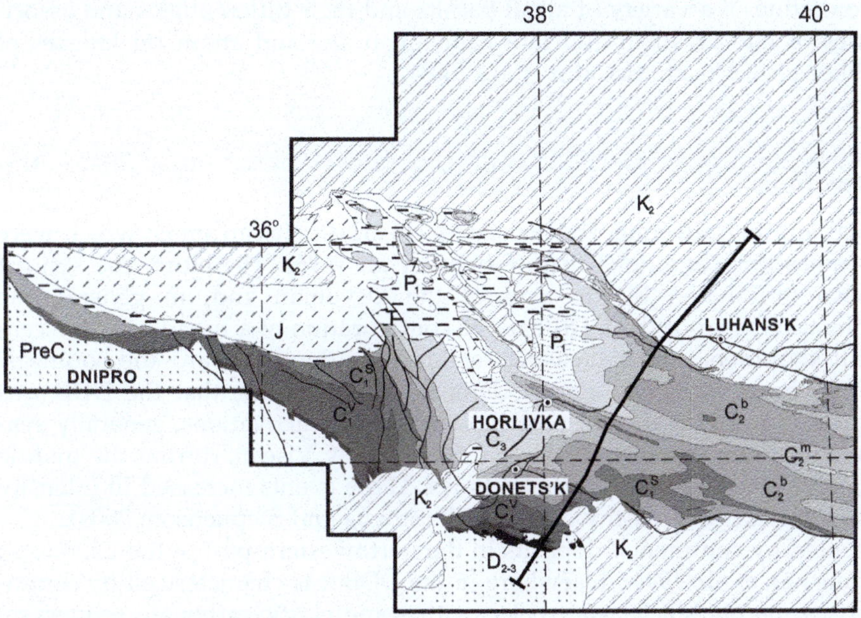

FIGURE 9.4 Cenozoic subcrop map of the Donbas Foldbelt and surrounding areas. For location, see Fig. 9.1. Stratigraphic labels as in Fig. 9.2 (PreC = Precambrian crystalline basement). Also shown is the location of the DOBRE deep seismic reflection profile seen in Fig. 9.2E. Source: *Modified from Stephenson, R.A., Yegorova, T., Brunet, M.-F., Stovba, S., Wilson, M., Starostenko, V., et al. (2006). Late Palaeozoic intra- and pericratonic basins on the East European Craton and its margins. In: Gee, D.G., Stephenson, R.A. (Eds.), European Lithosphere Dynamics, Geological Society of London, Memoir 32, 463-479.*

older than Middle Devonian have been encountered in any of the numerous boreholes that penetrate basement beneath the Paleozoic sediments of the DDB or adjacent Pripyat Trough (Chirvinskaya and Sollogub, 1980; Eisenverg, 1988). Rather, the Devonian–Carboniferous succession revealed by the reflection data is much thicker than inferred from the earlier velocity models and occupies the deeper parts of the section hitherto thought to be Riphean strata (cf. Stovba et al. 1996). Therefore, tectonic models suggesting that a precursor Riphean rift basin was reactivated during the Devonian cannot be supported.

Syn-rift structures and sedimentary succession

Although modified by postrift tectonic and especially salt movements (e.g., Stovba and Stephenson, 2003), the basic architecture of the DDB seen in Fig. 9.2 was developed during its Late Devonian rifting stage. High and laterally variable syn-rift subsidence rates, accompanied by the development of grabens and half-grabens, resulted in a wide range of local depositional environments and considerable paleogeographic heterogeneity (both in space and time), intense volcanism, and multidirectional tectonic movements (Stovba et al., 1996).

The main marginal faults, as well as numerous smaller faults of variable polarity, were activated during the initial rifting stages of the DDB and these cut the basement and overlying prerift Devonian sediments into 2- to 5-km-wide small blocks. Fault throws were in the range 50–300 m with fault orientations ranging from subparallel to highly oblique to the strike of the rift axis. Smaller faults are, in general, beyond the resolution of the seismic data. Subsequently, most syn-rift fault displacement appears to be confined to the main border faults (cf. Stovba et al., 1996) and two or more subparallel faults (cf. Stovba et al., 1996) subdividing the basin into an axial zone and so-called northern and southern "preflank" zones.

The rift margins are characterized regionally either by pervasive faults, exhibiting displacements from hundreds of meters to 2–4 km, or by sets of less-developed faults, displaying various displacements. Major faults dip at 70–80° and, based on DSS studies (Chekunov, 1994), may affect a significant part of the crust, as also seen in deep seismic reflection profiles crossing the related Pripyat Trough basin to the northwest (e.g., Juhlin et al., 1996).

Syn-rift deposits in the DDB reach a maximum thickness of about 4 km (e.g., Stovba et al., 1996; Ulmishek et al., 1994). They overlie the prerift sequence discordantly and are absent on some interior fault blocks due to reduced sedimentation and subsequent erosion. Much of the lower part of the syn-rift sequence consists of Frasnian salt, called the "lower salt," that alternates with clastics and carbonates rocks in a complex laterally variable pattern. The depositional thickness of the Frasnian series is at least 1000 m and reaches a maximum (up to 2 km) in the axial zone of the southeastern part of the DDB. The upper part of the syn-rift series consists of a thinner Famennian "upper salt" that thickens in the northwestern part of the DDB.

Syn-rift volcanic and intrusive rocks, consisting of a variety of alkali basalts and their differentiates and associated pyroclastics, occur in two main series of late Frasnian and late Famennian age and attain thicknesses of more than 2000 m (e.g., Wilson and Lyashkevich, 1996; cf. McCann et al., 2003).

Postrift succession and tectonic reactivations

There is no evidence of syn-depositional faulting affecting Tournaisian and earliest Visean units, which were apparently deposited under tectonically quiescent conditions (e.g., Stovba et al. 1996). Therefore, the syn-rift phase *sensu stricto* is considered to have terminated by the end of the Devonian and, in general, the Carboniferous and younger postrift sedimentary basin of the DDB has the configuration of a broad syncline centered on the rift axis, overlapping the rift shoulders, and increasing in thickness depth toward the southeast (Fig. 9.2). Seismic profiles published by Stovba et al. (1996) clearly demonstrate, however, that the DDB was affected during its Permo-Carboniferous evolution by a series of postrift extensional reactivations, generally synchronous with salt movements but tectonic in origin, namely, at the end of the early Visean, during the middle Serpukhovian, and during latest Carboniferous—earliest Early Permian times. These events increased in intensity toward the southeast, being minor to not observed in the northwestern DDB (Stovba and Stephenson, 2003).

The Carboniferous succession is represented by continental deposits in the northwestern part of the DDB (e.g., Dvorjanin et al., 1996; Izart et al., 1996; Ulmishek et al., 1994). Elsewhere in the DDB it is characterized by continuous rhythmic sedimentation and comprises mainly siliciclastic rocks (with some clastic-carbonate sequences) deposited in shallow marine and lagoonal environments (e.g., van Hinsbergen et al., 2015). There is little variation in the position of the basin depocenter. Only in the axial part of the southeastern DDB, where the Lower Carboniferous includes marine carbonates, did the depth of deposition exceed 200 m. Exposed and drilled Lower Carboniferous sediments in the DF are mainly marine limestones overlain by sandy-clay deposits interbedded by thin coal and limestone beds. Middle and Upper Carboniferous successions are exposed throughout most of the DF and consist mainly of arenaceous-argillaceous rocks interbedded with coal and limestone. With the exception of coal beds and sandy-clay continental intercalations, most were deposited in a shallow marine environment. Carboniferous sediments in the DDB reach a thickness of 11 km, with a maximum depth of their base at about 15 km (Stovba et al., 1996). The present-day maximum thickness of carboniferous sediments in the DF area is about 15 km based on the DOBREflection profile (Fig. 9.2).

Effects of postrift tectonic reactivation can be seen in the DDB in the regional cross-sections (Fig. 9.2). Normal faulting during the early Visean extensional reactivation was accompanied in places by the intrusion of Devonian salt into lower Visean strata (e.g., Stovba and Stephenson, 2003). There is no evidence of uplift, either of the basin or its flanks. Stovba and Stephenson (1999) documented faulting on and in the vicinity of the northern part of the profile seen in Fig. 9.2D, younger than Late Carboniferous and older than Triassic, interpreted to be Early Permian in age, reactivating the boundary fault of the rift. The same fault was active at the end of early Visean and in the middle of Serpukhovian. The stratigraphic thicknesses show that during the remainder of the Carboniferous and during the Mesozoic, no further conspicuous faulting took place. The intensity of each of the Permo-Carboniferous extensional events increases in the DDB southeastwards toward the DF but it is difficult to document in the DF. Saintot et al. (2003b) inferred a transtensional paleostress field in the DF that could be of this age. Additional evidence of Early Permian extensional deformation along the northern margin of the DF, documented widely but generally not in published literature, is presented and discussed by Stovba and Stephenson (1999). Elsewhere in the DF, Upper Cretaceous sediments directly overlie block-faulted and rotated Devonian and Carboniferous strata (Fig. 9.2). The lack of a Permian-Early Cretaceous sedimentary record prevents a definite interpretation of the age of these faults. However, there is no evidence for Mesozoic extensional deformation in the DDB, with the exception of minor, Late Cretaceous normal faults associated with salt movements (see below). Therefore, it is likely that faulting and block rotation seen along the southwestern margin of the DF are part of the widespread phase of Early Permian extension (transtension) seen throughout the DDB (Stovba and Stephenson, 1999; cf. McCann et al., 2003).

The lowermost Lower Permian sediments are represented by monotonous sand-shale series containing rare interbeds of limestones and coals that, similar to the Upper Carboniferous, reflect coastal-continental facies. Asselian sediments consist of five to seven layers of rock salt, separated by clastics and carbonates and also include numerous beds of gypsum, anhydrite, and dolomite. The thickness of salt layers and percent volume increase upward in the section (Eisenverg, 1988). The Sakmarian part of the series consists of a single salt layer likely representing redeposited Devonian salt dissolved from diapirs piercing the depositional surface in the

Early Permian (Stovba and Stephenson, 2003). In the southern preshoulder zone of the DDB, the Lower Permian sequence abruptly decreases in thickness and pinches out as a result of a decrease in depositional thickness as well as subsequent erosion. In contrast, its thickness decrease toward the northern shoulder of the basin is far more gradual. There are no Early Permian sediments preserved within the DF, although Upper Carboniferous and Lower Permian sediments are documented beneath the eastern extension of the northern margin of the DF.

A general absence of Upper Carboniferous and Lower Permian sediments in the northwesternmost part of the DDB can be explained by a decrease in the rate of postrift subsidence within a platform-wide regime of relative sea-level fall. Elsewhere within the DDB, the basin margins, particularly the southern one, were exposed during Early Permian times while the axial part of the basin continued to subside (cf. Fig. 9.2). Uplift of the southern margin of the DDB was very short-lived, lasting no more than 2–3 Myr between the late Asselian and early Sakmarian (Stovba et al., 1996). Extensive erosion occurred with progressively older sediments subcropping beneath the erosion surface in the direction of the Ukrainian Shield; by implication, considerable erosion of the Ukrainian Shield may also have occurred. Locally, more than 2 km of Upper and Lower Carboniferous were eroded at this time and during an ensuing dormant phase, which lasted until the Triassic. The widespread regional Permian unconformity observed throughout the DDB is therefore interpreted to be the result of the Early Permian event followed by a relative sea-level low stand during the later Permian.

Sedimentation resumed in the DDB in the Triassic, a time of tectonic quiescence, rising sea levels, and continuation of postrift subsidence. Most of the Mesozoic succession, comprising both marine and continental sediment sections, occurs throughout the area, overlying the rift axis as well as its flanks. Exceptions are the Upper Triassic and Lower Jurassic units, which occur only in the southeastern part of the DDB, and the Upper Cretaceous marls and chalks, which were eroded from large parts of the southern flank. The Upper Cretaceous succession was, as a whole, characterized by Chirvinskaya and Sollogub (1980) as "close to" platform type, although subsidence coincident with the Devonian rift axis exceeds that of the marginal zones (cf. Fig. 9.2). It is up to 2000 m thick in the central part of the DDB with maximum thicknesses of the Triassic, Jurassic, and Cretaceous units being 900, 700, and 1000 m, respectively.

An angular unconformity between the Triassic and Jurassic has been documented (Eisenverg, 1988). Triassic and Jurassic sequences thin smoothly in the direction of the DF, perhaps due to reduced subsidence, but also because of the later erosional truncation (Stovba and Stephenson, 1999). In the DF, Mesozoic sediments are preserved only along the basin margins. Along the northern margin, 150- to 200-m-thick Triassic sediments occur locally in the vicinity of exposed Paleozoic rocks, while the Jurassic is absent. East of Luhans'k (Fig. 9.4) the Mesozoic section contains only Upper Cretaceous units. Similarly, on the southern margin of the DF, only Upper Cretaceous sediments are preserved. They reach a thickness greater than 500 m and rest on eroded Paleozoic strata or the basement (Fig. 9.4). The present-day limits of the Mesozoic sequences in the DF are clearly erosional, indicating that the original depositional area may have been considerably larger (e.g., Ulmishek et al., 1994). Furthermore, no marginal facies or developments are observed near the erosional edges of the Mesozoic successions. It is therefore likely that the entire area of the DF underwent postrift subsidence during the Mesozoic and that, depending on the relative sea-level variations, the Mesozoic successions were deposited within its confines but were later eroded.

There is little evidence of postrift magmatic activity in the DDB; however, this is not the case for the DF, where igneous rocks of Early Carboniferous, Early Permian, and Mesozoic ages have been reported (Alexandre et al., 2004). Thermal modeling (Shymanovskyy et al., 2004) as well as detailed diagenesis studies in the DDB (Misch et al., 2018) and zircon and apatite fission track data from the DF (Danišík et al., 2009) all suggest moderate to low heat flow in the basin throughout the Mesozoic.

A widespread angular unconformity in the DDB developed at the end of Cretaceous-beginning of Paleogene (Kabyshev et al., 1998). The magnitude of inferred relative uplift increases toward the Ukrainian Shield and, as during the Early Permian, its maximum occurred in the area bordering the DF. In this area, Upper Cretaceous, Jurassic, and Triassic sediments were eroded as seen on the Cenozoic subcrop map (cf. Fig. 9.4). In the axial part of the southeastern DDB, there are local folds, domes, and salt diapirs defining linear trends, which correspond to the trends of the main folds of the DF. Within the DF itself, structural relationships determining the age of formation of folds, thrust, and reverse faults can be observed only near its margins, where Lower Permian, Mesozoic, and Cenozoic sediments are preserved. Stovba and Stephenson (1999) reported that no single geological section could be found in the published literature showing tightly constrained, structurally defined pre-Triassic folding or reverse faulting in the DF. In contrast, where Cretaceous sediments are present, for example, along the northern margin of the DF, reverse faults, and/or folds younger than the Cretaceous sediments and exposed at the surface are relatively common. Reverse faulting of Late Cretaceous age is also evident on the southern margin of the DF (Stovba and Stephenson, 1999; Saintot et al., 2003a,b) determined that the paleostress field associated with compressional structures found in the Cretaceous sediments on the margins of the DF is identical to that recorded by the outcropping Carboniferous sediments. Thus it can be concluded that the inversion of the DDB and formation of the DF occurred mainly in the Late Cretaceous (Stovba and Stephenson, 1999; Stephenson et al., 2001; cf. Danišík et al., 2009). The DOBREflection profile (Fig. 9.2E) shows that the shortening of the DF occurred at the crustal scale as a "mega–pop-up," which involved a major detachment fault through the entire crust and an associated back-thrust (Maystrenko et al., 2003; Stephenson et al., 2006, 2021).

The Cenozoic section of the DDB unconformably overlies Upper Cretaceous and older series and reaches a maximum thickness of 500 m to the northwest DDB (Eisenverg, 1988). The Paleogene consists mainly of sands, clays, and marls, and the Neogene mainly of sands with clayey interbeds.

Salt tectonics in the Dniepr-Donets Basin

The distribution and style of salt kinematics in the DDB and, importantly, its temporal relationship with tectonic events have been described in detail by Stovba and Stephenson (2003). Salt structures in the DDB formed episodically, with flow beginning during the deposition of syn-rift sediments directly overlying the Frasnian salt. At least half of all present-day known salt structures began forming during the rift stage (Stovba and Stephenson, 2003). Primary salt structures also controlled the subsequent development of many salt pillows and diapirs. The distribution of salt structures in the DDB is shown in Fig. 9.5.

Initiation of the postrift phases of salt movements in the DDB is intimately related to regional tectonic reactivations in the basin, as documented by seismic and other data. In particular, the onset of distinct episodes of salt movement coincides with the onset of distinct periods of extensional tectonic reactivation recorded in the late Visean, middle Serpukhovian, and Early Permian, and with compressional tectonic inversion at the end of Cretaceous. Some salt structures display phases of growth beginning with all of the tectonic reactivation events, whereas others display phases of growth beginning with only one or some of them. The intensity of tectonic reactivation in the DDB increases to the southeast toward the DF (Stovba and Stephenson, 1999) and the number of salt structures displaying growth during all periods of tectonic reactivation also increases in this direction. It should be noted, however, that the postrift succession thickness gradient follows the same trend.

Concordant (pillows and anticlines) and discordant (diapirs) salt structures formed during every phase of salt movement in the DDB. Though the onset of these phases during postrift basin evolution coincided with

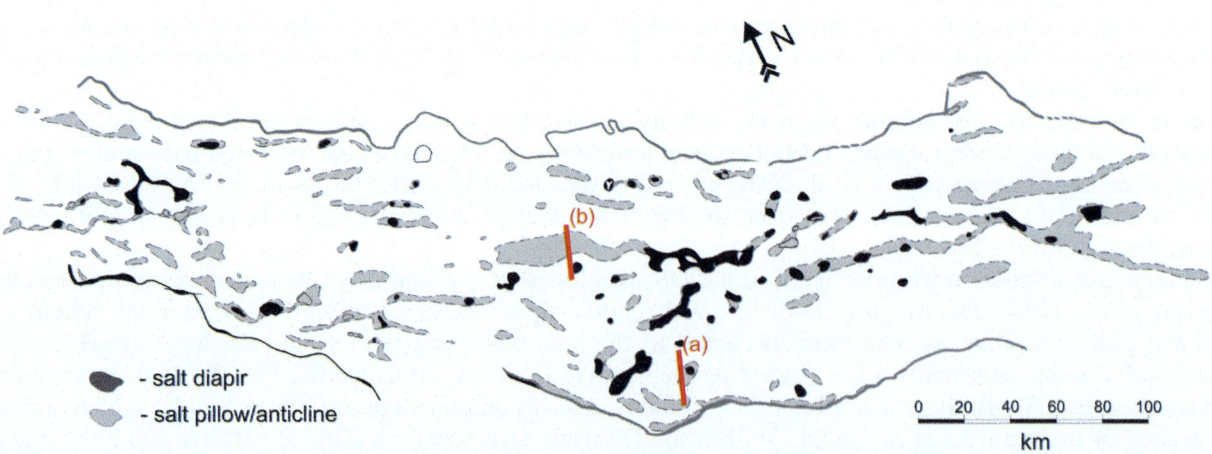

FIGURE 9.5 A simplified map showing the distribution of the main concordant (gray) and discordant (black) salt structures in the DDB, based on seismic and well data. The Devonian salt has been penetrated in boreholes for many of the structures shown. In the axial zone, concordant salt bodies are inferred to correlate with anticlinal structures seen in seismic data and are along strike of known salt diapirs. Red lines with numbers indicate the locations of seismic and geological sections displayed in Fig. 9.6 (panels A and B). *DDB*, Dniepr-Donets Basin. Source: *Modified from Stovba, S.M. & Stephenson, R.A. (2003). Style and timing of salt tectonics in the Dniepr-Donets Basin (Ukraine): implications for triggering and driving mechanisms of salt movement in sedimentary basins. Marine and Petroleum Geology, 19, 1160-1189.*

the onset of tectonic reactivations, the duration of periods of active halokinesis, especially when salt became extrusive, generally exceeded the duration of the corresponding regional tectonic event. Periods of halokinesis were followed by periods of quiescence, during which diapirism was frozen and up to several kilometers of sediments could be deposited above it before the next regional tectonic event triggered renewed salt movement (Stovba et al., 1996).

Fig. 9.5 demonstrates the relationship between salt diapir growth and regional tectonics. The salt structure shown in Fig. 9.6A formed episodically in the late Visean, middle Serpukhovian, and Early Permian. The growth of this structure was accompanied by the development of a rim-syncline. A significant proportion of the upper Visean sediments (up to one-third in the north and up to one-half to the south of the structure) was accumulated within this primary rim-syncline. The subsidence that provided the accommodation space for these Visean sediments was produced by the outflow of Frasnian salt from the salt series itself into the core of the salt anticline. The thinning of the Tournaisian–lower Visean succession toward the top of the structure is related to erosion in the earliest late Visean, occurring simultaneously with the formation of the rim-syncline. The late Visean growth phase was followed by a period of quiescence before renewed uplift after the deposition of the lower Serpukhovian succession. A period of quiescence also preceded the renewal of salt flow after the deposition of Late Carboniferous sediments, inferred to have occurred in the earliest Early Permian (Stovba and Stephenson, 1999). The timing of this tectonic event is more clearly demonstrated in Fig. 9.6B. This concordant salt structure is typical of those forming mainly at the end of Carboniferous–Early Permian. This is evinced by the erosive truncation of Carboniferous sediments on the crest of the structure and by the onlap of Early Permian deposits on the flanks of the structure. Quiescence is indicated until the latest Carboniferous but the structure was active in the Early Permian, the onset of growth coinciding with the onset of latest Carboniferous regional extension, although the duration of salt flow was longer than the duration of the short-lived tectonic event. After a period of quiescence in the Mesozoic, the growth of this structure was renewed in the latest Cretaceous-earliest Tertiary times, when the regional compressional tectonic event affected the DDB (leading to its inversion).

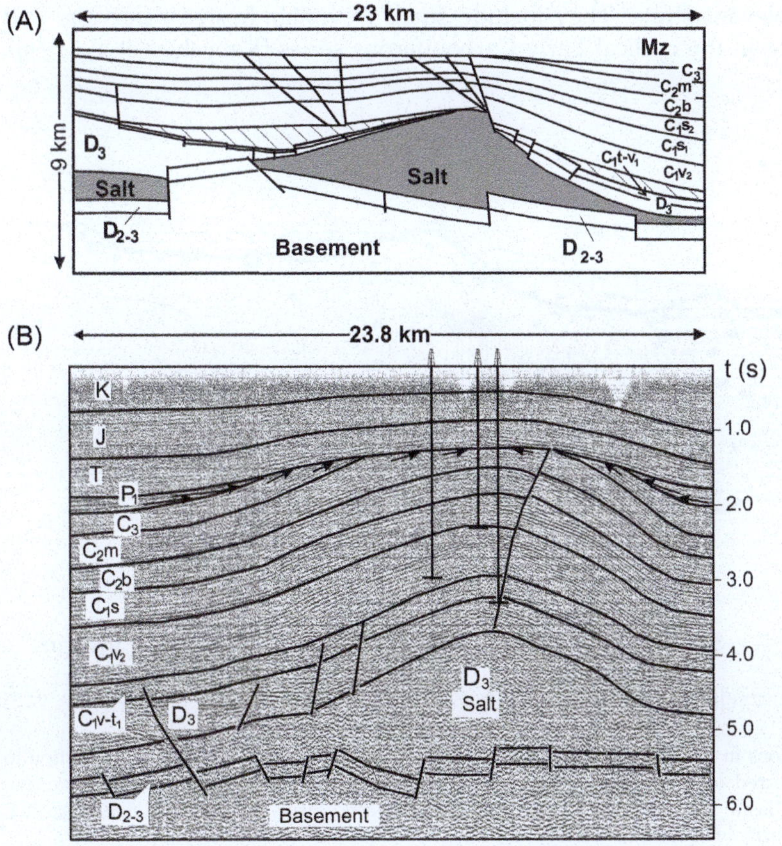

FIGURE 9.6 (A) Depth-converted seismic section (located in Fig. 9.5) showing a salt anticline that formed during three stages of tectonic reactivation in the DDB. The formation of this salt structure was accompanied by the development of primary rim-synclines in the late Visean, Serpukhovian, and at the end of Carboniferous (Early Permian). The crosshatched part of the upper Visean was deposited during outflow of Devonian salt from beneath the primary rim-syncline towards the core of the salt structure. (B) Seismic section (located in Fig. 9.5) showing a salt structure formed mainly at end Carboniferous–Early Permian times with renewed growth in latest Cretaceous-Early Cenozoic time. Stratigraphic symbols are as in Fig. 9.2. DDB, Dniepr-Donets Basin. Source: *From Stovba, S.M. & Stephenson, R.A. (2003). Style and timing of salt tectonics in the Dniepr-Donets Basin (Ukraine): implications for triggering and driving mechanisms of salt movement in sedimentary basins. Marine and Petroleum Geology, 19, 1160-1189.*

Hydrocarbon occurrence in the Dniepr-Donets Basin

Hydrocarbon exploration began in 1936 and has resulted in the discovery of about 200 oil and gas accumulations. Most of these are contained in Carboniferous to Permian series that were deposited during the postrift stage of the basin evolution (Fig. 9.7). The Devonian syn-rift series contains only a small part of the total reserves of the DDB, in contrast to the Pripyat Trough, where the Devonian syn-rift series is the main hydrocarbon producer (Ulmishek et al., 1994). Additional small oil and gas accumulations are stored in Precambrian basement and in Triassic and Jurassic reservoirs. Areas that were strongly affected by compressional tectonic reactivation contain only minor hydrocarbon reserves. The most significant fields are Shebelynka, Zakhidno-Krestyshche, Melychivka, and Jablunivka, the locations of which are shown in Fig. 9.7.

No oil- and gas-generating source rocks within the Devonian succession of the DDB have been reported, although their presence cannot be ruled out (Sachsenhofer et al., 2010). The strata sourcing hydrocarbons comprise mainly marine Tournaisian–early Visean, late Visean, and Serpukhovian shales, which have an average total organic carbon (TOC) content of 1.5%–2.3% and a maximum of up to 5%–6% (Kabyshev et al., 1998). TOC as high as 16% is reported by Sachsenhofer et al. (2010), cf. Schulz et al. (2010) in oil-prone Serpukhovian black shales in the northwestern DDB. The Early Carboniferous sequence in the northwestern part of the basin lies partly in the gas window, the rest being in the oil window, while all of this sequence is in the gas generation window in the southeastern part of the basin. Middle Carboniferous strata, comprising coals and shales, are ubiquitous, have a TOC content in the range of 0.6%–0.9%, and have entered the gas window in the basin center and the oil window along the basin flanks. In the southeastern part of the basin, near the transition to the DF, Lower Serpukhovian and Middle Carboniferous coals are oil-prone and gas-condensate-prone, respectively, and gas in this area may be sourced from the latter (Sachsenhofer et al., 2010). The Late Carboniferous and Early Permian sediments of the axial part of the DDB contain coal and shales with TOC of 0.2%–1.1% that have reached conditions for oil and gas generation only in the deepest, southeastern part of the DDB. Mesozoic source rocks are absent and are immature for the generation of hydrocarbons (Kabyshev et al., 1989, 1998).

"Primary" hydrocarbon accumulations of the DDB, meaning those associated with short lateral and vertical migration distances and having closely associated source rocks, occur in a variety of structural (fault blocks and anticlines), stratigraphic (Devonian reefs and sand pinch-outs), and combination (unconformity) traps that developed during and shortly after the deposition of the relevant strata. They include small Devonian-hosted reservoirs and stacked reservoir/seal pairs (2–6 levels) occurring in the cyclical Early Carboniferous series (Kabyshev et al., 1998).

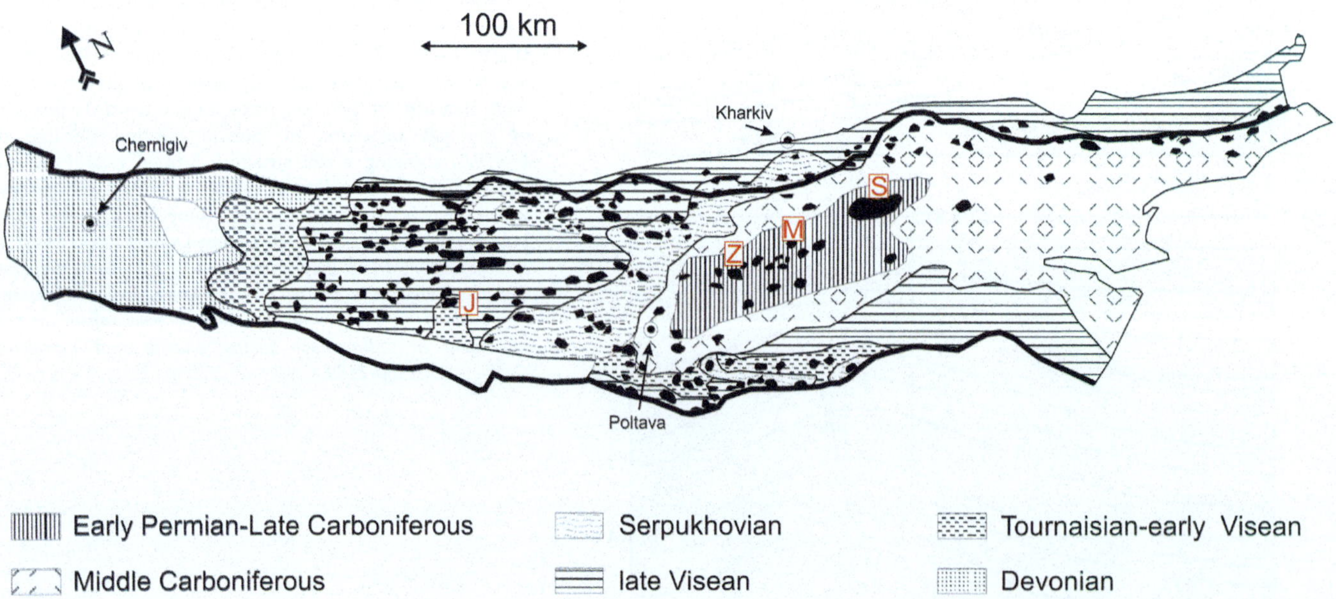

FIGURE 9.7 The distribution of hydrocarbon accumulations in the DDB (oil, gas-condensate, and gas fields shown in black), in relation to the age of the dominant productive formations. Fields referred to in the text are indicated as such: S—Shebelynka; M—Melykhivka; Z—Zakhidno-Krestyshche; J—Jablunivka. *DDB*, Dniepr-Donets Basin. Source: *Modified from Kabyshev, B., Krivchenkov, B., Stovba, S. & Ziegler, P.A. (1998). Hydrocarbon habitat of the Dniepr-Donets Depression. Marine and Petroleum Geology, 15, 177-190.*

"Secondary" accumulations, those associated with a greater degree of vertical migration from deeper levels, are trapped in structures related to the diapiric ascent of Devonian salt that occurred in conjunction with postrift tectonic movements (cf. Stovba and Stephenson, 2003). Salt diapirs associated with hydrocarbon reservoirs in the prolific Late Carboniferous–Early Permian series are broken by faults that have facilitated migration from below. Most of these traps developed during Mesozoic times and are characterized by a high fill ratio (Kabyshev et al., 1998). Mesozoic-hosted reservoirs contain oil and gas that have migrated from older formations through gaps in the regional Permian salt seal. Sealing conditions, in general, are most favorable in the middle parts of the DDB but deteriorate toward the shallower, northwestern parts of the basin, where shales are sandier, and toward its deeper southeastern parts, where shales are more brittle due to their greater degree of diagenesis.

Regional tectonic setting of the Dniepr-Donets Basin

Shatsky (1946) considered the DDB as being a type example of an aulacogen. Clearly being a "failed rift" in the sense that it did not itself lead to continental breakup and ocean crust formation, there remains disagreement on whether the DDB rift represents the failed third arm of a triple-rift system, the other arms of which ultimately formed a passive continental margin (e.g., Saintot et al., 2003b; Zonenshain et al., 1990) or other, subsequently closed, failed rifts (e.g., Stephenson et al., 2021). Indeed, subsequent to Shatsky's work and until quite recently, some authors (e.g., Chekunov et al., 1992; Volozh et al., 1999) have regarded the DDB to be part of a very long, singular intracontinental rift system, including the Karpinsky Swell (on the northern margin of the northern Caucasus domain; Fig. 9.1), that crosses the Caspian Sea and continues into western Asia. Such a model appears, at least in part, to be untenable within a plate tectonic framework.

Related to the question of whether the DDB is a failed arm of a triple-rift junction or not is the age of its initial rifting stage. Shatsky (1946) and later writers considered it to have been initiated as one of a set of Neoproterozoic (Riphean) rifts cross-cutting the EEC, some of which (including the DDB and the closely related Peri-Caspian Basin) were thought to have been very long-lived and only strongly reactivated in Late Paleozoic times. Scattered Neoproterozoic strata are exposed within the Ukrainian Shield (e.g., Chekunov et al., 1992) and Neoproterozoic sediments of continental affinity reportedly underlie Devonian sediments in the northwesternmost part of the DDB rift system and in its farthest southeastern extent, near the northeast end of the Azov Sea but nowhere else (e.g., Stephenson et al., 2001). However, Stovba et al. (1996) clearly showed from regional near-vertical reflection seismic profiling that what was thought to be Neoproterozoic is, in fact, part of the Late Paleozoic (middle Devonian and younger) platformal and rift-related sedimentary sequence. Thus the peri-cratonic rifting that formed the DDB is convincingly not older than Late Paleozoic, followed by several extensional reactivations in Permo-Carboniferous times, as described earlier.

In a pioneering plate tectonic synthesis of former Soviet Union geology, Zonenshain et al. (1990) represented the DDB as the failed arm of a Late Devonian rift system that led to the development of subsequently closed, small ocean basins along the southern margin of the EEC (see also Kostyuchenko et al., 2004; Saintot et al., 2003b; Saintot et al., 2006; Stephenson et al., 2006; Stephenson et al., 2021; Yegorova et al., 2004). In their paleogeographic and paleotectonic reconstructions, Nikishin et al. (1996) appear to hold to this view though it is not explicitly expressed as such. A critical issue in this respect would seem to be the tectonic origins and age of the oldest sediments in the Peri-Caspian Basin (Fig. 9.1). Some authors (e.g., Volozh et al., 2003) argue strongly and without equivocation that these are undoubtedly at least as old as Early Paleozoic based primarily on the presence of sediments of this age on the northern, peri-cratonic platformal margin of the basin. However, others suggest that definitive evidence of the age of these very deeply buried and inaccessible sediments simply does not exist and that a tectonic scenario in which the basin is due to Late Devonian rifting (contemporaneously with the DDB, therefore) fully and more satisfactorily fits available observations than other ones (e.g., Brunet et al., 1999). At best, this remains an open question.

Devonian rifting was accompanied by major magmatic activity (Wilson and Lyashkevich, 1996) and probably by the uplift of the Ukrainian Shield and the Voronezh Massif, which together form a large radius dome that is transected by the DDB (e.g., Stephenson et al., 1993) (Fig. 9.1). Rifting appears to cut across the structural grain of the underlying craton as determined by Shchipansky and Bogdanova (1996) although Stephenson et al. (2021) have inferred from gravity and deep seismic studies that a Precambrian basement boundary might lie in the vicinity of the rift. In any case, major boundaries within the basement are generally thought to provide conduits for magma expulsion during rifting. The geochemical signature of rift-related magma (Wilson and Lyashkevich, 1996) plus the sheer volume of magma suggests that the origin of the DDB is mantle plume related (e.g., Chekunov, 1994), as has also been inferred from various more recent subsidence-modeling studies mentioned earlier (e.g., Kusznir et al., 1996; Poplavskii et al., 2001; Starostenko et al., 1999; van Wees et al., 1996). However, there is no clear indication that the intensity of magmatism increases to the southeast in the direction of the supposed triple junction, although the depth of the basin increases significantly in this direction. Potential field anomalies indicate that mafic rocks have also intruded the crust beneath the sedimentary basin. Magnetic anomalies in the Dniepr segment, indicative of both supracrustal and crustal depth sources, correlate well with igneous bodies known to lie within the sedimentary sequence (Orlyuk and Pashkevich, 1994; Stephenson et al., 2001). Gravity modeling has long suggested the presence of a high-density lower crustal body along the rift axis, interpreted as an "axial dyke" related to rifting (Starostenko et al., 1986). Crustal velocity modeling from wide-angle seismic profiling along the axis of the rift in its northwesternmost segment clearly substantiates this (Starostenko et al., 2018) and Yegorova et al. (1999; 2004) show that this feature extends the entire length of the DDB to its culmination at the Karpinsky swell, taking a 120° bend to the northeast parallel to the present-day recognizable margin of the EEC at this culmination, essentially along the axis of the Peri-Caspian Basin (Fig. 9.1), providing some evidence that this could have been the triple-rift junction locus.

DSS (e.g., Chekunov et al., 1992; Ilchenko, 1996; cf. Stephenson et al., 2006) and more recent wide-angle reflection-refraction (WARR) seismic studies (DOBREfraction'99 Working Group, 2003) show that the amount of crustal thinning beneath the DDB increases to the southeast, concurrently with increasing sedimentary thickness (cf. Stephenson et al., 2001). The most recent profile is DOBRE (DOBREfraction'99 Working Group, 2003), crossing the inverted DF segment of the DDB. The sedimentary basin itself is well-defined, overlying a main crustal layer that thins significantly beneath the main sedimentary depocenter. In turn, a high-velocity lower crustal layer thickens significantly in the same part of the profile. The shape of the sedimentary basin is asymmetric, with the steepest crystalline basement surface on the southwestern margin of the basin, whereas the asymmetry of the high-velocity layer displays its steepest upper surface beneath the northeastern margin of the basin. The high-velocity layer has been interpreted as a zone of crustal underplating or "rift-pillow" (DOBREfraction'99 Working Group, 2003). The asymmetry may be an evidence of some degree of so-called "simple-shear" rather than "pure-shear" during rifting. The Moho depth is more or less constant across the DF (40 ± 2 km). The lower crustal high-velocity layer has presumably incorporated (and been thickened by) mantle-derived magmas during the rifting process. A minimum of 10% but as much as 50% of the high-velocity layer comprises mantle material, implying somewhat more than 50% thinning of the prerift crystalline crust beneath this part of the rift basin and a maximum crustal "stretching factor" of as much as 2.25–2.5 (DOBREfraction'99 Working Group, 2003), a figure that compares favorably with what has been inferred from subsidence modeling in the vicinity of the DOBREfraction profile (Stovba et al., 2003).

References

Alexandre, P., Chalot-Prat, F., Saintot, A., Wijbrans, J., Stephenson, R., Wilson, M., et al., 2004. $^{40}Ar/^{39}Ar$ Dating of magmatic activities in the Donbas Foldbelt and the Scythian Platform (Eastern Europe). Tectonics 23 (5), TC5002. Available from: https://doi.org/10.1029/2003TC001582.

Brunet, M.F., Volozh, Y.A., Antipov, M.P., Lobkovsky, L.I., 1999. The geodynamic evolution of the Precaspian Basin (Kazakhstan) along a north-south section. Tectonophysics 313, 85–106.

Chekunov, A.V., 1994. The geodynamics of the Dniepr-Donets rift syneclise. Geophysical Journal 16 (3), 3–13.

Chekunov, A.V., Gavrish, V.K., Kutas, R.I., Ryabchun, L.I., 1992. Dniepr-Donets paleorift. Tectonophysics 208, 257–272.

Chirvinskaya, M.V., Sollogub, V.B., 1980. Deep Structure of the Dniepr-Donets Aulacogen According to Geophysical Data. Naukova Dumka, Kiev, p. 178.

Danišík, M., Sachsenhofer, R.F., Frisch, W., Privalov, V.A., Panova, E.A., Spiegel, C., 2009. Thermotectonic evolution of the Ukrainian Donbas Foldbelt revisited: new constraints from zircon and apatite fission track data. Basin Research 22, 681–698. Available from: https://doi.org/10.1111/j.1365-2117.2009.00434.x. 1-18.

DOBREfraction'99 Working Group, 2003. "DOBREfraction'99" – velocity model of the crust and upper mantle beneath the Donbas Foldbelt (east Ukraine). Tectonophysics 371, 81–110.

Dvorjanin, E.S., Samoyluk, A.P., Egurnova, M.G., Zaykovsky, N.Y., Podladchikov, Y., Van Den Belt, F.J.G., et al., 1996. Sedimentary cycles and paleogeography of the Dniepr Donets Basin during the late Visean-Serpukhovian based on multiscale analysis of well logs. Tectonophysics 268, 169–187.

Eisenverg, D.E. (Ed.), 1988. Geology and Oil and Gas Occurrences of the Dniepr-Donets Depression: Stratigraphy. Naukova Dumka, Kiev, p. 148.

Ilchenko, T., 1996. The Dniepr-Donets Rift: deep structure and evolution from DSS profiling. Tectonophysics 268, 83–98.

Izart, A., Briand, C., Vaslet, D., Vachard, D., Coquel, R., Maslo, A., 1996. Stratigraphy and sequence stratigraphy of the Moscovian of the Donets Basin. Tectonophysics 268, 189–209.

Juhlin, C., Stephenson, R.A., Klushin, S., 1996. Reappraisal of deep seismic reflection Profile VIII across the Pripyat Trough. Tectonophysics 268, 99–108.

Kabyshev, B., Krivchenkov, B., Stovba, S., Ziegler, P.A., 1998. Hydrocarbon habitat of the Dniepr-Donets Depression. Marine and Petroleum Geology 15, 177–190.

Kabyshev, B.P., et al., 1989. Geology and Oil and Gas habitat in the Dniepr-Donets Depression. Academy Nauk Ukrainskoy CCP. Institute of Geology. Kiev Naukova Dumka 201.

Kostyuchenko, S.L., Morozov, A.F., Stephenson, R.A., Solodilov, L.N., Vedrentsev, A.G., Popolitov, K.E., et al., 2004. The evolution of the southern margin of the East European Craton based on seismic and potential field data. Tectonophysics 381, 101–118.

Kusznir, N.J., Stovba, S.M., Stephenson, R.A., Poplavskii, K.N., 1996. The formation of the northwestern Dniepr-Donets Basin: 2-D forward and reverse syn-rift and post-rift modelling. Tectonophysics 268, 237–256.

Maystrenko, Y., Stovba, S., Stephenson, R., Bayer, U., Menyoli, E., Gajewski, D., et al., 2003. Crustal-scale pop-up structure in cratonic lithosphere: DOBRE deep seismic reflection study of the Donbas Foldbelt, Ukraine. Geology 31, 733–736.

McCann, T., Saintot, A., Chalot-Prat, F., Kitchka, A., Fokin, P., Alekseev, A., et al., 2003. Evolution of the southern margin of the Donbas (Ukraine) from Devonian to Early Carboniferous times. In: McCann, T., Saintot, A. (Eds.), Tracing Tectonic Deformation Using the Sedimentary Record. Geological Society, 208. Special Publications, London, pp. 117–135.

Misch, D., Wegerer, E., Gross, D., Sachsenhofer, R.F., Rachetti, A., Gratzer, R., 2018. Mineralogy and facies variations of Devonian and Carboniferous shales in the Ukrainian Dniepr-Donets Basin. Austrian Journal of Earth Sciences 111, 15–25. Available from: https://doi.org/10.17738/ajes.2018.0002.

National Report of Ukraine. (2019). Extractive Industries Transparency Initiative, Kyiv, 511 pp. https://eiti.org/files/documents/ukraine_report-eiti_en_2019_final.pdf.

Nikishin, A.M., Ziegler, P.A., Stephenson, R.A., Cloetingh, S.A.P.L., Furne, A.V., Fokin, P.A., et al., 1996. Late Precambrian to Triassic history of the East-European craton: dynamics of sedimentary basin evolution. Tectonophysics 268, 23–63.

Orlyuk, M.I., Pashkevich, I.K., 1994. Evaluation of consolidated crust magnetization of the Dniepr-Donets aulacogen. Doklady of the Academy of Sciences if Ukraine 5, 125–128.

Poplavskii, K.N., Podladchikov, Yu, Yu, Stephenson, R.A., 2001. Two-dimensional inverse modeling of sedimentary basin subsidence. Journal of Geophysical Research 106 (B4), 6657–6672.

Sachsenhofer, R.F., Shymanovskyy, V.A., Bechtel, A., Gratzer, R., Horsfield, B., Reischenbacher, D., 2010. Palaeozoic source rocks in the Dniepr-Donets Basin, Ukraine. Petroleum Geoscience 16, 377–399. Available from: https://doi.org/10.1144/1354-079309-032.

Saintot, A., Stephenson, R.A., Brem, A., Stovba, S.M., Privalov, V., 2003b. Palaeostress field reconstruction and revised tectonic history of the Donbas fold-and-thrust belt (Ukraine and Russia). Tectonics 22, 1059. Available from: https://doi.org/10.1029/2002TC001366.

Saintot, A., Stephenson, R.A., Stovba, S., Brunet, M.-F., Yegorova, T. & Starostenko, V. (2006). The evolution of the southern margin of Eastern Europe (Eastern European and Scythian platforms) from the latest Precambrian-Early Palaeozoic to the Early Cretaceous. In: Gee, D.G., Stephenson, R.A. (Eds.), European Lithosphere Dynamics, Geological Society of London, Memoir 32, 481-505.

Saintot, A., Stephenson, R.A., Stovba, S.M., Maystrenko, Y., 2003a. Structures associated with inversion of the Donbas Foldbelt. Tectonophysics 373, 181–207.

Schulz, H.-M., Horsfield, B., Sachsenhofer, R.F., 2010. Shale gas in Europe: a regional overview and current research activities. Geological Society, London, Petroleum Geology Conference series 7, 1079–1085. Available from: https://doi.org/10.1144/0071079.

Shatsky, N.S., 1946. The great Donets basin and the Wichita system; comparative tectonics of ancient platforms. *Akademiya Nauk SSSR Izvestiya*. Seriya Geologicheskaya 6, 57–90.

Shchipansky, A., Bogdanova, S., 1996. The Sarmatian crustal segment: Precambrian correlation between the Voronezh Massif and the Ukrainian Shield across the Dniepr-Donets Aulacogen. Tectonophysics 268, 109–126.

Shymanovskyy, V.A., Sachsenhofera, R.F., Izart, A., Li, Y., 2004. Numerical modelling of the thermal evolution of the northwestern Dniepr–Donets Basin (Ukraine. Tectonophysics 381, 61–79. Available from: https://doi.org/10.1016/j.tecto.2003.10.013.

Starostenko, V., Janik, T., Yegorova, T., Czuba, W., Sroda, P., Lysynchuk, D., et al., 2018. Lithospheric structure along wide-angle seismic profile GEORIFT 2013 in Pripyat-Dniepr-Donets Basin (Belarus and Ukraine). Geophysical Journal International 212, 1932–1962. Available from: https://doi.org/10.1093/gji/ggx509.

Starostenko, V.I., Danilenko, V.A., Vengrovitch, D.B., Kutas, R.I., Stovba, S.M., Stephenson, R.A., et al., 1999. A new geodynamical-thermal model of rift evolution, with application to the Dniepr-Donets Basin, Ukraine. Tectonophysics 313, 29–40.

Starostenko, V.I., Kozlenko, V.G., Oganesyan, S.M., Shen, E.L., Oganesyan, M.G., Yegorova, T.P., et al., 1986. 3-D distribution of the density in the crust of the Dniepr Graben. Geophysical Journal 8, 3–19.

State Balance of Mineral Reserves of Ukraine (2019). Geoinform of Ukraine, Kyiv, Volumes 23, 47, 60, 1591 pp.

Stephenson, R., Yegorova, T., Stovba, S., 2021. An investigation of how intracratonic rifting is “seeded”: case study of the Late Devonian Dniepr-Donets Basin rift within the East European Craton. Precambrian Research 362, 106305. Available from: https://doi.org/10.1016/j.precamres.2021.106305.

Stephenson, R.A., Chekunov, A., Ilchenko, T., Kaluzhna, L., Baranova, Y., Starostenko, V., et al., 1993. Continental rift development in Precambrian and Phanerozoic Europe: EUROPROBE and the Dniepr-Donets rift and Polish Trough basins. Sedimentary Geology 86, 159–175.

Stephenson, R.A., Stovba, S.M., Starostenko, V.I., 2001. Pripyat-Dniepr-Donets Basin: implications for rift geodynamics and northern Peri-Tethyan tectonic history. Mémoires du Muséum National d’Histoire Naturelle 186, 369–406.

Stephenson, R.A., Yegorova, T., Brunet, M.-F., Stovba, S., Wilson, M., Starostenko, V., et al. (2006). Late Palaeozoic intra- and pericratonic basins on the East European Craton and its margins. In: Gee, D.G., Stephenson, R.A. (Eds.), European Lithosphere Dynamics, Geological Society of London, Memoir 32, 463-479.

Stovba, S.M., Stephenson, R.A., 1999. The Donbas Foldbelt: its relationships with the uninverted Donets segment of the Dniepr-Donets Basin, Ukraine. Tectonophysics 313, 59–83.

Stovba, S.M., Stephenson, R.A., Kivshik, M., 1996. Structural features and evolution of the Dniepr-Donets Basin, Ukraine, from regional seismic reflection profiles. Tectonophysics 268, 127–147.

Stovba, S.M., Stephenson, R.A., 2003. Style and timing of salt tectonics in the Dniepr-Donets Basin (Ukraine): implications for triggering and driving mechanisms of salt movement in sedimentary basins. Marine and Petroleum Geology 19, 1160–1189.

Stovba, S.M., Maystrenko, Yu.P., Stephenson, R.A., Kusznir, N.J., 2003. The formation of the south-eastern part of the Dniepr-Donets basin: 2-D forward and reverse syn-rift and post-rift modelling. Sedimentary Geology 156, 11–33.

Stovba, S.M., Tolkunov, A.P., Stephenson, R.A., 2005. Regional investigations of geological structure and evolution of the Donbas Foldbelt (DOBRE international project). Problems of the Petroleum and Natural Gas Industry. Kyiv 21–34.

Ulmishek, G.E., Bogino, V.A., Keller, M.B., Poznyakevich, Z.L., 1994. Structure, stratigraphy and petroleum geology of the Pripyat and Dniepr-Donets basins, Belarus and Ukraine. AAPG Memoir 59, 125–156.

van Hinsbergen, D.J.J., Abels, H.A., Bosch, W., Boekhout, F., van der Meer, D.G., Kitchka, A., et al., 2015. Sedimentary geology of the Middle Carboniferous of the Donbas region (Dniepr-Donets Basin, Ukraine. Scientific Reports (Nature Publishing) . Available from: https://doi.org/10.1038/srep09099.

van Wees, J.D., Stephenson, R.A., Stovba, S.M., Shimanovsky, V., 1996. Tectonic variation in the Dniepr-Donets Basin from automated modelling of backstripped subsidence curves. Tectonophysics 268, 257–280.

Volozh, Yu.A., Antipov, M.P., Brunet, M.-F., Garagash, I.A., Lobkovskii, L.I., Cadet, J.-P., 2003. Pre-Mesozoic geodynamics of the Precaspian Basin (Kazakhstan). Sedimentary Geology 156, 35–58.

Volozh, Yu.A., Antipov, M.P., Leonov, Yu.G., Morozov, A.F., Yurov, Yu.G., 1999. Setting of the Karpinsky Swell. Geotectonics 1, 28–43.

Wilson, M., Lyashkevich, Z.M., 1996. Magmatism and the geodynamics of rifting of the Pripyat-Dniepr-Donets rift, East European Platform. Tectonophysics 268, 65–81.

Yegorova, T.P., Stephenson, R.A., Kostyuchenko, S.L., Baranova, E.P., Starostenko, V.I., Popolitov, K.E., 2004. Structure of the lithosphere below the southern margin of the East European Craton (Ukraine and Russia) from gravity and seismic data. Tectonophysics 381, 81–100.

Yegorova, T.P., Stephenson, R.A., Kozlenko, V.G., Starostenko, V.I., Legostaeva, O.V., 1999. 3-D gravity analysis of the Dniepr-Donets Basin and Donbas Foldbelt, Ukraine. Tectonophysics 313, 41–58.

Zonenshain, L.P., Kuzmin, M.I., Natapov, L.M., 1990. Geology of the USSR: a Plate Tectonic Synthesis, Geodynamics Series, 21. American Geophysical Union, p. 242.

CHAPTER

10

Sedimentary basins of the late Mesozoic extensional domain of China and Mongolia ☆

Justin Edward Tully[1], *Cari Lynne Johnson*[1], *Stephan Alan Graham*[2], *Tim Cope*[3] *and Bradley David Ritts*[4]

[1]Department of Geology and Geophysics, University of Utah, Salt Lake City, UT, United States [2]Department of Earth and Planetary Sciences, Stanford University, Stanford, CA, United States [3]Department of Geosciences, DePauw University, Greencastle, IN, United States [4]Office of Research, University of Idaho, Moscow, ID, United States

A major intracontinental extensional province, more than twice the size of the Cenozoic Basin and Range Province of North America, developed in east Asia during late Mesozoic time. Like the Basin and Range, the east Asian province features numerous extensional basins that developed in both low-strain and high-strain settings within heterogeneous crust that had experienced prior shortening. The low-strain basins comprise isolated or linked half-grabens that are characterized by fluvial-lacustrine facies assemblages and internal drainage. In northeastern China, normal faults that bound numerous Early Cretaceous half-graben basins sole into Jurassic thrust faults at depth, demonstrating that the geometry of low-strain extension was strongly influenced by the prerift basement structure. Across Mongolia, from at least 98° longitude eastward, lacustrine-sourced petroliferous basin areas formed from Late Jurassic–Early Cretaceous syn-rift depocenters that locally coalesced due to progressive fault linkage. An Upper Cretaceous postrift succession was deposited following an episode of mid-Cretaceous transpression that inverted many of these basins. High-strain basins associated with Early Cretaceous metamorphic core complexes constitute supradetachment basin systems typically filled with thin (<1.5 km) successions of dominantly coarse-grained alluvial strata derived from the footwalls of successive detachment splays.

Multiple driving mechanisms have been implicated in this vast extensional event, the understanding of which remains an ongoing topic of research. However, it is generally recognized that some combination of at least three fundamental geodynamic processes were likely acting concurrently to enable such widespread extension. Briefly (as discussed further in the implications section), these include Late Jurassic–Early Cretaceous gravitational collapse of overthickened crust of early Mesozoic orogens in the continental interior; postcollisional dynamics of the Mongol-Okhotsk Ocean closure and relict slab break off beneath the resulting orogen/suture zone; and evolution of the Asian Pacific margin, particularly Paleo-Pacific flat-slab subduction and rollback cycles capable of driving complex deformation in eastern Asia since at least the Triassic, at times more than 1000 km inboard from the plate margin.

Introduction

Much of eastern China and eastern Mongolia was extended during the Late Jurassic and Early Cretaceous. As noted by various authors (Li et al., 1995; Ren and Li, 1998), the extended area of eastern Asia was comparable in several ways with the Cenozoic Basin and Range Province of North America: it was characterized by structurally controlled topography, dominated by normal faults and subsidiary strike-slip faults, and it featured basins dominated by

☆ In this chapter, outline to be updated with new chapter number.

Regional Geology and Tectonics: Volume 2
DOI: https://doi.org/10.1016/B978-0-444-64136-6.00002-6

facies characteristic of internal drainage. However, it differed in size, forming over an area more than twice as vast (Fig. 10.1). The Asian extensional basin system formed over an ~55 my period, between Kimmeridgian and Albian time, with a period of widespread high-strain extension that lasted for 15–20 my (~135–115 Ma). Within that greater period of extension, a general eastward-younging trend in basin formation is recognized (Liu et al., 2021a; Traynor and Sladen, 1995), whereas the North American Basin and Range system has been extending for over 40 my, with a period of widespread high-strain extension for c.20 my (Dickinson, 2002). Recent decades of investigation have revealed that the history and character of the North American Basin and Range system are complex. Extension and associated magmatism have been heterogeneous in time and space (Burchfiel et al., 1992; Dickinson, 2002; Long, 2019). Early in the history of the province, some regions were characterized by remarkable zones of high-strain extension, whereas other areas were substantially unaffected (McQuarrie and Wernicke, 2005). In the latest Cenozoic, high rates and magnitudes of strain have been less important, and more conventional rift structuring has been typical.

The region typified by late Mesozoic extension in eastern Asia is equally complicated. Basins of the region vary greatly in size and include conventional low-strain extensional basins, transtensional basins, and rapidly filled basins associated with high-strain extensional metamorphic core complexes (Figs. 10.1 and 10.2). Magmatism was widespread across the Mesozoic extensional province. Latest Jurassic–Early Cretaceous bimodal volcanism and intrusion complexes have been reported from the westernmost extensional basins in the Eastern Valley of Lakes in the north of the Gobi-Altai region, across the East Gobi-Hailar/Tamtsag-Erlian basin belts, and ultimately to the Paleo-Pacific margin-controlled basins (Fig. 10.2) (e.g., Songliao, Bohai Bay, and Sanjiang basins; Li et al., 2015; Wang et al., 2016; Zhang et al., 2015). Regionally, the most abundant magmatism is reported between 135 and 115 Ma, but overall magmatism was occurring from 180 to 80 Ma in various locations (Buriánek et al., 2008; Daoudene et al., 2012; Feng and Graham, 2023; Graham et al., 2001; Liu et al., 2021b, 2021a; Meng et al., 2003). Along the Paleo-pacific continental margin, an inboard sweep of magmatism occurred between 165 and 140 Ma, followed by a trenchward retreat from 140 to 110 Ma, and this is one of the main arguments in support of late Mesozoic flat-slab subduction along the entire Chinese portion of the Paleo-Pacific continental margin (Liu et al., 2021a).

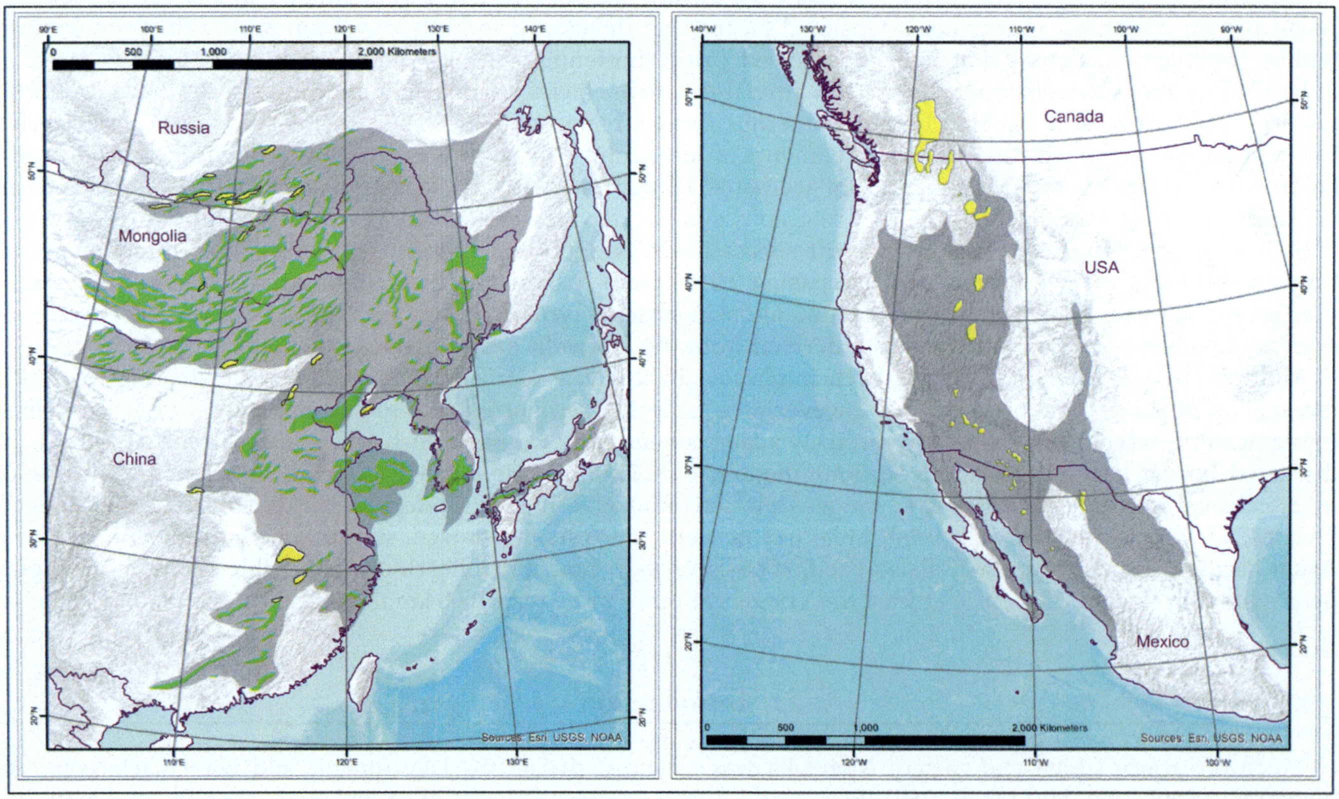

FIGURE 10.1 Same-scale comparison of the Mesozoic extensional domain in Asia (left, ~4.8 million km^2) and the North American Basin and Range (right, ~1.8 million km^2). Also shown are continental extension envelopes (grey), metamorphic core complexes (yellow), and syn-extensional basin depocenters of Asia (green). Asian basin outlines and metamorphic core complexes are modified from Wang et al., 2011 and Daoudene et al., 2017. North American metamorphic core complexes are modified after Gebelin et al., 2011.

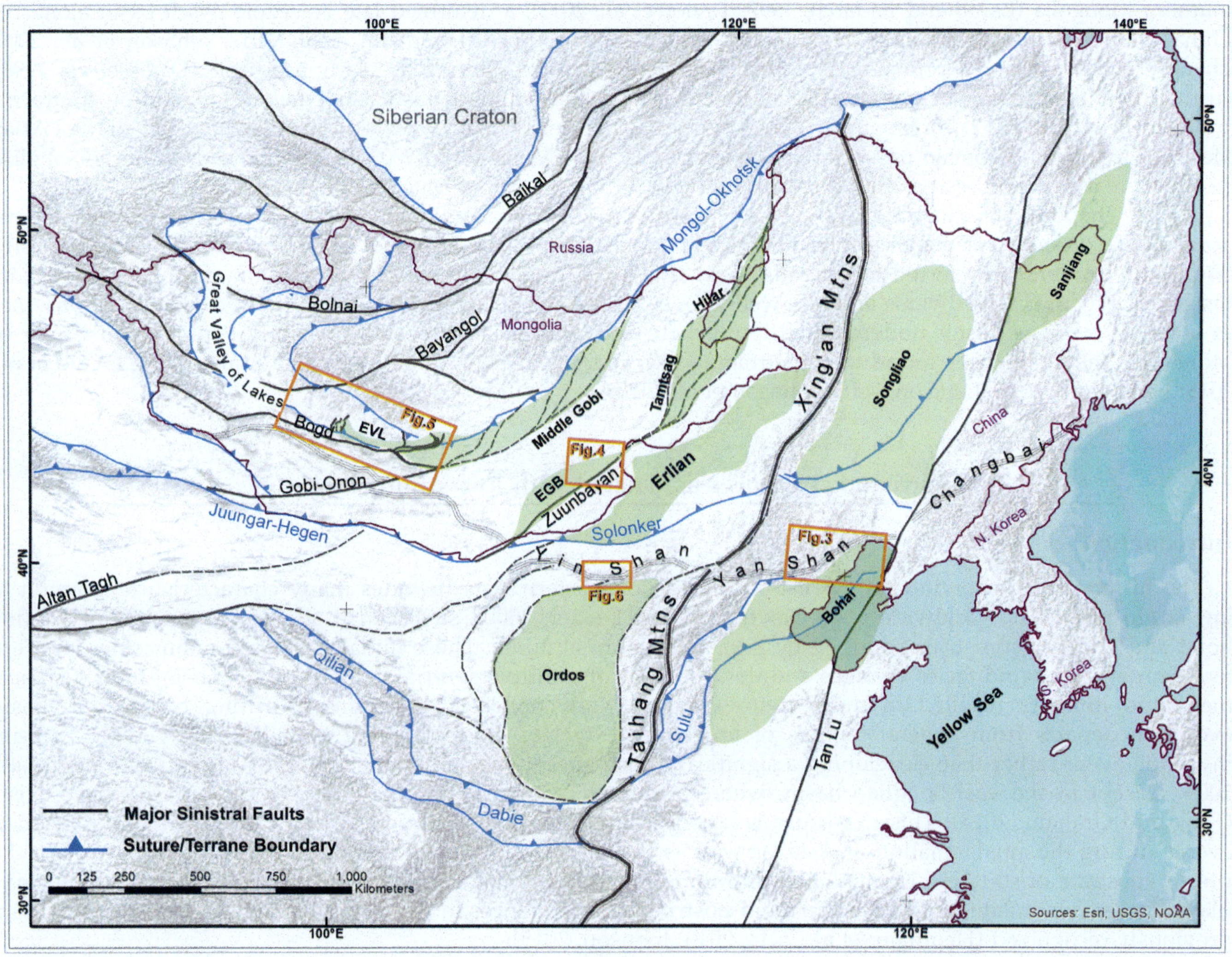

FIGURE 10.2 Reference map showing key structural elements and Jurassic-Cretaceous basins of northeastern Asia. International borders shown in purple.

Coeval with this (135–115 Ma) flux of magmatism was the widespread emplacement of metamorphic core complexes spanning eastern Asia from Siberia to South China and more than 2000 km E-W from the Pacific margin (Fig. 10.1). The core complexes were predominantly emplaced along major lithospheric boundaries of the Asian continental collage (Daoudene et al., 2017; Li et al., 2019; Wang et al., 2011). Break off of a relict subducted slab and the resulting aesthenosphere–lithosphere interactions are thought to have mobilized outward flow of the middle lithosphere along the Mongol-Okhotsk Suture Zone and its adjacent orogen (Meng, 2003; Van der Voo et al., 1999, 2015). This outflow of the mid-lithosphere combined with and/or aided by flat-slab foundering, rollback, and extension at the Pacific margin coincided with widespread metamorphic core complex emplacement over a narrow 15–20 my period (~135–115 Ma; Fig. 10.1).

Temporal variations in styles and rates of extension also characterize the east Asia extensional domain. The sheer geographic scale of the two intracontinental regions indicates that plate-scale tectonic processes have been operative. Nevertheless, important differences exist between the Asian and North American extensional provinces. The Asian basins display a remarkable diversity of sizes, in part reflecting an underlying control on basin size and bounding faults by preexisting crustal structure. Certainly, basin-bounding normal faults in America have been demonstrated to reactivate Mesozoic and early Cenozoic thrust faults (e.g., Gries, 1983). In contrast, in eastern Asia, the prerift substrate, an amalgam of crustal materials of variable dimensions, uncratonized and significantly shortened only a few tens of million years before extension (e.g., Johnson et al., 2001; Zheng et al., 1991), exerted a fundamental control on rifting. As a first approximation, the region is underlain, from north to south, by accreted Paleozoic arc terranes in southern Mongolia (Lamb and Badarch, 1997; Taylor et al., 2013); a deformed late Paleozoic flysch basin along the China–Mongolia border marking the Junggar–Hegen–Solon suture (Heumann et al., 2012; Johnson et al., 2008; Zhang et al., 1984); and North China cratonal basement tectonized during the early Mesozoic (Davis et al., 1998). Furthermore, the puzzling, close geographic, and temporal bracketing of the Asian extended domain by contractile orogenic belts (Fig. 10.2) differs from the American Basin and Range Province. Finally, the crust of the Asian Mesozoic extensional domain remains thick, as emphasized by Meng et al. (2003), except beneath the large Songliao basin where the Moho lies at about 30 km (including basin fill) (Fig. 16 of Feng and Graham, 2023).

In this chapter, we illustrate representative structures and basins of the east Asia Mesozoic extensional domain. We do not treat the largest basins of the region in detail, the Songliao, Erlian, and Hailar-Tamsag basins (Fig. 10.2). Readers are referred to Feng and Graham (2023) in this volume for a discussion of the Songliao basin. Instead, we focus on the diverse and less well-known basins and structural features of the province. The province as a whole remains vastly understudied, but where such basins have been sufficiently explored (in China or Mongolia), they have been found to be petroleum charged (e.g., Davaa, 2010; Feng and Graham, 2023; Guo et al., 2019; Jiang et al., 2016; Johnson, 2004; Johnson et al., 2015).

Basins and structures associated with low-strain extension

Introduction

The history of intracontinental extension in Mesozoic eastern Asia includes many characteristics common to low-strain provinces worldwide (e.g., Gawthorpe and Leeder, 2000), such as rift structures that broadly define horst and graben geometries bounded by high-angle normal faults. These initially isolated nonmarine depocenters demonstrate rapid facies changes and development of multiple syn-rift sequences. Increasing fault displacement and linkages resulted in more continuous and interconnected subbasins as rifting progressed. Basin evolution departs from general models of low-strain extension in that rifting ended abruptly with a basin-inversion event rather than developing a significant postrift sequence reflecting thermal subsidence (Meng et al., 2003), except in the vast Songliao basin, where a very thick postrift thermal subsidence megasequence exists (Feng and Graham, 2023). This departure is largely a function of the frequency of intracontinental deformation events during the amalgamation of Asia, as well as the heterogeneous basement structures that characterize the prerift substrate of southern Mongolia. In addition, the role of strike-slip faulting is poorly understood and may play a larger role in late Mesozoic rifting and postrift basin sequences in the China–Mongolia border region than previously recognized (Heumann et al., 2014, 2018; Johnson, 2015).

Northeastern China

The late Mesozoic extensional province in northeast China extends from the Mongolian border eastward to the Tan-Lu fault (Fig. 10.2). As many as 200 rift basins and subbasins, each bounded by northeast-trending normal faults, developed within this broad area during mid-Cretaceous time (Dou, 1997; Lin et al., 2001; Meng et al., 2003). Well-documented examples of late Mesozoic basins related to this rifting event include a number of coal-bearing half-graben basins in Inner Mongolia (Li et al., 1984) and the Erlian basin of northern China (Fig. 10.2, Lin et al., 2001; Guo et al., 2018). Over 70 Late Jurassic–Aptian half grabens (Fig. 10.5 of Feng and Graham, 2023) underlie the thick postrift thermal cooling and overlying foreland strata of the Songliao basin (Fig. 10.2; Dou, 1997; Xue and Galloway, 1993; Wang et al., 2016).

The rift system is superimposed upon a complex and heterogeneous basement fabric consisting of multiple Paleozoic suture zones and Mesozoic belts of intracontinental shortening, most notably the Yanshan fold-thrust belt (Fig. 10.2). The Yanshan belt and the coeval Yinshan fold-thrust belt farther west together comprise one of the largest and longest-lived zones of intracontinental shortening in Asia (Davis et al., 1998, 2001). Rift basins of Cretaceous age developed both within the Yanshan belt and along its northern periphery. Rifting closely followed, and perhaps overlapped, Late Jurassic–Early Cretaceous shortening in the Yanshan region. North of Beijing, major Yanshan thrust structures are intruded by a plutonic suite that yields ages of 132–129 Ma (Davis et al., 1998, 2001), indicating that the bulk of thrusting was completed by this time. Low-strain extension in the Yanshan region began c.130–121 Ma, the age range of volcanic units interbedded with and underlying syn-rift fill throughout the area. Lower Cretaceous strata associated with rifting in the Yanshan region comprise humid climate, coal-bearing fluvial, alluvial, and lacustrine facies associations (Fig. 10.3) that thicken and coarsen toward

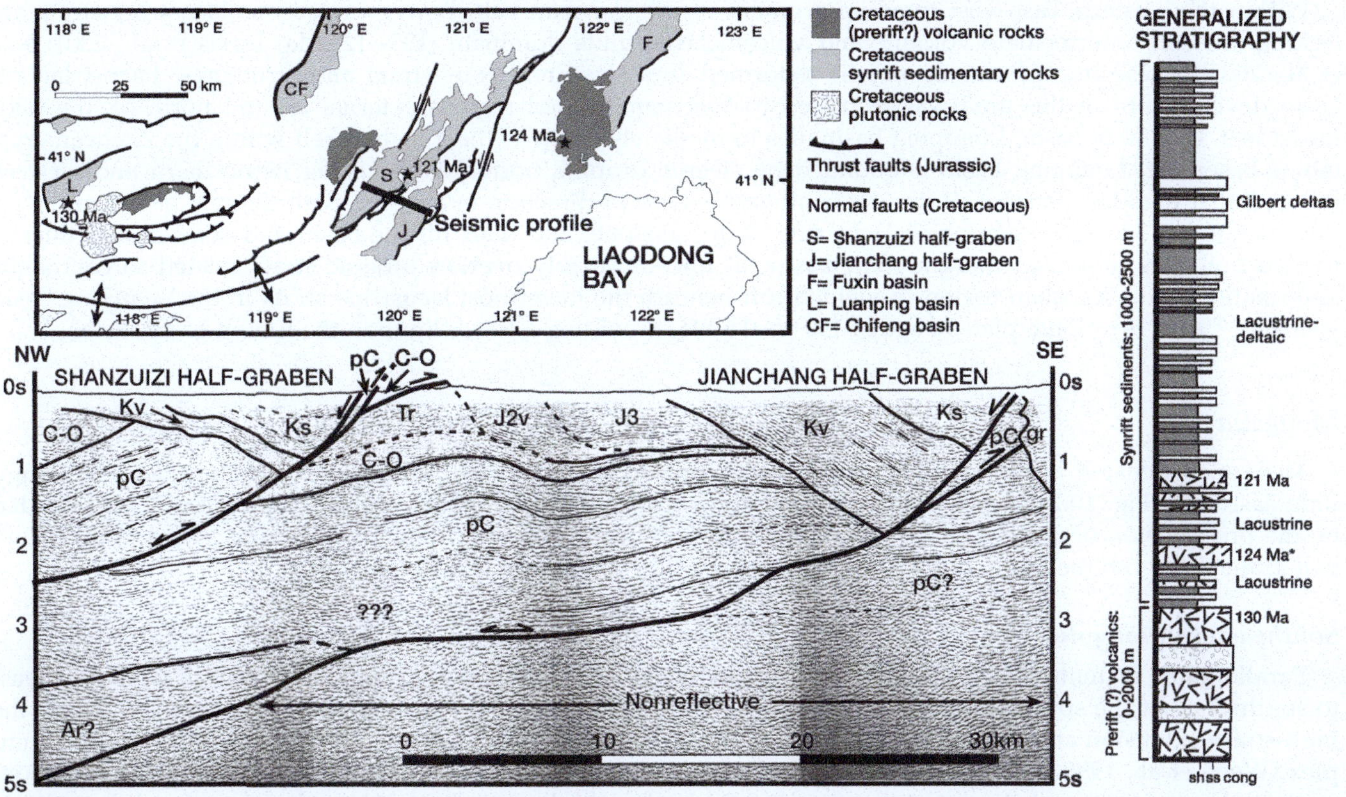

FIGURE 10.3 NW-SE time-migrated seismic line showing extensional basins developed atop the eastern Yanshan fold-thrust belt, Liaoning Province, China. Cretaceous normal faults bounding two half-graben basins (Shanzuizi basin and Jianchang basin) sole into Jurassic thrust faults at depth. Near-surface seismic interpretation is calibrated to surface exposures where these relationships are well exposed. Mesozoic structural trends plunge northeast, allowing for good control at depth. Geographic locations and stratigraphic position of $^{40}Ar/^{39}Ar$ ages shown; age denoted with asterisk from Swisher et al. (1999).

basin-bounding normal faults. These strata contain an abundant and diverse faunal assemblage, including the well-known Jehol biota that contains feathered dinosaur remains (Zhou et al., 2003).

The occurrence, geometry, and fill of rift basins in the Yanshan region were strongly influenced by the orientation of preexisting thrust trends relative to the Early Cretaceous stretching direction. Throughout most of its length, the Yanshan fold-thrust belt (and its Yinshan counterpart farther west, Fig. 10.2) trends strongly east-west and contains numerous large-offset features verging both north and south. In its easternmost extent, however, fold and thrust structures in the Yanshan belt trend northeast (Fig. 10.2) and are strongly southeast vergent. The Early Cretaceous stretching direction, as derived from the orientation of basin-bounding normal faults throughout northeast China, was to the northwest-southeast and apparently uniform throughout the extensional province. Early Cretaceous rift basins are widespread throughout the eastern segment of the Yanshan belt, where thrust structures were favorably oriented to be reutilized as normal faults in extension. Along western segments of the Yanshan belt, where the stretching direction was oriented oblique to basement fabrics, normal faults of Cretaceous age are common, but evidence for widespread rift basin development within the thrust belt is generally lacking.

Cretaceous rift basins developed atop the eastern, northeast-trending segment of the Yanshan belt are characterized by numerous southeast-deepening half-graben basins that occupy the core and periphery of the preexisting orogen and reactivate southeast vergent Jurassic–Cretaceous thrust structures as normal faults (Fig. 10.3). Normal faults that bound these half-graben uniformly dip to the northeast and sole into the older thrust faults in outcrop. Basin-bounding normal faults have shallow dips ($\sim 25°$in surface exposures), and the basins they bound seldom exceed 1–2 km in depth (Fig. 10.3). The fill of Jianchang and Shanzuizi basins (Fig. 10.3) comprises dominantly shallow lacustrine siltstone and sandstone punctuated by the occurrence of 2- to 6-m-thick progradational Gilbert delta complexes, fluvial sandstone, and coal. A tuff from the base of the Shanzuizi basin fill yields an $^{40}Ar/^{39}Ar$ biotite inverse isochron age of 121 ± 4 Ma; this is compatible with 124–125 Ma ages obtained from tuffaceous lacustrine strata elsewhere in Liaoning Province that broadly constrain the onset of rifting (Swisher et al., 1999).

Within the western, east-west trending segment of the Yanshan belt, Lower Cretaceous lacustrine strata are lacking. Instead, intermediate volcanic and volcaniclastic units dominate (135–122 Ma; Davis et al., 2001; Niu et al., 2003). These unconformably overlie deformed Jurassic–Cretaceous strata and structures. Lacustrine rift basin development in this area was restricted to basement-floored regions external to, and north of, the fold-thrust belt. One such basin, Luanping basin, lies immediately north of the Yanshan belt within the Archean crystalline basement. Luanping basin is bounded by steeply dipping normal faults along its northern and western margins that together define a northwest-southeast extension direction in keeping with the rest of the province. A minimum of 3 km of organic-rich lacustrine shale, siltstone, and marl fill the basin and coarsen and thicken toward both fault-bounded margins. The basin fill unconformably overlies Jurassic strata related to contractile deformation in the Yanshan region. A welded tuff beneath the lowermost lacustrine strata in the Luanping basin yields an $^{40}Ar/^{39}Ar$ biotite plateau age of 130 $\pm$ 0.5 Ma, providing a lower limit on the timing of extension.

Mongolia

Mesozoic extension in Mongolia, which spread across the eastern half of the country from at least 98° longitude eastward (Fig. 10.2), began in Late Jurassic time and continued through the Early Cretaceous, as recorded by the preservation of up to 3 km of nonmarine and volcaniclastic strata filling late Mesozoic-aged grabens and half-grabens in both the East Gobi Basin and Eastern Valley of Lakes (Figs. 10.4 and 10.5).

Southeastern Mongolia

Parallel fault systems broadly define the basin, which spans approximately 500 km in length from northeast to southwest. At the southwestern edge of the East Gobi Basin, Early Cretaceous rifting is characterized by the high-strain extension and formation of the Yagan-Onch Hayrhan detachment fault and metamorphic core complex (Webb et al., 1999). In contrast, the region near the active oil fields around Zuunbayan (Fig. 10.4) demonstrates characteristics of low-strain extension, including high-angle normal faults that border multiple subbasins (Johnson, 2004). Prerift, syn-rift, and postrift stratigraphic megasequences are well-expressed on seismic reflection profiles near Zuunbayan (Fig. 10.4; Heumann et al., 2018; Traynor and Sladen, 1995) and can be

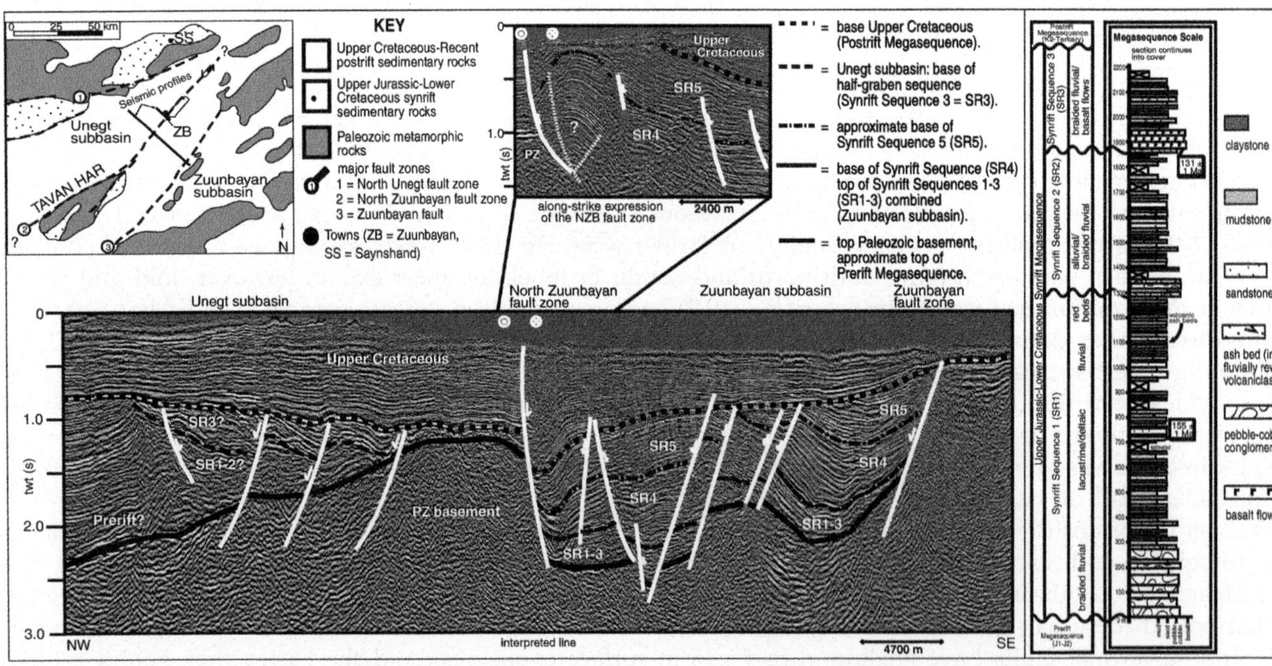

FIGURE 10.4 Regional seismic profile showing rift structure and prerift, synrift, and postrift megasequences of the East Gobi basin, Mongolia. Location of line shown in inset map. Vertical axis is in two-way travel time. Smaller seismic profile is from a parallel line several km away. Note high angle faults bounding major rift features, which were reactivated during a transpressional mid-Cretaceous basin inversion. Seismic data are from Johnson (2004). Stratigraphic section generalized from outcrops in the Unegt subbasin (Graham et al., 2001; Johnson and Graham, 2004a) is typical of Mongolian sector low-strain rift-basin fill.

correlated to outcrop exposures, particularly along the northern margin of the East Gobi Basin and at other basement uplifts (Graham et al., 2001).

Basement rocks in southeastern Mongolia mainly consist of Paleozoic-aged volcanic arc sequences (spanning Ordovician–Carboniferous sequences) that accreted before Mesozoic time as part of a closing ocean basin along the northern margin of the North China block (Lamb and Badarch, 1997). Although Precambrian rock units are mapped in parts of the East Gobi Basin (e.g., Tavan Har locality, Fig. 10.4; Yanshin, 1989), this interpretation is largely based on the high grade of metamorphism and inferred relative age relations between crystalline metamorphic rocks and "younger" Paleozoic units. Available radiometric dates indicate that many of these metamorphic basement rocks are Paleozoic and younger tectonite protoliths overprinted during Mesozoic shearing (Heumann et al., 2012; Taylor et al., 2013; Webb et al., 2010). Thus basement rocks underpinning late Mesozoic rift basins in southeastern Mongolia are relatively young and distinctly heterolithic, particularly compared to the more homogeneous "microcontinent" crust of the Precambrian North China block. Evidence for polyphase reactivation of major structures, particularly as long-lived strike-slip faults, may in part reflect this basement heterogeneity (Johnson, 2004, 2015; Johnson and Ritts, 2012; Heumann et al., 2018).

Prerift units also include Lower-Middle Jurassic strata of the Khamarkhavoor Formation. Locally, this unit is up to 1 km thick and it contains widespread coal beds that distinguish it from the rest of the Mesozoic sequence. These strata appear to predate (and are offset by) the main extensional structures of the East Gobi Basin, and they also lack the bimodal volcanic units common to the overlying syn-rift Mesozoic sequence (Graham et al., 2001). The basin setting for these units is poorly constrained but they correlate in age and depositional style to lower Mesozoic foreland and intermontane deposits found in western Mongolia (Hendrix et al., 2001) and may reflect a related prerift contractile event in the East Gobi Basin.

The syn-rift megasequence of southeastern Mongolia includes up to 3 km of preserved basin fill (Fig. 10.4). Five distinct syn-rift sequences are defined based on unconformable relations and indications of regional changes in depositional style (Graham et al., 2001). Nonmarine basin fill includes alluvial and fluvial deposits as well as widespread lacustrine and perilacustrine facies, particularly in Lower Cretaceous units at the basin center near Zuunbayan (Fig. 10.4; Johnson et al., 2003). Seismic stratigraphic units show asymmetric growth patterns that thicken into horst-bounding normal faults, supporting the interpretation of a syn-rift basin setting for these sequences. Rhyolitic ash beds and basalt flows are also common and provide radiometric constraints on the timing of rifting, from at least 155–126 Ma (Graham et al., 2001). Mapping of seismic reflection data indicates that the earliest phase of rifting (during Late Jurassic time) was characterized by isolated depocenters that coalesced into larger, more continuous subbasins during Early Cretaceous time as displacement on extensional faults increased, and fault systems linked to form more continuous structures (Johnson and Graham, 2004a,b). Syn-rift fluvial-lacustrine strata of the East Gobi Basin also form the major source, reservoir, and seal units for hydrocarbon accumulations in the subsurface near Zuunbayan (Johnson et al., 2003).

The main period of rift-related subsidence ended abruptly during approximately mid-Cretaceous time (early Cenomanian; Graham et al., 2001). A major basin-reorganization event is reflected by a regional angular unconformity between deformed syn-rift strata as young as Aptian–Albian, and relatively undeformed Upper Cretaceous strata of the postrift megasequence. Syn-rift normal faults were reactivated as thrust faults during this event, as indicated by the reverse drag of syn-rift strata seen on seismic data (Fig. 10.4). Mapping at Tavan Har and correlation to seismic data suggest that this basin-inversion event reflects transpressional movement along a major basin-dividing strike-slip fault (the North Zuunbayan fault; Heumann et al., 2014; Webb et al., 1999). Although poorly understood, outcrop and subcrop relations suggest that this was a long-lived structure with multiple phases of movement as old as Triassic and reactivated as recently as Cenozoic (Heumann et al., 2018; Johnson, 2004).

West-central Mongolia

In western and central Mongolia, a 50- to 150-km-wide, 1200-km-long belt of topographic lowlands hosts a string of internally drained lake basins, which form an arcuate (concave northeast) valley extending from the northwestern national border into central Mongolia (Fig. 10.2). Locally, this vast depression is referred to as the (Great) Valley of Lakes. This modern topographic basin is the most recent expression of a more regional plate-interior polyphase (PIP) basin trend (Johnson and Ritts, 2012): the present-day basin area is underlain by successive basin fill assemblages, separated by unconformities, that reactivated spatially similar depocenters through time. Active sinistral fault systems of the Gobi-Altai transpressional orogen and much of the surrounding Central Asian Orogenic Belt (Cunningham, 2005, 2010; Rippington et al., 2013) dominate the modern structural architecture of the region (Fig. 10.2). The Great Valley of Lakes is traversed by the Bolnai Fault in the north and bounded by the Bogd Fault in the south (Fig. 10.2), the latter of which ruptured along ~300 km of its trace in 1957 (>8 MM, Baljinnyam, et al., 1993; Cunningham, 2013).

From west to east, the Valley of Lakes transitions from north-south-trending to east-west trending (Fig. 10.2). This change occurs at approximately 95° longitude as it follows the southwest flank of amalgamated Precambrian cratonic blocks entrapped within the collage of terranes that make up the Mongolian continental assemblage. It is this eastern, east-west trending portion that contains Mesozoic extensional basins, which have been imaged in the subsurface by petroleum industry seismic reflection profiles (Fig. 10.5) and dated by outcrop studies and petroleum exploration drilling between 2010 and 2018 (Johnson et al., 2015; Petro Matad Group, personal communication).

This family of Mesozoic extensional basins in the Eastern Valley of Lakes formed upon Archean-Proterozoic cratonic basement of the Baidrag Terrane (Badarch et al., 2002). The major late Mesozoic depocenters, the Taats and Tugrug basins, are localized around the basement block's tapered and faulted southeast end (Fig. 10.5). The location and geometry of these basins suggest accommodation space was created as the cratonic block partially dislocated northwestward from its neighboring terranes during Mesozoic extension. The adjacent terranes include island arc and accretionary wedge complexes of the Lake Terrane to the south, and to a lesser extent, the subsumed Mongol-Okhotsk Ocean basin and the overlapping Khangai-Khenti terrestrial basin complex to the north (Badarch et al., 2002).

A third basin area was formed within the same cratonic basement farther west in the Eastern Valley of Lakes, the Baidrag Basin (Fig. 10.5). This local collection of subbasins exhibits structural trends that are suborthogonal (NNE-SSW) to the more E-W trends of the Tugrug and Taats basins and protrudes farther into the body of the Baidrag cratonic block, rather than along its margin. These Baidrag Basin structures are on trend with other rift basin systems (e.g., Mesozoic/Cenozoic evolution of the Khovsgol/Baikal lake basins) that have formed within a transition zone between the continent-wide dichotomy of contraction-dominated basins of west-central Asia and extension-dominated basins to the east. Such basin architectures, rifts at high angles to orogenic collisions, have been termed impactogens (Ingersoll, 2012). This tectonic transition zone occurs through the cusp of the continental, concave-north basement grain inherited during Phanerozoic accretion about the southern promontory of the Siberian Craton (Fig. 10.2) (Buchan et al., 2002; Cunningham, 1998, 2005; van Hinsbergen et al., 2015). This geometry is consistent with regional NW-SE-directed late Mesozoic extension and with deformation of the cratonic block during development of nearby basins. Like the other Eastern Valley of Lakes basins, despite some geometric differences, it too is a PIP basin with unconformities separating an underlying prerift section, and overlying postrift sequence (Fig. 10.5) and currently harbors the largest internally drained lake of the region.

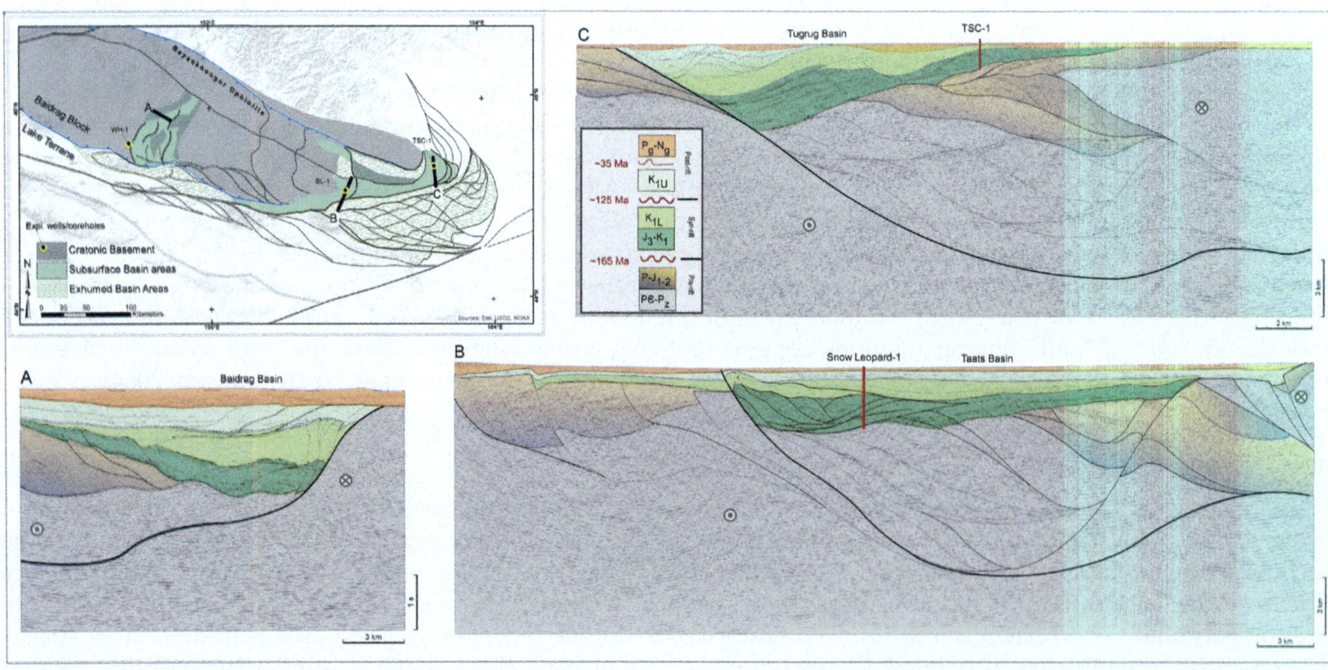

FIGURE 10.5 Reference map showing the distribution of Jurassic-Cretaceous basins of the Eastern Valley of Lakes with respect to relevant structures and terrane boundaries. Geoseismic dip-line examples from the (A) Baidrag Basin, (B) Taats Basin, and (C) Tugrug Basin. Note: Dip-line A vertical scale is in time (1 s scale bar); lines B & C are in depth and H = V (3 km scale bar). See Fig. 10.2 for location.

The Archean–Neoproterozoic basement rocks in this region are overlain by successions of Neoproterozoic–Early Cambrian marine carbonates and Silurian-Carboniferous arc successions of sedimentary and volcanic units. These older units are unconformably overlain by extensive and thick Permian–Lower Jurassic igneous and sedimentary sequences. Three main stratigraphic megasequences are recognized in outcrop and seismic data in the Eastern Valley of Lakes and surrounding region: prerift (pre-Kimmeridgian), syn-rift (Kimmeridgian–Barremian), and post-rift (Aptian–Albian to present). These megasequences are separated by prominent angular unconformities of Middle/Late Jurassic and mid-Cretaceous ages, which correlate with similar relationships recognized elsewhere in Mongolia and China. The associated angular unconformities are clearly visible on seismic data, in outcrop, and have been penetrated by the Tugrug Basin Stratigraphic Corehole (TSC-1), which recovered Triassic basalt below the Late Jurassic–Early Cretaceous basin fill (Johnson et al., 2015) and by the Snow Leopard-1 basin center exploration well in the Taats Basin, ~120 km to the west of Tugrug (Fig. 10.5).

The Eastern Valley of Lakes basins are bounded by high-angle to listric master normal faults, which are north-dipping in the Tugrug and Taats basins, and east and west dipping in the Baidrag Basin, but all preserve up to 3 km of petroliferous syn-rift fluvial-lacustrine basin fill. The Baidrag Basin, ~250 km west of the Taats Basin, represents the westernmost subsurface evidence of Mesozoic extension in the Eastern Valley of Lakes, with westernmost outcrop evidence just 60 km farther west. Beyond that, the Valley of Lakes then rests upon the Lake Terrane, not cratonic basement (Fig. 10.5), and basin geometry changes both at the surface and in the subsurface. To the west, sedimentary basins are less elongate and structurally asymmetric, becoming more ovoid and symmetrically filled, whereas seismic stratigraphic patterns appear more consistent with contractile or intramontane basins like those of central-western China.

Late Jurassic–Early Cretaceous rocks intermittently outcrop along the flanks of inverted ranges and basin margins of the Eastern Valley of Lakes. The syn-rift petroleum system is characterized by rich (5%–25% total organic carbon [TOC]) Upper Jurassic-Lower Cretaceous lacustrine source rocks, which act as both source and intraformational seals. Source rocks are generally within early oil window maturity based on source rock vitrinite reflectance values (0.5%–0.7% R_o) and RockEval Tmax values (>435°C). Stacked Upper Jurassic-Lower Cretaceous reservoir units were deposited in marginal lacustrine (fluvial-deltaic) and offshore (turbidite) lacustrine depositional environments (Johnson et al., 2015), and are directly comparable to those of the East Gobi Basin and eastern China (Graham et al., 2001; Johnson and Graham, 2004b; Wei et al., 2016). Hydrocarbon generation and expulsion have been proven in all three basin areas to varying degrees. The Tugrug Basin stratigraphic core (TSC-1) revealed a visually oil-stained sandstone interval and pervasive hydrocarbon fluid inclusions across a majority of the ~1600 m continuously cored interval (Johnson et al., 2015). The Snow Leopard-1 exploration well in the Taats Basin produced aromatic, light (~35 API) thermogenic oil on mud pits while drilling through the main exploration target, as well as elevated gas chromatography while drilling, and mud sample head-space gas during postwell analyses. The Wild Horse-1 exploration well, drilled on the SW flank of the Baidrag Basin, returned migrant hydrocarbon extractions from traces of stained sandstone cuttings near the base of the well (Petro Matad Group, personal communication).

Basins and structures associated with high-strain extension

Introduction

High-strain extensional styles are increasingly recognized across the northeast China–Mongolia Early Cretaceous extended region (Fig. 10.2). Structurally, high-strain extension is marked by the presence of detachment faults, low-angle normal faults of regional extent upon which large amounts of extension (at least tens of kilometers) are accommodated (Davis and Lister, 1988; Lister and Davis, 1989). The low-angle regional nature of detachment faults results in efficient accommodation of extension, and thus these faults are found in regions that undergo large magnitudes and high rates of extension (Friedmann and Burbank, 1995). In northeast China and southern Mongolia, many of these detachment faults control the formation of extensional metamorphic core complexes, which are created when tectonic removal of the upper plate of the detachment unloads the lower plate, resulting in isostatic uplift and doming. The resultant metamorphic core complex is characterized by a domal lower plate composed of mid- or lower crustal metamorphic rocks separated from a lower-grade upper plate by the detachment fault, which is usually characterized by a mylonitic shear zone overprinted by brittle fault rocks (Davis et al., 1986; Lister and Davis, 1989). The upper plate of detachment faults, whether in metamorphic core complexes or not, often contains syn-extensional sedimentary basins, termed "supradetachment basins" by

Friedmann and Burbank (1995). Supradetachment basins can be distinguished based on their structural setting in the upper plate of a detachment and on a variety of sedimentary characteristics, including dominance of footwall provenance, transverse deposystems, coarse sediment, proximal sedimentary facies distributed across the basin, and relatively thin sequences of strata (Friedmann and Burbank, 1995).

Documented examples of extensional detachment faults, metamorphic core complexes, and supradetachment basins include the northern Mongolia border (Daoudene et al., 2012, 2013, 2017); the Yagan-Onch Hayrhan metamorphic core complex in the southern Mongolia-China border region (Meng, 2003; Webb et al., 1999; Zheng and Zhang, 1993); the Ertomiao and Hohhot metamorphic core complexes in Inner Mongolia (Darby, 2003; Davis et al., 2002); and several examples in eastern Inner Mongolia and the Liaoning provinces (Davis et al., 1996; Darby, 2004). Other documented detachment systems in eastern China lie outside this area, including structures along the Mesozoic Qinling-Dabie orogen (e.g., Zhang and Zheng, 1999; Webb et al., 1999). Of all the high-strain extensional structures in this region, the Hohhot detachment is part of the most thoroughly documented metamorphic core complex and is associated with well-developed and well-exposed supradetachment basins (Davis et al., 2002). Because the structural characteristics and tectonics of the Hohhot detachment and Hohhot metamorphic core complex are well-constrained, and because associated supradetachment basins are widespread, we detail this system below as an excellent type example of east Asian high-strain extensional systems and sedimentary basins.

Hohhot detachment: structural setting

The Hohhot detachment system lies within the Daqing Shan north of Hohhot, the capitol of the Inner Mongolia Autonomous Region (Fig. 10.2). The detachment stretches along-strike in an east-west direction for more than 120 km (Fig. 10.6) and accommodates at least 40 km of top to the southeast extension in its central portion, where the Hohhot metamorphic core complex is developed (Davis et al., 2002). The age of extension on the Hohhot detachment is constrained to approximately 125–121 Ma by $^{40}Ar/^{39}Ar$ hornblende and biotite cooling ages in the footwall of the detachment, cross-cutting relationships, and $^{40}Ar/^{39}Ar$ dating of volcanics in synextensional sedimentary basins (Davis et al., 2002).

The Hohhot metamorphic core complex consists of the Hohhot extensional detachment fault, the Daqing Shan antiform, a roughly east-west trending culmination of metamorphic and plutonic rocks, and a folded upper plate of pre-Cretaceous crystalline and sedimentary units and Lower Cretaceous syn-extensional strata (Davis et al., 2002). On the south flank of the Daqing Shan antiform, the Hohhot detachment fault is a south-dipping low-angle (15–30°) mylonitic normal fault system with a brittle overprint (Fig. 10.6); this fault is the master detachment. It is corrugated into synforms and antiforms with hinges parallel to the NW-SE extension direction (Davis et al., 2002). Synformal corrugations in the detachment preserve numerous Lower Cretaceous nonmarine clastic basins (Ritts et al., 2010). On the northern flank of the Daqing Shan antiform, two detachment faults are stacked and synformally folded, with top-to-the-south slip on both splays (Fig. 10.6; Davis et al., 2002). The lower detachment separates mylonitic rocks from overlying nonmylonitic rocks, primarily Proterozoic crystalline rocks and Permian granitic gneisses. The upper detachment carries a succession of Cretaceous volcanic and sedimentary rocks (basin N2 in Fig. 10.6) that are highly deformed by normal faulting related to extension. The lower, oldest detachment was the original detachment, but with extension, the footwall was progressively unloaded, triggering isostatic uplift and bringing lower plate rocks to the surface. The resultant antiformally folded detachment was deactivated and a new splay propagated to the surface to accommodate further extension (Davis et al., 2002). The new splay, in turn, was also antiformally folded due to continued uplift with further unloading. The Hohhot detachment, on the southern flank of the Daqing Shan antiform, is the youngest fault splay and accommodates the remaining extension (Davis et al., 2002).

Hohhot detachment: supradetachment basins

The eastern Daqing Shan contains at least six distinct Lower Cretaceous basins that formed synchronously with extension along the underlying Hohhot detachment (Fig. 10.6). Because of their structural position above the Hohhot detachment and their synchronous evolution with the detachment (demonstrated by $^{40}Ar/^{39}Ar$ dating of intercalated volcanic rocks, cross-cutting relationships, and basin analysis), these basins are, by definition, supradetachment basins (Friedmann and Burbank, 1995). The six basins together comprise a *supradetachment basin system* that is defined here as a series of supradetachment basins that form in the upper plate of a detachment or linked detachments and that can be ascribed to a specific structural depozone within that system.

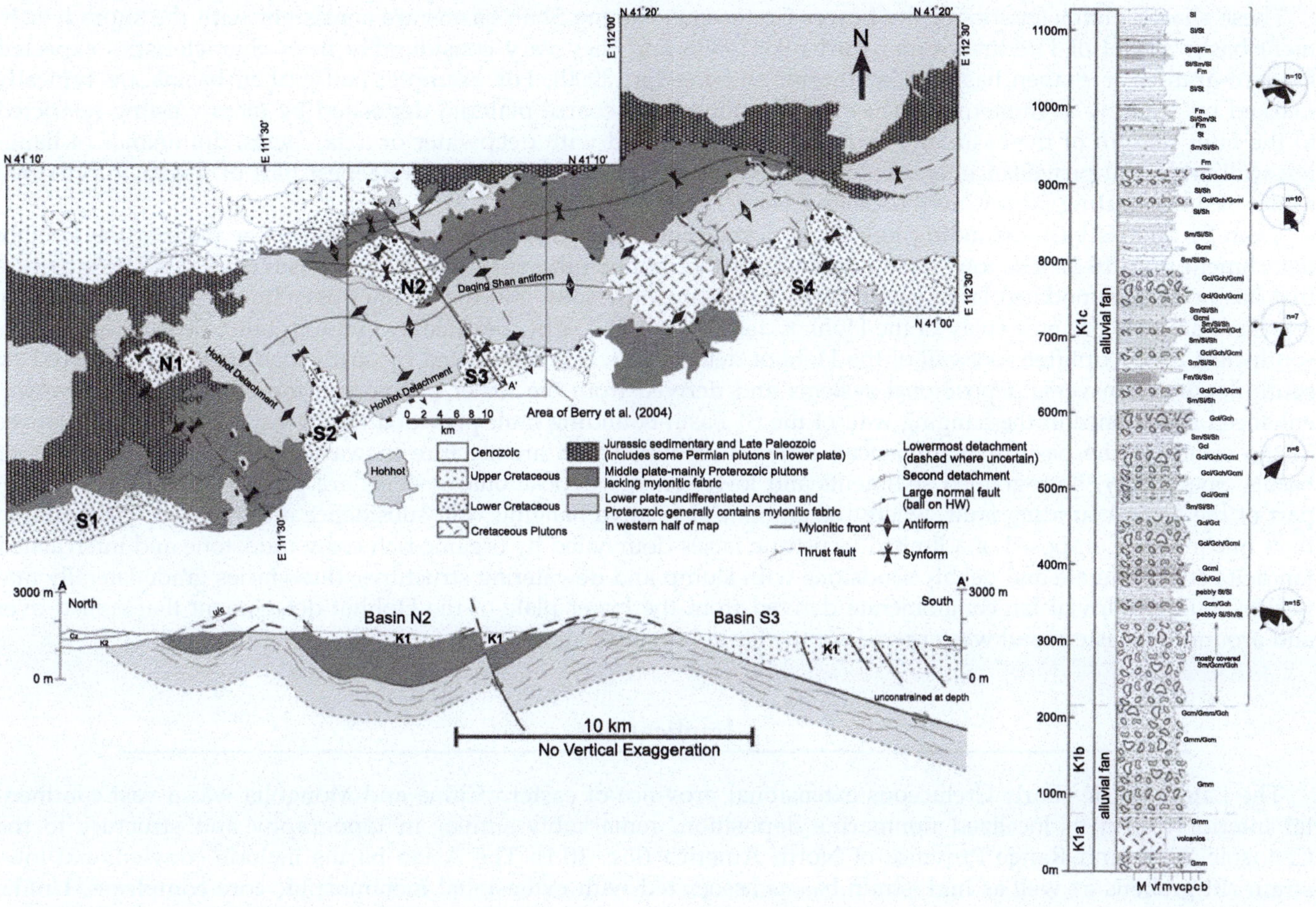

FIGURE 10.6 Structure, basin development, and typical stratigraphy associated with the Hohhot detachment fault and metamorphic core complex, near Hohhot, Inner Mongolia, China. See Fig. 10.2 for location of area. Note low angle normal faulting, folding of detachment fault, and doming of lower plate metamorphic rocks. Basin fill sequences, such as that shown at the right from basin S1 (north is to the top of paleocurrent diagrams), are dominated by very coarse grained alluvial and fluvial strata; limited accommodation precludes development of thick lacustrine sequences.

Basins N1 and S1–S4 (Fig. 10.6) all formed in the upper plate of the youngest splay of the Hohhot detachment, whereas basin N2 formed above an older splay of the detachment and is now located in the synformal keel of the Hohhot metamorphic core complex (Davis et al., 2002). Of the basins in the upper plate of the youngest splay of the Hohhot detachment, basins N1, S2, and S3 directly overlie and are cut by the Hohhot detachment, and basin S1 is bounded by a low-angle brittle normal fault that is wholly within the upper plate of the Hohhot detachment.

All of the supradetachment basins share the basic sedimentologic and stratigraphic characteristics of supradetachment basins as defined by Friedmann and Burbank (1995), although different depozones, particularly the upper plate basin depozone represented by basin S1, have some marked variation peculiar to their structural setting. All six basins are relatively thin; the maximum stratigraphic thickness is ~1200–1400 m and the thinnest basin is basin N2 with less than 400 m of strata. Strata in the basins are overwhelmingly dominated by conglomerate (Fig. 10.6), frequently cobble to boulder grade, that were deposited by streamflow and sheetflood-dominated alluvial fan depositional systems. Large (some more than 2 km long and 100 m thick) gravity-driven slide blocks are common in basins S2, S3, N1, and N2, also demonstrating the coarse-grained, proximal nature of the basins' depositional systems.

In each basin, sediment is derived from the lower plate, with only relatively minor amounts from the upper plate, and then only early in the basin's evolution. The fluvial and alluvial fan depositional systems that delivered the sediment filled the basin transversely from either the detachment breakaway or the domal core of the metamorphic core complex; paleocurrents in each basin are to the south-southeast, parallel to the extension direction. Finally, in each basin, facies are laterally uniform over the scale of the exposed basin, as is stratigraphic thickness.

These shared characteristics of the Lower Cretaceous Daqing Shan basins are consistent with the supradetachment basin model of Friedmann and Burbank (1995), and they vary considerably from characteristics expected for half-graben or graben basins (Gawthorpe and Leeder, 2000). For example, half-graben basins are typically marked by strongly asymmetrical facies distributions, with coarse material deposited by mass wasting restricted to the fault margin of the basin and immediately juxtaposed with deepwater or axial facies, dominance of hanging wall and axial provenance, and wedge geometry in cross section with the deepest part of the basin proximal to the basin-bounding fault (Gawthorpe and Leeder, 2000).

Basin S1 and its basin-bounding low-angle normal fault are wholly restricted to the upper plate of the Hohhot detachment (Fig. 10.6). The low-angle normal fault that bounds the northern edge of basin S1 is a brittle structure that has not accommodated large amounts of slip. It probably soles into the Hohhot detachment at depth and may represent an incipient new splay of the Hohhot detachment (B. Darby, personal communication). Because basin S1 is southward of the uplifted footwall of the Hohhot detachment, it is dominated by coarse conglomerate delivered by south-flowing transverse depositional systems and derived from the lower plate, as in the other basins. However, enhanced subsidence in the hanging wall of the S1 basin-bounding fault provided additional accommodation space in the center of the basin that influenced facies distribution and architecture in ways not observed in the other basins. Specifically, finer-grained sediments and lower-energy depositional systems are preserved along the central part of the basin-bounding fault, where the amount of relative hanging wall subsidence and fault slip rate were at their maximum. Composed of a limited lacustrine facies dominated by organic-rich coaly mudstone and intercalated fan delta conglomerate and pebbly sandstone with slump and dewatering structures, these facies grade laterally into pebble-boulder alluvial fan conglomerate derived from the lower plate of the Hohhot detachment that spilled over and around the eastern and western ends of the basin-bounding upper plate fault.

Implications

The Late Jurassic–Early Cretaceous extensional province of eastern China and Mongolia was a vast continental interior region of localized nonmarine deposition, remarkably similar in topography and structure to the Cenozoic Basin and Range Province of North America (Fig. 10.1). The Asian basins include conventional low-strain rift grabens as well as high-strain basins associated with extensional metamorphic core complexes. Unlike the North American Basin and Range Province, temporal and spatial distribution trends of the two end-member basin types have yet to be conclusively demonstrated, although Meng et al. (2003) proposed an evolutionary linkage between the two rift modes. All basins contain exclusively nonmarine fill, varying from relatively modest total thickness, very coarse-grained sequences in the high-strain setting basins, to thick accumulations of fine-grained lacustrine facies with coarse marginal facies in the low-strain basins. As in all nonmarine basins, climate is a major factor in determining the character of facies, whereas both tectonism and climate control the stratigraphic architecture (Carroll et al., 2010; Graham et al., 2001; Chamberlain et al., 2013; Johnson and Graham, 2004b). The deepest water lacustrine facies are best developed in the largest basins of the region, where accommodation is maximized along major basin-bounding faults. The larger basins tended to evolve from initially isolated sets of grabens, which over time became linked to form larger, lacustrine-dominated entities.

The Asian basins demonstrate pronounced structural control, including both localization of basins by the fabric of the underlying accretionary basement, as well as reactivation of specific faults. The regional pattern of extension abruptly terminated in the mid-Cretaceous with an episode of basin inversion. Only the largest of the basins, notably Songliao (Feng and Graham, 2023), display a clearly discernible pattern of postrift thermal subsidence (termed the "sag" phase in local literature).

The scale of the province and duration of extension suggest that extension was driven by plate-scale processes. No single extensional driver has yet been convincingly demonstrated, but a variety of mechanisms have been advanced (Fig. 10.7): (1) backarc extension behind the Pacific margin arc of Asia (Ren et al., 2002; Traynor and Sladen, 1995; Watson et al., 1987); (2) extension associated with the closing of the Mongol-Okhotsk seaway (Graham et al., 2001); (3) transtension associated with collisional tectonics (Kimura et al., 1990; Lamb et al., 1999); (4) gravitational collapse of orogenically thickened crust (Junggar–Hegen–Solonker [JHS] suture, Yinshan-Yanshan belt) (Graham et al., 2001; Traynor and Sladen, 1995) or orogenic collapse followed by relict Mongol-Okhotsk slab-break off and compensatory mid-lower crustal flow (Meng et al., 2003; Van Der Voo et al., 1999); and (5) Paleo-Pacific flat-slab subduction and foundering/retreat cycles, at least for the eastern Asian continental margin, and inboard for +/-1000 km, has been shown to be both likely and capable of generating the conspicuous patterns and timing of deformation, deposition, lithospheric thinning, and dynamic magmatism (Li et al., 2019; Liu et al., 2021a,b).

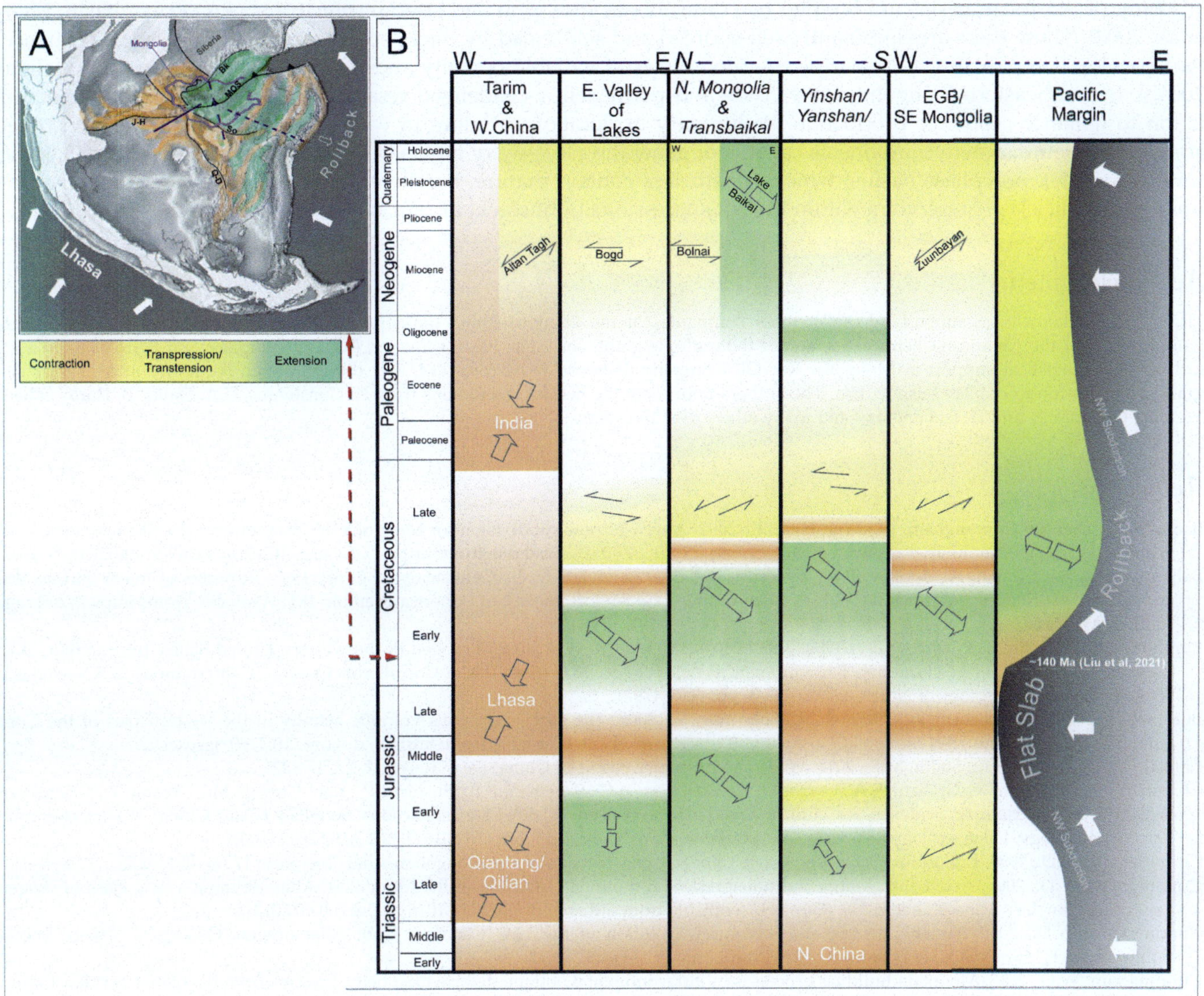

FIGURE 10.7 (A) Paleogeographic reconstruction of east-central Asia for the Early Cretaceous (~143 Ma), modified from C.R. Scotese, Paleomap Project (Moore and Scotese, 2021). Major tectonic sutures and regions of contraction and extension are highlighted. MOS = Mongol-Okhotsk; Bk = Baikalide; J-H = Jungar-Hegen; So = Solonker; Q-D = Qilian-Dabie. Blue transect lines refer to regions highlighted in 7B. (B) Mesozoic tectonic events chart for the same region. Major terrane collisions are represented in white, periods of contraction in orange, periods of transpression/transtension in yellow, periods of extension in green, and paleo-Pacific margin slab dynamics in grey.

Interestingly, the time periods attributed to flat-slab subduction/retreat driven events recognized at the paleo-Pacific margin closely align with similar events (basin subsidence vs inversion phases, magmatism, metamorphic core complex emplacement, etc.) farther inboard (>1000 km), where it is progressively less likely that the Pacific margin dominated the tectonic processes (Fig. 10.7). This coincidence in style of geodynamic pulsing suggests that the greater circum-Asian superconvergence regime (Paleo- and Neotethyan systems) of the Mesozoic may have worked together with relaxation of the interior orogenic belts (Mongol-Okhotsk, Jungar-Hebden, and Qilian-Dabie sutures) to manifest the vast extensional province (Li et al., 2019).

Many basins of the Asian extensional system are petroliferous (Fig. 10.7). Most notably, the Songliao basin contains Daqing oil field, the largest field in China (Feng and Graham, 2023). In general, the prospectivity of basins in the province is directly related to basin size and thickness of fill. Basins with a greater history of accommodation host better-developed lacustrine petroleum source rock facies and greater burial-facilitated maturation. High heat flow, typical of rift settings, somewhat offsets minimal burial in some basins: the deepest grabens of the East Gobi Basin contain at most only 3 km of fill, yet are petroliferous (Johnson, 2004; Prost, 2004).

Many of the Asian basins are typified by conventional structural and stratigraphic traps (e.g., Erlian basin, Meng et al., 2003). These traps are dominantly self-sourced and self-sealed by thick, organic-rich lacustrine shale sequences that envelop complex fluvial-lacustrine reservoir facies of variable porosity and permeability (Johnson and Graham, 2004b). In southeastern Mongolia, waxy crude and permeability-challenged reservoirs have required fracture stimulation to enhance economic production. Preliminary exploration activities in western Mongolia have demonstrated similar petroleum system components capable of generating less-waxy higher API crude from relatively small, structurally complex polyphase basins, which nonetheless contain mature, excellent-quality lacustrine source rocks and nonmarine siliciclastic reservoirs within the rift successions (Johnson et al., 2015).

Acknowledgments

The authors acknowledge support by the member companies of the Stanford-China Industrial Affiliates Program, the National Science Foundation, and the American Chemical Society for the multiple grants for them. Nescor Energy, Roc Oil Company, and Petro Matad Group Ltd provided logistical support in Mongolia. Roc Oil Company, Petro Matad Group Ltd, and the Petroleum Authority of Mongolia also granted permission to publish seismic data. The authors appreciate the helpful discussions from K. Constenius, G.A. Davis, B. Darby, Zheng Yadong, G. Badarch, and D. Badamgarav and many others over the years.

References

Badarch, G., Dickson Cunningham, W., Windley, B.F., 2002. A new terrane subdivision for Mongolia: implications for the phanerozoic crustal growth of Central Asia. Journal of Asian Earth Sciences 21 (1), 87–110. Available from: https://doi.org/10.1016/S1367-9120(02)00017-2.

Baljinnyam, I., Bayasgalan, A., Borisov, B.A., Cisternas, A., Dem'yanovich, M.G., Ganbaatar, L., et al., 1993. Ruptures of major earthquakes and active deformation in Mongolia and its surroundings. Geological Society of America Memoir 181. Available from: https://doi.org/10.1130/MEM181.

Burchfiel, B.C., Cowan, D.S., Davis, G.A., 1992. Tectonic overview of the Cordilleran orogen in the western United States. In: Burchfiel, B.C., Cowan, D.S., Davis, G.A., Burchfiel, B.C., Lipman, P.W., Zoback, M.L. (Eds.), The Cordilleran Orogen. Conterminous U.S. Geological Society of America Geology of North America G-3, pp. 407–479.

Buriánek, D., Hanžl, P., Erban, V., Gilíková, H., Bolormaa, K., 2008. The early cretaceous volcanic activity in the western part of the Gobi-Altay rift (Shiliin Nuruu, SW Mongolia). Journal of Geosciences 53 (2). Available from: https://doi.org/10.3190/jgeosci.026.

Carroll, A.R., Graham, S.A., Smith, M.E., 2010. Walled sedimentary basins of China. Basin Research 22, 17–32.

Chamberlain, C.P., Wan, X., Graham, S.A., Carroll, A.R., Doebbert, A.C., Sageman, B.B., Blisniuk, P., Kent-Corson, M.L., Wang, Z., Chengshan, W., 2013. Stable isotopic evidence for climate and basin evolution of the Late Cretaceous Songliao basin, China. Palaeogeography, Palaeoclimatology, Palaeoecology 385, 106–124. Available from: https://doi.org/10.1016/J.PALAEO.2012.03.020.

Cunningham, W.D., 1998. Lithospheric controls on late Cenozoic construction of the Mongolian Altai. Tectonics 17 (6), 891–902.

Cunningham, W.D., 2005. Active intracontinental transpressional mountain building in the Mongolian Altai: defining a new class of orogen. Earth and Planetary Science Letters 240, 436–444. Available from: https://doi.org/10.1016/j.epsl.2005.09.013.

Cunningham, W.D., 2010. Tectonic setting and structural evolution of the Late Cenozoic Gobi Altai orogen. Geological Society Special Publication 338, 361–387. Available from: https://doi.org/10.1144/SP338.17.

Cunningham, W.D., 2013. Mountain building processes in intracontinental oblique deformation belts: Lessons from the Gobi Corridor, Central Asia. Journal of Structural Geology 46, 255–282. Available from: https://doi.org/10.1016/j.jsg.2012.08.010.

Daoudene, Y., Gapais, D., Ruffet, G., Gloaguen, E., Cocherie, A., Ledru, P., 2012. Syn-thinning pluton emplacement during Mesozoic extension in eastern Mongolia. Tectonics 31. Available from: https://doi.org/10.1029/2011TC002926.

Daoudene, Y., Ruffet, G., Cocherie, A., Ledru, P., Gapais, D., 2013. Timing of exhumation of the Ereendavaa metamorphic core complex (north-eastern Mongolia) - U-Pb and 40 Ar/ 39 Ar constraints. Journal of Asian Earth Sciences 62, 98–116. Available from: https://doi.org/10.1016/j.jseaes.2011.04.009.

Daoudene, Y., Gapais, D., Cogné, J.P., Ruffet, G., 2017. Late Jurassic - early Cretaceous continental extension in Northeast Asia - relationships to plate kinematics. Bulletin de la Societe Geologique de France 188. Available from: https://doi.org/10.1051/bsgf/2017011.

Darby, B.J., 2003. Mesozoic intraplate deformation in the east Asian tectonic collage: the enigmatic northwest Ordos region, China. Ph.D. dissertation, University of Southern California, 174.

Darby, B.J., 2004. The newly discovered Waziyu metamorphic core complex. Yiwulu Mountains, western Liaoning Province, north China. Frontiers in Earth Science. 11, 145–155.

Davaa, B.-A., 2010. Geodynamic development and hydrocarbon potential of the Tamtsag Basin, Eastern Mongolia. Ph.D. Dissertation, University of Freiburg.

Davis, G.A., Lister, G.S., 1988. Detachment faulting in continental extension; perspectives from the Southwestern U.S. Cordillera. In: Clark, S. P., Burchfiel, B.C., Suppe, J. (Eds.), Processes in Continental Lithospheric Deformation, 218. Geological Society of America Special Papers, pp. 133–159.

Davis, G.A., Lister, G.S., Reynolds, S.J., 1986. Structural evolution of the Whipple and South Mountains shear zones, southwestern United States. Geology 14, 7–10.

Davis, G.A., Qian, X., Zheng, Y., Yu, H., Wang, C., Tong, H.M., et al., 1996. Mesozoic deformation and plutonism in the Yunmeng Shan: a Chinese metamorphic core complex north of Beijing, China. In: Yin, A., Harrison, T.M. (Eds.), The Tectonic Evolution of Asia. Cambridge University Press, pp. 253–280.

Davis, G.A., Cong, W., Zheng, Y.D., Zhang, J.J., Zhang, C.H., Gehrels, G.E., 1998. The enigmatic Yinshan fold-and-thrust belt of northern China: new views on its intraplate contractional styles. Geology 26, 43–46.

Davis, G.A., Zheng, Y., Wang, C., Darby, B.J., Zhang, C., Gehrels, G., 2001. Mesozoic tectonic evolution of the Yanshan fold and thrust belt, with emphasis on Hebei and Liaoning provinces, northern China. Geological Society of America Memoirs 194, 171–197.

Davis, G.A., Zheng, Y., Darby, B.J., Spell, T.L., 2002. Geometric and temporal evolution of an extensional detachment fault, Hohhot metamorphic core complex, Inner Mongolia, China. Geology 30, 1003–1006.

Dickinson, W.R., 2002. The Basin and Range Province as a composite extensional domain. International Geology Review. 44, 1–38.

Dou, L., 1997. The Lower Cretaceous petroleum system in NE China. Journal of Petroleum Geology. 20, 475–488.

Feng, Z.Q., and Graham, S.A., 2023. The Songliao Basin. in Chiarello, D., ed., Phanerozoic Regional Geology of the World, in press (this volume).

Friedmann, S.J., Burbank, D.W., 1995. Rift basins and supradetachment basins: intracontinental extensional end members, Basin Research, 7. pp. 109–127.

Gawthorpe, R.L., Leeder, M.K., 2000. Tectono-sedimentary evolution of active extensional basins. Basin Research 12, 195–218.

Gébelin, A., Mulch, A., Teyssier, C., Heizler, M., Vennemann, T., Seaton, N.C.A., 2011. Oligo-Miocene extensional tectonics and fluid flow across the Northern Snake Range detachment system, Nevada. Tectonics 30. Available from: https://doi.org/10.1029/2010TC002797.

Graham, S.A., Hendrix, M.S., Johnson, C.L., Badamgarav, D., Badarch, G., Amory, J., et al., 2001. Sedimentary record and tectonic implications of late Mesozoic rifting, southeast Mongolia. Geological Society of America Bulletin 113, 1560–1579.

Gries, R., 1983. North-south compression of Rocky Mountain foreland structures. In: Lowell, J.D. (Ed.), Rocky Mountain Foreland Basins and Uplifts. Rocky Mountain Association of Geologists, pp. 9–32.

Guo, Z.X., Shi, Y.P., Yang, Y.T., Jiang, S.Q., Li, L.B., Zhao, Z.G., 2018. Inversion of the Erlian Basin (NE China) in the early Late Cretaceous: implications for the collision of the Okhotomorsk Block with East Asia. Journal of Asian Earth Sciences v. 154, 49–66. Available from: https://doi.org/10.1016/j.jseaes.2017.12.007.

Guo, Z.-X., Yang, Y.-T., Zhao, X.-Z., Dan, W.-N., Wang, X., Du, L., 2019. Early Cretaceous tectonostratigraphic evolution of the Erlian Basin, NE China. A record of Late Mesozoic intraplate deformation in East Asia. Marine and Petroleum Geology v. 110, 539–564.

Hendrix, M.S., Beck, M.A., Badarch, G., Graham, S.A., 2001. Triassic synorogenic sedimentation in southern Mongolia: early effects of intracontinental deformation. Geological Society of America Memoirs. 194, 413–434.

Heumann, M.J., Johnson, C.L., Webb, L.E., Taylor, J.P., Jalbaa, U., Minjin, C., 2012. Paleogeographic reconstruction of a late Paleozoic arc collision zone, southern Mongolia. Geological Society of America Bulletin 124, 1514–1534.

Heumann, M.J., Johnson, C.L., Webb, L.E., Taylor, J.P., Jalbaa, U., Minjin, C., 2014. Total and incremental left-lateral displacement across the East Gobi Fault Zone, southern Mongolia: Implications for timing and modes of polyphase intracontinental deformation. Earth and Planetary Science Letters 392, 1–15.

Heumann, M.J., Johnson, C.L., Webb, L.E., 2018. Plate interior polyphase fault systems and sedimentary basin evolution: A case study of the East Gobi Basin and East Gobi Fault Zone, southeastern Mongolia. Journal of Asian Earth Sciences 151. Available from: https://doi.org/10.1016/j.jseaes.2017.05.017.

Ingersoll, R.V., 2012. Tectonics of Sedimentary Basins, with Revised Nomenclature. In Busby, C., and Azor, A., eds, Tectonics of Sedimentary Basins: Recent advances, p. 1–43. Doi: 10.1002/9781444347166.ch1.

Jiang, S., Xu, Z., Feng, Y., Zhang, J., Cai, D., Chen, L., Wu, Y., Zhou, D., Bao, S., Long, S., 2016. Geologic characteristics of hydrocarbon-bearing marine, transitional and lacustrine shales in China. Journal of Asian Earth Sciences 115, 404–418. Available from: https://doi.org/10.1016/J.JSEAES.2015.10.016.

Johnson, C.L., 2004. Polyphase evolution of the East Gobi basin: sedimentary and structural records of Mesozoic-Cenozoic intraplate deformation in Mongolia. Basin Research 16, 79–100.

Johnson, C.L., 2015. Sedimentary basins in transition: distribution and tectonic settings of Mesozoic strata in Mongolia. In Anderson, T.H., Didenko, A.N., Johnson, C.L., Khanchuk, A.I., and MacDonald, J.H., Jr., eds., Late Jurassic Margin of Laurasia—A Record of Faulting Accommodating Plate Rotation: Anderson, T., ed., Late Jurassic margins of Laurasia, Geological Society of America Special Paper, v. 513, p. 543–560. doi:10.1130/2015.2513(17).

Johnson, C.L., Graham, S.A., 2004a. Sedimentology and reservoir architecture of a synrift lacustrine delta, southeastern Mongolia. Journal of Sedimentary Research. 74, 770–785.

Johnson, C.L., Graham, S.A., 2004b. Cycles in perilacustrine facies of Upper Mesozoic rift basins, southeastern Mongolia. Journal of Sedimentary Research. 74, 786–804.

Johnson, C.L., Ritts, B.D., 2012. Plate Interior Poly-Phase Basins. Tectonics of Sedimentary Basins: Recent Advances 565–582. Available from: https://doi.org/10.1002/9781444347166.ch28.

Johnson, C.L., Webb, L.E., Graham, S.A., Hendrix, M.S., Badarch, G., 2001. Sedimentary and structural records of late Mesozoic high-strain extension and strain partitioning, East Gobi basin, southern Mongolia. In: Hendrix, M.S., Davis, G.A. (Eds.), Paleozoic and Mesozoic Tectonic Evolution of Central and Eastern Asia, 194. Geological Society of America Memoir, pp. 413–434.

Johnson, C.L., Greene, T.J., Zinniker, D.A., Moldowan, M.J., Hendrix, M.S., Carroll, A.R., 2003. Geochemical characteristics and correlation of oil and nonmarine source rocks from Mongolia. AAPG Bull 87, 817–846.

Johnson, C.L., Amory, J.A., Graham, S.A., Lamb, M.A., Badarch, G., and Affolter, M., 2008, Accretionary tectonics and sedimentation during late Paleozoic arc collision, China-Mongolia border region. In Draut, A., Clift, P., and Scholl, D., eds., Formation and application of the sedimentary record in arc collision zones, Geological Society of America Special Paper 436, p. 363–390.

Johnson, C.L., Constenius, K.C., Graham, S.A., Mackey, G., Menotti, T., Payton, A., et al., 2015. Subsurface evidence for late Mesozoic extension in western Mongolia: tectonic and petroleum systems implications. Basin Research v. 27, 272–294. Available from: https://doi.org/10.1111/bre.12073.

Kimura, G., Tasaki, T., Kono, M., 1990. Mesozoic collisional extrusion tectonics in eastern Asia. Tectonophysics 181, 15–23.

Lamb, M.A., Badarch, G., 1997. Paleozoic sedimentary basins and volcanic-arc systems of southern Mongolia; new stratigraphic and sedimentologic constraints. International Geology Review. 39, 542–576.

Lamb, M.A., Hanson, A.D., Graham, S.A., Badarch, G., Webb, L.E., 1999. Left-lateral sense offset of upper Proterozoic to Paleozoic features across the Gobi Onon, Tost, and Zuunbayan faults in southern Mongolia and implications for other Central Asian faults. Earth and Planetary Science Letters 173, 183–194.

Li, S., Li, B., Yang, S., Huang, J., Li, Z., 1984. Sedimentation and tectonic evolution of late Mesozoic faulted coal basins in north-eastern China. In: Rahmani, R.A., Flores, R.M. (Eds.), Special Publication of the International Association of Sedimentologists 7, pp. 387–406.

Li, S., Mo, X., Yang, S., 1995. Evolution of Circum-Pacific basins and volcanic belts in East China and their geodynamic background. Journal of China University of Geosciences 6, 48–58.

Li, S.-Q., Yang, Y.-Z., Xie, Q.-L., Wang, Y., Chen, F., 2015. Age constraints on late Mesozoic lithospheric extension and origin of felsic volcanism in the Songliao basin, NE China. The Journal of Geology 123, 153–175.

Li, S., Suo, Y., Li, X., Zhou, J., Santosh, M., Wang, P., et al., 2019. Mesozoic tectono-magmatic response in the East Asian ocean-continent connection zone to subduction of the Paleo-Pacific Plate. Earth-Science Reviews 192, 91–137.

Lin, C., Eriksson, K., Li, S., Wan, Y., Ren, J., Zhang, Y., 2001. Sequence architecture, depositional systems, and controls on development of lacustrine basin fills in part of the Erlian Basin, Northeast China. AAPG Bull 85, 2017–2043.

Lister, G.S., Davis, G.A., 1989. The origin of metamorphic core complexes and detachment faults formed during Tertiary continental extension in the northern Colorado River region, U.S.A. Journal of Structural Geology. 11, 65–94.

Liu, L., Peng, D., Liu, L., Chen, L., Li, S., Wang, Y., et al., 2021a. East Asian lithospheric evolution dictated by multistage Mesozoic flat-slab subduction. Earth-Science Reviews 217103621.

Liu, L., Liu, L., Xu, Y.-G., 2021b. Mesozoic intraplate tectonism of East Asia due to flat subduction of a composite terrane slab. Earth-Science Reviews 214103505.

Long, S.P., 2019. Geometry and magnitude of extension in the Basin and Range Province (39 N), Utah, Nevada, and California, USA. Constraints from a province-scale cross section. Bulletin 131, 99–119.

McQuarrie, N., Wernicke, B.P., 2005. An animated tectonic reconstruction of southwestern north america since 36 Ma. Geosphere 1, 147–172. Available from: https://doi.org/10.1130/GES00016.1.

Meng, Q.R., 2003. What drove late Mesozoic extension of the north China – Mongolia tract? Tectonophysics 369, 155–174.

Meng, Q.R., Hu, J.M., Jin, J.Q., Zhang, Y., Xu, D.F., 2003. Tectonics of the late Mesozoic wide extensional basins system in the China-Mongolia border region. Basin Research 15, 397–415.

Moore, T.L., and C.R. Scotese, 2021. Ancient Earth: Breakup of Pangea, Vers. 3.0.5, iOS Mobile Application, retrieved from https://apps.apple.com/us/app/breakup-of-pangea/id509943912on01/05/2023.

Niu, B., He, Z., Song, B., Ren, J., 2003. Zhangjiakou Group volcanic rock SHRIMP age and significance. Geological Bulletin of China 22, 140–141.

Prost, G.L., 2004. Tectonics and hydrocarbon systems of the East Gobi basin, Mongolia. AAPG Bulletin 88, 483–514.

Ren, J., Li, S., 1998. Comparison between the faulted basin system in northeastern Asia and the extension of the Basin and Range Province in western North America: Dizhi Keji Qingbao. Geological Science and Technology Information 17, 7–11.

Ren, J., Tamaki, K., Li, S., Zhang, J., 2002. Late Mesozoic and Cenozoic rifting and its dynamic setting in eastern China and adjacent areas. Tectonophysics 344, 175–205.

Rippington, S., Cunningham, D., England, R., Hendriks, B., 2013. The crustal assembly of southern Mongolia: new structural, lithological and geochronological data from the Nemegt and Altan ranges. Gondwana Research 23, 1535–1553. Available from: https://doi.org/10.1016/j.gr.2012.09.001.

Ritts, B.D., Berry, A.K., Johnson, C.L., Darby, B.J., Davis, G.A., 2010. Early Cretaceous supradetachment basins in the Hohhot metamorphic core complex. Inner Mongolia, China. Basin Research 22, 45–60.

Swisher III, C.C., Wang, Y., Wang, X., Xu, X., Wang, Y., 1999. Cretaceous age for the feathered dinosaurs of Liaoning, China. Nature 400 (6739), 58–61.

Taylor, J.P., Webb, L.E., Johnson, C.L., Heumann, M.J., 2013. The lost South Gobi microcontinent: protolith studies of metamorphic tectonites and implications for the evolution of continental crust in Southeastern Mongolia. Geosciences v. 3, 543–584.

Traynor, J.J., Sladen, C., 1995. Tectonic and stratigraphic evolution of the Mongolian People's Republic and its influence on hydrocarbon geology and potential. Marine and Petroleum Geology. 12, 35–52.

Van Der Voo, R., Spakman, W., Bijwaard, H., 1999. Mesozoic subducted slabs under Siberia. Nature v. 397, 246–249. Available from: https://doi.org/10.1038/16686.

Van der Voo, R., Van Hinsbergen, D.J.J., Domeier, M., Spakman, W., Torsvik, T.H., 2015. Latest Jurassic-earliest Cretaceous closure of the Mongol-Okhotsk Ocean: a paleomagnetic and seismological-tomographic analysis. Special Paper of the Geological Society of America 513, 589–606. Available from: https://doi.org/10.1130/2015.2513(19).

Wang, T., Zheng, Y., Zhang, J., Zeng, L., Donskaya, T., Guo, L., et al., 2011. Pattern and kinematic polarity of late Mesozoic extension in continental NE Asia. Perspectives from metamorphic core complexes. Tectonics 30. Available from: https://doi.org/10.1029/2011TC002896.

van Hinsbergen, D.J.J., Cunningham, D., Straathof, G.B., Ganerød, M., Hendriks, B.W.H., Dijkstra, A.H., 2015. Triassic to Cenozoic multi-stage intra-plate deformation focused near the Bogd Fault system, Gobi Altai, Mongolia. Geoscience Frontiers 6, 723–740. Available from: https://doi.org/10.1016/j.gsf.2014.12.002.

Wang, P.J., Mattern, F., Didenko, N.A., Zhu, D.F., Singer, B., Sun, X.M., 2016. Tectonics and cycle system of the Cretaceous Songliao Basin: an inverted active continental margin basin. Earth-Science Reviews 159, 82–102. Available from: https://doi.org/10.1016/j.earscirev.2016.05.004.

Watson, M.P., Hayward, A.B., Parkinson, D.N., Zhang, Z.M., 1987. Plate tectonic history, basin development, and petroleum source rock deposition, onshore China. Marine and Petroleum Geology. 4, 205–225.

Webb, L.E., Graham, S.A., Johnson, C.L., Badarch, G., Hendrix, M.S., 1999. Occurrence, age, and implications of the Yagan-Onch Hayrhan metamorphic core complex, southern Mongolia. Geology 27, 143–146.

Webb, L.E., Johnson, C.L., Minjin, C., 2010. Late Triassic sinistral shear in the East Gobi Fault Zone, Mongolia. Tectonophysics 495 (3–4), 246–255. Available from: https://doi.org/10.1016/j.tecto.2010.09.033.

Wei, W., Zhang, C., Zhang, S., Wu, C., Yang, K., 2016. Study on the Cretaceous turbidite and reservoir features in the Qingshankou Formation in northern Songliao Basin, NE China. Marine and Petroleum Geology 78, 797–806. Available from: https://doi.org/10.1016/j.marpetgeo.2016.01.016.

Xue, L., Galloway, W.E., 1993. Genetic sequence stratigraphic framework, depositional style, and hydrocarbon occurrence of the Upper Cretaceous QYN formations in the Songliao lacustrine basin, northeastern China. AAPG Bulletin 77, 1792–1808.

Yanshin, A.L., 1989. 1:1,500,000 Map of Geologic Formations of the Mongolian People's Republic. Academia Nauka USSR, 2 sheets.

Zhang, J., Zheng, Y., 1999. Multistage extension and age dating of the Xiaoqinling metamorphic core complex, central China. Acta Geologica Sinica 73, 139–147.

Zhang, Z.M., Liou, J.G., Coleman, R., 1984. An outline of the plate tectonics of China. Geological Society of America Bulletin 95, 295–312.

Zhang, F.Q., Chen, H.L., Batt, G.E., Dilek, Y., A, M.N., Sun, M.D., et al., 2015. Detrital zircon U-Pb geochronology and stratigraphy of the Cretaceous Sanjiang Basin in NE China: Provenance record of an abrupt tectonic switch in the mode and nature of the NE Asian continental margin evolution. Tectonophysics 665, 58–78. Available from: https://doi.org/10.1016/j.tecto.2015.09.028.

Zheng, Y., Zhang, Q., 1993. The Yagan metamorphic core complex and extensional detachment fault in Inner Mongolia. Acta Geologica Sinica 67, 301–309.

Zheng, Y., Wang, S., Wang, Y., 1991. An enormous thrust nappe and extensional metamorphic core complex in Sino-Mongolian boundary area. Science in China. Series B, Chemistry, Life Sciences & Earth Sciences 34, 1145–1152.

Zhou, Z., Barrett, P., Hilton, J., 2003. An exceptionally preserved Lower Cretaceous ecosystem. Nature 421, 807–814.

Further reading

Darby, B.J., Davis, G.A., Zheng, Y., 2001. Structural evolution of the southwestern Daqing Shan, Yinshan belt, Inner Mongolia, China. Geological Society Memoirs. 194, 199–214.

Donskaya, T.V., Windley, B.F., Mazukabzov, A.M., Kröner, A., Sklyarov, E.V., Gladkochub, D.P., et al., 2008. Age and evolution of late Mesozoic metamorphic core complexes in southern Siberia and northern Mongolia. Journal of the Geological Society 165, 405–421. Available from: https://doi.org/10.1144/0016-76492006-162.

Hendrix, M.S., Graham, S.A., Carroll, A.R., Sobel, E.R., McKnight, C.L., Schulein, B.J., et al., 1992. Sedimentary record and climatic implications of recurrent deformation in the Tian Shan: evidence from Mesozoic strata of the north Tarim, south Junggar and Turpan basins, northwest China. Geological Society of America Bulletin 104, 53–79.

Hendrix, M.S., Graham, S.A., Amory, J.Y., Badarch, G., 1996. Noyon Uul (King Mountain) syncline, southern Mongolia: early Mesozoic sedimentary record of the tectonic amalgamation of central Asia. Geological Society of America Bulletin 108, 1256–1274.

Li, P., Sun, M., Narantsetseg, T., Jourdan, F., Hu, W., Yuan, C., 2021. First structural observation around the hinge of the Mongolian Orocline (Central Asia): implications for the geodynamics of oroclinal bending and the evolution of the Mongol-Okhotsk Ocean. Geological Society of America Bulletin 1994–2006. Available from: https://doi.org/10.1130/b36200.1.

Sheldrick, T.C., Barry, T.L., Dash, B., Gan, C., Millar, I.L., Barfod, D.N., et al., 2020. Simultaneous and extensive removal of the East Asian lithospheric root. Scientific Reports 10. Available from: https://doi.org/10.1038/s41598-020-60925-3.

Van der Beek, P., Delvaux, D., Andreissen, P., Levi, K., 1996. Early Cretaceous denudation related to convergent tectonics in the Baikal region, southeast Siberia. Journal of the Geological Society. 153, 515–523.

Zorin, Y.A., 1999. Geodynamics of the western part of the Mongolia-Okhotsk collisional belt, Trans-Baikal region (Russia) and Mongolia. Tectonophysics 306, 33–56.

CHAPTER

11

Xialiao basin arm of the North China basin: a Paleogene rift basin modified by strike-slip deformation

Li-Yuan Hsiao[1] *and Stephan Alan Graham*[2]

[1]Chevron North America Exploration and Production, Covington LA, United States [2]Department of Earth and Planetary Sciences, Stanford University, Stanford, CA, United States

Introduction

A broad area of northeastern China was extended from Paleocene to Oligocene time, forming an intricate system of rifts and grabens known as the North China basin (hereafter NCB, Ye et al., 1985; also known as Bohai basin, Allen et al., 1997, 1998; Bohai Bay basin, Chang, 1991; Bohaiwan basin, Ren et al., 2002) (Fig. 11.1). The rifting involved faulting, differential tectonic subsidence, and voluminous volcanism (Chang, 1991), followed by diminished faulting, postrift thermal subsidence, and reduced volcanism in Neogene and Quaternary times. Over 12 km thick of terrestrial sediments accumulated in the NCB throughout the Cenozoic (Liu, 1987).

The NCB figures importantly in the long-standing debate about the origins of extensional basins, because it has variously been cited as a classic example of a passive rift, an active rift, and a transtensional basin. The passive rifting model of Ye et al. (1985) proposed that this basin was formed by uniform lithospheric stretching. Alternatively, Liu's (1987) active rifting model suggested that the basin formed due to asthenospheric upwelling, based on the observation of thinned crust and elevated heat flow/asthenospheric conductivity/gravity beneath the basin center. This interpretation is consistent with that deduced from subsidence analysis by Shedlock et al. (1985). However, in view of the prominence of the Tan-Lu strike-slip fault along the eastern margin of the basin (Fig. 11.1), a composite pull-apart basin model was subsequently proposed by Allen et al. (1997, 1998) to explain the formation of the NCB. This model suggests that the NCB was initially formed by E-W extension and was later pulled apart by the dextral movement of two basin-bounding strike-slip faults. This model thus represented the NCB as a classic example of a composite pull-apart basin (cf., Nilsen and Sylvester, 1995). The correctness of this model remains uncertain, although it can be easily assessed from structural maps of the basin, that is, the model requires a particular pattern of the rift-related faulting (Figure 11.1 in Allen et al., 1998). However, evaluation of this and other models of the NCB has been difficult, because of limited descriptions and documentation in the literature (e.g., Xu et al., 1983; Liu et al., 1987).

This chapter presents a summary of (1) basement geology, (2) rift stratigraphy and structural features, including the Tan-Lu strike-slip fault, as mapped from 2-D and 3-D seismic data sets, (3) igneous activities, and (4) the amount of extension of the Xialiao basin, which forms the NE rift arm of NCB. We then discuss the origins and lithospheric dynamics of the Xialiao basin, as well as the larger NCB, based on direct structural observations, subsidence analysis, crustal thinning, heat flow, and postrift thermal subsidence.

Regional Geology and Tectonics: Volume 2
DOI: https://doi.org/10.1016/B978-0-444-64136-6.00012-9

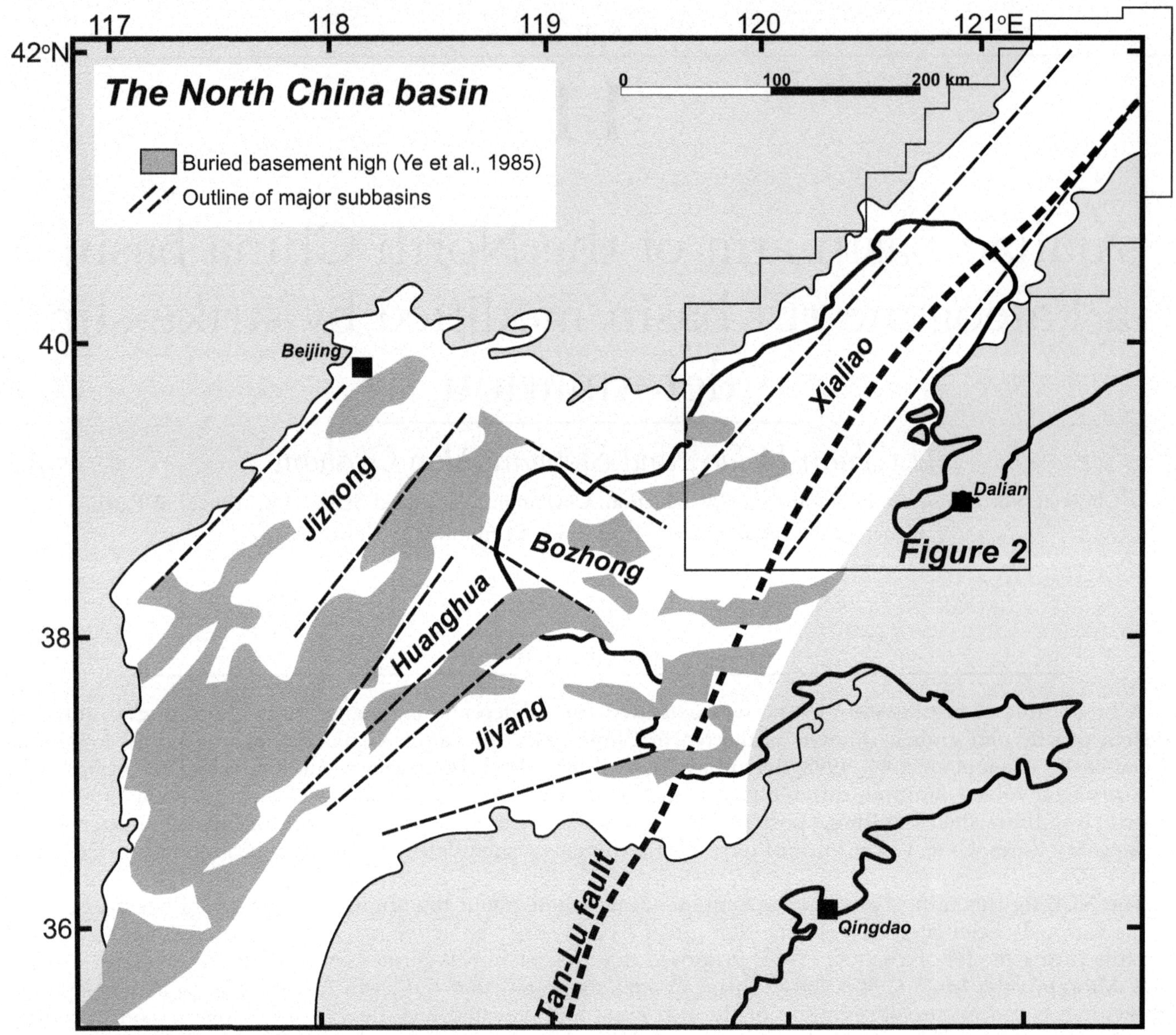

FIGURE 11.1 Tectonic setting of the North China basin.

Geologic setting

North China basin

Extending across 200,000 km^2 of NE China, the NCB is defined by rift structures (Fig. 11.1), postrift sediments, thinned continental crust, and high heat flow (Liu, 1987). The NCB mainly consists of five depressions (major subbasins; Fig. 11.1), composed of over 50 Paleogene half-graben and graben subbasins. These subbasins are mostly asymmetrical sedimentary troughs bounded by high-angle normal or strike-slip faults subparallel to the trend of depressions (Chang, 1991; Hsiao et al., 2004).

The Bozhong depression, along with Xialiao and Jiyang depressions, form a three-armed geometry (Fig. 11.1), reminiscent of basins of rift origins (Burke and Dewey, 1973). The Paleogene troughs and highs are buried by less-deformed Neogene/Quaternary strata of relatively uniform thickness, suggesting that the Paleogene rifting was followed in the Neogene by postrifting regional subsidence (Liu, 1987). Although postrift strata are less deformed, their distribution is also clearly controlled by the three-armed rift geometry.

Localization of Cenozoic NCB rifting might have been influenced by preexisting lithospheric heterogeneity, for example, the Xialiao basin is elongated along the Tan-Lu fault (TLF), which had a Mesozoic history of major strike-slip (e.g., Yin and Nie, 1993; Xu, 1980). Two other major depressions are superposed on prerift structural systems: a Cretaceous rift and older Mesozoic Yanshan orogen, along which the Jizhong (Tang et al., 1988) and Huanghua depressions (Wang et al., 1998) were developed, respectively. However, the detailed relationships between Cenozoic and Mesozoic structures remain undocumented.

Tan-Lu fault

The TLF system trends along the length of the Xialiao basin and parallel to the east margin of the NCB. As a major tectonic feature of east Asia (Fig. 11.1), this fault zone extends NNE for c.3500 km within China, is tens to 200 km wide, and has been depicted repeatedly in regional tectonic maps (e.g., Molnar and Tapponier, 1977; Tapponnier and Molnar, 1977). The TLF is a lithospheric-scale structure because it corresponds to earthquakes up to nearly 100 km deep (http://ciei.colorado.edu/~nshapirco/MODEL/); magnetic, gravity, and heat flow anomalies; and sudden changes in crustal thickness and lithospheric velocity structure (Huo et al., 2000). The TLF was active as a major left-lateral strike-slip fault between Late Triassic and Early Cretaceous, when it accrued an estimated 740 km of slip, based on the correlation of potential piercing points (Xu, 1980). Slip sense reversed in the Cenozoic, when an estimated 35 km of right-slip occurred in the Xialiao basin sector (Hsiao et al., 2004).

Xialiao basin

The Xialiao basin (Fig. 11.2; Shedlock et al., 1985) forms the northeastern arm of the composite NCB (Fig. 11.1). It extends NNE for nearly 400 km and is 80–120 km wide. Liaodong Bay and Liaohe basin comprise its offshore (southern) and onshore (northern) portions. The present-day rift deepens southward, and southerly axial drainage has been the primary sedimentary system since the late Oligocene (Liu et al., 1987).

Regional cross sections demonstrate that the rift valley has a classic continental rift, steer's head configuration: faulted prerift basement is overlain by Cenozoic rift-associated strata (Fig. 11.3; cf. White and McKenzie, 1988). Divergent-configured syn-rift strata overlie the prerift basement and are covered by less-tectonized postrift strata. Syn-rift strata were deposited in isolated half grabens, whereas postrift strata are regionally extensive, less-tectonized, and thicken toward the basin center. The otherwise typical rift cross section is modified by the Tan-Lu strike-slip fault system, which bisects the basin (Fig. 11.3).

Prerift basement

The onshore geology of Liaoning Province provides the best constraint on the nature of the basement under the Xialiao basin. The prerift basement records a long and complex geologic history, as described by the Bureau of Geology and Mineral Resources of Liaoning Province (1989) and summarized as below.

Liaoning Province is generally bisected by the Xialiao basin into two geologic terranes (Liaoxi and Liaodong, west and east of Liaoning, respectively). The two terranes share common features as well as notable differences. Precambrian metamorphic rocks and granite are covered by metamorphosed Ordovician/Silurian shallow-marine carbonate strata, which are overlain by Carboniferous/Permian terrestrial sandstone and metasandstone formations. The thickness of the carbonate section changes significantly on the opposite sides of the Xialiao basin (Liu et al., 1987).

During Mesozoic time, terrestrial basins developed, possibly related to the Yanshan orogeny. Triassic to Cretaceous strata, up to thousands of meters thick, include terrestrial sandstone, conglomerate, and shale. In addition to sedimentary rocks, Jurassic and Lower Cretaceous volcanic rocks, consisting of andesite and basalt, are interlayered in this sedimentary sequence.

The stratigraphy in the Liaodong area differs from that of the Liaoxi area in terms of thickness and rock type. The thickness of Paleozoic carbonate strata changes abruptly across the Liaodong Bay. One thousand meters of Cambrian shallow-marine carbonate strata are present, overlain by unmetamorphosed Ordovician carbonate. Devonian to Permian shallow-marine to terrestrial sandstone, shale, and conglomerate formations lie on the Silurian unconformity. Mesozoic terrestrial deposits also occur in Liaodong, but the thickness is significantly less than in Liaoxi. Another significant difference is that Jurassic andesitic magmatism is missing in Liaodong.

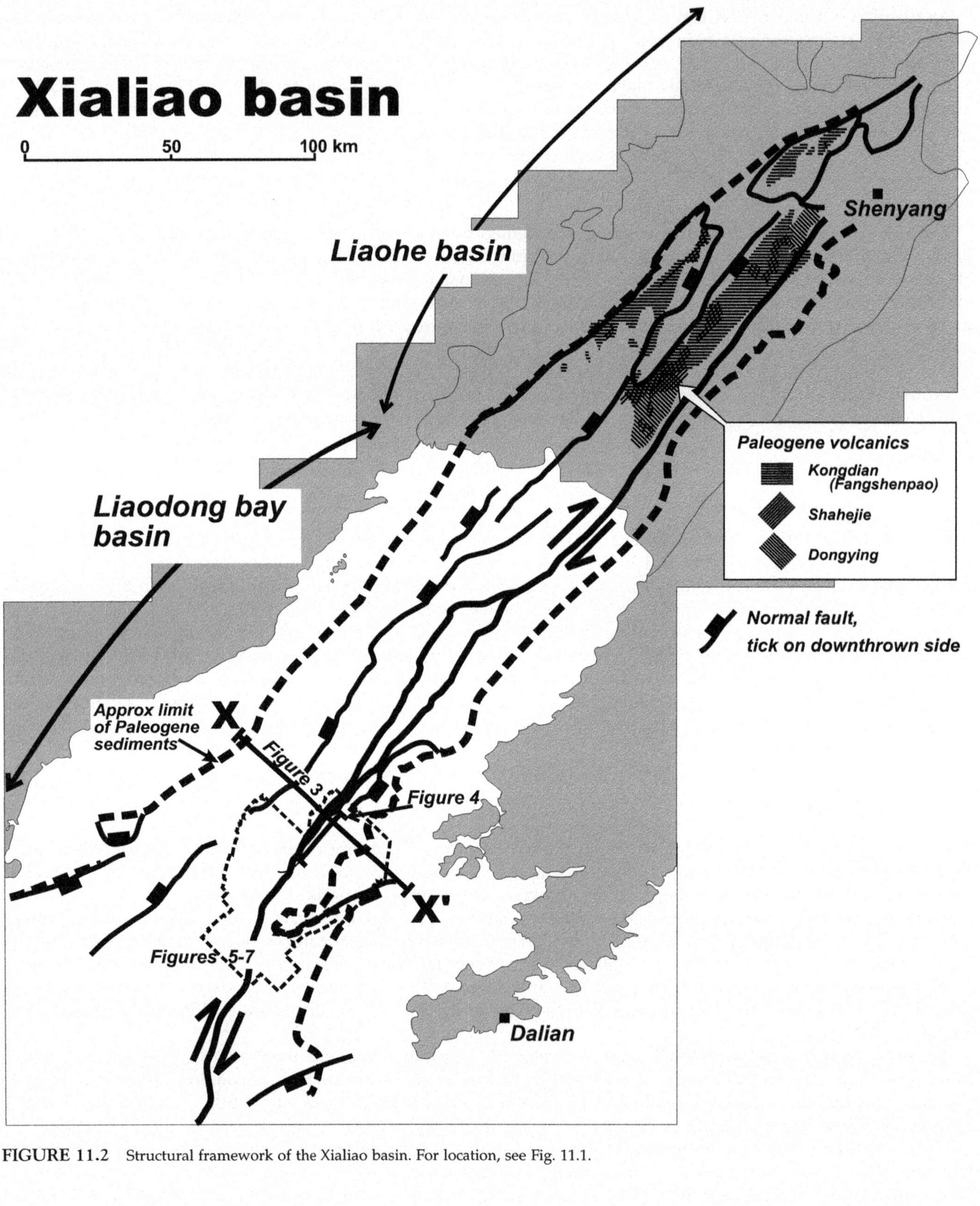

FIGURE 11.2 Structural framework of the Xialiao basin. For location, see Fig. 11.1.

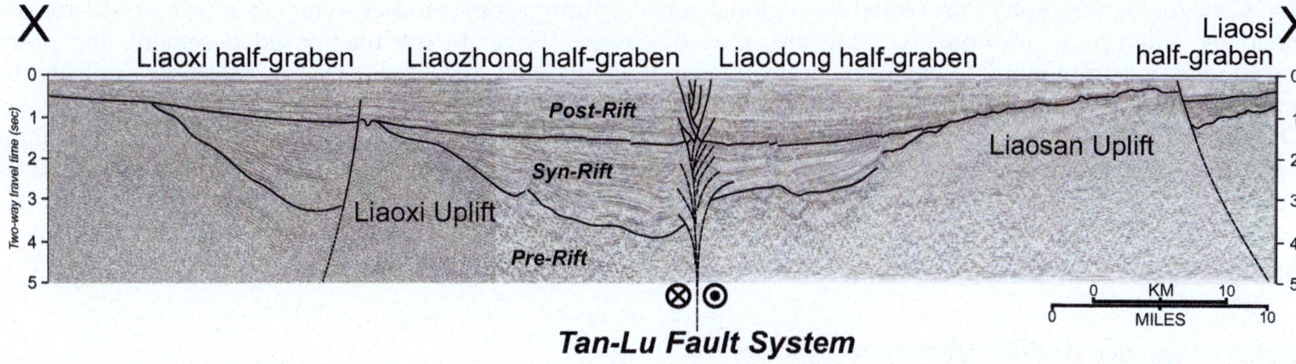

FIGURE 11.3 Regional cross section of the Xialiao basin. For location, see Fig. 11.2. Interpreted seismic reflection profile (above), and depth-converted geologic profile with using interval velocity of data processing (below). Circle with dot indicates motion toward the viewer; circle with x is away from the viewer. Source: *From Hsiao, L.Y., Graham, S.A., and Tilander, N. (2004). Reflection seismic imaging of a major strike-slip fault zone within a rift system: Paleogene structure and evolution of the Tan-Lu Fault system, Liaodong Bay, Bohai, offshore China. American Association of Petroleum Geologists Bulletin, 88, 71–98. AAPG© (2004), reprinted by permission of the AAPG whose permission is required for further use.*

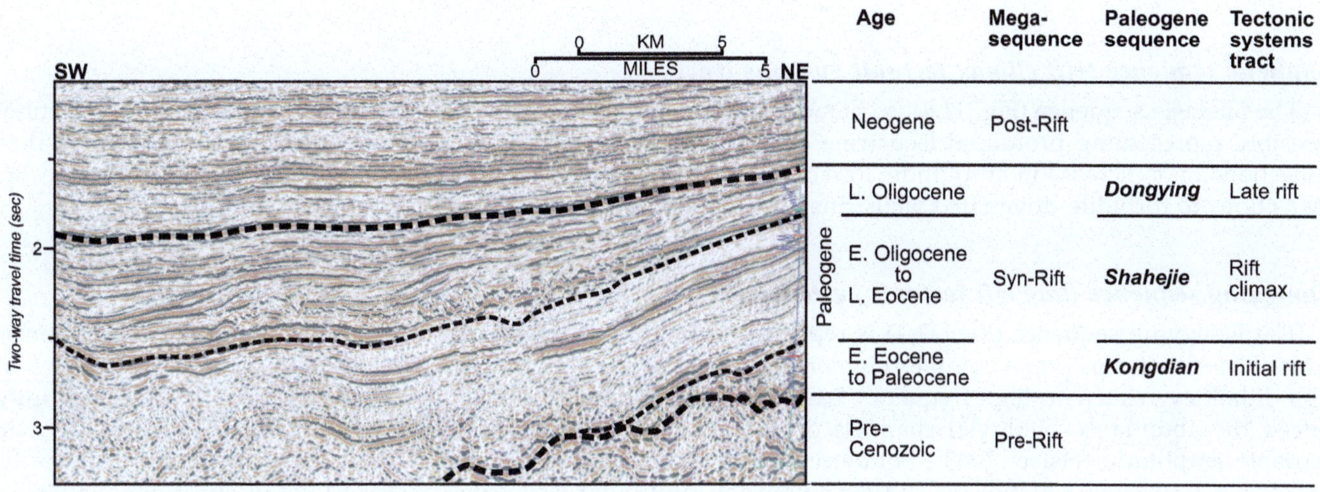

FIGURE 11.4 Seismic section longitudinal to the basin illustrating character of stratigraphic sequences. Well log is a shale-volume log, increasing sand plotted to the left. See Fig. 11.2 for location. Source: *From Hsiao, L.Y., Graham, S.A., and Tilander, N. (2004). Reflection seismic imaging of a major strike-slip fault zone within a rift system: Paleogene structure and evolution of the Tan-Lu Fault system, Liaodong Bay, Bohai, offshore China. American Association of Petroleum Geologists Bulletin, 88, 71–98. AAPG© (2004), reprinted by permission of the AAPG whose permission is required for further use.*

Mesozoic activity of the TLF likely accounts for contrasts on opposite sides of the rift. Strata in Liaodong apparently were tectonically transported northward for hundreds of kilometers along the TLF, accounting for the abrupt change in thickness of the Paleozoic carbonate and differences in Mesozoic stratigraphy (Bureau of Geology and Mineral Resources of Liaoning Province, 1989).

Rift stratigraphy

Faulted prerift basement is overlain by divergently configured strata deposited during basement tilting. The divergent strata are covered by a regional stratigraphic package, which is less tectonized and becomes thinner toward the basin margin. These two stratigraphic packages are regarded as syn- and postrift megasequences separated by a megasequence boundary (Figs. 11.3 and 11.4). The ages of the three megasequences are pre-Cenozoic, Paleogene, and Neogene/Quaternary, respectively (Fig. 11.4; Hsiao, 2003; Hsiao et al., 2004).

The Cenozoic stratigraphy is revealed by regional seismic stratigraphic studies, whereas prerift strata are constrained by outcrops in onshore Liaoning areas, as discussed above. In the unexposed basement, the prerift megasequence is commonly free of seismic reflections, but locally well-developed seismic reflections indicate that basement lithologies vary significantly.

The distribution of the syn-rift megasequence reflects the rift structure. Strata thicken toward the major faults but are thin toward or pinch-out on the ramp side of each half-graben (Fig. 11.3). This divergent configuration suggests that the basement tilted during deposition. The syn-rift megasequence consists of three sequences as tectonic systems tracts that reflect the mode of continental rifting (Fig. 11.4), as suggested by their stratal geometry and dominant environmental facies (Hsiao, 2003; cf. Lambiase, 1990; Prosser, 1993).

Kongdian sequence (initial rift tectonic systems tract)

The lowest Kongdian sequence (Fig. 11.4) displays a wide variety of seismic reflection features, including high to variable amplitude/continuity concordant, chaotic, and clinoform reflections, indicating its highly variable environment of deposition (cf. Mitchum et al., 1977). Reported rock types include alluvial conglomerate, lacustrine sandstone and siltstone, and volcaniclastics. In the Liaohe basin to the north, the voluminous volcanic layers (flood basalt of Zhou et al., 1988) occur at the bottom of the Cenozoic strata. Their thicknesses are generally hundreds of meters and locally more than 1200 m (Zhao et al., 1999).

Shahejie sequence (rift climax tectonic systems tract)

The Shahejie sequence (Fig. 11.4) is characterized by parallel, continuous, and consistent amplitude reflections, possibly representing profundal lacustrine environment near the basin center. Toward the basin margin, these reflections may increase in amplitude, lose continuity, or change shape from concordant to clinoform, suggesting transitions to turbidite-dominated lacustrine fans and exposed islands.

Dongying sequence (late rift tectonic systems tract)

The Dongying sequence (Fig. 11.4) is represented by a set of clinoforms and associated topset and bottomset units. The clinoforms are in complex sigmoidal-oblique configuration, representing deltaic sedimentation in variable lithology related to high-frequency base-level fluctuations (Hsiao, 2003). This base-level change also influenced the abundance of fluvial channels developed on the topsets, characterized by parallel reflections with variable amplitude (Hsiao, 2003). Bottomsets are parallel reflections with fairly consistent amplitude, possibly indicating lacustrine hemipelagic sedimentation (cf. profundal lacustrine facies of Carroll and Bohacs, 1999) of and/or turbidite-dominated facies of hyperpycnal flow (cf. Parsons et al., 2001).

Postrift megasequence

The post- and syn-rift megasequences are separated by a high-amplitude and continuous seismic reflection mappable throughout the entire NCB. The postrift megasequence is relatively uniform in distribution and untectonized. It is thickest near the basin center and thinner toward the basin margin (Figs. 11.3, 11.4). It also thickens toward SSW along the basin axis (Fig. 11.2), although there is another depocenter in northern Liaodong Bay as the result of local transtensional tectonics of the TLF (Hsiao et al., 2004).

The postrift megasequence is characterized by continuous and discontinuous seismic reflections in parallel configuration. The discontinuous reflections are regionally widespread, whereas continuous reflections are better developed to the south. Extending to the north, they die out or laterally become discontinuous reflections. The discontinuous reflections often correspond to sand and gravel deposited in fluvial environments (Liu et al., 1987). The continuous reflections, in contrast, often correlate with mud layers, the deposits of swamp and/or shallow lacustrine environments.

The basin filled during postrift time (Neogene and Quaternary). Due to limited accommodation, coarse-grained sediments were distributed throughout the basin. Around the center of Bohai to the south of the Xialiao basin, a shallow lake developed in the area of greatest subsidence (Liu et al., 1987). This depression likely connected with the open marine environment in late Neogene time, as suggested by the fact that upper Neogene strata overlie the structural high to the east of Bozhong depression that bounded the Bohai and the Yellow Sea (Jiaoliao uplift; N. Tilander, oral communication, 2001).

Rift structures

The Xialiao basin is characterized by three NNE-orienting fault systems: Liaoxi and Liaodong normal fault and Tan-Lu strike-slip fault systems (Fig. 11.2), all of which are high-angle and cut into basement subvertically. The Liaoxi fault system extends throughout the Xialiao basin subparallel to the coastline. The Liaodong normal faults trend NE, and one of those may extend onshore to the Liaodong Peninsula and die out near the bay center.

The normal faults are the boundaries between areas of subsidence and uplift on the hanging-wall and footwall sides of the fault, respectively. The magnitude of subsidence is commonly 4 + km toward the hanging wall and becomes zero or uplift toward the footwall. The magnitude of the uplift is uncertain, but erosional features on Liaosi fault block suggest that the uplift may have been at a scale of hundreds of meters (Hsiao, 2003; Hsiao et al., 2004).

The timing and intensity of the rifting are directly suggested by stratal geometry. The Paleogene strata are in divergent configuration in each half graben bounded by faults, whereas the Neogene/Quaternary strata are relatively uniform, that is, their thickness does not significantly change across the faults (Fig. 11.3). The tectono-stratigraphic relationship suggests that the faults were mostly active during the Paleogene time, particularly the period during which the Shahejie sequence was deposited because this sequence has the most significantly divergent geometry.

Tan-Lu fault

In Xialiao basin, the TLF zone is a through-going feature that bisects the rift system. In the southern basin, it is characterized by a flower structure in cross section (Fig. 11.3) and en echelon folds and faults in plain view (Figs. 11.5, 11.6), but northward it divides into multiple strands characterized by flower structures (Fig. 11.7). The strands merge again beneath the northern limit of Liaodong Bay, and a single major strand cuts Liaohe basin (Fig. 11.2). Regionally distributed focal solutions demonstrate that the fault is a right-slip fault (Chen and Li, 1998; Yang et al., 1989), consistent with the stepping sense of en echelon folds related to the fault in Xialiao basin (Fig. 11.5). Hsiao et al. (2004) inferred 30–40 km of post-Eocene right slip based on the separation of deformation zones and depocenters. The distribution of basement blocks related to rift-fault geometry forced a major restraining bend in the TLF early in its Cenozoic history of activity. This created a positive flower structure during Shahejie time, causing deformed growth strata (Fig. 11.7), influencing Shahejie facies patterns in the center of the rift, and ultimately deforming Shahejie strata in a pattern of en echelon folds (Fig. 11.5). By Dongying time, transtension dominated the TLF zone, all flower structures became negative, subsidiary normal faults were widespread along the fault zone (Fig. 11.6), and the earlier axial transpressional welt subsided (Fig. 11.7). Strike-slip continues to the present day, judging from stratal disruption upward all the way to the sea floor (Figs. 11.3, 11.7).

Igneous activity

Cenozoic igneous activity was widespread throughout the Xialiao basin. In onshore Liaohe basin, volcanism occurred throughout the Paleogene, as revealed by the extensive distribution of volcanic rocks (Fig. 11.2). In the lower Kongdian sequence, basaltic sections are generally 100–200 meters in thickness and locally exceed 1200 meters (Zhao et al., 1999). This thickness is comparable in scale to that of other basins in similar settings, such as Red Sea rift (Chazot et al., 1998). In the Shahejie sequence, basaltic layers are about 10–30 meters thick. In Dongying sequence, volcanism was localized to the north and south of the Liaohe basin, where the basaltic sequence thickness locally exceeds 800 m. The basalt is distributed along major faults and hence possibly originated from fissure eruptions.

Basaltic rocks from the Kongdian sequence yield K/Ar ages ranging from 45 to 65 Ma (Liang et al., 1992), and the geochemical signatures are fairly similar to basalt in other continental rift settings. The alkaline-dominated basaltic rocks mainly consist of basalt and trachybasalt, and some basaltic trachybasalt and basanite using the TAS classification scheme (Le Maitre et al., 1989). They mostly contain olivine, hypersthene, plagioclase, and augite (Zhao et al., 1999). The Nd-Sr isotopic field from lower Shahejie basalt indicates near standard mantle values (PREMA and HIMU zones of Zindler and Hart, 1986), whereas those from younger sequences suggest more lithospheric influences. Such a trend is also observed in Huanghua depression, where the alkalinity of the basalt decreases from the Paleogene to the Neogene systems (Gao, 1986). Paleogene tholeiitic basalt has been reported in Jizhong (Zhou et al., 1988). In Huanghua depression, voluminous magmatism has been imaged by seismic reflection profiling and confirmed by wells (Lu, 1990). In general, Liaohe volcanism was fed by the asthenosphere (cf. Fitton et al., 1998).

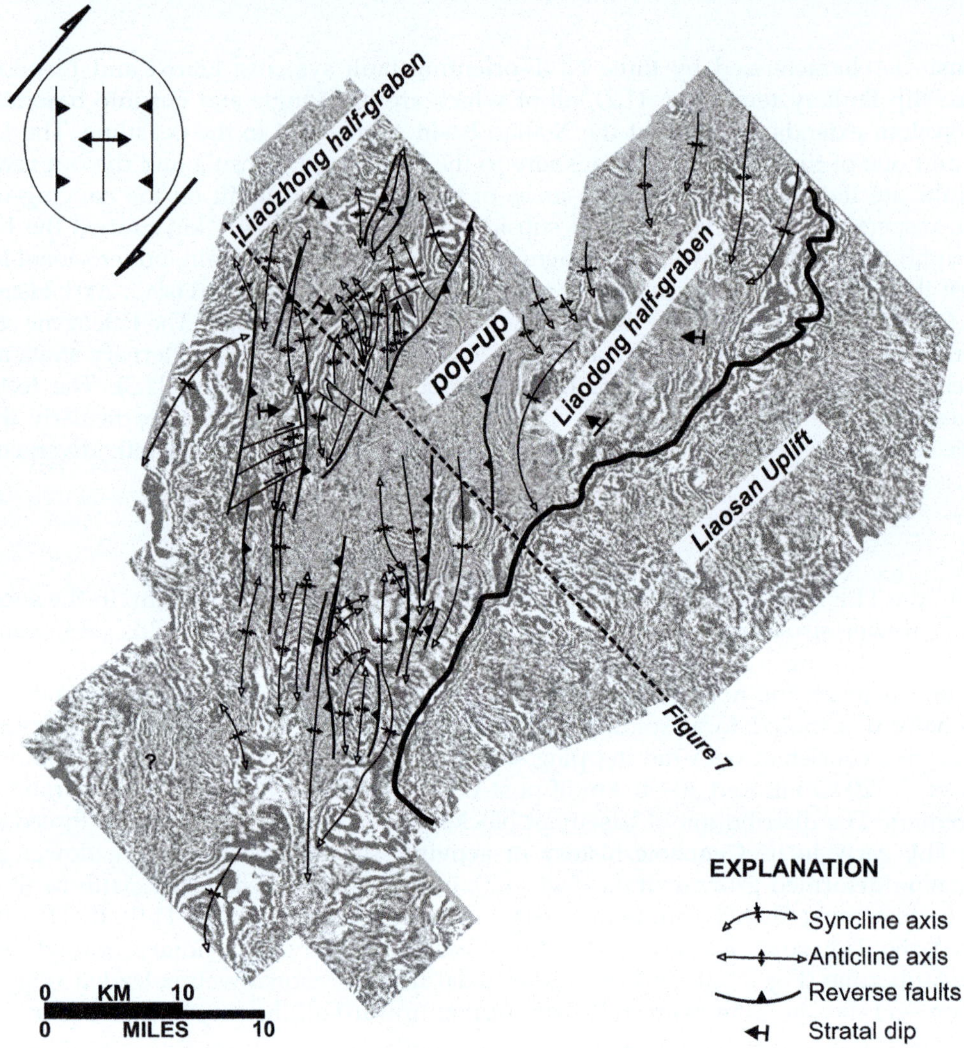

FIGURE 11.5 Interpreted time slice (2864 Ms, amplitude display) from 3-D seismic volume near the middle of the Shahejie sequence showing the distribution of shortening features at lower structural levels of the basin. See Fig. 11.7 for the position of time slice in the vertical section. Strain ellipse is from Wilcox et al. (1973). See Fig. 11.2 for location. Source: *From Hsiao, L.Y., Graham, S.A., and Tilander, N. (2004). Reflection seismic imaging of a major strike-slip fault zone within a rift system: Paleogene structure and evolution of the Tan-Lu Fault system, Liaodong Bay, Bohai, offshore China. American Association of Petroleum Geologists Bulletin, 88, 71–98. AAPG© (2004), reprinted by permission of the AAPG whose permission is required for further use.*

Lower Cenozoic basalt also exists in other northeastern China areas, including Hannuoba (Song and Frey, 1989; Zhi et al., 1990), Wangqing (Xu et al., 2001), and Jiyang and Huanghua depressions (Zhou et al., 1988) in the NCB. These rocks mostly consist of tholeiitic and subalkalic basalt, and minor alkalic basalt. In Hannuoba and Wangqing areas, the basalt includes mantle xenoliths that involve spinel peridotite with minor garnet peridotite, suggesting that the depth of the magmatic source exceeded 75 km. The whole-rock Nd-Sr isotope ratios scatter between zones of standard mantle and lithosphere (e.g., Smith, 1998; Zartman et al., 1991; Tatsumoto et al., 1992; Rollinson, 1993). These data suggest an asthenospheric source for the magma, an interpretation that is consistent with results from the Liaohe basin.

Extension

The amount of basin extension (i.e., extension factor β) can be estimated from direct measurement of heaves along basement faults (βf). By assuming that this basin was formed by lithospheric stretching, the extensional factors can also be evaluated by subsidence analysis (βs), crustal stretching (βc), amount of postrifting thermal subsidence (βt), and heat flux (βh) (Jarvis and McKenzie, 1980; McKenzie, 1978; White, 1988; Ziegler, 1983; Table 11.1). The extension factor βf from Fig. 11.3 is 6%, giving rise to βf 1.06, by directly measuring the heave of the two

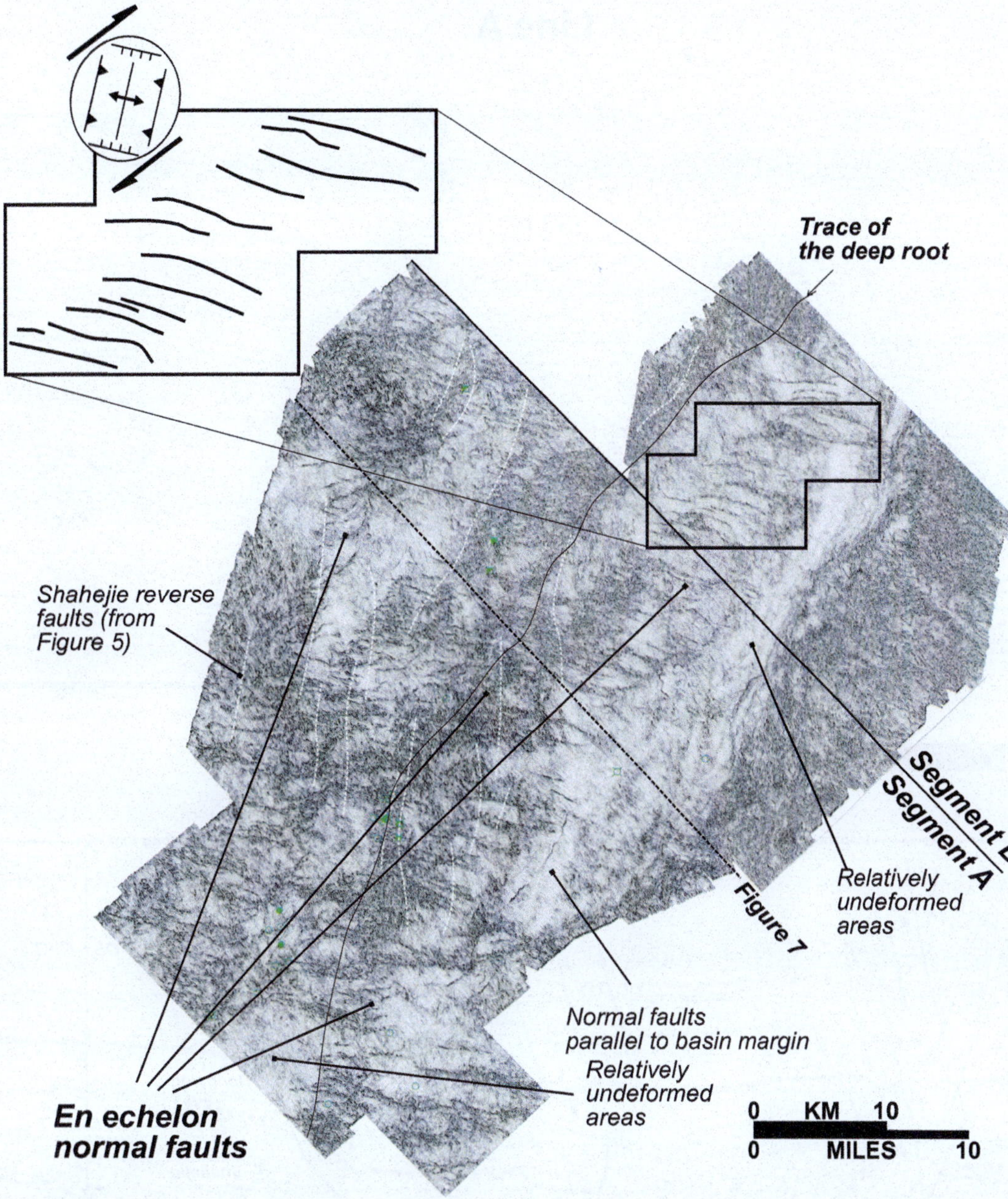

FIGURE 11.6 Time slice (1904 Ms, coherency display) from 3-D seismic volume near the top of the Dongying sequence showing the distribution of normal faults at upper structural levels of the basin. The left-stepping en echelon pattern of normal faults typical of the area is highlighted in the inset map. The traces of reverse faults at depth are shown as white lines. See Fig. 11.7 for the position of time slice in the vertical section. Strain ellipse is from Wilcox et al. (1973). See Fig. 11.2 for location. Source: *From Hsiao, L.Y., Graham, S.A., and Tilander, N. (2004). Reflection seismic imaging of a major strike-slip fault zone within a rift system: Paleogene structure and evolution of the Tan-Lu Fault system, Liaodong Bay, Bohai, offshore China. American Association of Petroleum Geologists Bulletin, 88, 71–98. AAPG© (2004), reprinted by permission of the AAPG whose permission is required for further use.*

normal faults. This is probably an overestimation because the effect of block rotation (cf. Sclater and Célérier, 1989) has not been considered. This low amount of extension is in good agreement with the Liaohe basin, where Chen and Li (1998) analyzed six sections throughout the basin and obtained an average βf of 1.08.

The observed extension, βf, is far less than βs, βc, βt, and βh values (Table 11.1). In Liaohe basin, Shedlock et al. (1985) derived average βs values of 1.3. Subsidence analysis in the same area yielded βs values ranging from 1.1 to 2.0 (Chen and Li, 1998). The βc is approximately 1.5–2 for Xialiao basin. The βt and βh are estimated as near 2 and >4, respectively, using the heat flux-time and sediment thickness-time plots provided by McKenzie (1978). The cause for the discrepancy between βf and other measures of extension is best viewed in the context of the rest of the NCB and other rift systems, as discussed below.

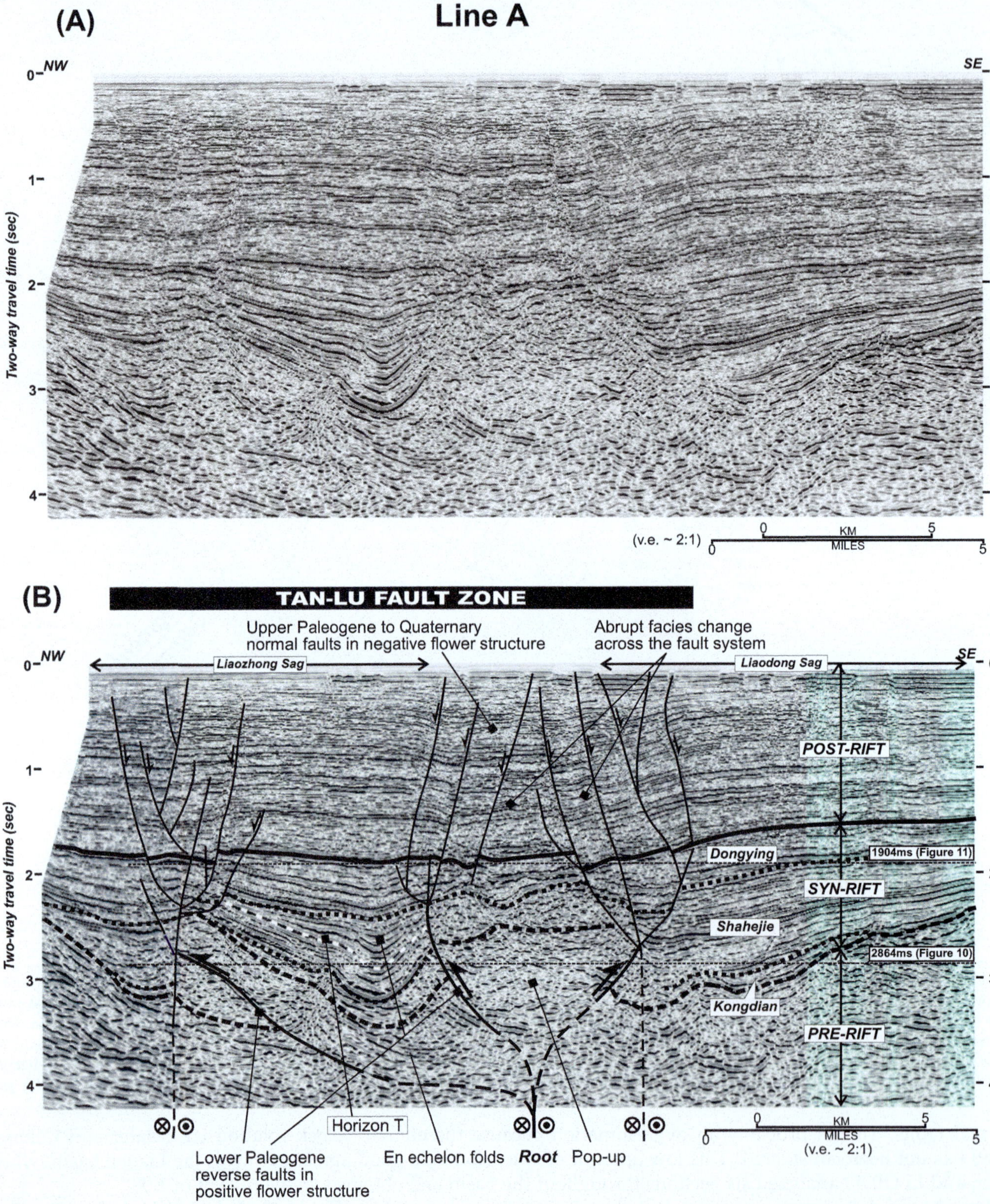

FIGURE 11.7 (A) Uninterpreted and (B) interpreted 2-D seismic line illustrating detailed structure in transverse section of flower structures of the Tan-Lu fault zone. The TLF strand in the center of the basin (and the center of the seismic line) features a positive flower structure at Shahejjie levels, characterized by a central pop-up structure and flanking compressional growth syncline (especially on its left flank). In contrast, from the Dongying level to the sea floor, both flower structures imaged in this section are negative. Positions of time slices in Figs. 11.5 and 11.6 are indicated at the right edge of the section. See Fig. 11.2 for the location of seismic line. *TLF*, Tan-Lu fault. Source: *From Hsiao, L.Y., Graham, S.A., and Tilander, N. (2004). Reflection seismic imaging of a major strike-slip fault zone within a rift system: Paleogene structure and evolution of the Tan-Lu Fault system, Liaodong Bay, Bohai, offshore China. American Association of Petroleum Geologists Bulletin, 88, 71–98. AAPG© (2004), reprinted by permission of the AAPG whose permission is required for further use.*

TABLE 11.1 Extension factors of the Xialiao and North China basins.

Subbasins						Extensional factors					
		βf	βs	Thermal subsidence (km)[d]	βt[e]	Moho depth (km)[f]	Total subsidence (km)[b]	Upper crustal thickness (km)[g]	βc[h]	Present heat flow (HFU)[f]	βh[e]
Xialiao basin	Liaohe basin	[a]**1.08**	[c]**1.3**	0.5–1.2	**<1.25**	29–33	[a]5–9	20–24	**1.5–1.8**	1.6–2.0	~4–∞
	Liaodong bay basin	[b]**1.06**	N/A	1.2–2.5	**<1.25–1.5~2**	~28	[b]6–10	18–22	**1.64–2**	1.8–2.0	∞
East Bozhong depression	**N/A**	[d]**>1.3**	3.7+	**2–4**	<28	[f]>10	~18	**~2**	2.36	∞	

[a]*Chen and Li, 1998.*
[b]*This study.*
[c]*Shedlock et al. (1985).*
[d]*Ye et al. (1985).*
[e]*Using plots of McKenzie (1978).*
[f]*Liu (1987).*
[g]*Moho depth—total subsidence.*
[h]*Prerift crustal thickness (36 km; Liu, 1987)/upper crustal thickness.*

Discussion

Lithospheric dynamics

Beyond the Xialiao rift, unexpectedly high βs, βc, βt, and βh values also have been derived from other parts of the NCB. The subsidence analysis of Hellinger et al. (1985) throughout the NCB derived βs values ranging from 1.26 to 1.72. The βt and βh are both estimated as >4. The mismatch between measurable fault extension and magnitude of extension predicted from other criteria is by no means unique to the NCB, as documented by others in various rift settings (e.g., Chenet et al., 1982; Karner et al., 1987; Moretti and Pinet, 1987). Nor can the discrepancy in extension estimates from NCB be explained by errors from estimation (cf. White, 1990); instead, it probably suggests that the lithospheric stretching of McKenzie (1978) is not the (only) mechanism contributing to basin formation because the basin subsided without sufficient upper crustal extension.

Instead, asthenosphere upwelling may have accompanied or caused the basin rifting (Shedlock et al., 1985; Liu, 1987), accounting for the discrepancy among extension indicators. The process of mantle penetration through the crust is unclear in detail but a preexisting zone of anomalously low-viscosity asthenosphere is required (Olsen et al., 1988). In the case of NCB, the zones of lithospheric weakness could have been provided by the TLF, Mesozoic orogenic belts, and Cretaceous rifting, as mentioned earlier.

We are uncertain about the origin of asthenospheric upwelling and its relationship to modest crustal extension in the NCB. Asthenospheric upwelling may result in continental extension; crustal fractures created by extension and preexisting zones of weakness may release mantle pressure and trigger the melting and upwelling of the upper mantle (Saunders et al., 1992). The former mechanism is similar to that of active rifting (e.g., Burke and Dewey, 1973), except that it is driven by the asthenosphere instead of a mantle plume. Under the active rifting model, the base of the lithosphere would be ponded by rising magma for long periods of time. The ponding results in conductive heating of the overlying lithosphere and the production of large amounts of melt (flood basalt; Farnetani and Richards, 1994). This scenario could account for the present high heat flow (Ye et al., 1985) and the low P-wave velocity under the NCB area (Zhang, 1998). Hence, either passive or active rifting processes remain the possible causes for NCB formation, like other well-studied cases of rift basins such as East Africa, Baikal, Rio Grande, and Rhine rifts, which have many similar geological features (e.g., magmatism, lithospheric thinning, and asthenospheric upwarping; Ruppel, 1995).

In either passive or active rifting scenario, prerift lithospheric heterogeneity may have played an important role in basin formation. Thinned continental crust is preferred for mantle upwelling, decompressional melting, magmatism, and/or crustal underplaing (e.g., McKenzie and Bickle, 1988; Huang et al., 1996). In the NCB case, the TLF changes orientation c.10 degrees near the NCB center (Fig. 11.1), which should have created great structural complexity. Resulting extension may have promoted mantle melting. At upper crust level, preexisting fabrics can control the geometry and location of rifts by initiating, diverting, or inhibiting fracture propagation (cf. Lezzar et al., 2002). At a smaller scale, the Xialiao, Huanghua, and Jizhong depressions might correspond to preexisting zones of weakness.

Evolution of the Xialiao basin

We infer the following evolution of the Xialiao basin:

Stage 1: Paleocene to Eocene: rift initiation

Due to either lithospheric extension or asthenospheric upwelling, abnormal mantle occurred under and along the present-day Xialiao basin area. Hot upper mantle material heated the lithosphere and reduced its thickness and density. Driven by isostatic buoyancy, lithospheric uplift produced a tensile stress field that triggered normal faulting and basin formation. Voluminous basaltic magma flooded the Liaohe basin to the north and may also have extruded along the major faults of the Liaodong Bay basin. Basin accommodation was limited and mostly distributed along the faults. Deposited sediments include alluvial conglomerate and sandstone, lacustrine mudstone, and possible volcaniclastics.

Stage 2: Eocene to Oligocene: rift climax

Asthenosphere upwelling persisted and reached its climax. The lithosphere was hottest and had lowest density, and therefore the maximum magnitude of uplift, normal faulting, and basin accommodation occurred. In addition, the TLF was reactivated as a right-slip fault soon after the onset of rifting. Rift-related normal faults are high angle. Between the faults, blocks were rotating, producing uplift and subsidence of the hanging and footwall sides, respectively. Like the normal faults related to rifting, the TLF formed an uplift zone due to the transpressional bending in southern Liaodong Bay and was subject to erosion (e.g., erosion on the order of 10^2 meters occurred on the Liaosi uplift in the southeast bay; Hsiao, 2003). Sublacustrine fans or exposed platform facies developed along uplifted basin margins, whereas fine-grained lacustrine sediments dominated the basin center.

Stage 3: Upper Oligocene to present: rift fading and postrifting thermal subsidence

Asthenosphere upwelling ceased and the lithosphere began cooling, but the TLF remained active as a transtensional system. Melt may have underplated the NCB, as suggested by minor later Tertiary magmatism, relatively high heat flow and intense thermal subsidence. The magnitude of normal faulting decreased. The region subsided and sediments filled the basin through deltaic progradation along the axis of the rift basin.

Evolution of the North China basin

We propose that asthenospheric upwelling may have played a key role in forming the NCB, and the upwelling may or may not have been triggered by lithospheric extension (e.g., active and passive rifting, respectively). The asthenospheric upwelling was probably located near the center of Bohai based on the geometry of three rift arms: Xialiao (NNE), Jiyang (SSW), and Bozhong (E-W) depressions. The upwelled asthenosphere heated the lithosphere along these rift arms and caused localized extension in the crust above. In each rift valley, Paleogene rift-related normal faults are parallel to subparallel to the rift valley, as well as Neogene to Quaternary pattern of sediments distribution. The Paleogene normal faulting reflects localized extension, and postrift sedimentation was controlled by basin accommodation created by thermal subsidence, the amount of which is directly related to the distribution of heated continental lithosphere. The three-arm geometry of asthenospheric upwelling corresponds to regional geophysical measurements, including thickness of crust and lithosphere, high heat flow, residual gravity anomaly, and low p-wave velocity (Liu, 1987).

The three-arm geometry still controls modern depositional patterns and coastline geomorphology. Relatively intense thermal subsidence localized the axis of subsidence along the center of the rift valleys. Each rift valley therefore corresponds to a major fluvial input system: Liaohe, Huanghe (Yellow), and Haihe rivers for Xialiao, Jiyang, and Bozhong depressions, respectively. Due to the limited amount of sediment input, the coastline of Xialiao and Bozhong depressions has concave-landward shapes, whereas the Jiyang rift has concave-seaward geometry, the latter probably reflects sediment input from the Huanghe River so voluminous that it exceeds basin accommodation from thermal subsidence.

Failure of the composite pull-apart basin model and the role of TLF in the NCB formation

Allen et al. (1997, 1998) proposed a composite pull-apart basin model for the formation of the NCB, based on recognition of right-lateral strike-slip movement of the TLF, and the overall "lazy z" shape in map view of the NCB (Fig. 11.1). If this 300-km wide (N-S direction) portion had been pulled apart by right-slip, the following would be required: (1) Paleogene rift-related normal faults in the Xialiao basin should be arranged in en echelon pattern, (2) TLF slip would have ceased near the Oligocene-Miocene boundary time, and (3) the TLF would have accrued 67–133 km of slip throughout the Paleogene (based on β ranging from 1.3 to 1.8; Allen et al., 1997).

However, these requirements are not met by the regional geologic relations derived from the structural analysis of Hsiao et al. (2004) and as summarized below. The Paleogene rift-related major normal faults in Liaodong Bay are mostly long and straight rather than en echelon (Fig. 11.2). The TLF has been active from Paleogene through Quaternary: stratal wedging signals Paleogene fault actively, yet most allied faults cut through the entire Cenozoic strata rather than terminate at the top-Oligocene horizon. Finally, total Cenozoic slip along the TLF is 30–40 km (Hsiao et al., 2004). We therefore conclude that although the TLF was active during rifting of the NCB, the magnitude of strike-slip is insufficient to account for the NCB. Instead of pulling apart the entire basin, the Cenozoic and Mesozoic movements of the TLF may have contributed to the subsidence of the NCB by weakening the lithosphere, as discussed earlier.

Conclusions

The Xialiao basin displays a classic continental rift, a symmetric "steer's head" configuration that consists of pre-, syn-, and postrift megasequences. The prerift basement consists of metamorphic, carbonate, sedimentary, and volcanic strata. The thickness and type of prerift rocks change abruptly across the Liaodong Bay, a change that is related to major Mesozoic strike-slip along the TLF. The syn-rift megasequences consist of three tectonic systems tracts, characterized by alluvial/fluvial/lacustrine, profundal lacustrine, and deltaic settings; the postrift megasequence consists of fluvial and shallow lacustrine deposits. Volcanic rocks are particularly abundant within the initial rift sequence and mainly consist of alkaline basalt, probably reflecting asthenospheric upwelling. The strata were cut by high-angle normal faults that accommodated initial basin rifting and extension. These basement-involved structures trend NNE subparallel to the basinal axis. The observed amount of extension is about 6%, which is markedly lower than anticipated by subsidence analysis, crustal thinning, thermal subsidence, and heat flow.

Abnormally high values of heat flow, thermal/initial subsidence, and crustal thinning of the NCB are probably the consequences of asthenospheric upwelling. This mantle event occurred along the major depressions of the NCB, among which the Xialiao, Bozhong, and Jiyang depressions comprise a three-armed geometry of continental rifting. The Huanghua, Jizhong, and Xialiao depressions correspond to preexisting zones of lithospheric weakness. The asthenospheric upwelling may have elevated the crust and provided tensile stress that rifted the continent, and extensional fractures by extension and TLF activity may have triggered the asthenospheric upwelling. The TLF was active as a right-lateral strike-slip fault system during basin formation, and it served as a weak zone for intense asthenosphere upwelling instead of directly causing NCB rifting.

Acknowledgments

This article was originally published as Chapter 10 in *Regional Geology and Tectonics: Phanerozoic Rift Systems and Sedimentary Basins*, by Hsiao, L.Y., and Graham, S.A., Xialiao, North China Basin, pages 237–257, Copyright Elsevier ISBN: 9780444563569, 2012. The authors acknowledge support by the member companies of the Stanford-China Industrial Affiliates Program. The authors thank Dennis Astilla, Attila Aydin, Edmund Chang, Tim Cope, Kuan Hsu, Yu Jin, Chuck Kluth, Hongde Liang, Nat Tilander, Maoshan Wang, Yongjun Yue, and Lori Zhang for discussions and/or assistance in preparing this chapter. ChevronTexaco, Petronas, and the China National Offshore Oil Company granted permission to publish the seismic data presented in this chapter, and the American Association of Petroleum Geologists Bulletin granted permission to reproduce Figures 11.3–11.7 of this chapter from Hsiao et al. (2004).

References

Allen, M.B., McDonald, D.I.M., Zhao, X., Vincent, S.J., Brouet-Menzies, C., 1997. Early Cenozoic two-phase thermal subsidence and inversion of the Bohai Basin, northern China. Marine and Petroleum Geology 14, 951–972.

Allen, M.B., McDonald, D.I.M., Zhao, X., Vincent, S.J., Brouet-Menzies, C., 1998. Transtensional deformation in the evolution of the Bohai Basin, Northern China. In: Holdsworth, R.E., Strachan, R.A., Dewey, J.F. (Eds.), Continental Transpressional and Transtensional Tectonics, 135. Geological Society Special Publication, London, pp. 215–229.

Bureau of Geology and Mineral Resources of Liaoning Province, 1989. Regional Geology of Liaoning Province. Geology Press, Beijing, p. 856.

Burke, K., Dewey, J., 1973. Plume generated triple-junctions: key indicators in applying plate tectonics to old rocks. Journal of Geology 81, 406–433.

Carroll, A.R., Bohacs, K.M., 1999. Stratigraphic classification of ancient lakes: balancing tectonic and climatic controls. Geology 27, 99–102.

Chang, C.Y., 1991. Geological characteristics and distribution patterns of hydrocarbon deposits in the Bohai Bay Basin, east China. Marine and Petroleum Geology 8, 98–106.

Chazot, G., Menzies, M.A., Baker, J., 1998. Pre-, syn-, and post-rift volcanism on the south western margin of the Arabian plate. In: Purser, B.H., Bosence, D.W.J. (Eds.), Sedimentation and Tectonics of Rift Basin- Gulf of Aden. Chapman & Hall, London, pp. 50–55.

Chen, Q., Li, Z. (Eds.), 1998. Structures of Eastern Depression, Liaohe Basin and Hydrocarbon Potential Analyses. Geology Press, Beijing, p. 156.

Chenet, P.Y., Montadert, L., Gairaud, H., Roberts, D., 1982. Extensional ratio measurements on the Galicia, Portugal and northern Biscay continental margins. In: Watkins, J.S., Drake, C.L. (Eds.), Studies in Continental Margin Geology, 34. American Association of Petroleum Geologists Memoir, Tulsa, pp. 703–715.

Farnetani, C.G., Richards, M.A., 1994. Numerical investigations of the mantle plume initiation model for flood basalt events. Journal of Geophysical Research 99, 13813–13833.

Fitton, J.G., James, D., Kempton, P.D., Ormerod, D.S., Leeman, W.P., 1998. The role of lithospheric mantle in the generation of late Cenozoic basic magmas in the western United States. Journal of Petrology, Special Lithospheric Issue 331–349.

Gao, Z., 1986. The Tertiary blind volcanic rocks in Huanghua Depression and their tectonic setting. Acta Petrologica Sinica 1986, 14–30.

Hellinger, S.J., Shedlock, K.M., Sclater, J.G., Hong, Y., 1985. The Cenozoic evolution of the North China basin. Tectonics 4, 343–358.

Hsiao, L.Y., 2003. Sequence Stratigraphy and Structure of the South Xialiao Basin, Bohai, Offshore China [Ph.D. Dissertation]. Stanford University, Stanford, p. 281.

Hsiao, L.Y., Graham, S.A., Tilander, N., 2004. Reflection seismic imaging of a major strike-slip fault zone within a rift system: Paleogene structure and evolution of the Tan-Lu Fault system, Liaodong Bay, Bohai, offshore China. American Association of Petroleum Geologists Bulletin 88, 71–98.

Huang, W., Gao, W., Ding, G., 1996. Neogene volcanism and Holocene earthquakes in the Tanlu fault zone, eastern China. Tectonophysics 260, 259–270.

Huo, G., Wang, X., Bai, J., Li, R., 2000. Geophysical features of the Tan-Lu fault zone. In: Wang, X. (Ed.), Tan-Lu Fault Zone. Geology Press, Beijing, pp. 74–107.

Jarvis, G.T., McKenzie, D.P., 1980. Sedimentary basin formation with finite extension rates. Earth and Planetary Science Letters 48, 42–52.

Karner, G.D., Lake, S.K., Dewey, J.F., 1987. The thermal and mechanical development of the Wessex Basin. In: Coward, M.P., Dewey, J.F., Hancock, P.L. (Eds.), Continental Extensional Tectonics, 28. Geological Society Special Publication, London, pp. 517–536.

Lambiase, J.J., 1990. A model for tectonic control of lacustrine stratigraphic sequences in continental rift basin. In: Katz, B.J. (Ed.), Lacustrine Basin Exploration; Case Studies and Modern Analogs, 50. American Association of Petroleum Geologists Memoir, Tulsa, pp. 265–276.

Le Maitre, R.W., Bateman, P., Dudek, A., Keller, J., Lameyre Le Bas, M.J., Sabine, P.A., et al., 1989. A Classification of Igneous Rocks and Glossary of Terms. Blackwell, Oxford, Boston, p. 193.

Lezzar, K.E., Tiercelin, J.J., Le Turdu, C., Cohen, A.S., Reynolds, D.J., Le Gall, B., et al., 2002. Control of normal fault interaction on the distribution of major Neogene sedimentary depocenters, Lake Tanganyika, East Africa rift. American Association of Petroleum Geologists Bulletin 86, 1027–1059.

Liang, H., Shen, S., Liu, X., Chen, W., Li, D., 1992. The age of the volcanic rocks and their geological time in Liaohe. Depression. Acta Petrolei Sinica 13, 35–41.

Liu, G., 1987. The Cenozoic rift system of the North China Plain and the deep internal processes. Tectonophysics 133, 277–285.

Liu, X., Wang, X., Wang, Y., Ho, B., Wang, L., Chen, H., 1987. Sedimentary basins in Bohai region. In: Wang, S., Zhao, L. (Eds.), Sedimentary Basins on Continental Shelf of North Offshore China; Petroleum Geology of China 16. Petroleum Industry Press, Beijing, pp. 3–262.

Lu, B., 1990. Typical Seismic Sections in China. Petroleum Industry Press, Beijing, p. 260.

McKenzie, D., Bickle, M.J., 1988. The volume and composition of melt generated by extension of the lithosphere. Journal of Petrology 29, 625–679.

McKenzie, D., 1978. Some remarks on the development of sedimentary basins. Earth and Planetary Science Letters 40, 25–32.

Mitchum Jr., R.M., Vail, P.R., Thompson III, S., 1977. Seismic stratigraphy and global changes of sea level, Part 2: the depositional sequence as a basic unit for stratigraphic analysis. In: Payton, C.E. (Ed.), Seismic Stratigraphy- Applications to Hydrocarbon Exploration, 26. American Association of Petroleum Geologists Memoir, Tulsa, pp. 53–62.

Molnar, P., Tapponier, P., 1977. Relation of the tectonics of Eastern China to the India-Eurasia collision: application of slip line field theory to large scale continental tectonics. Geology 5, 212–216.

Moretti, I., Pinet, B., 1987. Discrepancy between lower and upper crustal thinning. In: Beaumont, C., Tankard, A.J. (Eds.), *Sedimentary basins and basin forming mechanisms.* Canadian Society of Petroleum Geology Memoir, 12. pp. 233–239.

Nilsen, T.H., Sylvester, A.G., 1995. Strike-slip basins. In: Busby, C.J., Ingersoll, R.V. (Eds.), Tectonics of Sedimentary Basins. Blackwell Science, Malden, pp. 425–458.

Olsen, P., Shubert, F., Anderson, C., Goldman, P., 1988. Plume formation and lithospheric erosion: a comparison of laboratory and numerical experiments. Journal of Geophysical Research 93, 15065–15084.

Parsons, J.D., Bush, J.W.M., Syvitski, J.P.M., 2001. Hyperpycnal plume formation from riverine outflows with small sediment concentrations. Sedimentology 48, 465–478.

Prosser, S., 1993. Rift-related linked depositional systems and their seismic expression. In: Williams, G.D., Dobb, A. (Eds.), Tectonics and Seismic Sequence Stratigraphy, 71. Geological Society Special Publication, London, pp. 35–66.

Ren, J., Tamaki, K., Li, S., Zhang, J., 2002. Late Mesozoic and Cenozoic rifting and its dynamic setting in Eastern China and adjacent areas. Tectonophysics 344, 175–205.

Rollinson, H.R., 1993. Using Geochemical Data: Evaluation, Presentation, Interpretation. Longman, Singapore, p. 352.

Ruppel, C., 1995. Extensional processes in continental lithosphere. Journal of Geophysical Research 100, 24,187–24,215.

Saunders, A.D., Storey, M., Kent, R.W., Norry, M.J., 1992. Consequences of plume-lithosphere interactions. In: Storey, B.C., Alabaster, T., Pankhurst, R.J. (Eds.), Magmatism and the Causes of Continental Break-up, 68. Geological Society Special Publication, London, pp. 41–60.

Sclater, J.G., Célérier, P.A.F., 1989. Errors in extension measurements from planar faults observed on seismic reflection lines. Basin Research 1, 217–221.

Shedlock, K.M., Hellinger, S.J., Hong, Y., 1985. Evolution of the Xialiao basin. Tectonics 4, 171–185.

Smith, A.D., 1998. The geodynamics significance of the DUPAL anomaly in Asia. In: Flower, M.F.J., Chung, S.L., Lo, C.H., Lee, T.Y. (Eds.), Mantle Dynamics and Plate Interactions in East Asia, 27. American Geophysical Union Geodynamics Series, pp. 89–106.

Song, Y., Frey, F.A., 1989. Geochemistry of peridotite xenoliths in basalt from Hannuoba, Eastern China: Implications for subcontinental mantle heterogeneity. Geochimica et Cosmochimica Acta 53, 97–113.

Huabei Oilfields. In: Tang, Z., Wu, H., Gao, W., Qin, K., Liu, Y. (Eds.), Petroleum Geology of China 5. Petroleum Industry Press, Beijing, p. 569.

Tapponnier, P., Molnar, P., 1977. Active faulting and tectonics in China. Journal of Geophysical Research 82, 2905–2930.

Tatsumoto, M., Basu, A.R., Huang, W., Wang, J., Xie, G., 1992. Sr, Nd, and Pb isotopes of ultramafic xenoliths in volcanic rocks of Eastern China: enriched components EMI and EMII in subcontinental lithosphere. Earth and Planetary Science Letters 113, 107–128.

Wang, Z., Liu, H., Duan, Z., Wang, H., 1998. Analysis of Meso-Cenozoic tectonic inversion of Huanghua Depression, Hebei, China. Journal of China University of Geoscience 23, 289–293.

White, N. (1988). Extension and subsidence of the continental lithosphere [Ph.D. dissertation]. Cambridge: University of Cambridge (as quoted in White, 1990).

White, N., 1990. Does the uniform stretching model work in the North Sea? In: Blundell, D.J., Gibbs, A.D. (Eds.), Tectonic Evolution of the North Sea Rifts. Clarendon Press, Oxford, pp. 217–240.

White, N., McKenzie, D.P., 1988. Formation of the 'steer's head' geometry of sedimentary basins by differential stretching of the crust and mantle. Geology 16, 250–253.

Wilcox, R.E., Harding, T.P., Seely, D.R., 1973. Basic wrench tectonics. American Association of Petroleum Geologists Bulletin 57, 74–96.

Xu, J., 1980. The horizontal displacement of the Tancheng-Lujiang fault zone and its geologic significance, Scientific Papers on Geology for International Exchange, No. 1. Geological Publish House, Beijing, pp. 129–142.

Xu, H., Lu, W., Wang, S., Wan, J., 1983. Paleogene sedimentary facies systems and direction of searching for oil and gas in Bohaiwan basin. Acta Geologica Sinica 3, 243–253.

Xu, Y.G., Menzies, M.A., Ross, J.V., 2001. Thermo-tectonic destruction o fthe Archaean lithospheric keel beneath the Sino-Korean craton in China: evidence, timing and mechanism. Physics and Chemistry of the Earth (A) 26, 747–757.

Yang, M., Lu, P., Hua, X., Song, R., 1989. Focal mechanism solutions of earthquakes. In: Ma, X., Ding, G., Cao, W., Zhang, H., Zhang, B., Ma, Z. (Eds.), Lithospheric Dynmaics Atlas of China. China Cartographic Publishing House, p. 26.

Ye, H., Shedlock, K.M., Hellinger, S.J., Sclater, J.G., 1985. The North China basin: an example of a Cenozoic rifted intraplate basin. Tectonics 4, 153–170.

Yin, A., Nie, S., 1993. An indentation model for the North and South China collision and the development of Tanlu and Honam fault systems, eastern Asia. Tectonics 12, 801–813.

Zartman, R.E., Futa, K., Peng, Z.C., 1991. A comparison of Sr-Nd-Pb isotopes in young and old continental lithospheric mantle: Patagonia and eastern China. Australian Journal of Earth Sciences 38, 545–557.

Zhang, Y.S., 1998. Three-dimensional upper mantle structure beneath east Asia and its tectonic implications. In: Flower, M.F.J., Chung, S.L., Lo, C.H., Lee, T.Y. (Eds.), Mantle Dynamics and Plate Interactions in East Asia, 27. American Geodynamics Union Geodynamics Series, pp. 11–23.

Zhao, C., Meng, W., Jin, C., Cai, G., Zhao, S., Ji, H., 1999. Volcanic rocks and oil and gas in Liaohe basin. Petroleum Industry Press, Beijing, p. 101.

Zhi, X., Song, Y., Frey, F.A., Feng, J., Zhai, M., 1990. Geochemistry of Hannuoba basalts, Eastern China; constraints on the origin of continental alkalic and tholeiitic basalt. Chemical Geology 88, 1–33.

Zhou, X., Zhu, B., Liu, R., Chen, W., 1988. Cenozoic basaltic rocks in eastern China. In: Macdougall, J.D. (Ed.), Continental Flood Basalt. Kluwer Academic Publishers, Dordrecht, pp. 311–330.

Ziegler, P.A., 1983. Crustal thinning and subsidence in the North Sea: matters arising. Nature 304, 561.

Zindler, A., Hart, S.R., 1986. Chemical geodynamics. Annual Review of Earth and Planetary Sciences 14, 493–571.

Index

Note: Page numbers followed by "*f*" and "*t*" figures and tables, respectively.

O

P

Q

R

S

T

U

V

W

X

Y

Z

Printed and bound by CPI Group (UK) Ltd, Croydon, CR0 4YY
14/09/2026
02216388-0001